Training Engineering Students for Modern Technological Advancement

Anabela Carvalho Alves
University of Minho, Portugal

Natascha van Hattum-Janssen
Saxion University of Applied Sciences, The Netherlands

IGI Global
PUBLISHER of TIMELY KNOWLEDGE

A volume in the Advances in Higher Education and Professional Development (AHEPD) Book Series

Published in the United States of America by
IGI Global
Engineering Science Reference (an imprint of IGI Global)
701 E. Chocolate Avenue
Hershey PA, USA 17033
Tel: 717-533-8845
Fax: 717-533-8661
E-mail: cust@igi-global.com
Web site: http://www.igi-global.com

Library of Congress Cataloging-in-Publication Data

Names: Alves, Anabela Carvalho, 1972-editor. | van Hattum-Janssen, Natascha, 1971-editor.
Title: Training engineering students for modern technological advancement / Anabela Alves, and Natascha van Hattum-Janssen, editors.
Description: Hershey, PA : Engineering Science Reference, an imprint of IGI Global, [2022] | Includes bibliographical references and index. | Summary: "This book looks at the role of engineering teachers in preparing the next generation of engineers by presenting perspectives on and active learning methods for engineering education for a future generation of engineers"-- Provided by publisher.
Identifiers: LCCN 2021031962 (print) | LCCN 2021031963 (ebook) | ISBN 9781799888161 (h/c) | ISBN 9781799888178 (s/c) | ISBN 9781799888185 (ebook)
Subjects: LCSH: Engineering--Study and teaching.
Classification: LCC T65 .T69 2022 (print) | LCC T65 (ebook) | DDC 620.0071--dc23
LC record available at https://lccn.loc.gov/2021031962
LC ebook record available at https://lccn.loc.gov/2021031963

This book is published in the IGI Global book series Advances in Higher Education and Professional Development (AHEPD) (ISSN: 2327-6983; eISSN: 2327-6991)

British Cataloguing in Publication Data
A Cataloguing in Publication record for this book is available from the British Library.

For electronic access to this publication, please contact: eresources@igi-global.com.

Advances in Higher Education and Professional Development (AHEPD) Book Series

Jared Keengwe
University of North Dakota, USA

ISSN:2327-6983
EISSN:2327-6991

Mission

As world economies continue to shift and change in response to global financial situations, job markets have begun to demand a more highly-skilled workforce. In many industries a college degree is the minimum requirement and further educational development is expected to advance. With these current trends in mind, the **Advances in Higher Education & Professional Development (AHEPD) Book Series** provides an outlet for researchers and academics to publish their research in these areas and to distribute these works to practitioners and other researchers.

AHEPD encompasses all research dealing with higher education pedagogy, development, and curriculum design, as well as all areas of professional development, regardless of focus.

Coverage

- Adult Education
- Assessment in Higher Education
- Career Training
- Coaching and Mentoring
- Continuing Professional Development
- Governance in Higher Education
- Higher Education Policy
- Pedagogy of Teaching Higher Education
- Vocational Education

IGI Global is currently accepting manuscripts for publication within this series. To submit a proposal for a volume in this series, please contact our Acquisition Editors at Acquisitions@igi-global.com or visit: http://www.igi-global.com/publish/.

The Advances in Higher Education and Professional Development (AHEPD) Book Series (ISSN 2327-6983) is published by IGI Global, 701 E. Chocolate Avenue, Hershey, PA 17033-1240, USA, www.igi-global.com. This series is composed of titles available for purchase individually; each title is edited to be contextually exclusive from any other title within the series. For pricing and ordering information please visit http://www.igi-global.com/book-series/advances-higher-education-professional-development/73681. Postmaster: Send all address changes to above address.

Titles in this Series

For a list of additional titles in this series, please visit: http://www.igi-global.com/book-series/advances-higher-education-professional-development/73681

Advising Preservice Teachers Through Narratives From Students With Disabilities
Kimberly Dianne Cassidy (Shawnee State University, USA) and Beverly Sande (Prairie View A&M University, USA)
Information Science Reference • © 2022 • 235pp • H/C (ISBN: 9781799873594) • US $195.00

Global Trends, Dynamics, and Imperatives for Strategic Development in Business Education in an Age of Disruption
Anatoly Zhuplev (Loyola Marymount University, USA) and Robert Koepp (Loyola Marymount University, USA)
Business Science Reference • © 2022 • 288pp • H/C (ISBN: 9781799875482) • US $195.00

Handbook of Research on Future of Work and Education Implications for Curriculum Delivery and Work Design
Sunil Ramlall (Concordia University Wisconsin, USA) Ted Cross (Western Governors University, USA) and Michelle Love (Western Governors University, USA)
Information Science Reference • © 2022 • 640pp • H/C (ISBN: 9781799882756) • US $245.00

Handbook of Research on Credential Innovations for Inclusive Pathways to Professions
Yi Huang (Coppin State University, USA)
Business Science Reference • © 2022 • 492pp • H/C (ISBN: 9781799838203) • US $245.00

Global Perspectives on Quality Assurance and Accreditation in Higher Education Institutions
Hesham Magd (Modern College of Business and Science, Oman) and Siraj Kariyilaparambu Kunjumuhammed (Modern College of Business and Science, Oman)
Information Science Reference • © 2022 • 378pp • H/C (ISBN: 9781799880851) • US $195.00

Coaching Applications and Effectiveness in Higher Education
Ziad Hunaiti (Knowledge Well Limited, UK)
Information Science Reference • © 2021 • 274pp • H/C (ISBN: 9781799842460) • US $195.00

Partnership Motives and Ethics in Corporate Investment in Higher Education
Morgan R. Clevenger (Hiram College, USA) Cynthia J. MacGregor (Missouri State University, USA) and Paul Sturm (Kent State University, USA)
Information Science Reference • © 2021 • 471pp • H/C (ISBN: 9781799845195) • US $185.00

Handbook of Research on Leading Higher Education Transformation With Social Justice, Equity, and Inclusion
Clint-Michael Reneau (California State University, Fullerton, USA) and Mary Ann Villarreal (University of Utah, USA)

IGI Global
PUBLISHER of TIMELY KNOWLEDGE

701 East Chocolate Avenue, Hershey, PA 17033, USA
Tel: 717-533-8845 x100 • Fax: 717-533-8661
E-Mail: cust@igi-global.com • www.igi-global.com

Editorial Advisory Board

Table of Contents

Detailed Table of Contents

Solutions to global challenges need a range of engineers with diverse skills and attributes, and it is the responsibility of engineering educators to shape the engineering education landscape, using their problem-solving expertise to educate future engineers for modern technological advances. Project-based learning (PjBL) is an educational approach that can integrate such needed skills and attributes into the curriculum. However, delivering a truly effective PjBL approach can be quite difficult without considering a holistic approach encompassing three key pillars: PjBL curriculum and assessment, PjBL culture, and physical and online PjBL spaces. This chapter presents a comprehensive overview of how PjBL has been successfully deployed across the Chemical Engineering curriculum at the University of Nottingham, UK, through the lenses of those pillars, and in the form of design projects, with a progressive integration and development of diverse skills and competencies throughout the years.

Industrial firms increasingly concentrate on their core competences and outsource non-core activities, affecting the personal (soft) skills requirements of purchasing and supply chain management (PSM) personnel in their boundary-spanning roles. In parallel, machines take over processes but cannot replicate humans' soft skills such as creativity and strategic thinking. The literature shows that learning objectives in PSM courses in higher education are evaluated for not covering soft skills. Moreover, there is evidence that soft skills development is challenging. It is questionable which soft skills can be developed and which didactics are applicable. This study presents an educational soft skills experiment with IEM graduates, and it provides evidence that soft skills learning can effectively be introduced in existing courses. The graduates self-rated their competence levels of 36 soft skills before and after the course that provided soft skills workshops and a case study. In the first survey, "strategic thinking" ranked low and could be improved the most in the second survey.

The study aimed to map the critical success factors for the adoption of information and communication technologies (ICTs) in the educational process of educational institutions. Problems related to the adoption of ICT in the educational system stem from the need to adapt to the use of new technologies in the internal processes of institutions and in teaching and learning processes, common to different profiles of educational institutions including of engineering courses with them specificities. To meet the objective, a review of the existing bibliography in the Scopus database was carried out to highlight articles relevant to the topic. Based on the review, 31 articles identified the main factors and effects that influence and impact the process of implementation and continued use of ICTs. The survey generated a broader view of the challenges faced in different dimensions, from SWOT framework, involving different stakeholders. It is suggested in future studies to engineering analyze deeper the complex scenario that involves the theme.

Imparting quality technical education and training can be expected to be the vision of most engineering institutions globally to build a healthy society. The major stakeholders who contribute to this are students, teachers, industrialists, researchers, and institutes. The current scenario of rapid technological advancements demands engineering students to be dynamic and novel. Considering the heterogeneous intellectual ability of students, institutions frame time-restricted curriculums. Students who want to outperform have a challenge that they cannot be completely dependent on their academic curriculum. The objective of this chapter is to motivate and bring awareness among engineering students to adapt self-learning to excel in their professions. E-learning and project-based learning are identified as the two significant tools that could help students to self-learn. The influence of these tools on engineering students has been proved in this chapter with a case study, surveys, and feedback from students.

Preface

Engineers are indispensable in solving the world's challenges, especially the challenges related to sustainable development goals and unexpected events, as related to pandemic situations and/or climate changes among other potential catastrophes. Engineering education continues to be the crucial preparation for the next generation of engineers, while engineering practices rapidly evolve pressured by technological advancements. Engineering schools are integrated into large and rigid Higher Education Institutions (HEI) that are not known for their agility. Nevertheless, engineering teachers must have the agility to go beyond HEI structures and boundaries and close the gap between the needs of the professional practice and engineering education. Many engineering teachers are highly motivated to prepare engineering students for the quick change needed and technological, social, and environmental evolution.

In 2010, UNESCO published the first report on engineering called Engineering, Issues, Challenges and Opportunities for Development (UNESCO, 2010). The last chapter is dedicated to engineering education and the transformation needed in engineering education to provide answers to the challenges the world is facing to be solved by engineers. The authors identify challenges like outdating knowledge, innovation being based on collaborative knowledge, more complex construction of collaborative knowledge, and sustainability and social responsibility. They acknowledge the need for new skills for engineers, going beyond technical knowledge. In the 2010 report the word pandemic was only mentioned in a reference and online or hybrid engineering education was not addressed. Ten years later, UNESCO (2021) published a new report on engineering, building on the previous one and highlighting the urge to work on the Sustainable Development Goals, or as the report says, "It is therefore important for engineering education to find ways to educate engineers who can incorporate sustainable values into the development of technology" (UNESCO, 2021). Hadgraft and Kolmos (2020) distinguish three major challenges for the future of engineering education. First sustainability and the UN Sustainable Development goals that engineers play a large role in. Second the fourth industrial revolution that includes a strong integration of technologies such as Internet of Things, Artificial Intelligence, advanced materials, robotics and augmented realities. The third major challenge consists of employability and innovation competences that engineers need to be prepared for their professional life. Being ready for work is not automatically part of each and every engineering degree program. Engineers need to be trained to respond to problems of a high degree of complexity. This requires student driven learning environments with active learning methods (Freeman et al., 2014) like problem and project-based learning (PBL), for which the philosophical and pedagogical underpinnings are clearly explained by Christie and De Graaff (2017). Learning processes in which students are faced with complex challenges and work in teams on problems and solution are a characteristic of a curriculum that prepares future engineers for the challenges that society is facing. Engineering degree programs at many institutions are shifting from traditional to more problem and project-oriented approaches to learning (Bauters et al., 2020).

Mann et al. (2020) explore practice-based education (PBE), that is even more focused on complexity of engineering challenges. PBE is characterized by three elements: (1) the context of an authentic engineering practice, (2) supporting learners' agency in the process of becoming professionals, and (3) opportunities to work and learn simultaneously (Mann et al., 2020). It has a strong component of professional socialisation through the work context and as such prepares future engineers for their professional roles (Weidman et al., 2001). PBE is seen as an extension of PBL, but the preparation for the professional practice is more obvious in PBE as it takes place in the professional practice. Many institutions of higher education offering engineering degree programs are, in the light of the global challenges and the future role of their engineering graduates in these challenges, shifting from traditional approaches to learning and teaching to approaches that prepare students better for these challenges. They change from teacher-centered to student-centered learning and implement forms of active learning. Arrudo and Silva (2021) have identified key success factors of active approaches to learning and developed a maturity model containing 14 key success factors, divided into five dimensions: 1) Quality of content (Artifacts, Student assessment and Learning facilitation), 2) Organizational environment (Culture, Policy, Student feedback and Instructional design), 3) Organizational structure (Classrooms and Technology), 4) Professor/instructor (Knowledge, Skills and Attitude), and 5) Interactions (Between students and With instructors). The framework helps to outline the different maturity levels of active learning approaches and creates awareness of how student-centeredness can be enhanced. When considering the enhancement or implementation of active learning, a number of questions become apparent for institutions, course directors and especially for engineering teachers. In the first place the possible ways of implementing active learning in a degree program or in a single course. Although teaching staff, course directors and institutional leaders may be convinced of the necessity to change the teaching and learning, there are many possibilities to change to more active approaches to teaching and learning, with different degrees of impact on the students' preparedness for their future, but also with different degrees of impact on the organization and teaching staff. Changing implies making choices and deciding on what is needed for the future of engineering education, but also balancing with what is feasible within the organization. The readiness for change at institutional level does not necessarily coincide with the readiness for change of the teaching staff. This can influence the choices made for the new approaches and the scope within the organization, as described by Helle et al. (2003) with regard to different approaches to PBL. Insight in different approaches is an important first step in the change process. Experiences of colleagues at other institutions, including the path that led to the implementation of active learning approaches, are valuable for both individual teachers, teaching teams as well as institutions. Both the insights in possible active learning methods as well as the experiences of engineering teachers with these methods, including the hurdles they had to overcome, are important inputs for those who consider the implementation of active learning or who want to improve their current practice in active learning.

Another question is on the perception of students of these changes. Shifting from traditional engineering education that largely consists of lecture-based education to active learning approaches that increase the involvement and responsibility of the student implies a change in the involvement and behavior of students. The perception of students based on the experiences they go through is relevant in the decision making on a shift to active learning approaches.

Active approaches to learning that require more involvement of the professional practice ask for insight in the role and perception of the professional practice. Active approaches to learning can imply that students work for or in a company. Industry provides real assignments for students and includes individual student or student teams in their practice for a longer of shorter period of time. The experiences gained by companies and the perceptions of active learning are relevant in the search for ways of effective collaboration.

A major question is on the preparation of teachers. The role of teachers in active learning changes significantly, from a traditional lecturer who prepares lessons and works, in many institutions, rather individual and autonomous, to a coaching and facilitating role in a team of teachers that is often multidisciplinary and works together for a joint goal. Many engineering teachers do not have a pedagogical background and are especially trained for teaching their own disciplinary area. They face many challenges when their role is changing, not only with regard to the students they teach, but also with regard to the interaction with their colleagues and the way their teaching is embedded in their institution.

Finally, a rather recent question is on how to implement active approaches to learning into online and hybrid education. The COVID-19 pandemic changed the way higher education is organized and helped engineering teachers to rethink their teaching and learning approach. Active approaches to learning also became online or hybrid. In the past two years, teachers have gained experience in how to deal with online PBL and other active approaches to learning.

This book looks at the role of engineering teachers in preparing the next generation of engineers. It presents perspectives on and active learning methods for engineering education for a future generation of engineers. It provides engineering educators with examples of engineering education practices that could be followed, including the educational underpinnings. It contains practices focused on learning how to learn that enable students to be life-long learners. As such, it contributes to bypass the compartmentalized way of course organization so typical in many HEI and prepares students for more lean and agile engineering education. The book looks at the role of the teacher, especially the engineering teacher who does not necessarily have an educational background or is a researcher in the field of active learning, with best practices in engineering education. It aims to contribute with exploratory research, case studies, action-research projects, reviews, empirical cases that present and discuss active learning methodologies, hands-on activities, serious games, learning factories, and pedagogical labs among other approaches. Those are considered suitable approaches to train engineering students in onsite or online environments, in academia, or in an industrial environment. Such approaches are fundamental to train the students for modern technological advancement because they promote an environment where students can "learn by doing".

The book is organized in 17 chapters, following the questions as outlines above. Chapter 1 till 5 focus on experiences of engineering teachers in different contexts. Experiences from different countries are discussed, either on a general level or zooming in on specific degree programs. Chapter 1 starts with on overview of the implementation of PBL in a Chemical Engineering curriculum at a UK university, showing what is necessary from to make the implementation successful: curriculum and assessment, culture, and physical and online space. The chapter provides clear case studies from a chemical engineering context. Chapter 2 describes how soft skills are developed by students in an Industrial Engineering and Management degree program in a student-centered learning approach. The soft skills as identified in this study, play in important role in the preparation of future graduates for engineering challenges, although many engineering teachers feel a certain discomfort in the inclusion of these skills in the curriculum. This chapter presents an effective way to do so. The third chapter comes from a Mechanical Engineering context and explores the collaboration with external partners. It illustrates the role of local companies in the connection between higher education and the professional practice. In Chapter 4, a specific case of active learning, namely game-based learning, is explored. In a survey varied out in 19 countries, perceptions of learners involved in game-based learning methods were analyzed. The results show that a higher reported problem-solving ability and better reported second language learning. The authors explain how game-based learning can help to develop engineering competences. Chapter 5 proposes a framework that connects the agile Scrum framework to education, better known as EduScrum. The framework was applied to a Civil Engineering degree program and evaluated. The authors used EduScrum to bring the learning process closer to the professional practice.

The next chapter is one of the two chapters that are written from a student perspective. In Chapter 6, former students, currently working as engineers, look back on the experiences they had with PBL and the impact of these experience on their professional lives. They highlight the contribution of PBL to the development of transversal competences. In Chapter 7, a group of Industrial Engineering and Management students also emphasizes the transversal competences developed through PBL and furthermore acknowledges the resemblance of PBL with the professional practice, making them better prepared engineers.

Chapters 8 till 11 look at the role of industry in active learning methods. Firstly, the perspective of the industry itself is discussed. In Chapter 8, the translation of the advantages of PBL as an active learning approach is made to the professional practice. A Lean School was set up, based on the PBL principles of a university's experiences in PBL. Chapter 9 describes the close collaboration between a university and a car manufacturer in the region. A learning factory was created to simulate an authentic factory environment and to learn lean manufacturing. The second part of the industry perspective is on the competences defined by the industry that, in their view, students need to develop to be better prepared for future challenges. Chapter 10 is focus on Lean Leadership education and, through the analysis of two Lean Leadership programs, the author makes a case for the systematic inclusion of lean competences in the engineering curriculum as a way to make engineering students better prepared for their professional future. In the next chapter, written from an Industrial Engineering and Management perspective as well, the skills needed for Industry 4.0 and the challenges for industrial engineers are identified through a literature review. Different types of competences are identified and the implications for curricula in Industrial Engineering and Management degree programs are outlined.

Chapter 12 is also a literature review and has a focus on the meaning of active learning for the teaching staff. When student learning changes from an individual process to teamwork, the role of teachers and the way they work together also changes significantly. The chapter identifies the hurdles and challenges that teachers face, when doing team teaching in PBL.

The last set of chapters is dedicated to recent developments due to the COVID-19 pandemic. Starting in the spring of 2020, institutions of higher education were confronted with measures that inhibited students and teachers from meeting in classroom or other presential settings on-campus. New ways of teaching and learning had to be developed, almost overnight. Keeping the advantages of active approaches to learning while not being able to interact in the usual way, placed a number of challenges to teaching staff, institutions of higher education and of course to the students themselves. In a number of different contexts, the transition from presential active learning to online and hybrid forms of active learning is outlined. Chapter 13 describes the i9MASKS project, carried out as a summer program, where 21 students from different engineering degree programmes worked together in a project to create transparent facial masks for the prevention of virus spreading. The project showed that to the close link with an urgent societal problem, the engagement of students was high. In Chapter 14, the authors explore the impact of the COVID-19 pandemic on two degree programs at two different institutions and they conclude that new ways for teaching and learning and establishing relationships are developing. Students find new realities and the pandemic is of less influence in their academic life than in their personal and professional life. Chapter 15 zooms in on activities organized to work on digital skills for biomedical engineers and the tension between teachers' knowledge on the one hand and digital and other professional competences on the other hand that students need to develop for their participation in a digital and dynamic labor market. In Chapter 16, a literature study on the use of ICT in education is carried out in order to identify factors that influence the use of ICT in the educational process. A SWOT analysis gives insights in the implementation of ICT in education, which became, as the authors highlight, more relevant in the light of the COVID-19 pandemic and the shift to online education. The last chapter is on an experience that combines e-learning and project-based learning due to the pandemic and identifies the perceptions of students.

This book aims to support teaching staff, course directors, institutions of higher education but also students and the professional field that want contribute to engineering education that prepares students better for future challenges.

Anabela Carvalho Alves
University of Minho, Portugal

Natascha van Hattum-Janssen
Saxion University of Applied Sciences, The Netherlands

REFERENCES

Arrudo, H., & Silva, É. R. (2021). *Engineering Education Active Learning Maturity Model: a Conceptual Framework*. PAEE/ALE'2021, International Conference on Active Learning in Engineering Education, 13rd International Symposium on Project Approaches in Engineering Education (PAEE) and 18th Active Learning in Engineering Education Workshop (ALE). http://paee.dps.uminho.pt/proceedingsSCOPUS/PAEE_ALE_2021_PROCEEDINGS.pdf

Bauters, M., Holvikivi, J., & Vesikivi, P. (2020). *An overview of the situation of Project-Based Learning in Engineering Education*. SEFI 48th Annual Conference, Enschede, The Netherlands.

Christie, M., & de Graaff, E. (2017). The philosophical and pedagogical underpinnings of Active Learning in Engineering Education. *European Journal of Engineering Education*, *42*(1), 5–16. doi:10.1080/03043797.2016.1254160

Freeman, S., Eddy, S. L., McDonough, M., Smith, M. K., Okoroafor, N., Jordt, H., & Wenderoth, M. P. (2014). Active learning increases student performance in science, engineering, and mathematics. *Proceedings of the National Academy of Sciences of the United States of America*, *111*(23), 8410–8415. doi:10.1073/pnas.1319030111 PMID:24821756

Hadgraft, R. G., & Kolmos, A. (2020). Emerging learning environments in engineering education. *Australasian Journal of Engineering Education*, *25*(1), 3–16. doi:10.1080/22054952.2020.1713522

Helle, L., Tynjälä, O., & Olkinuora, P. (2003). Project-based learning in post-secondary education – theory, practice and rubber sling shots. *Higher Education*, *51*, 87–314.

Mann, L., Chang, R., Chandrasekaran, S., Coddington, A., Daniel, S., Cook, E., Crossin, E., Cosson, B., Turner, J., Mazzurco, A., Dohaney, J., O'Hanlon, T., Pickering, J., Walker, S., Maclean, F., & Smith, T. D. (2020). From problem-based learning to practice-based education: A framework for shaping future engineers. *European Journal of Engineering Education*, *46*(1), 27–47. doi:10.1080/03043797.2019.1708867

UNESCO. (2010). *Engineering: Issues, Challenges and Opportunities for Development*. UNESCO.

UNESCO. (2021). *Engineering for Sustainable Development*. UNESCO.

Weidman, J. C., Twale, D. J., & Stein, E. L. (2001). *Socialization of Graduate and Professional Students in Higher Education A Perilous Passage?* ASHE-ERIC Higher Education Report.

Acknowledgment

The editors would like to acknowledge the help of all the people involved in this project and, more specifically, to the authors and reviewers that took part in the review process. Without their support, this book would not have become a reality.

First, the editors would like to thank each one of the authors for their contributions. Our sincere gratitude goes to the chapter's authors who contributed their time and expertise to this book. In particular, the editors would like to acknowledge to the students, alumni and industry partners that have in this book their first experience as authors and their contribution is insightful for all interested in promoting active learning.

Second, the editors wish to acknowledge the valuable contributions of the reviewers regarding the improvement of quality, coherence, and content presentation of chapters. Most of the authors also served as referees; we highly appreciate their double task.

Third, the editors would like to acknowledge the support of their families.

Anabela Carvalho Alves
University of Minho, Portugal

Natascha van Hattum-Janssen
Saxion University of Applied Sciences, The Netherlands

Chapter 1
Project-Based Learning in Chemical Engineering:
Curriculum and Assessment, Culture and Learning Spaces

John Robinson
The University of Nottingham, UK

Daniel Beneroso
The University of Nottingham, UK

ABSTRACT

Solutions to global challenges need a range of engineers with diverse skills and attributes, and it is the responsibility of engineering educators to shape the engineering education landscape, using their problem-solving expertise to educate future engineers for modern technological advances. Project-based learning (PjBL) is an educational approach that can integrate such needed skills and attributes into the curriculum. However, delivering a truly effective PjBL approach can be quite difficult without considering a holistic approach encompassing three key pillars: PjBL curriculum and assessment, PjBL culture, and physical and online PjBL spaces. This chapter presents a comprehensive overview of how PjBL has been successfully deployed across the Chemical Engineering curriculum at the University of Nottingham, UK, through the lenses of those pillars, and in the form of design projects, with a progressive integration and development of diverse skills and competencies throughout the years.

INTRODUCTION

There is no secret anymore as to why engineering schools and departments need to teach differently. Students, industries, and institutions have been demanding different learning experiences over the last decade to produce graduates with diverse skills and attributes. Whilst student demands become more focused on novel teaching methods to help them design creative solutions to global challenges and become more employable, industries demand the development of competencies that can help graduates to

DOI: 10.4018/978-1-7998-8816-1.ch001

work across a multiplicity of boundaries and with people whose specialisation and cultural frameworks differ from their own (Mellors-Bourne, May, Haynes, & Talbot, 2017). On the other hand, higher education providers are demanding programmes to be designed around dynamic institutional strategies that represent the essence of their mission and values.

The most natural place in the engineering curriculum where these demands can be met and addressed is in engineering design, which is typically implemented as a standalone, capstone module or - more effectively - as the heart of the overall programme through all the course stages. In the realm of engineering, design can be defined as a "systematic, intelligent process in which designers generate, evaluate, and specify concepts for devices, systems, or processes whose form and function achieve clients' objectives or users' needs while satisfying a specified set of constraints" (Dym, Agogino, Eris, Frey, & Leifer, 2005). This process requires design teachers to look into the identification of competencies that promote expert-thinking and performance attitudes in design, and into the evaluation of how effective design courses are at developing such attributes. Adams and Atman greatly distilled the essence of design teaching as an iterative process of interpretation, selection, and planning (Adams & Atman, 1999), whereby students first access information to identify a design objective and monitor and search strategies to identify alternative options. This is then followed by an examination of the options and selection of a solution that best meets the design problem, followed by the implementation of the solution to verify that it meets the original objectives. This cyclic, non-linear iterative process reflects a natural feature of the engineer's competency (Bucciarelli, 1996) and leads to better quality solutions (Gero, 1996), offering plenty of opportunities to develop graduates who are able to value the broader socio-economic and environmental contexts, through an appreciation of sustainability, ethics, innovation and even entrepreneurship.

This chapter draws on the experiences over the past ten years in the department of chemical and environmental engineering at the University of Nottingham, UK, with the ultimate objective of providing a holistic view of how engineering students can be successfully trained for modern technological advancement through such iterative design-thinking approach. Particularly, the chapter discusses how and why diverse engineering skills and competencies are horizontally integrated through the duration of the engineering course alongside the importance of a collaborative departmental culture and suitable learning spaces.

BACKGROUND

Although embedding student, industry and institutional demands into a confined design module or programme seems to be the way forward for a sustained acquisition of professional competences, its implementation in educational practice has several challenges. Perkins developed a theory of knowledge whereby a distinction between knowledge-as-information and knowledge-as-design is made (Perkins, 2013). He claimed that in most engineering courses, knowledge is conveyed as information, relying on standard textbook problems and exercises that do not capture the broad pool of skills leading to expert-thinking of the discipline. This traditional approach to knowledge-as-information leads to the view that students cannot think or do engineering unless they know a lot of engineering, unlike the concept of knowledge-as-design, which is active and purposive, with students being able to break away from familiar frames of reference. Engineering design practised by professional engineers is eminently built upon knowledge-as-design and views the pieces of acquired knowledge as the foundational structures to develop the art and practice of changing the physical world through engineering (Moran, 2019). The deeply grounded view in many institutions of engineering education as knowledge-as-information represents a big obstacle to those advocating the teaching of design through the lenses of engineering practice and knowledge-as-design. But what do knowledge-as-design approaches look like in educational practice? The answer is simple: through inductive teaching (Graham, 2018; Prince & Felder, 2006). This type of teaching begins with tangible, authentic case studies, real-world problems, or real observations as the principal motivation to generate agency within students through a need for theories, facts, methodologies and models upon which they are either introduced with the essential information or facilitated to learn it for themselves. In this regard, project-based learning (PjBL) has been widely endorsed as the mechanism for teaching design (Graham, 2010; Heywood, 2005; Kolmos & de Graaff, 2014). PjBL is underpinned by a project brief from which students identify the need to generate a solution to a problem, perform a series of tasks leading to the production of a product that can be either a design, a device, a model, or a computer simulation. The product is normally presented as a written report summarising the identified problem and its constraints, the possible solutions, and the criteria used to select among them the most suitable solution. In the particular case of engineering design, PjBL usually begins with groups of students that need to deal with a client brief including an open-ended engineering problem full of constraints and that calls for the groups to formulate solution strategies and to iterate and re-evaluate their approach in response to their outcomes based on, economic, sustainability, social, ethical and safety criteria (Bissett-Johnson & Radcliffe, 2021). In reality, one could say that professional engineering design is PjBL, where professional engineers learn through acquisition of knowledge-as-design. The steps involved in PjBL are basically the same to those necessary to accomplish a design solution through an iterative process of interpretation, selection, and planning.

The learning and development of required engineering competences embedded within PjBL are rather complex phenomena, with students being required to embrace a composite set of inter-related skills such as systems thinking, critical thinking, dealing with uncertainties, making estimates and decisions based on conflicting information, communicating their technical knowledge to professional standards, or thinking as part of a diverse team (Chen, Kolmos, & Du, 2021). Supporting a sustained approach to consciously address these skills poses an additional challenge to the use of design as the means for integrating student, industry and institutional demands (Todd & Magleby, 2004). Such broad suite of skills to be learnt make PjBL implementation more time-consuming at the pre-project stage and at the active involvement of teaching staff and industries to ensure quality resources and strategies (Vasiliene-Vasiliauskiene, Vasiliauskas, Meidute-Kavaliauskiene, & Sabaityte, 2020). Unsurprisingly, a single member of staff could hardly teach and assess design to meet such demands without a strong supporting structure made up of more academics involved in the pedagogy of design teaching. However, the expertise of staff across engineering departments is usually quite diverse and many academics may not have a strong background on design teaching philosophies and tools, let alone the time to learn and use them to support the teaching delivery (Mitchell & Rogers, 2020). In this sense, a collaborative culture with departments becomes essential.

Whilst the educational literature has a large number of studies regarding how competencies can be integrated within a course by means of PjBL (Li & Antiohos, 2021; Thomassen & Stentoft, 2020), little is discussed about the importance of setting a continuous acquisition of such competencies over the programme years (Guerra & Rodriguez-Mesa, 2021), and how this can be supported by a collaborative departmental culture and well-designed learning spaces. This chapter covers this identified gap in the literature in the following three sections, each describing one of the pillars required to sustain the development of engineering skills using PjBL through all the years of an engineering degree.

In summary, the literature points towards three core elements for successful PjBL outcomes: A curriculum that is built around PjBL; people with sufficient skill and experience to lead and support students in PjBL activities; and a place where PjBL activities can be carried out. In this chapter, the authors present a holistic approach to engineering design teaching through project-based learning (PjBL), underpinned by a successful deployment across the chemical and environmental engineering curricula at the University of Nottingham, UK, over the past ten years. The chapter is structured in a background section and the three pillars that the authors believe are critical to the success of their approach: PjBL curriculum, collegiate department culture, and physical and online learning spaces where students can 'learn-by-doing'. Learnt lessons, including dealing with the management and leadership for a cultural change across the department, will be thoroughly discussed, providing rich perspectives from the student, academic and industry communities through a series of case studies embedded within the pillars sections.

PILLAR 1: CURRICULUM AND ASSESSMENT

The professional process design engineer is able to establish a number of alternative concepts, evaluate these relative to a given design basis or brief, make decisions about the most suitable approach, develop and optimise the chosen approach and communicate the outcomes such that the final design can be realised (Pahl & Beitz, 2013). The ambition of the PjBL design curriculum is to produce engineering graduates who are able to operate to a commensurate level, however it is not possible to achieve this outcome if design is covered in one single year of the curriculum. Take the decision-making component for example; in order to train, support and evaluate decision-making students first need the ability to be able to confidently and reliably obtain the information on which those decisions are based, and this information is typically in the form of material and energy balances, sizing calculations, hazard studies and economic analysis. To assess decision-making, it needs to be communicated in the form of drawings and standard professional report and specification templates, and these skills need to be embedded prior to the decision-making element.

The decision-making component can often be taken for granted, but there are multiple elements that form part of the process and it takes time for students to recognise these and learn how to deal with them. Underpinning design decisions is the evidence base, and it is here that conflicting sources of information can often occur. Students are commonly torn between approaches based on physics, chemistry and engineering principles, and those based on heuristics – the latter being extensively discussed by Moran (Moran, 2019) – . Different design outcomes can arise depending on the approach taken, but the professional design engineers use a balance of both heuristic and principle-based approaches through the duration of a design project. The decision-pathway of a typical project is illustrated in Figure 1, and a key component of PjBL in design is for students to recognise that their project is effectively a journey of decisions taken from two different perspectives. The role of the educator is to support students in recognising this journey, and to help them understand when principle-based or heuristic-based decisions may be the most appropriate. For example, how to deal with uncertainty, or interpret sets of codes and standards that apply to one industry but not others. This is not a trivial exercise and can only be taught effectively once students are able to reliably assemble the key information and communicate design outcomes to an appropriate standard. Getting students to this stage takes much more than one academic year, and it is important that a curriculum is established over a number of years so that students can gain the required knowledge and skills prior to focussing on the decision-making components. Foundations need to be laid in earlier years of the degree programme before PjBL can be used effectively, which at the University of Nottingham comes in year 3.

Figure 1. Process design decision-making pathway

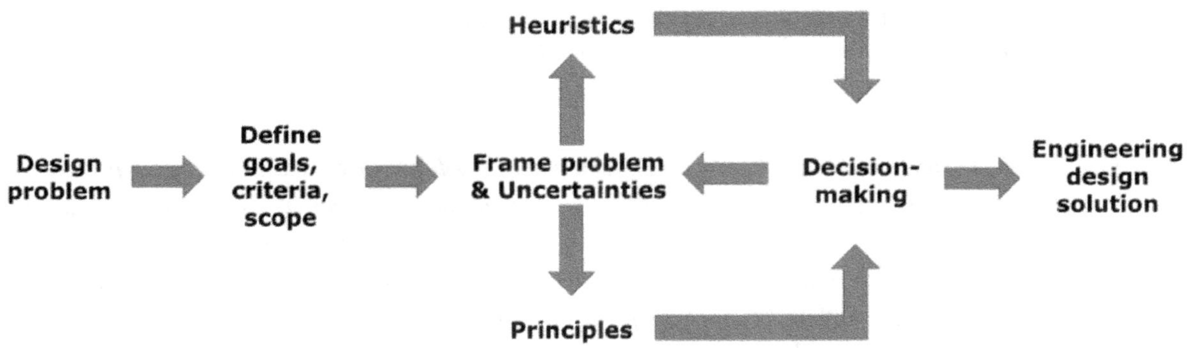

Each year of the degree programme contains one module dedicated to design, with initial teaching focused on the underpinning skills of obtaining the information needed for decision-making and communicating design outcomes. Each module complements the other by further developing earlier-learned skills and competencies, often within a more complex and more realistic context, whilst also introducing more advanced aspects of process engineering design. By years 3 and 4 students have the necessary competence in many areas, at which point teaching and support is concentrated on the newly introduced topics, with areas of competence revised and refreshed as required.

Modules are led by academic staff and supported by industrial design tutors, who guide students to develop design documents that are in line with those that they will typically encounter within an industrial setting. The industrial tutors reflect on Case Study 1 about their role at supporting the PjBL curriculum at the University of Nottingham.

Case Study 1

As adjuncts to academic teaching, the Department has a team (currently three) of Industrial Tutors, who are retired chemical engineers employed on a part time basis. Their experience covers a range of industries from oil and gas to water and pharmaceuticals and roles from conceptual design through Front End Engineering design and detail design to project management and commissioning. Their role is to show students how engineering design operates in the industrial environment so that, on graduation, they will be familiar with the tools that they will use in the workplace: Block Flow Diagram, Process Flow Diagram, Material and Energy Balance, Piping & Instrumentation Diagram, layout drawings, schedules and data sheets. They also teach the all-important safety aspects of HaZId (Hazards Identification) and HazOp (Hazard and Operatibility Study). In short, their role is to show the students how to apply the theoretical knowledge gained in their academic learning.

The Industrial Tutors have all had careers in industry and are motivated by the desire to pass on this experience to the next generation of chemical engineers, and seek to promote a relationship with the students that is more "trainer/apprentice" than "teacher/pupil". One of the most important lessons to be learned by design students is that there is never only a single answer to a problem, and they will frequently hear, and be involved in, the Industrial Tutors "brainstorming" design questions to find the best solution, just as happens in the engineering workplace. The students also learn, from the Industrial Tutors, that they will often work with insufficient information when "rules of thumb" and estimates need to be applied. To assist with this, the Industrial Tutors generated a series of design monographs on specific subjects such as how to prepare piping specifications, how to complete HazOp records, and how to specify steam distribution systems. These have grown to a library of some twenty-five design guidance notes.

Since most of the students have little or no experience of the hardware used in chemical plants, the Industrial Tutors also endeavour to give them an idea of what such equipment looks like using photographs, videos, "fly through" models and, where practicable, actual items of equipment.

Design projects are continually evolving, and the Design Tutors have a role in ensuring that they are based on current technology and reflect, as far as possible, current industrial practice. In addition to their design support role, the Industrial Tutors are also available to teach their own engineering specialism, using case studies from their own experience.

Alex Kulbicki, Richard Hill and Bruce Quare, Industrial Tutors

The overall design approach is supported then using a number of departmental design guidance documents and templates, which complement existing industry standards. Figure 2 depicts the development of skills and competencies throughout the four-year programme.

Figure 2. Vertical integration of engineering competencies through curriculum

Skill/Competence	Year 1	Year 2	Year 3	Year 4
Drawing				
Process Flow Diagrams				
Sizing, Specification and Standards				
HAZID				
Monitoring and Control				
HAZOP				
Economic and Sustainability Metrics				
Scheduling and Commissioning				
Uncertainty				
Concept selection				
Technology Limitations				
New Technologies and Applications				

Legend:

introduced
developed
competent

Design teaching in **year 1** takes place in the first term that students join the university. Prior to this the majority will have studied pre-university courses, which are heavily teacher-led. Students learn drawing as a means of communication, and the use of CAD. Methodologies for equipment sizing and specification are introduced but based around simple flow and separation systems under ambient conditions where limited prior knowledge is needed. The use of standards is also introduced alongside sizing and specification, and HAZID is taught as a methodology for quantifying hazards and to improve process safety awareness at the design stage. Students carry out their design in groups, with the briefs being extensive and well-defined. Process routes and equipment are pre-selected, and extensive design-data is provided. Year 1 design is largely based on heuristics, as students at this point have not had sufficient time to acquire the knowledge needed for principle-based decision-making.

Design teaching in **year 2** takes place in the second semester, at which point students will have covered mass and energy balances, fluid dynamics, process chemistry, material selection, heat transfer, binary separation processes and phase equilibria. Equipment sizing and specification skills are developed through the use of processes with more units, non-ambient conditions, chemical reactions and multiple phases. Heat balances and heat recovery are key elements to the year 2 project. Different costing techniques are introduced, and students are supported to include key instrumentation and control for their process. HAZOP is introduced to compliment the HAZID methodology learned in year 1 to include process safety improvements within all stages of the design process. The brief is well-defined. A range of design data is available, with selected data withheld to develop literature searching skills. The process route is pre-determined; however students have some flexibility to choose different equipment and configurations using both heuristic and principles-based approaches.

Design in **year 3** takes the form of a year-long 40-credit project. At this stage the core competencies of drawing, sizing and the use of standards are already in place, and students are challenged to demonstrate these for a complex, system-level process with inherent hazards and non-ambient conditions through PjBL. The tools and techniques learned in years 1 and 2 form the basis of decision-making in year 3. A brief is provided that includes large uncertainties in process conditions, however there is no stipulation as to the process route that should be adopted nor the equipment/configuration that should be used. The brief is carefully chosen so as to provide multiple process pathways as potential solutions, and where decision-making cannot be avoided. This presents the ideal context for PjBL. Students identify three different process concepts, then evaluate them based on economic, safety and environmental criteria. A single concept is taken forward to a Front-End Engineering Design (FEED) phase, which includes a functional specification and project schedule as newly introduced elements.

Approximately 50% of students graduate with a BEng degree after year 3. Design in **year 4** is included within the year-long 60-credit project and delivered through PjBL. The design element of this project serves to contextualise the research component by allowing students to understand the limitations of current technologies and the rationale for subsequent technology development. A well-defined brief is provided that typically stipulates the use of a new technology or presents a set of constraints that cannot be met with the best current available technology. Students produce a conceptual design to meet the given brief which will either contain significant uncertainty (new technology) or will identify the technology limitation that prevents the brief from being met, with a corresponding recommendation for alternative technologies that should be investigated.

The transition from year 1 through to 4 is not without significant challenges. The heuristic approach used in year 1 is very effective in allowing students to rapidly develop a number of core competencies, but one of the unintended consequences is that they do not develop an inherent sense of why a particular process or piece of equipment is used, nor what the credentials of potential alternatives might have been. Perhaps more significant is that it tends to lead to a view amongst students that there is a 'right answer' to the design problem, or a 'correct method' needed to approach it. This is the antithesis of PjBL, but an intended step in the journey to supply students with the skills needed to undertake PjBL in subsequent years. Also important is that many of the core skills and competencies that underpin later PjBL are acquired individually, i.e. through the use of CAD to produce engineering drawings or conducting a series of hydraulic calculations, but important elements of PjBL are group-based. The lifecycle of a PjBL project includes multiple components, some of which are more likely to be better carried out individually and some that conducted as a team based on our experience with PjBL over the last ten years (see Table 1).

Table 1. PjBL components in engineering design

PjBL Project Component	Primary Way of Working
Clarifying the brief/problem	Group
Identifying options and constraints	Group
Establishing the technical, economic, and environmental characteristics of each option	Individual
Evaluating the safety credentials of design options	Group
Process/concept selection decisions	Group
Communication of design decisions	Individual

Team components are those where multiple viewpoints are beneficial, for example when understanding the trade-offs between economic, environmental and safety criteria, and rationalising these against an ethical framework in order to make a decision. This is very different to simply sharing tasks amongst individual group members, for example where each group member might produce CAD drawings individually for inclusion in a group portfolio. The drawings in this case are done individually, but the drawings are ultimately communicating design decisions that have already been made by the team. Focussing on the underpinning skills in year 1 means a strong emphasis on project tasks that are best carried out individually. This presents an added challenge with PjBL, because not only are new skills and competencies being introduced but the ways of working needed to achieve them are also relatively new to students at the PjBL stage of the programme.

It is important that the project briefs and structures are carefully chosen in Year 3 to recognise the heuristic model that has been used before but drive the decision-making and thought processes that are characteristic of professional design engineers. Students are not well-placed to propose projects that fit these criteria, and for this reason the design problem/brief is set by the lead academic at the start of the academic year. The brief is typically based around scenarios of replacing life-expired process plant for industrial clients in the UK. Introducing inherent choice in concept selection is a key part of the PjBL process, so the project brief must be structured in a way that it can be satisfied using multiple design options. Students need to be supported to recognise the different options, and to work through the decision-making process in a logical manner. The assessment structure is therefore a vital part of the PjBL process, as student activity (and to a large extent their learning) is typically driven by the assessment criteria. Assessment criteria that focus on the perceived 'correctness' of design outcomes or the amount of work produced will not drive students to follow the required thought or decision-making processes needed for PjBL and will not encourage them to work together effectively. It is important therefore that the grading criteria promote the design thought processes over and above the design outcomes, and that they promote the right balance of individual/group ways of working shown in Table 1.

Assessing the Thought Process

A single set of grading criteria that covers both the thought processes and the design outcomes is not workable in practice, as students inevitably tend to default toward striving for the 'right answer'. In order to focus assessment on the thought and decision-making process in PjBL then references to the design outcomes need to be minimised or removed altogether. This may seem counter-intuitive for a student design project, but with a curriculum spanning over four years it affords the opportunity to separate design outcomes and thought processes from assessment and grading criteria. By assessing learning outcomes related to the design outcomes (e.g. P&ID drawings, HAZOP studies, hydraulic calculations) in years 1 and 2 there is no need to duplicate such assessments in later years. Consequently, the year 3 and 4 grading criteria can focus on the alternative options and justification that underpin the decision-making process. To achieve this, it is often necessary to deviate from the standard suite of design project deliverables. For example, a Piping and Instrumentation Diagram is a professional-standard engineering design output but gives no indication of the alternative configurations nor why students chose the process depicted in the drawing. For the professional engineer the thought-process is inherent, and there is no need to document this in the design output. Students are in a different stage of their careers, and consequently there is a need to document and assess the decision-making elements. To this end many of the deliverables are adapted to include a narrative element, and a conceptual design phase is introduced which drives students to use robust evaluation methods to compare different process options and document the reasons for their final decision. As an example, below are the assessment criteria for one of the deliverables (Optioneering Report) in the year 3 design project:

2.2 class: A safety evaluation is presented that identifies the relevant hazards that are inherent within the different design concepts. Quantitative economic evaluation and comparison of each option based on revenue generation, capital and operating costs for a single scenario. The environmental aspects of each concept are identified in terms of feedstock and utilities consumption per tonne of product.

2.1 class: A safety evaluation is presented that identifies and compares the relevant hazards that are inherent within the different design concepts. Quantitative economic evaluation and comparison of each option based on revenue generation, capital and operating costs for a limited range of scenarios. Qualitative evaluation of the environmental aspects of each concept, including feedstock, utilities and discharges per tonne of product.

1st class: A comprehensive safety evaluation is presented that identifies and compares the relevant hazards that are inherent within the different design concepts highlighting potential actions for reducing risks for consideration in Task 2. Quantitative economic evaluation and comparison of each option based on revenue generation, capital and operating costs for the full range of scenarios specified within the brief. Quantitative evaluation of the environmental aspects of each concept in terms of feedstock and utilities consumption per tonne of product, discharges compared with the Environmental Permit for the site and a brief explanation of how the permit values will be achieved.

Assessing Group Working

Team-based ways of working are essential elements of the decision-making process in engineering design, and are therefore fundamental components of PjBL. If the assessment and grading criteria do not effectively promote groupworking then many students will default to more individual ways of working, the benefits of team-based engineering design will not be realised and the overall outcomes for students will be lower as a result. It is important that assessment criteria are geared towards team-based ways of working, and that sufficient weighting is applied so that students understand the importance and adapt their behaviour accordingly. The traditional structure for process engineering design projects is to include well-defined group and individual components, however this approach tends to lead to a bias for the individual component and does not effectively promote the right ways of working. In many cases with this approach students do not support each other effectively; they compartmentalise their individual element and do not engage in the iterative and collaborative processes that are key to system-level design. Achieving the desired level of team working means moving away from the traditional split individual/group model for design projects. PjBL projects can be 100% group, in that the project deliverables are assessed and attributed to the group as a whole rather than to individual students as explained in the previous subsection. Individual marks are then assigned as a percentage of the group mark by using a metric that is related to team-based ways of working over the duration of the project, and that is confirmed at the end of the academic year by means of individual viva-voce examinations. This metric can be established using a balance of teacher and peer assessment, with students evaluated according to the following six criteria:

- Contributed effectively to group meetings, sub-tasks and milestone review meetings
- When leading tasks, coordinated and organised effectively, and was receptive to opinions and ideas from the support team
- When leading tasks, provided documents in advance of group meetings and formal hand-ins, and communicated task outcomes and key decision-points effectively with the rest of the group
- When supporting tasks, offered an appropriate and timely level of ideas and opinions
- When supporting tasks, provided effective review of outcomes and deliverables prior to submission/presentation
- Provided effective feedback and appraisal to task leaders during group meetings.

The criteria are indicative of effective teamworking, recognising that students will lead some tasks, support on others, need to be organised and communicate regularly and effectively. There is no reference to the quality of design outcomes in these criteria, nor is there reference to 'volume' of contribution. Without effective support and assessment frameworks students undertaking group projects tend to experience two extremes of behaviour within their groups; students that do not engage, and those that exhibit controlling behaviours. With the set criteria the expectation for students is clear, and this effectively manages the two behavioural extremes and results in better experience for the students. With improved groupworking driven by the assessment structure and grading criteria it naturally results in positive ways of working, good decision-making and better overall PjBL project performance and outcomes.

PILLAR 2: COLLEGIATE DEPARTMENT CULTURE

Although curriculum is the cornerstone of PjBL in design delivery, a key element that is often overlooked, yet critical to its success, is the development of a collegiate culture around design teaching. *Culture* in this context can be defined as 'shared systems of meaning and practice emerging from collective learning and taught to a group's newcomers as the correct way to think and behave' (Baba & Pawlowski, 2001). In this regard, a single member of staff could hardly teach design in a sustainable manner given the wide breadth of multidisciplinary competencies that students need to learn and master throughout. Taking such a flawed approach would lead to a lack of opportunities to daily engage with a professional dialogue between engineering practitioners and academics, so necessary for the constructive evolution of educational practice in design teaching. Rather, department and faculty leaders should prioritise the development of a shared meaning around what design teaching should be by listening to industry, students, and institutional demands, and how it should look like in practice, embracing the diverse backgrounds and competencies of their staff.

The former department head at the University of Nottingham reflects in case study 2 on the cultural changes that were needed to sustain a transition towards a model that can now be considered as exemplary within the UK, with a strong vertical integration of key engineering competencies through years 1 to 4. The case exemplifies the major role of department leaders at recognising and prioritising enhanced PjBL design pedagogy in order to ensure that all graduate students develop a unique set of skills to professional standards highly valued by employers.

Case Study 2

At the heart of a collegiate department is a common purpose grounded in pride in the quality of our graduate product. In saying this, it is important that the department is not just viewed as the staff, but also by association students, alumni and employers. If all of them recognise and support the common purpose and feel a tangible sense of pride, in what they are, what they produce or what they employ, then collegiality is coupled to a sustainable reputation.

Our department in 2011 was barely collegiate, if at all. Graduates were good, but undifferentiated from the crowd. Staff, most of whom were not from engineering backgrounds, survived on the academic freedom to deliver their own material within their allocated modules in their own way. A ground change was required. A change that would start with the quality of our graduate product. The obvious way of doing that was to improve the quality of design teaching, an employer's measure of an excellent engineer that was frequently bemoaned in industrial fora. Our aim was to ensure that all students would leave Nottingham with a set of skills that, irrespective of their final degree classification, would make them highly employable. We devised a strategy underpinned by a series of key elements:

- *Re-design the curriculum on the principle that design, delivered to industrial standards, was the backbone of the degree programme, where no module is greater than the programme it serves. In other words, staff would be instructed as to what content had to be in their modules and how that fitted into the programme structure.*
- *All staff must be involved in delivery and assessment of Chemical Engineering design teaching irrespective of their academic background; the reality being that everyone needed to be able to understand and read a P&ID. We were all in this together.*

- *Employ staff from industry. Academics are professional academics yet what we wanted to produce were professional engineers. So, to guarantee confidence in our graduate quality, raise the standard of design teaching and support the staff and students, non-academic, professional engineers on staff were essential. Typically, these were engineers towards the end of long careers who were keen to give something back.*
- *Degrees with a year in industry. This would also raise the standard via peer learning. It would also give students pride in what they had learned and prove to staff, all of whom were expected to visit students on placement, that the curriculum was delivering a quality product. It would also give us feedback as to where further improvements could be made.*
- *Appoint new academic staff that understood what would be expected and would enhance the evolving collegiate culture.*
- *Celebrate and promote success, biannual staff meals, and an annual networking dinner for students, staff, alumni and industry.*
- *Give oversight of the strategy to the right people.*

To make this work everyone needed to understand what was happening and why. To reap the benefit, it had to be done quickly generating anticipated dissatisfaction. Students on the old curriculum realised they had missed out; staff were annoyed at being forced outside their disciplines or not being able to teach their own subjects or run their own research projects. But within 3 years the tables had turned, reputation was building, graduate employability and student satisfaction were high, undergraduate recruitment was growing and a collegiate department was forming.
Professor David Large, Former Head of Department

One of the key elements to support this collegial and collaborative department culture around design teaching is the embodiment of curriculum skills and competencies through a single project brief. In other words, not letting students choose the project topic. A single brief keeps the overall physics of the process unaltered but is chosen by teachers to give enough breadth for groups of students to produce multiple solutions, with hundreds of different permutations in the process routes and configurations. The pre-PjBL design teaching components lead to a student tendency to default to a process route that they find in the literature, but without considering the client's needs and constraints or alternative approaches. The freedom of choice in this instance inevitably leads to significant bias on the part of the students, devising projects that they believe are interesting or easy. At this stage in their careers, students lack the holistic understanding to be able to select a project that builds on previous heuristic approaches but introduces elements of decision-making in process selection. With a common brief it is possible to involve more department staff, as non-engineers can be trained by senior colleagues, and it leads to much more consistency in how projects are assessed between different groups. 'Choice' and 'quality' in this instance are mutually exclusive – short-term dissatisfaction from students in the project allocation is significantly outweighed by the overall project experience and outcomes.

A single brief also provides a unique level of support focused on developing a system-level thinking approach in students, which would not be practical had groups selected different topics (or briefs) for their design projects involving completely different physics and process chemistry. A single project brief serves as a buffer element for the academic staff, providing a highly consistent student experience regardless of the process routes and configurations that groups decide to select. Based on the authors' experience over the years, a single brief tremendously helps in providing timely, reliable, and actionable feedback over the course of the project and during formal group design review meetings.

A single brief also provides a framework for consistency in assessment of both individual and collective team development of expert-thinking abilities. Given the large number of deliverables typically involved in a design project (e.g., basis of design, PFDs, P&IDs, material and energy balances, equipment specifications, functional design specifications, HAZID/HAZOP safety reports, economic analysis, environmental impact and sustainability assessment, plot plan layout and elevation drawings, EPC schedules or design proposals), a team of academics is usually needed to undertake the assessment. These academics usually have different level of expertise and therefore it is important that they are properly briefed and supported by the design leads through the entirety of the project. Splitting the deliverables – rather than overall group submissions – and ensuring that each deliverable is assessed by the same academic inherently produces a consistent assessment throughout the whole cohort of students. This framework based on a single client brief also facilitates the typical individual viva-voce examinations happening after the project deliverables are submitted, creating a consistent model to assess whether individuals have a thorough understanding of the entire process and alternative options, and whether they have understood the key decisions made during each design phase, relating design decisions to the project brief and to safety, economic and environmental criteria.

As one would expect, the composition of the student groups can really influence the quality of their outputs. In this regard, discarding self-sorting approaches for group formation becomes essential. Design leads have to carefully think about students' allocations into groups well in advance, ensuring that each group consists of students with stronger and weaker competencies – based on their previous course performance –, and mixed nationalities, genders and perhaps course strands. This make-up ensures that all the groups are diverse and skills-balanced and have the same opportunities to produce a high-quality design. It also stretches students to work on a wider range of abilities than their technical knowledge and helps them to become confident communicators and team-workers readily available to be deployed across different parts of a business and to fit into diverse teams. Whilst teachers can select based on available student characteristics and data there is still a very large variable that is not (and likely cannot) be controlled when establishing groups – the way students will tend to behave once the group is formed. The pre-PjBL curriculum is generally focussed on the individual student, with individual ways of working being prevalent in the early stages of their university career and long before that in their secondary/tertiary education settings. Many students transition with relative ease to the team ways of working needed in PjBL design projects, but inevitably there are those who take longer to recognise the challenges and adapt to the team environment. Typically, there are two behavioural extremes that teachers frequently observe in group-based projects; (1) Those students that do not engage sufficiently with the team, (2) Those students that exhibit controlling behaviours. There are a multitude of factors that underpin these behaviours, for example a lack of understanding of the importance of team ways of working, a previously-successful approach of them working as individuals, their reticence to engage in social situations generally or issues with their mental health. In both cases the behaviour can have a significant negative effect on the team dynamic and the overall team/individual outcomes. Students that do not engage sufficiently result in unequitable division of tasks by the group, a weaker evidence base and more restricted opinions for the decision-making process. Students that exhibit controlling behaviours tend to want to lead, but they do not recognise the inherent skill sets available within the team, they struggle to establish trust and can often set unrealistic expectations of other team members.

The behavioural extremes cannot always be predicted in advance of forming the teams, so a mechanism is needed to identify and modulate these behaviours during the PjBL project itself. This takes the form of academic staff mentors - staff who are not allocated any design delivery or assessment duties

– or even those without a strong chemical engineering background (e.g. chemists or geologists) – acting as group mentors. These staff do not provide any technical support, but they carry out observations on group meetings and have access to group shared spaces, identifying the behavioural extremes and working with both individuals and teams to recognise these traits and establish more positive team ways of working. This new supporting element became a reality, once again, owing to the collegiate culture developed over the years in the department. Dr Ian Lowndes, current Director of Education & Student Experience at the department, shares his reflections on the role of academics as mentors in case study 3.

Case Study 3

To replicate the mentoring that a graduate engineer would normally receive, the department has introduced the Design Project mentors to help support the development of the vital group dynamic students require to deliver a successful group student design project. The department carefully manages the allocation of students to each project team, to deliver groups that have a similar mix of academic abilities, cultural & educational diversity & gender.

To prepare students for this major group project exercise the students undertake an introductory group skills workshop where they are introduced to the different character types and skills competencies that contribute to the successful delivery of group projects. The student groups are introduced to the principles of Honesty, Trust & Respect that underpin the establishment of effective and successful working relationships. Students are also introduced to the Belbin Team Roles and are encouraged to share their personal Belbin assessment profiles, generated from an online survey, with their group members. The nine Belbin team roles are: Resource Investigator, Teamworker and Co-ordinator (the Social roles); Plant, Monitor Evaluator and Specialist (the Thinking roles), and Shaper, Implementer and Completer Finisher (the Action or Task roles) (Belbin, 2012). Belbin reported that the most successful teams were made up of a diverse mix of behaviours, and that teams need to be able to access to each of the nine Belbin Team Role behaviours to become a high-performing team. The first role assigned to the mentor is therefore to discuss the survey results with the group, assist them to identify the strengths and gaps in the Belbin Team roles present, and introduce the concept of partnership to either compensate any deficit roles or design competencies to deliver assessed project outcomes.

*The role of a mentor is to foster the courage and confidence of the student within a group, to enable them to trust and respect each other, to admit and confide their design skills strengths and weaknesses, to recognise and accept the adoption of shared ownership of tasks in the delivery of the project. The latter skills recognition is perhaps the most challenging for the mentor to develop within any team. This is not surprising given that all of our students are the products of our public educational systems that has been regularly and rigorously testing them as an individual since they entered primary education. Our students enter University benchmarking themselves against others on the individual public examination results they have achieved. The involvement in shared group activity is viewed by many as an anathema and an unnecessary dilution of their personal academic achievements. The most satisfying experience comes for a mentor when you suddenly realise on attending a group progress meetings that the reporting language has changed from **I** to **we, we** have planned/executed/supported each other… the language of a successful team built upon Honesty, Trust, Respect & Partnership.*

An under-estimated and under used mentor tactic that can be effectively used to develop and reward good group dynamic – Use of PRAISE!!! Verbally and by written reports to the group recognise, praise and celebrate any instances of collegiate actions, deeds, comments demonstrated by students by both

praising the individuals and group during group meetings. Identify the good practice with the group to encourage more participation and engagement.
Dr Ian Lowndes, Director of Education and Student Experience

PILLAR 3: LEARNING SPACES

As discussed in Pillar 1 (Curriculum) a PjBL project requires both individual and team-based ways of working at different stages. To facilitate the appropriate way of working it is important that students have access to suitable spaces, both physical and virtual. Where collaborative elements are needed then there is a corresponding need for spaces where information can be shared, ideas discussed, and decisions made. Drawings, calculation sheets, process models and a multitude of other project milestones need to be shared in an environment where the whole group can observe them, and this can take the form of physical paper-based drawings or information that can be displayed on a screen. The same environment for information sharing needs to be conducive to group discussion, where ideas can be exchanged, annotations can be made on physical paperwork and on screen, and where subsequent decisions can be documented. Group interactions of this nature occur once or twice weekly and are typically up to two hours in duration. The traditional academic lecture theatre falls a long way short of the requirement for a group PjBL design space, both in terms of its configuration for group activities but also its availability to be used by multiple groups for the required duration. Other disciplines recognise the need for collaborative spaces, particularly architecture and product design, where studio-style layouts are an integral part of both education and professional practice. The same model works equally effectively for PjBL when applied to process engineering design projects (see Figure 3).

Figure 3. Group of students working in the design studio

'Design studios' are provided for students in all year groups of the degree programme. The studio consists of clusters of tables and chairs, with an integrated large screen at each station. Groups of up to 7 students can occupy a single cluster, and the studio spaces can typically accommodate 25 groups at any one time. Access is regulated through the student timetable, with unfettered access permitted outside of the formally timetabled periods. The studios are not used for formal teaching purposes, but the spaces are ideal for teaching staff to support student groups. With a clearly defined space it is possible for staff to observe and interact with all groups over the course of a two-hour period. With no studio space groups would typically meet in various locations around the university, and in that case, it becomes more difficult to provide adequate levels of support.

Groupworking spaces need not be exclusively physical spaces. With COVID-19 there was a need to move PjBL design projects to a 100% online model, and this meant creating an online space to support team-based ways of working in the same way a design studio does. MS Teams was used as a platform, with group channels created for each team so they could store and share files, and work on them collaboratively. The meeting functionality meant they could meet to coordinate group activity, share drawings on screens and mark them up in real time. Academic staff and mentors could also join MS Teams meetings as required in order to observe, support and assess. Aside from tactile interactions with physical paper drawings, the online MS Teams platform was a very effective substitute for the design studio as a space for team-based ways of working. What the online space lacks in face-to-face interaction it makes up in that fact that it is not constrained by university timetables and fosters higher levels of team interaction than is the case with the physical space alone. For this reason, the online groupworking space will become a regular feature of future PjBL activities, but one that will complement the design studio rather than replace it.

The physical studio space is very popular with students and well-used. As well as the support for teamworking it provides engineering students with a space that they feel belongs to them, something that is often lost where universities take a corporate or centralised approach to space allocations and timetabling. The popularity of the spaces amongst students leads to a number of unintended consequences, one of which is that the space can often be used as the 'default' for all PjBL. Soon after the studio spaces were established teams of students could be seen in groups around a workstation trying to draw a Piping & Instrumentation Diagram, and other groups were observed attempting to develop a spreadsheet to perform material & energy balance calculations. P&ID drawings are a means of communicating design decisions. The decisions are undertaken collaboratively, but the drawing itself is an individual activity. Material balance calculations quantify the functionality of a particular concept, and underpin subsequent safety, economic and environmental/sustainability metrics. Team collaboration is essential to review the calculation outcomes and decide on subsequent design iterations but is not appropriate for establishing the evidence on which the review and decisions are based. The studio space was driving students to inappropriate ways of working, working collaboratively on tasks that should be carried out individually. In vastly improving the environment for teamworking, the space had inadvertently weakened the PjBL components where individual ways of working were necessary. The solution was to signpost the ways of working approaches more effectively to students, to undertake regular observations of teamwork to ensure that groups are focussing on the right activities and to ensure adequate learning spaces exist for more individual ways of working.

Individuals within teams will typically lead specific tasks or project deliverables, will be tasked with obtaining key pieces of information, perform calculations and establish metrics for candidate design concepts and communicate findings for review by the rest of the group or for final assessment by academic staff. Many of the skills students need to do this are already embedded from taught design components from earlier years of the programme, but students will typically need support from teachers to demonstrate these skills within the PjBL framework. The design studio spaces provide one route to this level of support, where staff can be available in the space at defined periods for formal or ad-hoc meetings with individual students. An open-door policy amongst teaching staff also facilitates this type of support, but in PjBL projects with a single brief it is often the case that many individuals will ask broadly the same questions of staff. This is inefficient, but also very frustrating when supporting large student cohorts. The COVID-19 pandemic meant that individual face-to-face support could no longer be offered, and it is here that the online platform really showed its effectiveness. Students were signposted to 'open tutorials', which were timetabled meetings on MS Teams with up to 150 participants. Staff have camera and microphones enabled, and students are instructed to keep their microphones on mute. Students ask questions using the 'chat' function, which are then answered by teaching staff verbally, through the chat or by screen-sharing other content. The benefit of this approach is that students experience improved levels of peer learning, by observing the questions that others are asking and the corresponding answers that are given. Sessions are recorded and made available for students to watch later. The open tutorial approach not only improves the efficiency of student support, but the peer learning aspect significantly improves the rate at which students understand the different aspects of the project and its constraints. A further benefit of the online open tutorial is that multiple staff can engage effectively with the student cohort. Staff-staff discussions can be used to kick-start the tutorial process with pre-empted questions or themes, and student questions can be responded to by multiple staff in cases where multiple opinions are relevant. This dovetails very well with the objectives of PjBL in process engineering design but is a feature that is very difficult to incorporate in face-to-face student support. As well as open tutorials a web forum is also used to respond to student queries. Instead of the timetabled tutorial using MS Teams the questions can be posted at any time, and staff will typically respond to the post within 1 working day. This approach allows questions to be grouped into themes, stops duplication of questions and responses, allows students to solicit help outside of the timetabled tutorial periods and allows teaching staff to see how their colleagues are responding to queries. The latter is important as it allows a consistency of response in cases where this is important (e.g., deliverable/assessment queries) and also allow multiple teacher opinions to be posted where appropriate. Both the open tutorial and online forum spaces are invaluable for training junior teaching staff who may be supporting design projects for the first time, offering them the opportunity to learn effectively from senior colleagues.

Not all students are comfortable interacting in open online forums, in much the same way that many students choose not to speak up in a crowded lecture room. The open tutorial platform and to some extent the design studio provides spaces for these students to observe, but for their own queries it is important that suitable opportunities and spaces for 1-1 meetings are maintained. This can be as simple as ensuring availability of staff for surgery sessions, a system for booking specific slots or ad-hoc meetings, and in all cases, they can be online or in physical office spaces or meeting rooms.

CONCLUSION

Project-based learning in engineering design has proved to efficiently develop the required skills and attributes of the 21st century engineering graduates through consistent vertical and horizontal integration into the engineering curriculum. However, the authors of this chapter claim that this outcome would not have been possible to achieve if design had been covered in a single year of the curriculum as a compartmentalised module. To develop engineering expert-thinking – i.e., graduates being able to make and communicate engineering decisions underpinned by a robust body of evidence – students first need to be able to obtain the information upon which decision-making is based, and this requires a continuous training and support on a variety of engineering tools through the course stages. To achieve this, a holistic approach supporting the logistics and infrastructure of PjBL must be adopted. Educators should think about how they will be explicitly assessing the development of expert-thinking abilities and group working beyond the conventional individual assessment on the use of different engineering tools. Yet, this is not enough to ensure successful learning through PjBL. Department culture becomes an essential pillar; ensuring the buy-in from all staff to support the delivery, assessment and/or mentoring of groups can be really challenging and educational leaders within schools and departments should prioritise and allocate appropriate resources to sustain high-quality PjBL if engineering expert-thinking is the ultimate quality they wish their graduate product to be known for. Finally, learning spaces are also critical to the success of PjBL in engineering design. Providing equipped physical and online spaces for group working are ideal for teaching staff to support student groups. With a clearly defined space it is possible for staff to observe and interact with all groups with different degrees of constraints (e.g., online spaces are not constrained by university timetables like physical spaces are, and they tend to foster higher levels of team interaction than is the case with the physical space alone). However, support is usually required to help students understand how to use both spaces in an efficient way, as they tend to naturally default the use of these spaces to be working on every single individual task that does not require group inputs (see Table 1). The presented three-pillar framework has supported and sustained a 21st century model for PjBL that can be considered as exemplary within the UK and has produced more than 1000 highly employable chemical engineering graduates over the last ten years.

REFERENCES

Adams, R. S., & Atman, C. J. (1999). *Cognitive processes in iterative design behavior.* Paper presented at the FIE'99 Frontiers in Education. 29th Annual Frontiers in Education Conference: Designing the Future of Science and Engineering Education 10.1109/FIE.1999.839114

Baba, M. L., & Pawlowski, D. (2001). *Creating culture change: An ethnographic approach to the transformation of engineering education.* Paper presented at the International Conference on Engineering Education.

Belbin, R. M. (2012). *Team roles at work.* Routledge. doi:10.4324/9780080963242

Bissett-Johnson, K., & Radcliffe, D. F. (2021). Engaging engineering students in socially responsible design using global projects. *European Journal of Engineering Education, 46*(1), 4–26. doi:10.1080/03043797.2019.1674785

Bucciarelli, L. L. (1996). *Designing engineers.* MIT Press.

Chen, J., Kolmos, A., & Du, X. (2021). Forms of implementation and challenges of PBL in engineering education: A review of literature. *European Journal of Engineering Education, 46*(1), 90–115. doi:10.1080/03043797.2020.1718615

Dym, C. L., Agogino, A. M., Eris, O., Frey, D. D., & Leifer, L. J. (2005). Engineering design thinking, teaching, and learning. *Journal of Engineering Education*, *94*(1), 103–120. doi:10.1002/j.2168-9830.2005.tb00832.x

Gero, J. S. (1996). Creativity, emergence and evolution in design. *Knowledge-Based Systems*, *9*(7), 435–448. doi:10.1016/S0950-7051(96)01054-4

Graham, R. (2010). *UK approaches to engineering project-based learning*. White Paper sponsored by the Bernard M. Gordon/MIT Engineering Leadership Program.

Graham, R. (2018). *The global state of the art in engineering education. Massachusetts Institute of Technology*. MIT.

Guerra, A., & Rodriguez-Mesa, F. (2021). Educating engineers 2030 – PBL, social progress and sustainability. *European Journal of Engineering Education*, *46*(1), 1–3. doi:10.1080/03043797.2020.1828678

Heywood, J. (2005). *Engineering education: Research and development in curriculum and instruction*. John Wiley and Sons. doi:10.1002/0471744697

Kolmos, A., & de Graaff, E. (2014). Problem-based and project-based learning in engineering education. Cambridge handbook of engineering education research, 1, 141-161.

Li, R., & Antiohos, A. (2021). *Problem Based Learning (PBL) in four-week term block mode teaching*. Paper presented at the 2021 IEEE Global Engineering Education Conference (EDUCON). 10.1109/EDUCON46332.2021.9453991

Mellors-Bourne, R., May, T., Haynes, K., & Talbot, M. (2017). *The state of engineering*. Retrieved from https://www.engineeringuk.com/media/1355/enguk-report-2017.pdf

Mitchell, J. E., & Rogers, L. (2020). Staff perceptions of implementing project-based learning in engineering education. *European Journal of Engineering Education*, *45*(3), 349–362. doi:10.1080/03043797.2019.1641471

Moran, S. (2019). *An applied Guide to process and plant design*. Elsevier.

Pahl, G., & Beitz, W. (2013). *Engineering design: a systematic approach*. Springer Science & Business Media.

Perkins, D. N. (2013). *Knowledge as design*. Routledge. doi:10.4324/9781315803067

Prince, M. J., & Felder, R. M. (2006). Inductive teaching and learning methods: Definitions, comparisons, and research bases. *Journal of Engineering Education*, *95*(2), 123–138. doi:10.1002/j.2168-9830.2006.tb00884.x

Thomassen, A. O., & Stentoft, D. (2020). Educating students for a complex future–why integrating a problem analysis in problem-based learning has something to offer. *The Interdisciplinary Journal of Problem-Based Learning*, *14*(2). Advance online publication. doi:10.14434/ijpbl.v14i2.28804

Todd, R. H., & Magleby, S. P. (2004). Evaluation and rewards for faculty involved in engineering design education. *International Journal of Engineering Education*, *20*(3), 333–340.

Vasiliene-Vasiliauskiene, V., Vasiliauskas, A. V., Meidute-Kavaliauskiene, I., & Sabaityte, J. (2020). Peculiarities of educational challenges implementing project-based learning. *World Journal on Educational Technology: Current Issues*, *12*(2), 136–149.

Chapter 2
A Soft Skills Experiment in an Industrial Engineering and Management Academic Course:
A Demonstration of How to Develop Soft Skills

Klaas Stek
University of Twente, The Netherlands & Graz University of Technology, Austria

ABSTRACT

Industrial firms increasingly concentrate on their core competences and outsource non-core activities, affecting the personal (soft) skills requirements of purchasing and supply chain management (PSM) personnel in their boundary-spanning roles. In parallel, machines take over processes but cannot replicate humans' soft skills such as creativity and strategic thinking. The literature shows that learning objectives in PSM courses in higher education are evaluated for not covering soft skills. Moreover, there is evidence that soft skills development is challenging. It is questionable which soft skills can be developed and which didactics are applicable. This study presents an educational soft skills experiment with IEM graduates, and it provides evidence that soft skills learning can effectively be introduced in existing courses. The graduates self-rated their competence levels of 36 soft skills before and after the course that provided soft skills workshops and a case study. In the first survey, "strategic thinking" ranked low and could be improved the most in the second survey.

INTRODUCTION

Preface

In this chapter, the role of higher education lecturers in preparing the next generation of industrial engineering managers is addressed by presenting an active learning method for the training of (1) knowledge

DOI: 10.4018/978-1-7998-8816-1.ch002

and theory, (2) professional and interpersonal skills, and (3) intrapersonal traits. It presents the results of an educational experiment in an academic master's course with students of Industrial Engineering and Management (IEM) and Business Administration (BA), more precisely, in the field of purchasing and supply chain management (PSM) which belongs to the domain of Operations Management.

The challenges of further digitalisation and the circular economy need to be addressed with technological advancement, which is the focus of this book. Technological advancement is a human effort and requires competent humans to combine hard skills ([technological] knowledge, theory and professional skills) and soft skills (interpersonal skills and intrapersonal traits or attitudes), as shown in this study.

The recommendation is to distinguish between (1) knowledge and theory, (2) professional and interpersonal skills, and (3) intrapersonal traits and to design for each of these three (I) intended learning outcomes, (II) didactical approaches and (III) assessment methods. This study distinguishes between hard and soft skills and found evidence in the literature that soft skills learning objectives are absent in academic (PSM) curricula. Therefore the following recommendation is to formalise soft skills learning objectives. Educators need to understand that soft skills cannot be assessed the same way as knowledge. For instance, to test the progress in soft skills, the students were evaluated with two surveys before and after a cognitive and soft skills training course in which they worked on a real-life case study in an experimental educational design.

The Lack of Soft Skills in Higher Education

The experiment in this chapter is performed in a PSM course. The importance of the PSM function in organisations increased in the past decades. Although make-or-buy decision-making leading to outsourcing has a long tradition in organisations and academic writings (Ammer, 1983; Gross, 1966; Jauch & Wilson, 1979), from the 1990s, organisations increasingly started outsourcing non-core activities, meaning that those were not produced within the organisation but purchased from suppliers (Cousins, Lamming, Lawson, & Squire, 2008; Luzzini & Ronchi, 2016).

The significant increase of outsourcing by organisations is affected by political-economic, technological and demographic developments. From the 1980s, the political-economic systems have been affected by the trade tariffs elimination efforts of the GATT and WTO (Narlikar, 2003), which stimulated globalisation and led to increased global sourcing activities accelerated by the Chinese economic reform from the late 1970s (Logan, 2011). Organisations increasingly concentrated on their core competences, as underlined by Prahalad and Hamel (1990), and outsourced non-core activities. The outsourcing of non-core activities led to increased supplier management, supply chain management and strategic PSM decision-making, and the scope of PSM objectives (Luzzini & Ronchi, 2016; Schoenherr, 2010).

As a result, PSM has shifted from an operational, transactional, and highly strategic function (e.g. Bals, Schulze, Kelly, & Stek, 2019; Tassabehji & Moorhouse, 2008). Bals et al. (2019) confirm the PSM function's strategic focus and point at the effects of the Internet-of-Things or the 4th Industrial Revolution and "moving towards a circular economy and circular supply chains" (Bals et al., 2019, p. 10). In the 4th Industrial Revolution affects PSM, or "Procurement 4.0," machine-to-machine communication will take over operational tasks (Bals et al., 2019) and is "influencing the digitisation of procurement and supply chains" (Bienhaus & Haddud, 2018, p. 965). Moreover, it has led to another palette of required competences. i.e. a balanced mix of hard and soft skills, especially intrapersonal traits, like 'strategic thinking' (e.g. Bals et al., 2019) and 'creativity' (e.g. Kiratli, Rozemeijer, Hilken, de Ruyter, & de Jong, 2016).

There is evidence that formal learning objectives regarding soft skills, i.e. interpersonal human-to-human skills and intrapersonal character traits, are primarily absent in academic PSM courses and tracks (Birou, Lutz, & Zsidisin, 2016; Stinenbosch, 2017; Wong, Grant, Allan, & Jasiuvian, 2014). Bals et al. (2019) argue that PSM lecturers in higher education should introduce learning objectives that cover the context of future requirements caused by the challenges of sustainability and the Internet of Things. Nonetheless, Fawcett and Rutner (2014) have found that PSM higher education is "not evolving at the pace and in the way expected by professionals" (Fawcett & Rutner, 2014, p. 181).

The literature on organisational knowledge and skills provides different taxonomies. A body of literature distinguishes between hard skills and soft skills (e.g. Andrews & Higson, 2008; Bailly & Léné, 2013; Heckman & Kautz, 2012; Laker & Powell, 2011). Heckman and Kautz (2012, p. 451) define soft skills as: "personality traits, goals, motivations, and preferences". Laker and Powell (2011, p. 113) distinguishes "hard-skills or technical training (working with equipment and software) and soft-skills training (interpersonal or intrapersonal focus)".

Thus, in higher education courses, a significant role is given to the transfers of knowledge and theory, and in parallel, academic courses are not equipped to apply those hard skills in practice, i.e. via the development of soft skills in the curricula. Ahmed, Fernando Capretz, Bouktif, and Campbell (2012) provided evidence that soft skills (i.e. interpersonal skills and intrapersonal traits) are as crucial as knowledge factors and professional skills (or hard skills) for professionals. Moreover, the lack of soft skills is more likely to be the reason for ending a labour relationship than a lack of knowledge (Ahmed et al., 2012). Employers highly value soft skills necessary to carry out professional tasks or hard skills (Ahmed et al., 2012).

Stek and Schiele (2021) provided quantitative evidence that soft skills are necessary but not sufficient conditions to carry out hard skills. The literature differentiates between "sufficient" or "necessary" conditions. Sufficient conditions lead to outcomes, whereas a necessary condition will prevent an effect from taking place (Van der Valk, Sumo, Dul, & Schroeder, 2016). Therefore, Stek and Schiele's (2021) finding that soft skills are necessary conditions for hard skills execution means that the absence of soft skills predicts the absence of effectiveness in carrying out hard skills.

The increasing importance of intrapersonal character traits such as 'strategic thinking' and 'creativity' as proposed by Bals et al. (2019) and Kiratli et al. (2016) is in line with the findings Von der Gracht, Giunipero, and Schueller (2016). They researched future PSM skills of purchasers and foresaw existential threats in organisations. When in a "talent war", competitors recruit the most "creative and innovative minds" (Von der Gracht et al., 2016, p. 30). This finding is confirmed by Von der Gracht et al. (2018, p. 9), who found that machine-to-machine communication personnel must be "creative, productive and innovative" in the future era of machine-to-machine communication personnel. Hence, the right brainpower will be decisive in the age of the Internet-of-Things and artificial intelligence (AI): "To the extent that digital transformation is also transforming our society into a knowledge society, our economy could likewise change into a knowledge economy, or even into a 'human economy', where not only intellect but especially creativity, passion, character and team spirit will make the difference" (Von der Gracht et al., 2016, p. 10).

The conclusion is that humans are distinct from machines since humans possess creative, inventive skills that machines lack. The issue of whether machines or AI can be 'creative' is countered by Du Sautoy (2019) in 'The Creativity Code: Art and Innovation in the Age of AI' with the understanding that instead of a replacement of human creativity by machines, AI will support and accelerate human creativity (Du Sautoy, 2019). Hence, the creativity and inventiveness of the human workforce will stay an essential factor.

Bals et al. (2019) underlined the importance of fully integrating all competences needed for developing PSM function into higher and professional education and industrial training programmes: knowledge, professional and interpersonal skills, and a substantial set of intrapersonal traits. Moreover, Bals et al. (2019) highlighted that student-centred teaching methods should replace traditional, frontal teacher-centred methods "as current training and teaching methods are not necessarily suitable for developing all types of competencies, and the pedagogy needs to be adapted to reflect these requirements" and suggest: "in-class training formats such as role-plays and the potential for online courses and more interactive formats, e.g. blended learning or flipped classroom approaches" (Bals et al., 2019, p. 11).

In general, the question is how PSM research could address the complexity of the 21st century, especially those that concern students' education and training practitioners. Can lecturers in academia influence students to develop these soft skills actively, or are these interpersonal skills and intrapersonal traits innate character features? If soft skills and, specifically, intrapersonal traits can be actively influenced, the question raises how these could be developed in higher education, i.e. with which didactics. Therefore, it leads to the following research question: (RQ) which interpersonal skills and intrapersonal traits can be actively developed in PSM in higher education with which didactics?

Since the PSM competences, literature mainly focuses on listing competences and has neglected to research the testing, experimenting and describing how future PSM education should be organised. This research fills this gap by presenting an educational experiment. The student-centred, learning-by-doing approach of Scholten and Dubois (2017) is adopted as part of this experiment and will be elaborated on in the theory section. In this research, the evidence is given that in a time frame of 8 weeks, with a study load of 140 hours, interpersonal skills and intrapersonal traits can be developed within the context of a real-life case.

BACKGROUND

Theory on Soft Skills Development and Student-Centred Learning

A keyword search in Google Scholar and Scopus was performed to conduct a systematic literature review, following Durach, Kembro, and Wieland (2017). The keyword search generated an initial set of 1,007 articles; a list of PSM job requirements was gradually extracted (displayed in Table 1). Articles regarding organisational (i.e., non-individual) competences or capabilities in the described fields and articles that focused on consumer skills of family household purchasing budget use were discarded, likewise minor citations. In the end, from 33 studies published between 1987 and 2020, skills requirements could be extracted (see: Appendix 1).

Table 1 shows the PSM skills' occurrences stated at least once per article and showed little unanimity. Interestingly, a more significant proportion of the studies agreed upon the need for 'PSM knowledge' (82 per cent), as well as 'business knowledge' (76 per cent) and 'negotiation skills' (73 per cent), about one-third of the studies present items that the other studies did not propose, such as industry knowledge' (33 per cent), 'creativity' (33 per cent), 'or 'legal knowledge' (30 per cent).

Table 1. PSM requirements mentioned in the scientific PSM job requirements literature (1987-2020)

PSM knowledge	80%	Curiosity / the will to learn	37%	Entrepreneurial attitude	10%
Business knowledge	73%	Industry knowledge	33%	Result-driven	10%
Negotiation skills	73%	Creativity	37%	Consultancy Advisory skills	10%
Leadership skills	70%	Legal knowledge	30%	Power handling	10%
Relationship management	63%	Quality management	30%	Discipline	10%
Analytical thinking	63%	Process management	30%	Conscientiousness	10%
Holistic thinking	63%	Blueprint reading	33%	Will to compromise	7%
Computer literacy	63%	Logistic knowledge	27%	Self-confidence	7%
Team ability	63%	Project management	33%	Perseverance	7%
Communication skills	57%	Written proficiency	27%	CAD	3%
Problem solving skills	57%	Tactfulness	27%	Own initiative	3%
Strategic thinking	50%	Motivate skills	27%	Empathy	3%
Decisiveness / decision making	50%	Time management	23%	Entreprise Resource Planning	3%
Organisational skills	43%	Information management	20%	Patience	3%
Risk management	40%	Presentation skills	20%	Common sense	3%
Persuasive skills	43%	Assertiveness	20%	Conference skills	3%
Conflict resolution	43%	Flexibility	20%	Loyalty	3%
Customer orientation	37%	Mathematics/numeral skills	13%	Pro-activity	3%

Note: PSM requirements that are mentioned at least once per article in the scientific PSM job requirements literature (1987 – 2020) (see: Appendix 1)

The data were exposed to further analyses. Since Fawcett and Rutner (2014, p. 180) acknowledged that PSM practices in firms have "evolved dramatically over the past generation", the scientific sample was divided into '20th century' (1987-2001; $n = 16$) and '21st century' (2003-2020; $n = 17$) sub-samples. Interestingly, no significant differences among the specific requirements were found with a t-test, indicating certain academic development stagnation. The advent of the Fourth Industrial Revolution might change the picture, however.

Table 1 displays PSM job requirements and the percentage of times they occurred for requirements mentioned at least once in a single scientific PSM article in the entire set of articles. The top rankings indicate a profile of a PSM professional who possesses accumulated knowledge (i.e., knowledge and experience) in PSM and business; who can negotiate, network and communicate; who can be both a team leader and member; who can think both analytically and holistically; and who can work with computer systems.

The EU Directives on Higher Education's Role in Soft Skills Development

According to the European ministers of education, institutions for higher education have to anticipate future competences since fostering innovation and creativity in society is a task for these institutions (Leuven/Louvain-la-Neuve Communiqué, 2009). Moreover, institutions in higher education endorsed the shift towards student-centred methods via the European Association of Institutions in Higher Education

(EURASHE) and the European University Association (EUA). They co-developed with the European ministers of education the standards and guidelines for quality assurance in the European higher education area (ESG Report, 2015).

The ESG Report state that: "Institutions should ensure that the programmes are delivered in a way that encourages students to take an active role in creating the learning process and that the assessment of students reflects this approach (…) Student-centred learning and teaching plays an important role in stimulating students' motivation, self-reflection and engagement in the learning process" ESG Report (2015, p. 12).

Hence, student-centred approaches are preferred, especially for the training of soft skills, attitudes or traits (e.g. Bals et al., 2019). Nevertheless, teacher-centred, frontal, and classical lectures can be considered the dominant higher education design for 'transferring' knowledge and theory (Hoidn, 2017). Multiple barriers cause the reason why universities are dominantly teacher-centred. Firstly, since dozens or hundreds of students can attend the same lecture, frontal teacher-centred methods are highly efficient, though not as effective as student-centred methods (Hannafin & Land, 2000). However, it is doubtful whether a professional activity, such as education, can be efficient when effectiveness levels are low. According to Drucker (1977), a state of efficiency can be reached before effectiveness is established: "Effectiveness is the foundation of success – efficiency is a minimum condition for survival after success has been achieved. Efficiency is concerned with doing things right. Effectiveness is doing the right things" (Drucker, 1977, p. 33).

The learning effectiveness is hidden in learning, consisting of reconstructing bits of knowledge by students themselves. The learning process is not a 'transfer' of information from a lecturer directed towards the student; it is a genuine, personal process in the individual student's mind (Land & Hannafin, 2000).

The shift from frontal, classical teaching towards student-centred didactics profoundly changes the working modus (Anthony & Kadir, 2012). Traditional lecturers who change their didactics to student-centred approaches report "feelings of guilt" because the student-centred method seems to be initially "just guiding and supporting the students in the learning processes" and knowledge is no longer "transferred" in a classical, frontal mode (Anthony & Kadir, 2012, p. 57). The guilt-feeling is caused by the wrongly expected loss of the lecturers' authority in the classroom when the lecture leaves behind the frontal, classical method (Anthony & Kadir, 2012).

Moreover, despite their importance, the learning and teaching of soft skills are more complicated. Laker and Powell (2011, p. 113) distinguish "hard-skills or technical training (working with equipment and software) and soft-skills training (interpersonal or intrapersonal focus)" and found evidence that the soft skills learning process is associated with higher levels of resistance from both, the students/trainees and the management. Thus, soft skills training has not as direct applicability to the job as hard skills training. Therefore, soft skill learning results in a lower degree of achieved proficiency and self-efficacy. Moreover, the preciseness of identifying soft skills training objectives is lower (Laker & Powell, 2011). This lower level of preciseness might also be the case for the identification of training methods.

Moreover, the PSM competence literature has presented important competences (e.g. Bals et al., 2019; Giunipero & Pearcy, 2000; Knight, Tu, & Preston, 2014; Tassabehji & Moorhouse, 2008), but has failed to present best practices on how these competences could be best taught. Whereas Feisel, Hartmann, and Giunipero (2011) found that these intrapersonal traits of experienced professionals and their strategic behaviour are difficult to influence, the research of Scholten and Dubois (2017) showed, though, the positive outcomes of an educational experiment in cohorts of students in a PSM course as is shown after the next section.

The PSM competence literature hardly provides outcomes of didactical experiments on applying the necessary soft skills, especially intrapersonal character traits in PSM courses in higher education. The question arises whether the prevailing conceptualisation of PSM is capable of comprising the 21st century's complexity since the research community seems not to address the suitable didactical instruments for soft skills development to cope with the contemporary and future challenges in the field (e.g. C. R. Carter, Rogers, & Choi, 2015; Darby, Fugate, & Murray, 2019; Knight, Meehan, Tapinos, Menzies, & Pfeiffer, 2020).

The literature might fail to address soft skills development in PSM courses in higher education because formalised soft skills intended learning outcomes are almost absent at European, US and UK universities. These are predominantly focused on just the transfer of knowledge and theory (Birou et al., 2016; Hoidn, 2017; Stinenbosch, 2017; Wong et al., 2014).

Interestingly, employers explicitly demand from institutions for higher education to formalise soft skills learning objectives (Mursion, 2021). Soft skills have an essential role in applying knowledge and cognitive skills in daily practice. Soft skills are crucial or even more important than cognitive and professional skills or 'hard skills' for a professional (Ahmed et al., 2012). That is because soft skills are necessary conditions to carry out hard skills, as explained in the introduction (Stek & Schiele, 2021).

The presence of soft skills is an excellent forecaster to success in life, and an absence appears to be causing the ending of a labour relationship rather than a lack of cognitive skills (Ahmed et al., 2012; Heckman & Kautz, 2012; Zunk & Sadei, 2015). In line with that, Forrest and Swanton (2021, p. 1) have found evidence in a longitudinal study of secondary school pupils that "Self-reported problem solving, creativity, teamwork, and verbal communication were alternately associated with later high school performance, hourly wage, and employment status". Tuononen, Parpala, and Lindblom-Ylänne (2019, p. 581) found that "having diverse competences and an ability to recognise them at the time of graduation is important for later career success and may also be related to what kind of challenges graduates face in working life".

Didactical Approaches in Higher Education

Due to changing labour market demands, higher education must anticipate adjusting the learning objectives early (Hoidn, 2017). Expectations are that in 2025 about 45 per cent of the European jobs will require high-level qualifications, and another 45 per cent will need medium-level capabilities. After graduation, students need the "ability to apply knowledge and skills flexibly in different contexts", and academia has to prepare "students with the subject-based know-how as well as with high-level transversal competences and skills such as joint problem solving, critical thinking, and self-regulated learning" (Hoidn, 2017, p. 2).

As abovementioned, formalised soft skills learning objectives are absent in higher PSM education. Several explanations are given in the literature. The training method or didactics of soft skills training differs from hard skills didactics. Soft skills development can be influenced by external stimuli but is nevertheless a personal effort, and it requires a student-centred approach. However, the dominant design at universities is teacher-centred, frontal, and classical lecturing for hard skills (Hoidn, 2017). Teacher-centred techniques are more cost-efficient but less effective than student-centred methods (Hannafin & Land, 2000). A shift from teacher-centred to student-centred didactics is a profound change of working modus that causes guilt when losing authority when not following frontal, classical didactics (Anthony & Kadir, 2012).

The development of soft skills is more complicated than that of hard skills. It is associated with higher resistance levels from the learners because it does not have direct applicability as hard skills development (Laker & Powell, 2011). Moreover, soft skill development leads to a lower degree of achieved proficiency and self-efficacy than hard skills development. Another issue is that the preciseness of distinguishing soft skills training objectives and methods is lower (Laker & Powell, 2011).

Formalising soft skills learning objectives is further obstructed by the negative association lecturers in academia have. Chamorro-Premuzic, Arteche, Bremner, Greven, and Furnham (2010, p. 238) found that "IQ was negatively associated with soft skills ratings, such that individuals with higher cognitive ability were less likely to believe that soft skills were important for outstanding academic achievement or desirable job after graduating". Lecturers in academia believe that "lower ability students may use soft skills to compensate for their poorer analytic/reasoning skills, just as conscientious students are more likely to use soft skills to improve their academic performance".

Traditionally, lecturers at (European) universities use classical, frontal lecturing for 'transferring' knowledge and theory (Hoidn, 2017). The design of frontal teaching sets students in a passive listening role, which has the lowest effect on retaining knowledge (Masters, 2013), for which Poh, Swenson, and Picard (2010) even has provided empirical evidence.

Poh et al. (2010) measured a student's neurological activity over seven days with a device, a 'wearable sensor for unobtrusive, long-term assessment of electrodermal activity.' Poh et al. (2010) show that most neurological activity is found with the student during self-study, doing homework, doing laboratory work, exams, and sleeping (dreaming). Lesser brain activity is seen during watching television, relaxing and remarkably when following lectures in the classroom. Hence, for the student, having a classical, frontal lecture in most cases has the same neurological impact as watching television or remaining in a relaxing state of mind.

Poh et al. (2010) illustrate that learning consists of reconstructing knowledge by students themselves; the learning process is not a 'transfer' of information from a lecturer directed towards the student (Land & Hannafin, 2000). De Houwer, Barnes-Holmes, and Moors (2013, p. 633) define learning as 'ontogenetic adaption', i.e. "as changes in the behavior of an organism that are the result of regularities in the environment of that organism".

Nevertheless, Hannafin and Land (2000) found that many lecturers in higher education are convinced that they could transmit the knowledge they possess to the individual students. Yet, there is broad agreement that the students have to reconstruct knowledge individually (Hannafin & Land, 2000). In line with that, the European Commission in 2008 already noted that "traditional teaching approaches based on direct instruction or lecturing are no longer adequate" and that they have to be "replaced by more learner-focused models that are based on the learner's active involvement in the process of reflection and interpretation" (Hoidn, 2017, p. 5). Active involvement in experiments is a way to activate students in their learning process to reach a comprehensive learning effect. Active involvement has better learning results than students who passively watch the same similar experiment demonstrated by a lecturer (Bonwell & Eison, 1991).

Towards a Student-Centred Approach

Student-centred learning environments are a better alternative to the dominant design of classical, frontal instruction. The introduction refers to the ESG-standards from the year 2005 (ESG Report, 2015). The conclusion is that European higher education's educational methods still do not align with the ESG-standards (Hoidn, 2017; Stinenbosch & Stek, 2017), although academia and higher education institutions were involved via the EURASHE and the EUA.

As mentioned, in PSM, not many experiments with knowledge transfer, interpersonal skills and intrapersonal traits development are described in the academic literature. A rare example is Scholten and Dubois' (2017) attempt to train students' competences. In subsequent cohorts at a Swedish and a Dutch university, case study projects have been performed in which supply chain management master students were assigned to project groups on writing a joint e-book on PSM.

Scholten and Dubois' (2017) describe the courses aimed to comprehend the state of the art PSM theoretically and practically. "The courses lie within the domain of project-based learning, one approach within the constructivist theory, where students are given an assignment to carry out one or more tasks that lead to the production of a final product" (Scholten & Dubois, 2017, p. 1687).

The main conclusion is that "the teaching context influenced the learning process and the learning outcomes. Active involvement, self-directed learning, collaborative learning and learning from practice enabled by the set-up of the course are identified as key mechanisms for the learning outcomes in relation to skills and content" (Scholten & Dubois, 2017, p. 1683).

Scholten and Dubois' (2017) approach develops students' and lecturers' cognitive, interpersonal and intrapersonal skills. Scholten and Dubois (2017, p. 1696) conclude that for a student-centred approach, educators need other lecturing skills: "due to students' active involvement in decisions regarding content and process, a lot of flexibility and creativity is required from individual lecturers," and they add: "Our results show that students who are actively involved in the learning process by taking responsibility, engaging in collaborative learning and by taking the chance to learn from practice are able to develop higher-order learning in relation to content as well as skills that are needed in today's job environment" (Scholten & Dubois, 2017, p. 1696).

Compared to the traditional, classical way of lecturing methods, the method, as explained by Scholten and Dubois (2017), appears to be more in line with the Dublin descriptors (Dublin Descriptors, 2004). These descriptors are part of the Bologna Process and should be endorsed by all European higher education institutions. The Dublin descriptors state that learning objectives in general consist of both hard and soft skills: "(i) knowledge and understanding, (ii) applying knowledge and understanding, (iii) making judgements, (iv) communication skills and (v) learning skills" (Dublin Descriptors, 2004; Leoni, 2014, p. 4).

In higher education, classical, frontal teacher-centred lecturing is the dominant design, and the teaching is mainly focused on transferring knowledge and hard skills (Masters, 2013). However, the literature underlines the importance of soft skills development at universities and institutions for higher education. Therefore, based upon Scholten and Dubois (2017) and Laker and Powell (2011), the following is hypothesised:

Hypothesis: soft skills (interpersonal skills and intrapersonal traits) can be developed in PSM courses in higher education with learning-by-doing practices.

METHODOLOGY

A Quasi-Experiment with Intrapersonal Skills in a PSM Course

In this study, an experimental approach is pursued. It is following the call of Pettigrew (2001) for a form of 'management research after modernism' to "be prepared for a period of experimentation and learning"

(Pettigrew, 2001, p. 69). Darby et al. (2019, p. 1) calls for an "expanding the methodological toolbox" of PSM and alert not to use just a sheer observative, sociological positivist approach. This plea of Darby et al. (2019) can be associated with Hacking (1984, p. 154), who stated that: "no field in the philosophy of science is more systematically neglected than experiment."

The design of measurement in education is normally a pre-experimental design that is "the exposure of a group to an experimental variable or event, the effects of which are to be measured," (X) followed by "some process of observation or measurement" (O) (Campbell & Stanley, 1966, p. 6). Hence, the standard set-up in education is X-O, primarily a series of lectures followed by a test referred to by Campbell and Stanley (1966, p. 6) as "the one-shot case study".

For this research, an O_1-X-O_2 design has been set up, or "the one-group pre-test – post-test design", which is preferred over an X-O design and "to be worth doing where nothing better can be done" (Campbell & Stanley, 1966, p. 6), which is the case, although a design with a control group would have been a better alternative. In that case, the focal group would perform the surveys and follow the lectures (O_1-X-O_2), and the control group would only perform the surveys (O_1-O_2). Since there was no access to a control group, "pre-test – post-test control group design" could not be completed (Campbell & Stanley, 1966, p. 6). The O_1-X-O_2 design in this study consisted of two identical surveys O_1 and O_2, and the experiment X: a 5 ECTS course with (online) lectures, workshops, case study, academic writing and self-study in between both surveys.

The used method to assess the difference between O_1 and O_2 is the Paired-Samples or Dependent t-test, for which the Confidence Interval Percentage is set at 95 per cent. The missing values are set to be "excluded by analysis". The students filled out the first survey before the end of the first week of the ten-week course. The same students filled out the second survey before the end of the last week of the course. The 'mean' results are derived from subtracting the different items' outcomes in the second survey from the first survey. Both surveys were measured on a 5-point-Likert scale from "fully disagree" (1) to "fully agree" (5). Moreover, Cohen's *d* effect sizes are calculated. The effect size are considered to be 'small' ($.2<d<.5$); 'medium' ($.5<d<.8$); or 'large' ($.8<d<1.2$) by Cohen (1988, pp. 25-26).

Campbell and Stanley (1966) note that an O_1-X-O_2 design comes with internal validity problems. Primarily, 'history' forms a threat to validity. "Between O_1 and O_2, many other change-producing events may have occurred in addition to the experimenter's X" (Campbell & Stanley, 1966, p. 6). In this study, a one-group pre-test – post-test design experiment is performed in a cohort of students in an elective, introductory PSM course for the master curriculum Industrial Engineering Management and master students in the Business Administration track are enrolled. Parallel to the course, the participating students followed in the same period, on average, two other courses that might have affected the second survey outcomes (O_2).

Consequently, the quasi-experiment was not performed with a parallel group of students in another traditional course that has not incorporated interpersonal skills and intrapersonal traits development in the learning objectives. Nevertheless, the results show an increase in some interpersonal and intrapersonal skills. It is uncertain whether this increase was gained due to its specific learning objectives and training methods. Further, the increase between survey O_1 and O_2 could be due to other parallel courses in the curriculum or other private life circumstances.

Finally, some students stated that becoming conscious of the competence level caused differences between the two surveys, known as the Dunning-Kruger effect (Kruger & Dunning, 1999), i.e. unskilled persons tend "to hold overly favourable views of their abilities in many social and intellectual domains"

(Kruger & Dunning, 1999, p. 1121). However, skilled persons have the tendency "to underestimate their performance relative to their peers" (Kruger & Dunning, 1999, p. 1126).

There is a possibility that this effect occurred that student overestimated their perceived competence levels in the first survey and understated those in the second. The study is an experiment that measured in a timeframe of eight weeks twice the perceived competence levels. Between both surveys, the student got trained and skilled and might have developed from unaware to aware of their incompetence. There is a risk that the students filled out the first survey overstating competence levels and understating those in the second.

Course Design in Three Lines: Knowledge, Skills and Attitudes/Traits

The course attracted 95 students (30 females and 65 males; 82 Dutch students, seven citizens from other EU-countries and four from non-EU countries) in three subsequent cohorts in 2018-2019 (30 students: 8 females and 22 males; 24 IEM and 6 BA students; average age 23.7, δ=1.75); in 2019-2020 (26 students: 12 females and 14 males; 19 IEM and 7 BA students; average age 23.0, δ=1.06); and in 2020-2021 (39 students: 10 females and 29 males; 28 IEM and 11 BA students; average age 24.3, δ=2.52). About 75 per cent of the students were enrolled at the master track IEM and 25 per cent at BA studies. Students from both tracks were randomly assigned to working groups. The course consisted of about 16 lectures and workshops of 90 minutes and was divided into a knowledge line, a skills line, and an attitude or intrapersonal traits line, which formed the whole construct of the experimental approach of the course (X).

The researched sample consists of IEM and BA master students of an elective, introductory course, Purchasing Management, at a Dutch polytechnical university. The master course has been designed in such a way that it offered education in (1) knowledge and theory, (2) professional and interpersonal skills, and (3) intrapersonal traits. The didactical construct provided a mix of frontal, classical instructions, a Massive Open Online Course (MOOC), practical workshops, storytelling and learning-by-doing in two larger projects with tutoring meetings. The course philosophy is that competence consists of knowledge and theory, professional and personal skills and intrapersonal traits or attitudes (e.g. Campion et al., 2011; Delamare-Le Deist & Winterton, 2005). The knowledge and soft skills training served the case-solving (professional and personal skills) part, and the case gave meaning to the course's content.

The course syllabus stated: "Aims - To acquire 'success skills' from practice [knowledge (hard skills) as well as interpersonal as intrapersonal skills] to analyse the performance of the purchasing function in an organisation and make well-founded, inventive decisions on complex strategic purchasing issues" (…). "The didactic of this course is (…) entirely student-centred: PSM professionals are invited to perform active workshops on the content and the personal development, and the further lectures are built upon student participation. Each student group works out one of a dozen subjects and are subsequently digesting the textbook, journal articles and reports into a working paper and presentation. The students are encouraged to work in teams, communicate in a cross-cultural setting, consider calculated risk, think out-of-the-box, and find creative and inventive solutions to complex solutions problems to access the PSM content and theories. The aim is to (further) develop individual traits that are indispensable for a future career as a professional".

De Houwer et al. (2013, p. 633) defined learning as "ontogenetic adaption", i.e. "as changes in the behavior of an organism that are the result of regularities in the environment of that organism". Hence, effective learning leads to a sustained, long-term impact on the learner. Therefore, in the course, 'Constructive alignment' is applied, an educational approach introduced by Biggs (1996, p. 347) that combines two ways of thinking: "the first derives from constructivist learning theory and the second from the instructional design literature". Central in constructivism is creating the meaning of the learner's activities, impacting the teaching and assessment methods. The "Instruction design" underlines the alignment between a course's learning objectives and the student's performance assessment methods.

Biggs (1996, p. 347) combines both ways to "Constructive alignment". Constructivism is applied as the instructional design framework to create curriculum objectives "in terms of performances that represent a suitably high cognitive level, in deciding teaching/learning activities judged to elicit those performances and to assess and summatively report student performance" (Biggs, 1996, p. 347). Whereas the definition of competence by Campion et al. (2011) or Delamare-Le Deist and Winterton (2005) reveals a triangle of (1) knowledge and theory, (2) professional and interpersonal skills, and (3) intrapersonal traits, Biggs' (1996) Constructive alignment alerts educators to align (I) intended learning outcomes, (II) didactical approaches and (III) assessment methods. Therefore, for each of the three lines in the course, Constructive alignment is applied. The learning of knowledge and theory, professional and interpersonal skills, and intrapersonal traits differ in intended learning outcomes and hence also in didactics and assessments methods, as is shown below.

The knowledge line's practical substance consisted of an assignment to the students to co-author a book entitled 'State of the Art of Purchasing and Supply Management', inspired by Scholten and Dubois (2017). Teams were formed of about three students and were assigned to write a scientific paper, i.e. book chapter on a PSM topic. At Dutch universities, lecturers usually leave the initiative with the students to team up in groups. In most cases, this appears to lead to mono-cultural, uni-gender teams of acquainted students, which would not necessarily represent the daily practice in these graduates' future professional lives; usually, professionals cannot pick their fellow team members.

For each book chapter assignment, a topic and some guidance were given, such as two or three crucial articles and how to start academic writing. The book chapter project urged them to use the most recent literature and cite a minimum of 15 peer-reviewed articles. The use of the annotation program EndNote was set obligatory. Each student group was invited at least twice to meet the lecturer for 30 minutes to structure and improve the paper writing process. After the deadline and the lecturer's final editing, the book with ten chapters was made available in pdf format via intranet and was handed out in hard copy during the open-book exam. The student groups presented their chapter in a 15-minutes PowerPoint presentation during the final lectures.

However, the lecturer again teamed up the students for the skills line in other groups than for the book chapter writing assignment. The reason for that is twofold. Firstly, it is unlikely that professionals can form their teams or be consigned to two identical groups in professional life. Secondly, teaming up in different teams avoided exchanging tasks between students, leading to freeriding in one of the projects.

Real-life case studies were co-developed with the purchasing department of the university. The cases regarded the public tender procedures of tenders that would be carried out a few months later by the purchasing department. In the different lectures in the subsequent academic years, workshops were organised with guest speakers, such as chief purchasing officers (CPO) and purchasing experts.

In the case of kick-offs, the university purchasers, contract-owners and contract managers had a role. The case studies have been designed to be easily explained: e.g. 'the university needs new faculty housing' or 'the contract of the hot beverages vending machine is ending' and 'a tender procedure is upcoming'. Nevertheless, these 'simple' problems were hard to solve due to all the facets, like stakeholders' interest, sustainability issues, et cetera.

The case studies were subtitled 'talking with real people'. Therefore the written info in the case study assignments was limited. Instead, the students were invited to raise oral questions to obtain more information from the purchasers, contract-owners and contract managers, who are the real people who will work on the same case in real life in the following months. These practitioners, who work for the university, cooperate for different reasons. One reason is the willingness to contribute to the university's educational process and stay connected to the employers' core business. Another critical reason is exchanging ideas with the students, the guest speakers and the lecturers to get insights from a new angle and obtain a synergy advantage in the upcoming tender procedures.

For both the knowledge and the case line, in 2018 and 2019, classical and in 2020, due to the Covid-lockdown, online, frontal instruction lectures were provided on topics like public procurement procedures and the selection and awarding; purchasing management; supplier selection; and innovation sourcing. In addition, in the attitudes or intrapersonal traits line, several workshops were provided, such as a CPO's workshop on ethical behaviour and sound leadership, an interim management and consultancy agency workshop on consultancy skills, a creativity activation workshop and other necessary skills in a purchasing consultancy job; a negotiation lecture and workshop. Hence, the students are engaged with stories, experiences and workshops of several practitioners in the attitudes/traits line that served their performance in the case study. The cognitive lectures and paper writing in the knowledge line also partly served the solving of the case studies.

The third line in the course regarded the development of attitudes or intrapersonal traits. For this line, the one-group pre-test – post-test design experiment is performed in an O_1-X-O_2 design, whereas O_1 (observation 1) represents the first survey that was held in the first week of the course; X (exposure) represents the exposure to soft skills training; and O_2 (observation 2) represents the second survey that is identical to the first and was held in the last week of the course.

The survey consisted of 36 interpersonal skills and intrapersonal traits with a five-point Likert scale ("fully disagree" [1] to "fully agree" [5]). The 36 interpersonal and intrapersonal skills (see: Appendix 2) were derived from Stek and Schiele (2021), who performed an extensive PSM competence survey for the Erasmus+ PSM education project PERFECT, for which they selected 88 hard and soft skills items. Stek and Schiele (2021) derived these items from PSM competence literature (e.g. Giunipero & Pearcy, 2000; Tassabehji & Moorhouse, 2008) and depended for most of the soft skills items on KODE®X, which is a tool for the exploration of strategic corporate competency requirement profiles of requirement- or task-specific skills-set profiles and skills of potential employees and managers used. KODE®X was developed for an assignment to change the professional roles of 140 engineers and architects towards a job as project manager (Erpenbeck & Scharnhorst, 2005; Heyse, Erpenbeck, & Max, 2004).

The first survey in the experiment in the three cohorts was taken before the second lecture, and the second survey was taken after the final lesson. Thus, the first survey's results were kept unrevealed for the individual students until after filling out the second survey. Therefore, when filling out the second survey, the individual students were unaware of their initial scores in the first survey about eight to ten weeks earlier. After the course, students compared their scores and handed in a reflection on the course, including a personal development plan. For the analysis, Paired Samples or Dependent t-tests were performed. The four main assumptions of Paired Samples t-test have been met. The dependent variable is continuous, are normally distributed and does not contain any outliers. Moreover, the observations were independent of one another (Field, 2009). For the surveys, ethical approval was received from the university's Ethics Committee, and all students approved the use of anonymised data for scientific use.

RESULTS

An Increase of Levels of Interpersonal Skills and Intrapersonal Traits

The results of both surveys or observations O_1 and O_2 as per Campbell and Stanley (1966) were subject to Dependent *t*-tests. In total, 26 items showed a significant difference, and ten did not, as shown in Table 2. Herewith an answer is given to the research question on which soft skills can be developed. Moreover, there is enough evidence not to reject the hypothesis; indeed, the evidence is found for a substantial number of soft skills that these can be developed in a relatively short time frame of a ten-week course of 5 ECTS with the described didactics: a mix of frontal, classical instructions, a MOOC, practical workshops, storytelling and learning-by-doing in two larger projects with tutoring meetings.

The course was successful in increasing 'strategic thinking', 'negotiation skills' and 'salesperson skills' with significant p-values ($p < 0.01$) and 'medium' Cohen's *d* effect sizes ($.5<d<.8$) (Cohen, 1988, pp. 25-26; Sawilowsky, 2009). 'Strategic thinking' has the strongest effect size, meaning that the progress that the student sample made for this survey item is significant ($p < .000$) and with a 'medium' effect size (Cohen, 1988, pp. 25-26; Sawilowsky, 2009). 'Strategic thinking' was part of the case study's learning objectives; a lecture and workshop in strategic management were part of the case. The first assignment in the case consisted of forming a vision on higher education development in the following decades to define the building's purpose on the university campus.

'Negotiation skills' improved significantly, and the effect size is 'medium' ($p=.000$; $d=.636$) (Cohen, 1988, pp. 25-26; Sawilowsky, 2009). 'Negotiation skills' are explicitly practised during the negotiation workshop. 'Salesperson Skills' has a significant outcome with a medium effect size ($p=.000$; $d=.538$).

Furthermore, the course improved networking, teamwork, leadership, problem-solving, cross-functional management, communication skills, and empathic capacity to a lower extent, showing lower significances ($.01<p<.05$) and smaller effect sizes. Cohen's *d* effect sizes in these cases are 'small' ($.2<d<.5$) (Cohen, 1988; Sawilowsky, 2009). In the one-group pre-test – post-test, the students filled out identical surveys before and after the course and self-reported their skills levels in 36 skills as displayed in Table 1.

The course caused a significant difference in 'strategic thinking,' a relatively lower-ranked item in the students' sample means. On a 5-point Likert scale, the students' mean in O_1 was 2.70, and in O_2, it increased to 3.21. The impact of the course is considerable but should not be exaggerated. The course has played a role in developing students' cohorts from a lower level towards a more average strategic thinking level and is undoubtedly not excellent. For example, the listing of 36 competence items 'strategic thinking' was ranked 29 in O_1 and O_2. It increased to rank 15 (see: Appendix 2). 'Salesperson skills' and 'negotiation' remained at respectively rank 35 and 36.

Therefore, the initially lower-ranked competences are more likely to be subject to improvement. It is not surprising that the survey items with a higher self-rated competence level cannot be improved as those with lower self-ratings since the first set of items are developed, and the latter are underdeveloped. In other words, the proverbial low-hanging fruit is found in lower self-rated competences. Figure 1 the two rankings of the 36 competence items displayed in a scatter plot: the ranking in t-values from low to high and the perceived competence level from low to high on the 5-point Likert scale. The trendline reveals a negative relationship between both. Hence, the lower the initially perceived competence level, the higher the chance of a significant improvement. Table 3 shows evidence that the slope of the regression is significant ($p=.000$). The R^2 is .402.

Table 2. Effect sizes of significant differences O_1 and O_2 interpersonal and intrapersonal skills levels in a paired-samples or dependent t-test

		Survey scores		*Paired Differences*		*t-value*	*Sig. (p-value 1-tailed)*	*Cohen's d*
		Survey 1	*Survey 2*	*Mean*	*SD*			
1	Strategic Thinking	2,70	3,21	.518	.713	5.433	.000	.726
2	Negotiation skills	2,21	2,71	.500	.786	4.759	.000	.636
3	Salesperson skills	2,36	2,80	.446	.829	4.028	.000	.538
4	*Networking / Building Relations*	2,49	2,86	.351	.772	4.409	.000	.455
5	*Communication skills*	3,06	3,34	.362	.914	3.837	.000	.396
6	*Decision-making*	2,83	3,16	.309	.804	3.722	.000	.384
7	*Cross-functional management*	2,86	3,23	.391	.828	3.772	.000	.472
8	*Ability to Solve Problems*	3,13	3,38	.277	.768	3.493	.000	.360
9	*Comprehension of Complexity*	2,95	3,19	.245	.683	3.474	.000	.358
10	*Capacity to be empathetic*	3,34	3,66	.339	.721	3.524	.000	.470
11	*Leadership/community management*	2,51	2,89	.359	.824	3.491	.000	.436
12	*Flexibility and agility*	2,79	3,07	.277	.835	3.212	.001	.331
13	*Creativity*	2,60	2,83	.213	.670	3.079	.001	.318
14	*Inventiveness*	2,58	2,86	.277	.873	3.073	.001	.318
15	*Persistence*	2,70	3,04	.339	.837	3.033	.002	.405
16	*Proactivity*	2,87	3,10	.213	.731	2.821	.003	.279
17	*Teamwork*	3,17	3,47	.313	.889	2.813	.003	.351
18	*Cross-cultural awareness*	2,86	3,06	.245	.876	2.708	.004	.279
19	*Willingness to take risks*	2,83	3,14	.191	.766	2.424	.009	.250
20	*Customer-oriented*	3,14	3,27	.191	.846	2.195	.015	.226
21	*Stress management*	2,98	3,16	.297	1.122	2.116	.019	.265
22	*Willingness to Learn*	2,87	3,02	.160	.780	1.983	.025	.204
23	*Capacity to Advice*	2,87	3,03	.181	.904	1.941	.028	.200
24	*Holistic Thinking*	3,11	3,28	.149	.747	1.933	.028	.199
25	*Poise*	2,66	2,80	.149	.747	1.933	.028	.199
26	*Task management*	3,38	3,47	.219	.917	1.909	.030	.239
27	Power of Persuasion	2,95	3,12	.143	.699	1.530	.066	.204
28	Result-orientated action-taking	2,94	3,05	.128	.688	1.422	.079	.147
29	Self-assurance	2,73	2,94	.106	.809	1.274	.103	.131
30	Inter-generation skills	2,96	3,09	.156	.996	1.256	.107	.156
31	Honesty	3,67	3,74	.096	.804	1.154	.126	.119
32	Conscientiousness	3,47	3,50	.125	.833	1.124	.133	.150
33	Social Manners	3,32	3,32	.096	.881	1.054	.147	.108
34	Ability to Resolve Conflicts	3,21	3,27	.096	.928	1.000	.160	.103
35	Critical thinking	3,15	3,25	.107	.867	0.925	.180	.124
36	Loyalty	3,69	3,77	.053	.884	0.583	.281	.060

See Appendix 2 for the item's definitions. The sample consists of Business Administration and Industrial Engineering and Management Master students (n=94) in the cohorts 2018-2019 (n=30), 2019-2020 (n=26), and 2020-2021 (n=38; one student resigned during the course) of the introductory course Purchasing Management at the University of Twente. Used method: Paired t-test; Confidence Interval Percentage: 95 per cent; Missing Values: Exclude cases analysis by analysis (Cohen, 1988, pp. 25-26; Sawilowsky, 2009).

Figure 1. Graph displaying the relationship between perceived competence levels in O_1 and the significance of the progress of the personal skills in the course (t-values)
Note: This figure displays the relationship between the perceived competence levels in survey 1 (O_1) and the significance of the progress in perceived competence levels from survey 1 (O_1) to survey 2 (O_2) (t-values) (R^2 Linear = .402). In other words, skills items with lower scores in the first survey are more likely to improve during the course than skills items with higher scores.

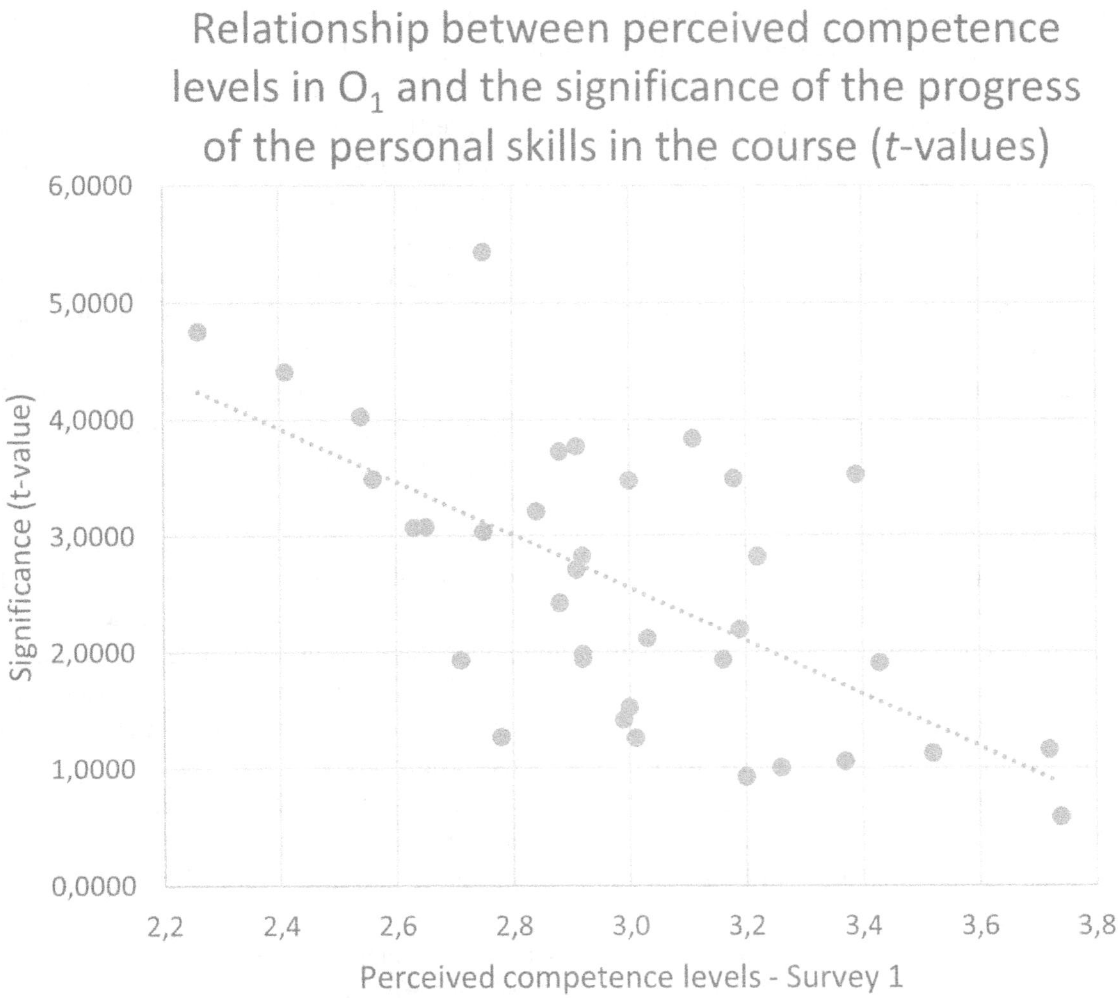

Table 3. Regression output belonging to Figure 1

Coefficients [a]						
		Unstandardised Coefficients		Standardised Coefficients	*t*	Sig.
		B	Std. Error	Beta		
(Constant)		9.265	1.410		6.573	.000
Perceived competence level (survey 1)		-2.276	.476	-.634	-4.781	.C00
a. Dependent Variable: Significance of progress (t-values)						

SOLUTIONS AND RECOMMENDATIONS

Contextual Fundament for Attitudes / Intrapersonal Traits Development

The research question is: which interpersonal skills and intrapersonal traits can be actively developed in PSM in higher education with which didactics? The skills items are displayed in Table 1; however, the second part is not answered explicitly. The didactical approaches of the experiment are multiple. The knowledge and theory line was dominated by (1) frontal, classical instruction, (2) the use of a MOOC, (3) the literature review writing assignment. This assignment led to the writing of chapters of a joint book used for the exam. The lecturer facilitated the writing of literature reviews with at least two 30-minutes tutoring meetings with the groups on the form and content of the chapter. The exam and the literature review were graded. For the literature review, a scoring rubric is developed.

The professional and interpersonal skills lines didactics were focused on learning-by-doing in practical workshops given by experts on negotiation and advisory or consultancy skills. Moreover, the students were assigned randomly to groups and could not form their team. It led to diverse, non-mono-cultural groups that would not usually appear in master courses, i.e. teams consisting of members with the same nationality and mother tongue and teams formed by students from the same gender. Moreover, the students had to present the results of the case study and literature review for their cohorts. The case study reports and presentations were graded following a rubric.

The intrapersonal traits or attitudes didactics were offered via storytelling and learning-by-doing. A CPO shared experiences in cross-cultural settings, with ethical issues and personal leadership. In a creativity lecture and workshop, the students were instructed how creative team processes occur and were encouraged to practice this in the case study groups. The assessment of this line was handing in a personal reflection and development plan based upon the differences in both surveys. The plans formed a necessary condition to receive the final grade and were not graded; handing in a “serious” plan was enough, and all students in all cohorts handed in “serious” reflections.

Hence, as shown in Table 2 and Figure 1, the students could acquire lesser developed soft skills. The three lines in the experiment formed an intertwined construct of learning knowledge and theory, professional and interpersonal skills, and intrapersonal traits or attitudes with multiple didactics and assessment methods.

Remarkably, two forms of thinking, ‘critical thinking’ and ‘strategic thinking’, represent the second lowest and the highest significance and effect size. Moreover, ‘critical thinking’ is perceived as a Top-10 ranked competence with the surveyed students, whereas ‘strategic thinking’ is ranked amongst the lowest ten. The students stated that they developed critical thinking in the four to five years of academic training (and over ten years of elementary and secondary school education). Strategic thinking, however, appeared to be underdeveloped.

Therefore, students could be more familiar with the term ‘critical thinking’ since they might have been primed with the notion that ‘critical thinking’ is a desirable objective for a student in general. In literature, there is a discourse on ‘thinking’-learning objectives, like ‘critical thinking’, ‘academic thinking’, et cetera. The discourse is led by Willingham (2008), who questions whether critical or other kinds of specific thinking are skills on its own: “If you remind a student to ‘look at an issue from multiple perspectives’ often enough, he will learn that he ought to do so, but if he doesn’t know much about an issue, he can’t think about it from multiple perspectives” (Willingham, 2008, p. 21). Willingham (2008) pleas for critical or other specific thinking in a given context, which is in line with Delamare-Le Deist and Winterton's (2005) definition of competency as a construct of three elements, knowledge, skills, abilities and other characteristics.

'Strategic thinking' is a skill that employers appreciate (Bals et al., 2019). The literature shows that attitudes and intrapersonal traits as 'strategic thinking' are hardly taught in academia (Birou et al., 2016; Wong et al., 2014). Thus, the course was directed on the strategic purchasing management theory and urged the students to think about strategic PSM-related problems. The case study provided a context to the thinking process.

Competence items that were part of the survey but were not formalised in the learning objectives for the introduction course to Purchasing Management, such as 'salesperson skills' and the 'willingness to take risks', showed a remarkable improvement. However, the modern purchaser needs these entrepreneurial skills, which already was confirmed by Giunipero and Pearcy (2000). Other significant unintended improvements are 'persistence', 'proactivity', 'teamwork' and 'cross-cultural awareness'. The evidence is provided with the sample consisting of business master students who are used to work in student groups, 'teamwork' skills can be improved significantly when these students are assigned to groups instead of form groups, i.e. to continue in old structures.

As stated in the introduction, the transfer of knowledge and theory has a significant role in higher education. Indeed, the importance of knowledge and theory is undeniable. However, soft skills (i.e. interpersonal skills and intrapersonal traits) are necessary to carry out professional tasks (Stek & Schiele, 2021). Moreover, employers highly value soft skills, and the lack of soft skills is more likely to be the reason for ending a labour relationship than a lack of knowledge (Ahmed et al., 2012).

For the operationalisation of strategic management knowledge and theory, strategic thinking or strategic handling is needed. In strategic management courses, students are taught strategic management knowledge and theory (which was also done in the course). However, in most of these courses, the students are not taken to the next level of carrying out strategic management and lack competences in strategic thinking. Delamare-Le Deist and Winterton (2005) showed evidence that knowledge, skills, and attitudes form a construct and Willingham (2008) expresses that attitudes and traits should be developed within a context (of knowledge and theory).

Giunipero (2000) distinguishes between hard skills (PSM knowledge and skills) and soft skills (intrapersonal traits and intrapersonal traits in PSM). A "world-class purchaser" is "continuously improving his/her skills; (…) is focused on professional development and education; (…) is willing to change and adapt; is a problem solver seeking the best solution; is flexible; (…) is ethical; (…)" and "adapts well to change" (Giunipero, 2000, p. 8), within the PSM context, as suggested by Willingham (2008).

RQ1 is focused on which soft skills 'could' be developed. Like is hypothesised, the evidence is shown that soft skills can be trained, which is confirmed by Laker and Powell (2011), although they mention that soft skills training comes with higher levels of resistance from participants and their managers. Students characterised the course as "different" from other courses and confirmed in their reflections that they could improve their soft skills in the course.

A male IEM-student in the cohort 2020-2021 noted: "The case we got to work on was very interesting and had a lot of room for creativity. Because it was a group assignment, you could also learn a lot from the creativity of your project members. This has had a positive effect on my own creativity skill. The assignment that came with the case had almost no guidelines. This also meant that you had to estimate for yourself what was asked". A female BA-student in the 2020-2021 cohort: "I thought the creativity lecture and workshop were interesting! It showed and taught me more ways of how to think of creative ideas and concepts. I think these might be very interesting for my future work-life".

A female IEM-student in the 2020-2021 cohort elaborate on what is referred to above as the Dunning-Kruger effect: "For the skill creativity, i.e. being creative in professional life / having creative ideas, the difference was -1 (i.e. the score of survey O_1 minus that of O_2). The reason behind this is that after attending the lecturer's creativity lecture, I had a different view on creativity. During the lecture, the lecturer mentioned that "creativity leads to innovativeness when it is novel and useful". During my working experience, I had many business ideas that were not that useful. (…) I had a very poor view of what this skill actually meant, but after the course, I could reflect on my working experience and noticed that I actually do manage this skill very well".

CONCLUSION

Hence, the question of whether soft skills 'could' be taught is answered, i.e. that the self-perceived competence levels of most soft skills have improved significantly. Nevertheless, Laker and Powell's (2011) findings trigger whether, in academia, soft skills should be taught. As shown by Ahmed et al. (2012) and many scholars in the PSM competence field, employers would agree (e.g. Bals et al., 2019; Feisel et al., 2011; Giunipero & Pearcy, 2000). The European ministers of Education also would agree, according to the statement that: "Higher education should be based at all levels on the state of the art research and development thus fostering innovation and creativity in society" (Leuven/Louvain-la-Neuve Communiqué, 2009, p. 4).

Interestingly, many parties would agree that soft skills, more precisely intrapersonal traits learning objectives, should be applied in academic courses, most notably by employers, politicians, and PSM scholars. Nevertheless, academia failed to offer a balanced volume of knowledge and theory, professional and interpersonal skills and intrapersonal traits. Hence, soft skills 'could' and 'should' be taught in higher education, but they are absent in intended learning outcomes. It raises the question of whether soft skills 'would' be taught in academia. Moreover, soft skills education is less attractive for all stakeholders. Students, trainees, and management prefer hard skills over soft skills education (Laker & Powell, 2011), and lecturers feel guilty when shifting to student-centred methods (Anthony & Kadir, 2012).

The student evaluations made it clear that the course was mostly positively evaluated and was experienced as 'different from other courses'. Remarkably, the students in the cohort 2020-2021 that, due to the Covid-measurements, followed most of the lectures online, seem to have appreciated the course better than the preceding cohorts. Probably, the reason why the students enjoyed the course in times of Covid lockdowns and distant learning was because of the student-centred approach. The course demanded many (online) interactions. The students worked in two separate teams on the book chapter and the case, and per group, multiple mentor meetings were planned with the lecturer.

The students revealed that they would usually start studying the lecture notes and PowerPoint presentations a week or two before the exams. In this course, the students studied their lecture notes earlier because the knowledge formed the case context or the book chapter. The evidence shows that the course followed Willingham's (2008) call for a contextual basis to develop specific thinking forms. Moreover, with the outcomes of the research of Poh et al. (2010) in mind, regarding the intrapersonal traits, the course content seems to have caused 'neurological activity' within the students' brains.

The recommendation of this study for educators is to consider the competence construct of knowledge, skills and attitude (e.g. Campion et al., 2011; Delamare-Le Deist & Winterton, 2005). Competent persons tend to possess the three elements of this construct. The course in this study offered the learning knowledge, skills and traits were facilitated, whereas regular classes focus only on knowledge and theory (Birou et al., 2016; Hoidn, 2017). This study illustrates how a construct is implemented and meant to inspire other educators and formalise soft skills learning objectives.

FUTURE RESEARCH DIRECTIONS

Limitations and Further Research

Indeed, the course might have caused 'neurological activity', i.e. some form of learning. The first limitation of this study is that it is questionable whether learning as 'ontogenetic adaption' occurred as meant by De Houwer et al. (2013, p. 633), who defined learning "as changes in the behavior of an organism that are the result of regularities in the environment of that organism".

The course consisted of three cohorts of only 95 students of an elective introduction course to PSM. Fifth, after getting information about the course's first lecture, a dozen students usually sign out for different reasons. Finally, in some cases, the required courses' timetables interfered with this elective course, and students expressed that they did not like the course design. Hence, only interested students stayed in the course after the introduction, which may be seen as a respondent's bias and probable convenience sampling.

The majority, 82 of the 95 students (86 per cent), has Dutch nationality, which might have caused cultural bias. Seventh, another potential limitation could be a social-desirability bias, i.e. that (some) students might have anticipated and replied having developed (some) skills in the course and therefore replied differently in the second survey. However, as mentioned, the first survey's outcomes were not disclosed before filling out the second survey.

Further research is suggested by replicating this study to test if the same results can be reached in other (cultural) circumstances. Moreover, it would be recommendable to distinguish in further research the most critical interpersonal and intrapersonal skills that lead to PSM success and replicate this study with a redesigned course and subsequent learning objectives.

The 'methodological toolbox' should not be restricted to a sheer observative, positivist research (Chicksand, Watson, Walker, Radnor, & Johnston, 2012; Darby et al., 2019; Pettigrew, 2001); PSM and PSM educational research would benefit from active scholarly involvement in the complex challenges the field is facing regarding digitalisation and circularity.

REFERENCES

Ahmed, F., Fernando Capretz, L., Bouktif, S., & Campbell, P. (2012). Soft skills requirements in software development jobs: A cross-cultural empirical study. *Journal of Systems and Information Technology*, *14*(1), 58–81. doi:10.1108/13287261211221137

Ammer, D. S. (1983). Purchasing and Materials Management for Health-Care Institutions (2nd ed.). Lexington Books.

Anderson, M. G., & Katz, P. B. (1998). Strategic sourcing. *International Journal of Logistics Management*, *9*(1), 1–13. doi:10.1108/09574099810805708

Andrews, J., & Higson, H. (2008). Graduate employability, 'soft skills' versus 'hard' business knowledge: A European study. *Higher Education in Europe*, *33*(4), 411–422. doi:10.1080/03797720802522627

Anthony, E. M., & Kadir, Z. A. (2012). A road not taken: A breakthrough in English for specific purposes via problem-based learning. *Journal of Technical Education and Training*, *4*(1).

Bailly, F., & Léné, A. (2013). The personification of the service labour process and the rise of soft skills: A French case study. *Employee Relations*.

Baily, P., Farmer, D., Crocker, B., Jessop, D., & Jones, D. (2008). *Procurement principles and management* (10th ed.). Pearson Education.

Bals, L., Schulze, H., Kelly, S., & Stek, K. (2019). Purchasing and supply management (PSM) competencies: Current and future requirements. *Journal of Purchasing and Supply Management*, *25*(5), 100572. doi:10.1016/j.pursup.2019.100572

Bienhaus, F., & Haddud, A. (2018). Procurement 4.0: Factors influencing the digitisation of procurement and supply chains. *Business Process Management Journal*, *24*(4), 965–984. doi:10.1108/BPMJ-06-2017-0139

Biggs, J. B. (1996). Enhancing teaching through constructive alignment. *Higher Education*, *32*(3), 347–364. doi:10.1007/BF00138871

Birou, L., Lutz, H., & Zsidisin, G. A. (2016). Current state of the art and science: A survey of purchasing and supply management courses and teaching approaches. *International Journal of Procurement Management*, *9*(1), 71–85. doi:10.1504/IJPM.2016.073388

Bonwell, C. C., & Eison, J. A. (1991). *Active Learning: Creating Excitement in the Classroom*. 1991 ASHE-ERIC Higher Education Reports: ERIC.

Burt, D. N., Dobler, D. W., & Starling, S. L. (2003). *World class supply management: The key to supply chain management*. McGraw-Hill Irwin.

Campbell, D. T., & Stanley, J. C. (1966). *Experimental and quasi-experimental designs for research*. Houghton Mifflin Company Boston.

Campion, M. A., Fink, A. A., Ruggeberg, B. J., Carr, L., Phillips, G. M., & Odman, R. B. (2011). Doing competencies well: Best practices in competency modeling. *Personnel Psychology*, *64*(1), 225–262. doi:10.1111/j.1744-6570.2010.01207.x

Carr, A. S., & Smeltzer, L. R. (2000). An empirical study of the relationships among purchasing skills and strategic purchasing, financial performance, and supplier responsiveness. *The Journal of Supply Chain Management*, *36*(3), 40–54. doi:10.1111/j.1745-493X.2000.tb00250.x

Carter, C. R., Rogers, D. S., & Choi, T. Y. (2015). Toward the theory of the supply chain. *The Journal of Supply Chain Management*, *51*(2), 89–97. doi:10.1111/jscm.12073

Carter, J. R., & Narasimhan, R. (1996). Is purchasing really strategic? *International Journal of Purchasing and Materials Management*, *32*(4), 20–28. doi:10.1111/j.1745-493X.1996.tb00216.x

Cavinato, J. L. (1987). Purchasing performance: What makes the magic? *Journal of Purchasing and Materials Management*, *23*(3), 10–16. doi:10.1111/j.1745-493X.1987.tb00186.x

Chamorro-Premuzic, T., Arteche, A., Bremner, A. J., Greven, C., & Furnham, A. (2010). Soft skills in higher education: Importance and improvement ratings as a function of individual differences and academic performance. *Educational Psychology*, *30*(2), 221–241. doi:10.1080/01443410903560278

Chicksand, D., Watson, G., Walker, H., Radnor, Z., & Johnston, R. (2012). Theoretical perspectives in purchasing and supply chain management: An analysis of the literature. *Supply Chain Management*, *17*(4), 454–472. doi:10.1108/13598541211246611

Cohen, J. (1988). *The effect size index: d* (Vol. 2). Lawrence Erlbaum Associates, Publishers.

Cousins, P. D., Lamming, R., Lawson, B., & Squire, B. (2008). *Strategic supply management: principles, theories and practice*. Pearson Education.

Cousins, P. D., & Spekman, R. (2003). Strategic supply and the management of inter-and intra-organisational relationships. *Journal of Purchasing and Supply Management*, *9*(1), 19–29. doi:10.1016/S1478-4092(02)00036-5

Cruz, C., & Murphy, E. (1996). Purchasing's new importance requires a broader education. *Purchasing*, *12*(9), 46–49.

Darby, J. L., Fugate, B. S., & Murray, J. B. (2019). Interpretative research: A complementary approach to seeking knowledge in supply chain management. *International Journal of Logistics Management*. Advance online publication. doi:10.1108/IJLM-07-2018-0187

De Houwer, J., Barnes-Holmes, D., & Moors, A. (2013). What is learning? On the nature and merits of a functional definition of learning. *Psychonomic Bulletin & Review*, *20*(4), 631–642. doi:10.375813423-013-0386-3 PMID:23359420

Delamare-Le Deist, F., & Winterton, J. (2005). What is competence? *Human Resource Development International*, *8*(1), 27–46. doi:10.1080/1367886042000338227

Dowd, K. O., & Liedtka, J. (1994). What corporations seek in MBA hires: A survey. *The Magazine of the Graduation Management Admission Council, 2*.

Drucker, P. F. (1977). *People and performance: The best of Peter Drucker on management*. Harper & Row, Publishers, Inc.

Du Sautoy, M. (2019). *The Creativity Code: Art and Innovation in the Age of AI*. Harvard University Press.

Dublin Descriptors, J. Q. I. (2004). *Dublin descriptors. The Joint Quality Initiative Working document*. Shared Dublin descriptors for the Bachelor's, Master's and Doctoral awards.

Durach, C. F., Kembro, J., & Wieland, A. (2017). A new paradigm for systematic literature reviews in supply chain management. *The Journal of Supply Chain Management*, *53*(4), 67–85. doi:10.1111/jscm.12145

Eltantawy, R. A., Giunipero, L. C., & Fox, G. L. (2009). A strategic skill based model of supplier integration and its effect on supply management performance. *Industrial Marketing Management, 38*(8), 925–936. doi:10.1016/j.indmarman.2008.12.022

Erpenbeck, J., & Scharnhorst, A. (2005). Modellierung von Kompetenzen im Licht der Selbstorganisation. In T. Meynhardt & E. J. Brunner (Eds.), *Selbstorganisation Managen - Beiträge zur Synergetik der Organisation*. Waxmann Verlag GmbH.

ESG Report. (2015). *Standards and Guidelines for Quality Assurance in the European Higher Education Area (ESG)*. Retrieved from https://www.enqa.eu/index.php/home/esg/

Faes, W., Knight, L., & Matthyssens, P. (2001). Buyer profiles: An empirical investigation of changing organizational requirements. *European Journal of Purchasing & Supply Management, 7*(3), 197–208. doi:10.1016/S0969-7012(00)00031-9

Fawcett, S. E., & Rutner, S. M. (2014). A longitudinal view of supply chain education: Assessing the challenge of retaining relevance in today's dynamic marketplace. *International Journal of Logistics Management, 25*(1), 180–201. doi:10.1108/IJLM-07-2012-0061

Feisel, E., Hartmann, E., & Giunipero, L. C. (2011). The importance of the human aspect in the supply function: Strategies for developing PSM proficiency. *Journal of Purchasing and Supply Management, 17*(1), 54–67. doi:10.1016/j.pursup.2010.07.001

Field, A. (2009). *Discovering statistics using SPSS* (Vol. 3). Sage publications.

Flöthmann, C., Hoberg, K., & Wieland, A. (2018). Competency requirements of supply chain planners & analysts and personal preferences of hiring managers. *Supply Chain Management, 23*(6), 480–499. doi:10.1108/SCM-03-2018-0101

Forrest, C. J., & Swanton, T. (2021). Longitudinal associations between soft skills, education and labour market outcomes: evidence from a survey of young Australians. *Education + Training*. doi:10.1108/ET-10-2020-0325

Giunipero, L. C., Denslow, D., & Eltantawy, R. (2000). *A skills-based analysis of the world class purchaser*. Academic Press.

Giunipero, L. C., Denslow, D., & Eltantawy, R. (2005). Purchasing/supply chain management flexibility: Moving to an entrepreneurial skill set. *Industrial Marketing Management, 34*(6), 602–613. doi:10.1016/j.indmarman.2004.11.004

Giunipero, L. C., & Handfield, R. B. (2004). Purchasing. *Education + Training*, 2.

Giunipero, L. C., Handfield, R. B., & Eltantawy, R. (2006). Supply management's evolution: Key skill sets for the supply manager of the future. *International Journal of Operations & Production Management, 26*(7), 822–844. doi:10.1108/01443570610672257

Giunipero, L. C., & Pearcy, D. H. (2000). World-class purchasing skills: An empirical investigation. *The Journal of Supply Chain Management, 36*(3), 4–13. doi:10.1111/j.1745-493X.2000.tb00081.x

Gross, H. (1966). Make or buy decisions in growing firms. *The Accounting Review, 41*(4), 745–753.

Hacking, I. (1984). Experimentation and scientific realism. In *Science and the Quest for Reality* (pp. 162–181). Springer. doi:10.1007/978-1-349-25249-7_7

Hannafin, M. J., & Land, S. M. (2000). Technology and student-centered learning in higher education: Issues and practices. *Journal of Computing in Higher Education*, *12*(1), 3–30. doi:10.1007/BF03032712

Heckman, J. J., & Kautz, T. (2012). Hard evidence on soft skills. *Labour Economics*, *19*(4), 451–464. doi:10.1016/j.labeco.2012.05.014 PMID:23559694

Heyse, V., Erpenbeck, J., & Max, H. (2004). *Kompetenzen erkennen, bilanzieren und entwickeln*. Waxmann Verlag.

Hoidn, S. (2017). *Student-centered learning environments in higher education classrooms*. Palgrave Macmillan. doi:10.1057/978-1-349-94941-0

Jauch, L. R., & Wilson, H. K. (1979). A strategic perspective for make or buy decisions. *Long Range Planning*, *12*(6), 56–61. doi:10.1016/0024-6301(79)90056-6

Keough, M. (1993). Buying your way to the top. *The McKinsey Quarterly*, *1993*(3), 41–63.

Kern, D., Moser, R., Sundaresan, N., & Hartmann, E. (2011). Purchasing Competence: A Stakeholder-Based Framework for Chief Purchasing Officers. *Journal of Business Logistics*, *32*(2), 122–138. doi:10.1111/j.2158-1592.2011.01011.x

Killen, K. H., & Kamauff, J. W. (1995). *Managing purchasing: Making the supply team work*. Irwin Publishing.

Kiratli, N., Rozemeijer, F., Hilken, T., de Ruyter, K., & de Jong, A. (2016). Climate setting in sourcing teams: Developing a measurement scale for team creativity climate. *Journal of Purchasing and Supply Management*, *22*(3), 196–204. doi:10.1016/j.pursup.2016.04.006

Knight, L., Meehan, J., Tapinos, E., Menzies, L., & Pfeiffer, A. (2020). Researching the future of purchasing and supply management: The purpose and potential of scenarios. *Journal of Purchasing and Supply Management*, *100624*. Advance online publication. doi:10.1016/j.pursup.2020.100624

Knight, L., Tu, Y.-H., & Preston, J. (2014). Integrating skills profiling and purchasing portfolio management: An opportunity for building purchasing capability. *International Journal of Production Economics*, *147*, 271–283. doi:10.1016/j.ijpe.2013.06.013

Kolchin, M. G., & Giunipero, L. C. (1993). *Purchasing education and training: Requirements and resources*. Center for Advanced Purchasing Studies.

Kruger, J., & Dunning, D. (1999). Unskilled and unaware of it: How difficulties in recognizing one's own incompetence lead to inflated self-assessments. *Journal of Personality and Social Psychology*, *77*(6), 1121–1134. doi:10.1037/0022-3514.77.6.1121 PMID:10626367

Laker, D. R., & Powell, J. L. (2011). The differences between hard and soft skills and their relative impact on training transfer. *Human Resource Development Quarterly*, *22*(1), 111–122. doi:10.1002/hrdq.20063

Land, S., & Hannafin, M. (2000). Student-Centered Learning Environments. In D. Jonassen & S. Land (Eds.), *Theoretical foundations of learning environments* (pp. 1–24). Lawrence Erlbaum Associates, Inc. Publishers.

Leoni, R. (2014). Graduate employability and the development of competencies. The incomplete reform of the "Bologna Process". *International Journal of Manpower, 35*(4), 448–469. doi:10.1108/IJM-05-2013-0097

Leuven/Louvain-la-Neuve Communiqué. (2009). *The Bologna Process 2020: the European Higher Education Area in the new decade*. Communiqué of the Conference of European Ministers Responsible for Higher Education, Leuven and Louvain-la-Neuve.

Logan, J. (2011). *The new Chinese city: globalization and market reform* (Vol. 52). John Wiley & Sons.

Luzzini, D., & Ronchi, S. (2016). Cinderella purchasing transformation: Linking purchasing status to purchasing practices and business performance. *Production Planning and Control, 27*(10), 787–796. doi:10.1080/09537287.2015.1137986

Masters, K. (2013). Edgar Dale's Pyramid of Learning in medical education: A literature review. *Medical Teacher, 35*(11), e1584–e1593. doi:10.3109/0142159X.2013.800636 PMID:23758180

McKeefry, H. (1998, Aug.). Opportunity Knocks-Education and Technical Skills Will Open the Door to Lucrative Job Offers. *CMP Net*, 3.

Mulder, M., Wesselink, R., & Bruijstens, H. C. (2005). Job profile research for the purchasing profession. *International Journal of Training and Development, 9*(3), 185–204. doi:10.1111/j.1468-2419.2005.00230.x

Muller, E. W. (2001). *Report on the job analysis to update the certified purchasing manager and accredited purchasing practitioner examinations*. Center for Advanced Purchasing Studies.

Murphy, E. (1995). Half the battle is knowing what skills to acquire. *Purchasing, 119*(9), 49–54.

Mursion. (2021). *From Skill to Instinct: How Higher Education Can Bridge the Gap Between Classroom and Career*. Academic Press.

Narlikar, A. (2003). *International Trade and Developing Countries: Bargaining coalitions in the GATT & WTO* (Vol. 13). Taylor & Francis.

Pagell, M., Das, A., Curkovic, S., & Easton, L. (1996). Motivating the purchasing professional. *The Journal of Supply Chain Management, 32*(3), 27.

Pettigrew, A. M. (2001). Management research after modernism. *British Journal of Management, 12*(s1), S61–S70. doi:10.1111/1467-8551.12.s1.8

Poh, M.-Z., Swenson, N. C., & Picard, R. W. (2010). A wearable sensor for unobtrusive, long-term assessment of electrodermal activity. *IEEE Transactions on Biomedical Engineering, 57*(5), 1243–1252. doi:10.1109/TBME.2009.2038487 PMID:20172811

Prahalad, C. K., & Hamel, G. (1990). The core competence of the corporation. *Harvard Business Review, 68*(3), 79–91.

Sawilowsky, S. S. (2009). New Effect Size Rules of Thumb. *Journal of Modern Applied Statistical Methods; JMASM*, *8*(2), 26. doi:10.22237/jmasm/1257035100

Schoenherr, T. (2010). Outsourcing decisions in global supply chains: An exploratory multi-country survey. *International Journal of Production Research*, *48*(2), 343–378. doi:10.1080/00207540903174908

Scholten, K., & Dubois, A. (2017). Advancing the skill set of SCM graduates–An active learning approach. *International Journal of Operations & Production Management*, *37*(11), 1683–1699. doi:10.1108/IJOPM-11-2015-0724

Schulze, H., Bals, L., & Johnsen, T. E. (2019). Individual competences for sustainable purchasing and supply management (SPSM) A literature and practice perspective. *International Journal of Physical Distribution & Logistics Management*, *49*(3), 287–304. doi:10.1108/IJPDLM-01-2018-0036

Stek, K., & Schiele, H. (2021). (in press). How to train supply managers – necessary and sufficient purchasing skills leading to success. *Journal of Purchasing and Supply Management, 100700*. Advance online publication. doi:10.1016/j.pursup.2021.100700

Stinenbosch, M. (2017). *Analysis of represented purchasing skills in academic literature and in current Dutch education provision*. University of Twente.

Stinenbosch, M., & Stek, K. (2017, July). The gap between academic purchasing education and the market needs. *In-Procurement*, 24-25.

Tassabehji, R., & Moorhouse, A. (2008). The changing role of procurement: Developing professional effectiveness. *Journal of Purchasing and Supply Management*, *14*(1), 55–68. doi:10.1016/j.pursup.2008.01.005

Tatham, P., Wu, Y., Kovács, G., & Butcher, T. (2017). Supply chain management skills to sense and seize opportunities. *International Journal of Logistics Management*, *28*(2), 266–289. doi:10.1108/IJLM-04-2014-0066

Trent, R. J., & Monczka, R. M. (2003). Understanding integrated global sourcing. *International Journal of Physical Distribution & Logistics Management*, *33*(7), 607–629. doi:10.1108/09600030310499286

Tuononen, T., Parpala, A., & Lindblom-Ylänne, S. (2019). Graduates' evaluations of usefulness of university education, and early career success–a longitudinal study of the transition to working life. *Assessment & Evaluation in Higher Education*, *44*(4), 581–595. doi:10.1080/02602938.2018.1524000

Van der Valk, W., Sumo, R., Dul, J., & Schroeder, R. G. (2016). When are contracts and trust necessary for innovation in buyer-supplier relationships? A necessary condition analysis. *Journal of Purchasing and Supply Management*, *22*(4), 266–277. doi:10.1016/j.pursup.2016.06.005

Von der Gracht, H., Giunipero, L. C., & Schueller, M. (2016). *Future-proof procurement; Now or never: the big procurement transformation*. Retrieved from https://advisory.kpmg.us/articles/2017/future-proof-procurement.html

Von der Gracht, H., Huber-Straßer, A., Schueller, M., Müller, N., Lichtenau, P. E., & Zühlke, H. (2018). *Rethinking the value chain. A study on AI, humanoids and robots - Artificial Intelligence: Possible business application and development scenarios to 2040*. Retrieved from https://assets.kpmg/content/dam/kpmg/xx/pdf/2018/09/rethinking-the-value-chain.pdf

Willingham, D. T. (2008). Critical thinking: Why is it so hard to teach? *Arts Education Policy Review, 109*(4), 21–32. doi:10.3200/AEPR.109.4.21-32

Wong, C. Y., Grant, D. B., Allan, B., & Jasiuvian, I. (2014). Logistics and supply chain education and jobs: A study of UK markets. *International Journal of Logistics Management, 25*(3), 537–552. doi:10.1108/IJLM-01-2013-0003

Zawawi, N. A. W. A., Umar, A. A., Aziz, A. R. A., Crouzier, O., Khamidi, M. F., & Idrus, A. (2014). PFI procurement skills: A cross-sector survey of practitioners. *Procedia Engineering, 77*, 170–178. doi:10.1016/j.proeng.2014.07.016

Zunk, B. M., & Sadei, C. (2015). Sharpening the Industrial Engineering and Management qualification profile: Research findings from Austria. *International Journal of Industrial Engineering and Management, 6*(3), 109–120. https://www.scopus.com/inward/record.uri?eid=2-s2.0-84947928261&partnerID=40&md5=b0f42724b97636d4a62cfbdbc4defdd2

ADDITIONAL READING

Bals, L., Schulze, H., Kelly, S., & Stek, K. (2019). Purchasing and supply management (PSM) competencies: Current and future requirements. *Journal of Purchasing and Supply Management, 25*(5), 100572. doi:10.1016/j.pursup.2019.100572

Biggs, J. B. (1996). Enhancing teaching through constructive alignment. *Higher Education, 32*(3), 347–364. doi:10.1007/BF00138871

Du Sautoy, M. (2019). *The Creativity Code: Art and Innovation in the Age of AI.* Harvard University Press.

Giunipero, L. C., & Pearcy, D. H. (2000). World-class purchasing skills: An empirical investigation. *The Journal of Supply Chain Management, 36*(3), 4–13. doi:10.1111/j.1745-493X.2000.tb00081.x

Kiratli, N., Rozemeijer, F., Hilken, T., de Ruyter, K., & de Jong, A. (2016). Climate setting in sourcing teams: Developing a measurement scale for team creativity climate. *Journal of Purchasing and Supply Management, 22*(3), 196–204. doi:10.1016/j.pursup.2016.04.006

Scholten, K., & Dubois, A. (2017). Advancing the skill set of SCM graduates–An active learning approach. *International Journal of Operations & Production Management, 37*(11), 1683–1699. doi:10.1108/IJOPM-11-2015-0724

Stek, K. (2021). Personality Development in Higher Education in the Era of Industry 4.0 - Comparing educational practices and philosophies in Industry 1.0 and Industry 4.0. In M. Olivas-Lujan & T. Bondarouk (Eds.), *Smart Industry - Better Management.* Emerald Publishing.

Stek, K., & Schiele, H. (2021). How to train supply managers – necessary and sufficient purchasing skills leading to success. *Journal of Purchasing and Supply Management, 100700.* Advance online publication. doi:10.1016/j.pursup.2021.100700

Von der Gracht, H., Giunipero, L. C., & Schueller, M. (2016). *Future-proof procurement; Now or never: the big procurement transformation*. Retrieved from https://advisory.kpmg.us/articles/2017/future-proof-procurement.html

Willingham, D. T. (2008). Critical thinking: Why is it so hard to teach? *Arts Education Policy Review*, *109*(4), 21–32. doi:10.3200/AEPR.109.4.21-32

KEY TERMS AND DEFINITIONS

Globalisation: The process of purchasing on the worldwide market that has been promoted by the United Nations World Trade Organisation (WTO) and its predecessor, the General Agreement on Tariffs and Trades (GATT).

Hard Skills: Knowledge and skills applicable for a specific job.

Internet of Things: Or machine-to-machine communication or the 4th Industrial Revolution is the process of autonomic production systems.

Procurement 4.0: The term for the implications of Industry 4.0 on the procurement function.

Quasi-Experiment: An experiment with a single group with observations before and after without a control group.

Soft Skills: Interpersonal or human-to-human skills and intrapersonal character traits.

Student-Centred Approach: The student acquires knowledge, skills, and attitudes through active, productive and effective learning.

Teacher-Centred Approach: The teacher instructs students in a straight classical frontal fashion setting students in a passive, consuming role.

APPENDIX 1

Table 4. PSM competences literature 1987-2020

authors	*methodology*
Anderson and Katz (1998)	Conceptual paper
Baily, Farmer, Crocker, Jessop, and Jones (2008)	Qualitative, interview-based (*n*=46)
Bals et al. (2019)	Qualitative, interview-based (*n*=46)
Burt, Dobler, and Starling (2003)	PSM textbook
Carr and Smeltzer (2000)	Quantitative, survey-based (*n*=85)
J. R. Carter and Narasimhan (1996)	Quantitative, survey-based (*n*=302)
Cavinato (1987)	Qualitative, interview-based (*n*=50)
Cousins and Spekman (2003)	Mixed, interview-based (*n*=23) and survey-based (*n*=310)
Cruz and Murphy (1996)	Practitioners article
Dowd and Liedtka (1994)	Quantitative, survey-based
Eltantawy, Giunipero, and Fox (2009)	Quantitative, survey-based (*n*=152)
Faes, Knight, and Matthyssens (2001)	Quantitative, survey-based (*n*=183)
Flöthmann, Hoberg, and Wieland (2018)	Quantitative, survey-based (*n*=243)
Giunipero and Pearcy (2000)	Mixed, interview-based (*n*=23) and survey-based (*n*=310)
Giunipero, Denslow, and Eltantawy (2005)	Mixed, interview-based (*n*=41) and survey-based (*n*=73)
Giunipero and Handfield (2004)	Mixed, interview-based (*n*=41) and survey-based (*n*=73)
Giunipero, Handfield, and Eltantawy (2006)	Qualitative, focus group meeting (*n*=53)
Kern, Moser, Sundaresan, and Hartmann (2011)	Quantitative, survey-based (*n*=148)
Keough (1993)	Conceptual paper
Killen and Kamauff (1995)	PSM textbook
Kolchin and Giunipero (1993)	Quantitative, survey-based (*n*=131)
McKeefry (1998)	Practitioners article
Knight et al. (2014)	Quantitative, survey-based (*n*=72)
Mulder, Wesselink, and Bruijstens (2005)	Quantitative, survey-based (*n*=261)
Muller (2001)	Quantitative, survey-based (*n*=2,416)
Murphy (1995)	Practitioners article
Pagell, Das, Curkovic, and Easton (1996)	Qualitative, interview-based (*n*=14)
Schulze, Bals, and Johnsen (2019)	Qualitative, interview-based (*n*=46)
Trent and Monczka (2003)	Quantitative, survey-based (*n*=216)
Tassabehji and Moorhouse (2008)	Qualitative, interview-based (*n*=18)
Tatham, Wu, Kovács, and Butcher (2017)	Quantitative, survey-based (*n*=216)
Zawawi et al. (2014)	Quantitative, survey-based (*n*=43)

APPENDIX 2

Table 5. Rankings of the items in the two student surveys

		Mean O_1	δ	Mean O_2	δ
1	**Loyalty** - Being loyal in professional life	3.69	.813	3.74	.829
2	**Honesty** - Being trustworthy in professional life	3.67	.706	3.77	.739
3	**Conscientiousness** - Conscientiousness implies a desire to do a task well and to take obligations to others seriously	3.38	.776	3.50	.874
4	**Result-orientated action-taking** - Aiming on effectiveness	3.34	.738	3.47	.813
5	**Capacity to be empathetic** - Capacity to listen and understand	3.32	.741	3.66	.769
6	**Social Manners** - Being tactful, diplomatic and having organisational sensitivity	3.21	.849	3.32	.806
7	**Ability to Resolve Conflicts** - Being able to avoid and resolve conflicts	3.17	.808	3.27	.894
8	**Teamwork** - Being able to work in a group of persons, acting together as a team	3.15	.755	3.47	.755
9	**Critical thinking** - Having the skills and knowledge of how to assess problems or issues in a critical manner	3.14	.819	3.25	.837
10	**Willingness to Learn** - Being professionally curious, motivation to learn continuously	3.13	.775	3.27	.764
11	**Ability to Solve Problems** - Being able to solve problems in a systematic way	3.11	.722	3.38	.705
12	**Task management (priority management)** - Being able to make a prioritisation in business-related tasks	3.06	.882	3.28	.745
13	**Communication skills** - Having the skills and knowledge of how to communicate	2.98	.825	3.34	.849
14	**Capacity to Advice** - Having consultancy skills	2.96	.886	3.16	.766
15	**Inter-generation ability** - Being aware of and able to work with people from different generations	2.95	.909	3.09	.988
16	**Comprehension of Complexity** - Being able to understand and solve complex problems	2.95	.674	3.19	.692
17	**Self-assurance** - Being assertive and having self esteem	2.94	.783	3.05	.872
18	**Cross-cultural awareness** - The ability to become aware of cultural values, beliefs and perceptions of yourself and other cultures	2.88	.955	3.12	.914
19	**Holistic Thinking** - Holistic thinking involves understanding a system by sensing its large-scale patterns and reacting to them	2.87	.640	3.02	.747
20	**Poise** - Being (self) confident	2.87	.802	3.03	.822
21	**Proactivity** - Being anticipatory, change-oriented and self-initiated behaviour in situations	2.87	.789	3.10	.804
22	**Customer-oriented** - being oriented on the end user	2.86	.766	3.06	.814
23	**Cross-functional management** - Being able to work with people from other professions and functions	2.86	.704	3.23	.707
24	**Decision Making** - Being able to make decisions	2.83	.767	3.16	.807
25	**Stress management** - Know how to manage stress at home and work using a variety of techniques	2.83	.977	3.14	.833
26	**Flexibility and agility** - Being able to adjust one's behaviour to new information or changing circumstances	2.79	.742	3.07	.737
27	**Willingness to take risks** - Taking well-considered risks	2.73	.805	2.94	.773
28	**Persistence** - Continuing in an opinion or course of action despite difficulty or opposition	2.70	.829	3.04	.738
29	**Strategic Thinking** - Strategic thinking is a process that defines how people think about, assess, view, and create the future for themselves and others	2.70	.829	3.21	.780
30	**Power of Persuasion** - Having influential skills	2.66	.721	2.80	.699
31	**Creativity** - Being creative in professional life / having creative ideas	2.60	.880	2.83	.771
32	**Inventiveness** - Being able to convert creative ideas in practice	2.58	.752	2.86	.756
33	**Leadership/community management** - Being able to manage employees in teams	2.51	.812	2.89	.819
34	**Networking / Building Relations** - Networking and relations management	2.49	.836	2.86	.863
35	**Salesperson skills** - Having acquisition strength and having canvassing ability	2.36	.883	2.80	.942
36	**Negotiation skills** - Being able to negotiate the specific commercial and legal terms in a contract needed to be settled in a satisfactory way for your organisation	2.21	.825	2.71	.780

Notes: O_1 is the first survey held in the first week of the course; O_2 is the second survey taken after the course; the items were measured on a 5-point Likert scale - fully disagree (1) to fully agree (5). The items are derived from Stek and Schiele (2021).

Chapter 3
Mechanical Engineering Students Project-Based Learning in OUAS:
Learning by Doing

Mira Kekkonen
Oulu University of Applied Sciences, Finland

Ville Isoherranen
https://orcid.org/0000-0003-3696-391X
Oulu University of Applied Sciences, Finland

ABSTRACT

This chapter introduces project-based learning approach which is used in the Oulu University of Applied Sciences (OUAS), School of Engineering and Natural Resources, Mechanical Engineering Department to get local companies to offer project works to mechanical engineering students. The concept is based on organizing a local event or online event for the companies to come to OUAS campus to present their challenges needing engineering students to solve. The companies are then competing, selling, or pitching their problem for engineering students as the engineering students will then individually select the most interesting cases to be solved, and which has linkage to potential summer job and thesis work opportunities if projects are successful. The concept has proven to be successful, and it has been established as traditional event with many companies returning to the pitching event annually to get their industry problems solved by group of motivated engineering students.

INTRODUCTION

Learning-based projects provides mechanical engineering students with the skills and experience to prepare them for the demands of working life. Project-based work is currently one of the most important and desired engineering skills in business life. Similar to learning to ride a bike, making mistakes

DOI: 10.4018/978-1-7998-8816-1.ch003

is an integral part of the process of gaining new experience as a professional in the field of mechanical engineering. It is thus important to create an environment where it is safe to students act and where an experienced mentor can support professional growth of students to help students dare to make mistakes and learn from them. Universities are able to provide students with an education on the theoretical foundations of engineering through classroom instruction, but as this education progresses to more specific vocational instruction, classroom instruction and lab exercises are no longer sufficient to satisfy students' thirst for knowledge.

Universities must offer students the opportunity to learn through experience as a part of their curricula. This is especially true for mechanical engineering students, who are typically interested in very practical work. This pragmatic perspective thus also influences student's intrinsic motivation to study in general. Therefore, it is imperative that universities also have well-equipped, modern laboratory facilities to support students in their learning and, specifically, include project-related work for mechanical engineering students. This practical work improves both students' theoretical knowledge and their professional career' after graduation. These abilities ultimately also impact the problem-solving abilities of students, and problem-solving is a fundamental part of the working life of an engineer.

Students aim to acquire skills in class that promote their employment after graduation. The competence that is valued in the "the real world" of post-graduate work/careers is created through experience. In the current world of work, it is not always sufficient to possess higher education in a particular field. From the employer's point of view, in addition to education, a student must have experience in the field to make them an ideal employee choice. The jobseeker who has completed university training and has also relevant practical experience is able to better demonstrate their motivation in a given profession and reduces the perceived risk in their hiring as an employee. If the candidate already has experience, this provides the foundation for immediate productivity and suggests developed problem-solving skills, which in general is the goal of education. The main task of engineers in working life is to solve problems with a team. A proficient engineer should first know the basics of design and then learn to fix problems that result from deviations from these basics. In addition to this, engineers need to identify the root-causes of problems that arise from processes of the companies. Often, if only the visible problem is solved, it may repeat itself or move the problem to another process phase, in which case the root cause has not been solved correctly. Engineers need to learn systematic thinking—that is, the overall impact of systems on each other. Knowledge in this domain is gathered through education and experience. However, a single engineer is not able to fully perceive or understand complex systems; therefore, a good engineer has good human skills in addition to problem-solving skills. Thus, overall, the ability to work on a team and identify and solve problems together are the key skills for a professional engineer in the engineering industry.

The goal of the project-based learning course case presented in this paper is to facilitate the employment of the students to the work that corresponding to their education after graduation, which is the goal of all vocational students. Graduation itself is not a sufficiently valuable reason for students for pursuing a profession; obtaining employment makes education valuable to the individual.

This chapter addresses how the Department of Mechanical Engineering of the School of Engineering and Natural Resources of the Oulu University of Applied Sciences (OUAS) has approached educating engineers to better serve and encourage closer cooperation with local companies while helping new engineering professionals obtain their first engineering job. This innovative solution is the annual Mechanical Engineering Pitching Event (MEPE).

Research Process and Method

In this chapter, the research method used is the case study (Eisenhard, 2007; Yin, 2009). The background and literature review are then introduced to build the background for the research in relation to previous research. Then, the overall project-based learning concept at OUAS, the MEPE, and its related process is explained, providing a framework for the cases. Specifically, this chapter includes three cases of student projects that were executed in the past years and details how these projects motivated the professional growth of young engineers. From the cases, the results and the feasibility of the concept are evaluated, and future research directions are presented. The main research question (RQ) can be formulated as the following: *Is the MEPE a feasible concept for providing project-based learning opportunities for engineering students?*

BACKGROUND

In Finland, the Universities of Applied Sciences (UAS) offer a Bachelor in Engineering education degree, the curriculum of which is designed to serve the needs of private and public companies, organizations, and society. In particular, the engineering degree is designed from the perspective of the realities of working life that companies require to ensure that new engineers have the optimal skillsets to address the existing needs of companies. In the Department of Mechanical Engineering at Oulu University of Applied Sciences (OUAS), special focus is placed on networking and communication skills, as these skills are essential in current, global working life. The work of engineers is project-based in many cases, and these projects are often multidisciplinary. The Mechanical Engineering degree program aims to bridge the transition from academia to industry to facilitate an easy and fast transition from school to working life.

Learning does not happen, and projects are not completed without realistic target-setting. Some students are sufficiently motivated to complete study tasks at hand, but, for many, an essential factor in their motivation is the acquisition of skills that are directly linked to employment after the completion of their education (Kekkonen & Juntunen, 2019). This means that the Universities of Applied Sciences (OUAS) play a specifically important role in ensuring the fit of the engineering education curriculum to the needs of business life. Additionally, engineering education faces many demands, as working life is also changing rapidly (Kropsu-Vehkaperä et al., 2013). Notably, project-based learning offers flexibility in curriculum design.

The UAS and the related industry have many collaboration programs. For example, the research, development, and innovation (RDI) projects funded by the European Union (EU) offer an effective platform for impactful development projects. These programs allow the UAS to develop their activities and laboratories and support the deployment of new technologies to companies. Additionally, the UAS can then offer new engineers who possess the relevant skills necessary for new technology to the industry. All members of the Finnish UAS group cooperate with local companies, and each university has its own process for executing projects in collaboration with these companies. Specifically, the publishing arm of Laurea University of Applied Sciences has produced an edited book that details how the members of the Finnish UAS have provided education in cooperation with local companies (Helariutta et al., 2021).

Literature Review

Work in practice is the core of the engineering profession, which requires certain skills, such as making judgements under conditions of uncertainty and learning from experience. This requires that professional engineering education prepare students to face real-life conditions and challenge them in regard to the integration of knowledge, practical skills, and ethical judgement in a setting often outside of actual practice. Engineering education must also support students' creativity and curiosity while simultaneously developing their problem-solving skills. It has also been studied that graduates nowadays not only need specific knowledge in different fields but also must be able to share this knowledge to solve complex problems in their working lives. Sheppard et al. (2006) state that engineering is centered around resolving undesirable conditions through technological applications and that engineering problem-solving will affect change and continuously create improvements in the world. For this reason, teaching methods have been developed that promote students' ability to solve problems and prepare them for their current and future working lives the needs of current and future working life.

Bédard et al. (2012) state that, in project-based learning, the learners control the learning process while the teachers advise from a distance. They also have noted that, during the curriculum, students alternating between studying at the university and working in the industry should not be considered project-based learning, as it takes place outside the curriculum and representatives from the university do not supervise it; this is considered problem-based learning (Bédard et al., 2012). Project-based learning practice procedures are recognized as consisting of eight phases: (1) identifying the creative final project; (2) clarifying the target audience; (3) exploring the connotation of the project; (4) designing the project; (5) making a schedule for the project; (6) beginning work on the project; (7) solving any problems and disputes; and (8) completing the project (Hong, 2007).

Project-based learning is widely used in different schools and undergraduate programs, combining two broad pedagogical principles—learning-by-doing and student-centered teaching—which help students implement their knowledge in a real-life context (Kokotsaki et al., 2016). The importance of the teacher's role in selecting driving questions for projects; offering support via complex instruction during project-based learning has been noted; and their role as a mentor has been found significant in expanding students' understanding (Krajcik & Blumenfeld, 2006). In project-based learning, concrete contributions are end products representing students' new understanding and attitudes that have grown during the project, with primary focus on the process of learning. Project-based learning is an active student-centered form of instruction that is characterized by students' autonomy, constructive investigation, goal-setting, and collaboration. It provides, through authentic questions and problems within real practices, opportunities for such, communication, and reflection, leading to meaningful learning experiences aiming to help students on the path to employment. It has been studied that project-based learning has a very positive impact on learners through, for example, helping students become better problem-solvers. Specifically, five essential characteristics of effective projects have been identified—(1) centrality, (2) a driving question, (3) constructive investigation, (4) autonomy, and (5) realism—with the importance of student collaboration, reflection, redrafting, and presentations also emphasized in many publications (Kokotsaki et al., 2016). In the end, even though many studies have noted the positive impacts of project-based learning in helping to develop students' skills toward becoming productive members of a global society, these developed skills are not measurable through standardized tests (Bell, 2010). This is challenging the teacher's evaluation process of students' project-based learning results to become more holistic.

Many schools and universities have implemented project-based learning in their education and sometimes also in cooperation with companies. One successful example presented by Lima et al. is from Portugal with industrial engineering students. In the Portugal example, local companies look forward to receiving fresh opinions from students as well as in the Oulu region (Lima et al., 2018). Experiences of success with every participant in the MEPE projects will specifically increase students' self-esteem, starts for them a cycle of success (Bell, 2010).

For several decades, the technology industry has emphasized engineers and other workers' learning skills and problem-solving abilities and attitudes by implementing well-proven Lean thinking, which is closely related to project-based learning. Learning has been at the center of Lean from its conception. Learning to recognize problems and their root causes and then solving them supports the major Lean idea of continuous improvement. One of the primary goals of Lean thinking at the companies is to create a learning organization and culture. Previous study said that adopting Lean requires change a way of thinking to succeed. In project-based learning as well, the goal is to teach individuals to recognize and solve real-life problems. In the literature it's well-known that the key to creating a truly Lean learning culture within companies is Lean management (Ballé et al., 2015). The main similarity between project-based learning and Lean is that both are mostly based on working in teams, with one experienced member (e.g., a teacher, sensei, or coach) working as a mentor who guides the team to work more efficiently and effectively. These mentors' values are related to team learning, creating a culture of problem-solving, and continuous improvement (Ballé et al., 2019; van Dun & Wilderom 2015). Also similar to project-based learning, teams are at the center of the Lean leadership model, which is based on self-development and a culture of improvement (Bhasin, 2013; Dombrowski & Mielke, 2015). The main traits of successful Lean plants also have many similarities with project-based learning, including high humane orientation, the use of small-group problem solving, and employee training to perform multiple tasks (Bortolotti et al., 2015).

Alves et al. (2012) found, in a literature review of industrial cases, that companies' workers are the source of capital for Lean and learning organizations. In studies, has noticed that, in organizations, the maturity of the lean concept can be likened to that of organizational learning (Hines et al., 2004). Bhasin and Found (2019) conducted an expansive literature review on sustaining the Lean ideology, with building a learning organization found to be a center principle of Lean. Moreover, the most asset for the companies is having a motivated, learning-minded, and solution-oriented staff, which is why universities need to train new professionals to meet these needs.

Overall, in reference to the above such successful experiences can be found from MEPE projects for both students and participating companies.

The OUAS RDI Program Initiated Systematic Project-Based Collaboration with Small and Medium Sized Companies (SMEs)

The mechanical engineering program within the OUAS, The European Regional Development Funded (ERDF) project called TEHOJA, which was executed together with companies from 2017–2020, is one example of a successful EU-funded program. The program began in 2017 with the aim of helping local, Oulu area, companies operationalize new automation technologies that could be expected to bring about competitive advantages in terms of costs, delivery schedules, and product quality. The program focused on finding suitable application environments for collaborative robots (called "cobots") for local small- and medium-sized (SME) companies. These cobot applications were piloted during the program at existing companies. The main goal of the program was to expand the knowledge of local SMEs in regard to collaborative robotics and its application in enhancing production as well as to train the personnel within these companies to apply these new collaborative robotics-related technologies in their everyday work (Broström et al., 2019).

Prior to the implementation of this program, the collaboration between local SMEs and the mechanical engineering education program of OUAS was not systematic, with one of the issues needing to be addressed being the initiation of the engagement of local SMEs in collaboration OUAS. The Mechanical Engineering Pitching Event (MEPE) was established to resolve this need for systematic and extensive collaboration. This event annually brings local SMEs and third-year mechanical engineering students together, where the local SMEs then pitch their current challenges to the students. Companies compete to obtain the best students to work on their challenges, "selling" their challenges to the students via a pitching event, which is a well-known mechanism used by start-up companies to sell their business ideas to investors. Students then select their project work topic for the spring semester based on the presentations delivered by the SMEs. This also provides participating companies with the opportunity to obtain summer trainees for the following summer (or "thesis workers", as the project-related work team members are known within the companies). The teams of students will also learn about the companies' processes, products, and development opportunities, thus making it easier for a company to employ the students for value-adding work within the company.

MECHANICAL ENGINEERING STUDIES AND THE PITCHING EVENT (MEPE)

From the very beginning of the program, the School of Engineering aims to teach students how to master product design and project management activities. First-year students gain knowledge through basic courses on mathematics and physics, but, additionally, they enroll in a course called Innovative Product Development in their first semester. In this course, students learn about the product development process phases; they then design and build their own products. At the end of the course, there are product exhibitions. Some new products developed through this course have been patented and licensed based on their novelty and innovativeness.

During their second year of study, the mechanical engineering students deepen their knowledge of theory as well as their professional understanding of mechanical engineering. Finally, students' studies are completed in their third year. Additionally, students gain practical experience from summer jobs undertaken between their semesters of study. They then possess the necessary basic understanding to undertake more demanding assignments within the industry (Kekkonen & Juntunen, 2018).

Preparing for the MEPE

The MEPE is organized annually at the beginning of the autumn semester. Preparatory work begins in the spring semester to ensure the participation of companies in the MEPE. In many instances, the final meeting of projects from the previous event offers the best occasion to recruit companies to continue to work with OUAS via the MEPE. Once companies have experienced the value of the MEPE, it is easier to convince them to continue with it the following year. New companies, especially micro companies or SMEs are also welcomed to the MEPE and are specifically actively recruited to increase the number of MEPE participants.

At first, companies might experience difficulties in defining their project topics, since development resources are especially scarce at small companies (Isoherranen & Ratnayake, 2018). The education staff from the School of Engineering and Natural Resources visits companies on-site to better understand their operations and business concept. This enables them to support companies in defining their development project topics, which the companies then present at the MEPE. The MEPE can be a virtual or locally held event on the OUAS campus. After the pitching portion, the following day, students can select the project topics they find most interesting. Their selection is then approved by the education staff via a specific interview process to ensure study progress as well as the fit and skills of the students for the project. It is essential that the students' previous studies are completed before their entering the project-based learning phase partners to ensure they have the necessary background knowledge to enable their successful project completion.

Executing Projects from the MEPE

Once the projects have been selected, the students start planning with their teacher's guidance and then proceed to a kick-off meeting with their respective companies. After the kick-off meeting, the project plan is developed, with all parties needing to approve the project plan before the start of the project. The project-related work then starts at the company at the beginning of the January.

Depending on the needs of the project, the project group will work on the company's work premises or at laboratories within the OUAS, utilizing its available equipment and design software. For example, to execute management-related Lean philosophy projects, such as the implementation of 5S, it may be crucial that project work occurs in the "Gemba", the actual environment or place where the value is created (Kekkonen et al., 2021). Otherwise, product design projects can be executed in laboratories (if all the necessary working equipment and materials are available) until the product testing phase. Nevertheless, the working premises of the companies are required to provide support and resources when needed for the student groups during the project. Every project group also has a supervising teacher who works as a mentor and helps complete the project.

During the spring semester, the execution of the project is followed by teachers rigorously to ensure its successful development for the company as well as an excellent learning experience for the students. Two intermediate seminars are held during the project execution. At these seminars, the project groups present their projects, their current progress, and their subsequent plans to the rest of the class and all supervising teachers. Students will then receive comments, recommendations, and advice from a very experienced teachers in audience. The supervising teachers also follow up on the projects by visiting the participating companies to obtain insight into the project and the company's needs; during this, they may also discover new development project options for following years. The project is finalized by the end of May, meaning that the project group completes their final reports and deliver a final presentation on their conclusions to the company representatives and their supervising teachers.

Process Map of the MEPE

Step 1: The supply of information (via email, websites, and other project and thesis meetings) to local companies about participating in the MEPE.

Step 2: Connection with companies between May to August. The OUAS staff contacts companies and discusses students' project options and the needs of the companies.

Step 3: The Mechanical Engineering Pitching Event (held in September). Companies deliver 15-minute pitches on their companies, offer specific projects, and try to impress the engineering students. The companies can participate in a local event or via a virtual portal. The entire event is streamed on YouTube, where students and teachers can follow the presentations if they are unable to participate in the local event.

Step 4: Selection. After the event, all project options will be listed on a virtual platform called Moodle. The mechanical engineering students then choose projects and create their own project teams based on their interests.

Step 5: Verification. Before the projects begin, the teachers interview the student groups and check the students' current study progress before giving their permission to start the project to ensure they possess the necessary background knowledge to enable successful project completion. Verification also serves the purpose of motivating students to complete courses in the years prior to their third-year project work, and this ensures the graduating numbers during the fourth and final study year.

Step 6: Kick-off meeting. The student team, supervising teacher, and client (the company representative) participate in an initial meeting, during which the project is launched. The supervising teacher sets a kick-off meeting with the company and the students to facilitate a coterminous project start. After this meeting, the students are responsible for setting up future project meetings with the company representative and other project members.

Step 7: Execution of the project. The project teams start their work in January either at the company premises or at OUAS laboratories. Regardless of the premises where the project is carried out, the company provides guidance, resources, and support to the student team throughout the entire project to ensure the project is executed successfully.

Step 8: Monitoring of the project. During the project, students participate in two intermediate seminars, during which they present their project progress to other students and their teachers.

Step 9: Follow up. The supervising teachers visit the companies to obtain insight into the projects and provide guidance to the students on practical issues that arise during the projects.

Step 10: End of the project. After an entire semester of work, the projects are completed. The students complete a final report and presentation about the project for the company representatives as well as the supervising teachers.

Step 11: After the project. The companies typically have the opportunity to observe successfully executed projects. Students may be able to continue their project work within their thesis or as a summer job.

EXAMPLES OF STUDENT PROJECTS

The professional interests of the students have an impact on the project topics, which the students select and carry out during the project course in their third year. Mechanical engineering students at the School of Engineering and Natural Resources at OUAS choose their study options during their first year of study. The options within the study program are Automotive Engineering, Machine Automation Engineering, and Production Engineering. In their third study year, Machine Automation engineering students complete their project studies with the course called Product Development Project (20 European Credit Transfer and Accumulation System [ECTS] credits), and Automotive and Production Engineering students complete the course, Practical Project Training (30 ECTS credits).

The Automotive Engineering students specifically concentrate on the automotive aftermarket, automotive electronics, the principles of combustion engines, and the basic structures of hybrid powertrain systems and modern automotive control systems in their studies. The project subjects for the automotive engineering students are usually related to different kinds of vehicles (e.g., the development of automotive service plans, the building of various prototypes, the planning of projects that are related to these vehicles, the planning of the layout of automotive workshops, or the planning of automotive testing equipment).

The Machine Automation students specialize in the planning of devices, machines, and control systems. When studying product development, they utilize modern software, such as SolidWorks. These students are interested in carrying out machine design projects that require the brainstorming and planning of the control system of a given device. Projects that offer the opportunity to verify the function of the device with the aid of a proof-of-concept prototype are specifically desired. Additionally, devices that are built for the connection of collaborative robot cells have been very successful projects in prior years.

The Production Engineering students study optimal methods to produce different devices and products more efficiently and economically. They also thoroughly study industrial maintenance, quality for safety management, and environmental management. Furthermore, they study Lean philosophy and management. Examples of excellent project focuses for production engineering students include factory layout planning, the development of production work adhering to the Lean philosophy and the commissioning of Lean tools (e.g., 5S), the determination of production turnaround times, the development of the initial designs/processes of robots, and the development of the operations, production control, and HSEQ systems of warehouses.

Generally, the authors have noticed that the most motivating and interesting projects for Mechanical Engineering students are pragmatic and include planning and implementation elements. The students feeling that the companies that participate in the MEPE appreciate their work contribution is also considered important. Furthermore, the reward of a summer job, which is often offered for successful project work that has been carried out well, is also motivating. The course also expands their job references on their curriculum vitae, which helps them in later finding jobs after graduation.

The following project examples will present three different engineering student projects carried out during the 2019–2020 school year. The first case study was implemented at a local small enterprise. This successful product development project with a proof-of-concept aspect was carried out by a group of Machine Automation engineering students. The second case study is a project that was conducted at a metal workshop that operates in Oulu. A Production Engineering student group carried out this production development project. This project was implemented using the principles of Lean philosophy and management, and it also included a staff training aspect. Third project is implemented by Production Engineering student during the exchange studies.

Case 1: Plastic Bottle Clamp

Head Recycle Systems (HRS) develops innovative plastic recycling equipment in the Oulu area. In 2019, this company participated in the MEPE for the second time, encouraged by their positive experience the previous year. They came to the event seeking enthusiastic mechanical engineers for a product development project. The project topic focused on the development of a new type of mechanical plastic recycling equipment for crushing plastic bottles. The goal was that the machine would be able to handle the storage and recycling of plastic materials as well as independently handle fault situations when machine jams occurred. This project was taken on by four mechanical engineering students with focuses on machine automation engineering and machine building.

The project team was largely given free rein to innovate and develop. In addition to this primary machine design process, the project group also designed several other sub-devices, including those that, for example, delimited the trajectories of moving parts, guided the goods to be pressed, and secured the motor. The team of students was given certain boundaries within which they could work, but there was also plenty of room for creativity and "outside-the-box" thinking. This freedom enabled the invention and development of a new method of constructing the device, which will be utilized in the future development of machinery and equipment for clients of HRS.

The result of the project was a fully functional device that met the client's requirements (see Figures 1 and 2) (i.e., the goals of the project were satisfactorily and completely achieved). The project team systematically worked as a team and gained significant new knowledge in several sub-areas of product design. Moreover, the project team revealed that the best moment of the project was the completion of the prototype as well as its initial tests. This culmination of the project was the moment when everyone could see the impact of their own contributions as well as the realization of common visions shared among each other (Heinonen et al., 2020).

Figure 1. 3D-printed proof of concept
Source: Heinonen et al., 2020

The result of this project was a successfully constructed machine using new innovative concepts. The entire team of engineering students continued to work with this company as thesis workers after the project completion. One of the students also secured his first engineering job with HRS after graduation. The project provided significant opportunities for the students to test their engineering skills in a real environment. Also, this project helped them obtain valuable feedback on the content of their mechanical engineering courses and their relevance to the real demands of working life within companies.

Figure 2. Photo of the working prototype of the product
Source: Heinonen et al., 2020

As a result of the project, the HRS company obtained an innovative prototype that was flexibly produced at a reasonable cost by the team of engineering students. The company was satisfied with the competence, innovation, and cooperation skills of the project team.

Case 2: 5S Implementation in a Metal Workshop

This case project was presented at the MEPE in 2019. In 2019, JMC Engine Oy underwent a 5S pilot project for one machine cell. The pilot 5S project earned positive feedback from the operators and resulted in improved worksite tidiness (see Figure 3). As a result, the company decided to expand this project to cover the entire production site and expand personnel understanding related to the 5S method via practical training. This project was carried out by a group of mechanical engineering students enrolled in the Production Engineering option.

The goal of this project was to expand the 5S methodology to the rest of the factory site, promote a Lean culture, and build a quality and measurement control tool to sustain the changes made within 5S. At this company the project also aimed to improve job satisfaction, safety, and the work environment as well as to eliminate waste from production.

Figure 3. Photo of the MCM 510 workstation before and after the implementation of organization based on 5S
Source: Tuomivaara & Laakko, 2020

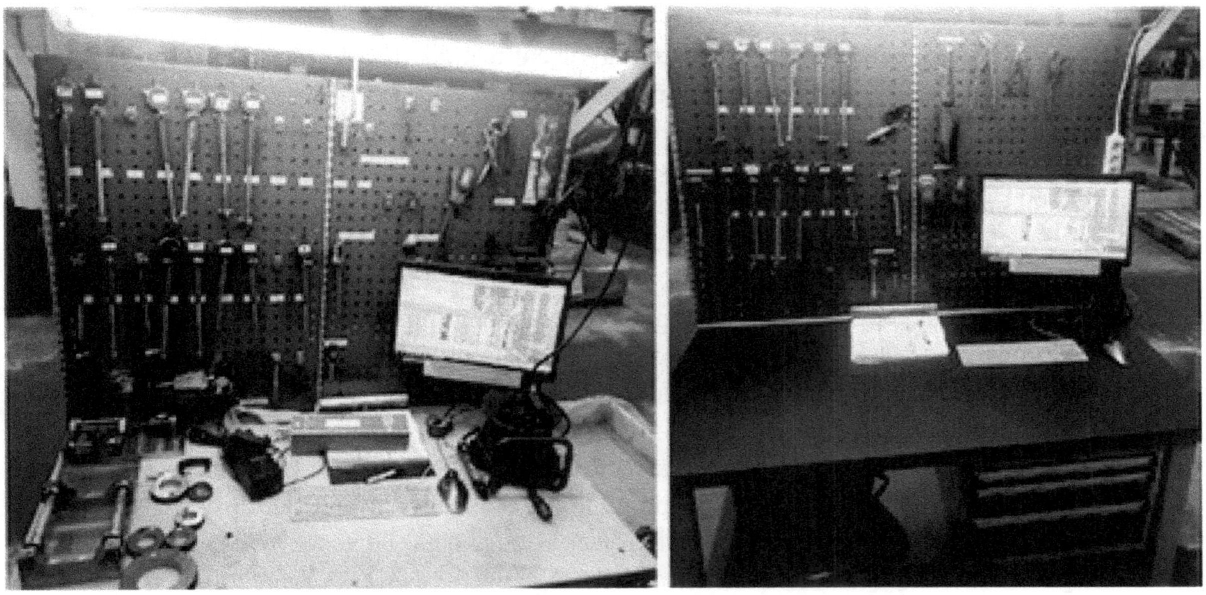

At the end of the project, a 5S training package for company personnel was implemented in collaboration with the staff of the project called POTKUA project, funded by the European Social Fund. The POTKUA project has been in existence from 2018 to 2021, and it has aimed to train and expand managements awareness of the Lean philosophy in SMEs and develop companies' competitiveness in commercial markets. Lean is a well-known method of managing production in factories and thus increase competitiveness; large companies have successfully implemented this approach in their activities. Meanwhile, micro-, small-, and medium-sized companies have not widely implemented Lean because of a lack of resources, a lack of support from management, and a lack of worker knowledge. The POTKUA project is another excellent example of how these kinds of EU-funded (European Regional Development Fund, ERDF) projects within the Department of Mechanical Engineering are able to educate both engineering students and SME companies and bring relevant participants together (Kekkonen et al., 2021).

The aim of the training was to teach employees through 5S theory and practice. At first, a formal teaching session was held to introduce the theory, where the employees were educated on the meaning of 5S and its benefits. The second part of their education was a workshop in which the employees organized their own workstations following the 5S approach learned from the theoretical section.

The student project group considered the training of the employees to be the most important milestone of this project. The project members also believed that training the employees was a significant step toward promoting a Lean culture and increasing a "working discipline" at the company. When everyone is trained in the principles of 5S, everyone understands the rules and what 5S stands for. The project members agreed that this would be key to maintaining the changes made at the worksite (Tuomivaara & Laakko, 2020).

Case 3: International Project Training

Mechanical engineering students at the OUAS also have the opportunity to complete their project studies via an exchange program in Germany at Ulm University of Applied Sciences or in the Netherlands at Hanze University in Groningen. This exchange program also entitles students to an international double degree in mechanical engineering. The program lasts an entire academic year and consists of an autumn semester of study at a university and a spring semester spent at an international project internship at a local company. When a student who has completed the exchange program graduates, he or she will receive diplomas from both the Finnish and exchange universities.

A successful project internship is also taken into account at international companies, and the student can continue working at the company that they worked with during their study abroad during their thesis. Production engineering student Petteri Tyni is one example of a student who was able to successfully complete an international project internship as part of the exchange program. This student first completed an exchange program at Ulm University of Applied Sciences and then completed a project internship in Switzerland for a company called Bystronic Laser AG. He also completed his thesis based on his work for the company. He found working on his thesis very instructive, because it was not written in his mother tongue, and it included two projects that were executed abroad. These projects were well-suited to the topic of his thesis and provided several professional challenges. Moreover, this student indicated that he had learned a significant amount about project management, language skills, problem solving, and working in an international work environment through his thesis work (Tyni, 2021).

The international aspect of mechanical engineering is an integral part of the industry, as the size of the market often requires that companies function as international partners or subcontractors. It is therefore important to prepare at school for international collaboration, which may be part of the working life of an engineer. To engineers, communication with international companies from different cultures is crucial for successful projects and companies. The double degree provides students with a competitive advantage that has a positive impact on their employment and also indirectly affects SMEs in the Oulu region, which like to employ engineers with a double degree. These engineers provide a competitive advantage to local companies, which may not already have significant international expertise.

LEGACY OF THE TEHOJA PROJECT

During the TEHOJA project, cooperation was established with 18 different companies in Northern Finland. As a result of the project, an innovative platform for collaborative robotics was built for OUAS, which enabled the construction of products and applications for needs of companies. A total of 42 products were developed and implemented as part of the TEHOJA project. The original goal of the project was to implement only 10 developed products. Perhaps the most successful robotic project was carried out for the world's northernmost ceramics factory, Pentik, in Posio, Finland. The factory wanted to participate in the project to improve the ergonomics, resilience, and well-being of its employees during work.

As a result of the TEHOJA project, several projects have been implemented in the Department of Mechanical Engineering at OUAS: Potkua, Roboreel, Roboboost, Roboedu, and Kotu. The TEHOJA project has also contributed to the employment of more than 10 students in local companies, such as JMC Engine Oy, Sähkö-Rantek Oy, and Pentik Oy.

Another result of TEHOJA is that the MEPE has become an integral part of mechanical engineering curriculum, which has made it possible to teachers work closely and network with local companies year after year— even after the TEHOJA project, which led to the creation of the MEPE. Through the event, OUAS has been able to improve the visibility of its training and RDI activities, and, over time, the MEPE has clarified a unified operating model that has created credibility for the operations of the department. Raising awareness among management of companies and through word-of-mouth thus contributes to the development of the credible image of OUAS in mechanical engineering education, which, in turn, further enhances student employment and also develops the reputation of OUAS as a meaningful and innovative workplace that actively develops the economy of the Oulu region.

During the first year of the MEPE (2016–2017), the event only involved the TEHOJA project. From 2018 onward, more local SMEs and other projects within the Department of Mechanical Engineering at OUAS also started to participate (see Table 1). Like the TEHOJA project, new projects have been introduced in cooperation with the business community in the area. Most of these new projects will further enhance the competitiveness of companies by designing new products and increasing the efficiency of their production, mainly via the Lean philosophy and the introduction of the capabilities of interoperable robots as in the TEHOJA project. In conclusion, the MEPE has widely increased the awareness of SMEs and their interest in cooperating with the Department of Mechanical Engineering at OUAS in Northern Finland (Autio, 2020).

In 2021, it has been observed that (in addition to the SME companies) large companies have expressed interest in the MEPE. These include SSAB, Nokia, and iLOQ, which have recently invested in and extended their operations into Oulu. As a result, the need for engineers has increased. However, the focus of the program is—and will continue to be in the future—supporting SME companies via the MEPE, because these companies provide the basis for the area's economy and thus make the comprehensive development of Oulu possible. Nevertheless, large companies employ numerous engineers, and they are also valuable for the Oulu area.

Even though the MEPE event has become a fixed part of the education program in the mechanical engineering department and of the operationalization of cooperation with local companies, its development work has not ended. The school is constantly contacting new SME companies and other new local companies that have not yet discovered the program that may be interested in offering students development project opportunities and cooperating with the school each year; this ensures that future engineers and employers will continue to meet and develop the local economy. This requires contacting local companies, demonstrating the activities and operations of the school, and marketing to companies that are not yet aware of the program. Over the years, individuals previously employed at companies that had participated in the program have occasionally changed employers and initiated new cooperation opportunities for the program because of their prior favorable experiences.

We have also improved the accessibility of the event. Up until 2019, the event was organized as a local event in which the representatives of companies met, networked with, and listened to presentations from other companies in a large concert hall. In 2020, as a result of COVID-19, the event had to be organized as a virtual event for students, and only business representatives were invited to the local event, which was streamed to students on YouTube. In the past years, all participants in the event were physically present on site. The hybrid event was found to be a well-functioning solution, saving entrepreneurs time and lowering the attendance threshold for the event. In the past, participating in the event required as much as half a working day and traveling to the site, since some of the participating companies are located quite far from Oulu, Linnanmaa.

Table 1. Number of companies participating in the MEPE and project cases between 2016–2020

Year	Number of Companies	Number of Project Cases
2016	10	10
2017	12	17
2018	22	50
2019	20	53
2020	17	34

The MEPE Concept Provides Inspiration

The implementation of business experiments and the teaching of collaborative robotics, as well as experimentation involving companies in the local area, have been integrated into the basic engineering of education due to the MEPE. Its operations will continue among local companies if there is sufficient demand and need for expertise and cooperation are considered useful.

The impact of the MEPE has also been identified in other OUAS collaborations with educational vocational institutions (called 6Aika) as the best practice to follow. These institutions have been told the purpose of the project and the operating principle of the business experiments, but the same operating principles are not possible at all educational institutions, as they are vocational training institutions and do not have the same capabilities as polytechnic institutions (Autio, 2020).

On the other hand, this kind of cooperation with companies also requires a special relationship of trust between all participants. Universities and other schools have to trust in their own work as educating bodies that the students they educate are able to execute project work independently for the participating companies and also trust that they are able to find necessary information on their own and ask for help when needed. Additionally, companies must trust that the engineering students are willing to do their best work for the projects they have selected. They must also trust that the university has sufficiently educated these near-graduation engineers. Moreover, the students have to trust that their teachers can guide and support them during their project work. They must also believe that the companies will provide all the necessary support for their work and that the representatives of companies appreciate their work for the company, which is unpaid.

This collaboration with companies and the integration of their needs into the training of mechanical engineers underscores the fact that the UAS responds to the needs of working life and the industry. Since the UAS supports the growth of the professional identity of its engineering students by training them in cooperation with companies regarding the realities of working life and utilizing the expertise of experienced lecturers, the UAS significantly contributes to the success of companies in the region by providing experts who meet the needs of the labor market. Students' employment immediately following graduation from OUAS can be seen from Figure 4, which shows a high level of employment in engineering jobs. This indicates the successful implementation of project-based learning for building industry-relevant engineering skills via the MEPE concept and process.

Figure 4. Employment, after graduation, %. school of engineering average between 9/2017–12/2019

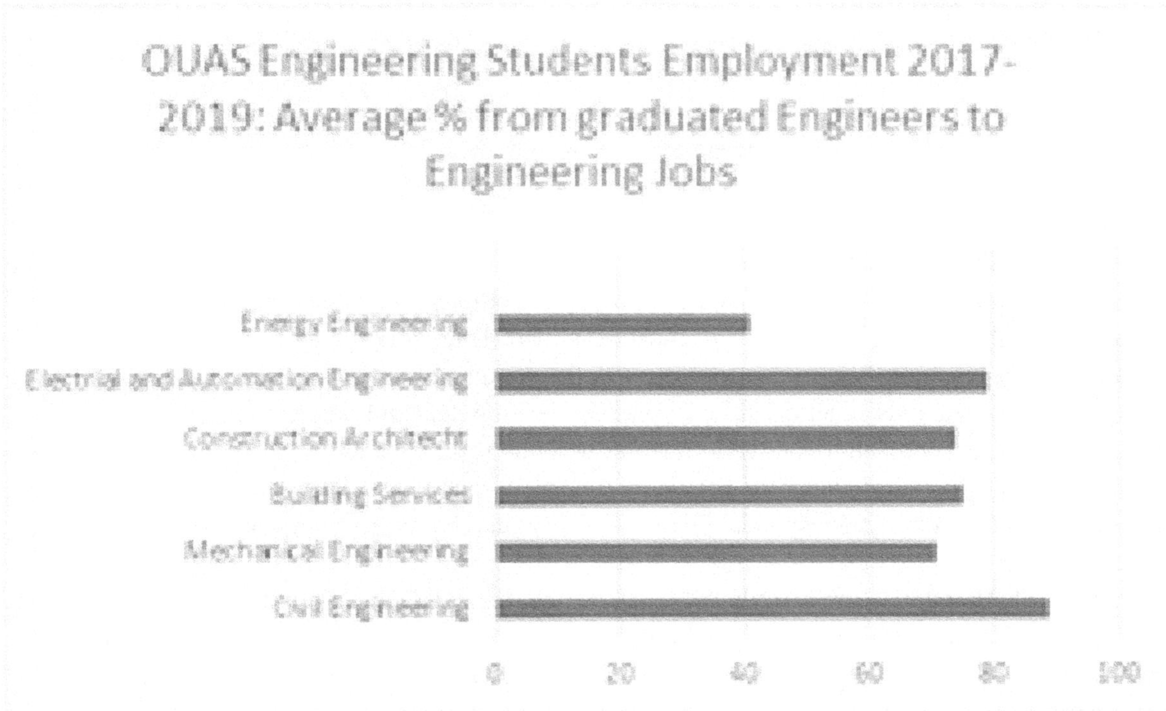

FUTURE RESEARCH DIRECTIONS

In the future, we will continue to systematically develop the MEPE and further clarify its concept. Specifically, the MEPE should target other study options for mechanical engineering students to better serve the OUAS students. In previous years, the project topics included in the MEPE have not equally represented all mechanical engineering study options. After targeting companies that were previously invited to the MEPE, the special interests of individual mechanical engineering fields were previously identified to provide equal opportunities for all students. The future direction of the program will include comprehensive service for the entire mechanical engineering program and will continue to emphasize cooperation with local companies.

The MEPE also presents research opportunities for evaluating the impact of the program on local businesses and post-graduate employment for students. As has been previously mentioned, student projects implemented through the MEPE have been meaningful for students and have met their ambition to undertake relevant project work, but their effects on businesses have not been studied. It would therefore be promising to investigate this domain, because, if the good practices of this event are extended to other organizations, universities, and curriculums, there should be evidence of the impact of MEPE events on local businesses and the professional development of students.

CONCLUSION

The research question regarding the feasibility of the *MEPE concept to provide project-based learning opportunities for engineering students* developed by the School of Engineering and Natural Resources at the Oulu University of Applied Sciences, Oulu, Finland. The needs of local companies regarding the development of their operations have been harnessed to support mechanical engineering training within the institution. In this paper, actual project-based learning MEPE project examples have been presented via case studies.

One of the goals of engineering education is to teach students to be self-directed and ideally, students would look for internships at companies themselves and complete the course independently. In order, to achieve the right learning goals, students must be guided. Therefore, the project topics being defined in advance by the teaching staff in cooperation with the companies enables the realization of the goals of every parties. In teaching practical skills in product development and project management, a lecture-driven, independent approach to studying may not produce results when the goal is to train engineers who possess practical skills. Thus, it is necessary to learn through motivation on an encouraging topic and in an environment that corresponds to real work assignments in the companies of a graduate mechanical engineer.

The aim of higher education institutions is to train engineers and develop the professional identity of graduates of the engineering profession through theorical classroom education and real-world development projects to applicate their theories. Therefore, as an educator, the School of Engineering and Natural Resources promotes the self-confidence and thus employment of students when the step from student life to working life is not too wide, because students school life has included the completion of real-world industry projects. For many engineering students, the first step in their engineering career and experience working as an engineer may be a project internship or product development opportunity through the MEPE, including typical or even business-critical challenges for real companies, that are tailored to suit graduate engineering engineers. The job of a mechanical engineer involves dealing with real-world problems, and this is best learned by doing, searching, or asking for advice from more experienced professionals, and, ultimately, through firsthand personal experience. For this reason, the work of each project and product development course is supervised by an experienced lecturer who has worked in the industry for several years. As an experienced engineering professional, this lecturer can guide, consider potential risks, and act as a mentor for engineering students starting their careers and working on a project—both in terms of project management and task development work. When supervising students, these lecturers are also able to observe the activities of local companies and expand their assessment of their needs and areas for development. This enables the continuous development of educational content and the emergence of new project ideas that would better be able to serve local companies (e.g., by providing a low-risk opportunities) to test the impact of robotics and other new technologies on a company's business.

ACKNOWLEDGMENT

The authors would like to thank the staff of the School of Engineering and Natural Resources for their valuable input for this book chapter as well as the Council of Oulu Region for the ERDF funding, which has supported the MEPE development projects.

REFERENCES

Alves, A. C., Dinis-Carvalho, J., & Sousa, R. M. (2012). Lean production as promoter of thinkers to achieve companies' agility. *The Learning Organization*, *19*(3), 219–237. doi:10.1108/09696471211219930

Autio, T. (2020). *Loppuraportti*. TEHOJA-hanke.

Ballé, M., Chaize, J., & Jones, D. (2015). Inclusive versus exclusive learning: The secret ingredient to creating a truly "lean" and "learning" culture. *Development and Learning in Organizations*, *29*(1), 20–23. doi:10.1108/DLO-10-2014-0080

Ballé, M., Chaize, J., & Jones, D. (2019). Lean as a learning system: What do organizations need to do to get the transformational benefits from Toyota's method? *Development and Learning in Organizations*, *33*(3), 1–4. doi:10.1108/DLO-11-2018-0147

Bédard, D., Lison, C., Dalle, D., Côté, D., & Boutin, N. (2012). Problem-based and project-based learning in engineering and medicine: Determinants of students' engagement and persistance. *The Interdisciplinary Journal of Problem-Based Learning*, *6*(2). Advance online publication. doi:10.7771/1541-5015.1355

Bell, S. (2010). Project-based learning for the 21st century: Skills for the future. *The Clearing House: A Journal of Educational Strategies, Issues and Ideas*, *83*(2), 39–43. doi:10.1080/00098650903505415

Bhasin, S. (2013). Impact of corporate culture on the adoption of the Lean principles. *International Journal of Lean Six Sigma*, *4*(2), 118–140. doi:10.1108/20401461311319329

Bhasin, S., & Found, P. (2019). Sustaining the lean ideology. *Management Decision*. Advance online publication. doi:10.1108/MD-09-2019-1254

Bortolotti, T., Boscari, S., & Danese, P. (2015). Successful lean implementation: Organizational culture and soft lean practices. *International Journal of Production Economics*, *160*, 182–201. doi:10.1016/j.ijpe.2014.10.013

Broström, T., Kaivosoja, L., & Kekkonen, M. (2019). Robotiikalla joustavaa automaatiota Oulun talousalueelle. *ePooki, Oulun ammattikorkeakoulun tutkimus- ja kehitystyön julkaisut.*

Dombrowski, U., & Mielke, T. (2013). Lean Leadership fundamental principles and their application. *Procedia CIRP*, *7*, 569–574. doi:10.1016/j.procir.2013.06.034

Eisenhardt, K. M., & Graebner, M. E. (2007). Theory building from cases: Opportunities and challenges. *Academy of Management Journal*, *50*(1), 25–32. doi:10.5465/amj.2007.24160888

Heinonen, H., Törmänen, J., Soukkio, J., & Riihinen, M. (2020). *Projektin HRS: loppuraportti*. Academic Press.

Helariutta, A., Fred, M., Kangastie, H., Merimaa, M., & Päällysaho, S. (2021). *Avoin TKI-integroitu oppiminen - toteuttamistapoja ja oppimisympäristöjä. Laurea julkaisut.* Academic Press.

Hines, P., Holweg, M., & Rich, N. (2004). Learning to evolve a review of contemporary lean thinking. *International Journal of Operations & Production Management, 24*(10), 994–1011. doi:10.1108/01443570410558049

Hong, J.-H. (2007). The Comparison of Problem-based Learning (PmBL) Model and project-based learning (PtBL) model. *International Conference on Engineering Education–ICEE.*

Isoherranen, V., & Ratnayake, R. C. (2018). Performance assessment of microenterprises operating in the Nordic Arctic region. *Journal of Small Business and Entrepreneurship, 30*(5), 431–449. doi:10.1080/08276331.2018.1438769

Kekkonen, M., Jokinen, T., & Isoherranen, V. (2021) Potkua Pohjois-Suomen PKT-yrityksiin Lean-toimintamallia soveltamalla. Avoin TKI-integroitu oppiminen - toimintamallit ja hyvät käytänteet. (pp. 43–48). Laurea julkaisut.

Kekkonen, M., & Juntunen, T. (2018). Tuottava tapa oppia projektityöskentelyä. *Toolilainen, ammattikorkeakoulujen tekniikan ja liikenteen alan järjestölehti.*

Kokostaki, D., Menzies, V., & Wiggins, A. (2016). Project-Based Leaning. *Improving Schools, 19*(3), 267–27. doi:10.1177/1365480216659733

Krajcik, J. S., & Blumenfeld, P. C. (2006). Project-based learning. The Cambridge handbook of the learning sciences, 317-333.

Kropsu-Vehkapera, H., Isoherranen, V., & Kess, P. (2013). Company assessment in production management education. Education conference, Tampere Finland.

Lima, R. M., Dinis-Carvalho, Sousa, R., Arezes, P., & Mesquita, D. (2018) Project-based learning as a bridge to the industrial practice. In Closing the gap between practice and research in industrial engineering, lecture notes in management and industrial engineering. doi:10.1007/978-3-319-58409-6_41

Sheppard, S., Colby, A., Macatangay, K., & Sullivan, W. (2006). What is engineering practice? *International Journal of Engineering Education, 22*(3), 429–438.

Tuomivaara, A., & Laakko, M. (2020). *5S Project JMC Engine Oy* [Final report]. Academic Press.

Tyni, P. (2021). *New standards for development of laser cutting machines.* Oulu University of Applied Sciences. https://www.theseus.fi/handle/10024/356966

van Dun, D. H., & Wilderom, C. P. M. (2016). Lean-team effectiveness through leader values and members' informing. *International Journal of Operations & Production Management, 36*(11), 1530–1550. doi:10.1108/IJOPM-06-2015-0338

Yin, R. K. (2009). Case study research: Design and methods. *The Canadian Journal of Action Research, 14*(1), 69–71.

Chapter 4
Influence of Game-Based Methods in Developing Engineering Competences

Helder Gomes Costa
https://orcid.org/0000-0001-9945-0367
Universidade Federal Fluminense, Brazil

Frederico Henrichs Sheremetieff
Universidade Federal Fluminense, Brazil

Elaine Aparecida Araújo
Universidade Federal Fluminense, Brazil

ABSTRACT

This research aims to understand the influence of game-based learning methods on engeineer competences. Competencies expected from an engineer, which competencies are commonly explored by game-based learning methods, and perceptions from a sample composed of 92 respondents about the question that drives the research are explored. All competencies analyzed had more positive influence responses than negative ones, or non-impact responses. The competence analyzed most positively is "problem solving"; the one with the most negative impact responses is "second language learning," and the one with the most non-impact responses is "continuous search for career improvement." This study fills the following gaps: compiles and analyzes articles on game-based learning methods and carries out unprecedented research regarding the influence of game-based learning methods on the professional competences of graduates of engineering courses.

INTRODUCTION

One of the main goals of undergraduate institutions is to improve the quality and efficiency of education and to improve the professionals who will graduate. So moreover than developing just technical and

DOI: 10.4018/978-1-7998-8816-1.ch004

theoretical knowledge required of graduating professionals, it is necessary the activation of cognitive and learning competences (Liuta et al., 2019) - according to (De Jesus & Costa, 2013), it is also a need in forming engineers. In the other hand, according to (Hamari et al., 2016), it can also be noted that student disinterest is an international problem because, once 20-25% of them classified themselves as having low participation and a low sense of belonging, as discovered in a research in 28 countries members of the Organization for Economic Co-operation and Development (OECD). Taking in mind that more than 2.5 billion people in the world spent time playing video games (SuperData Research, 2019), many have wondered why not change this problem to this solution.

So, there is an expectative about the game-based teaching and learning contribution to improve student engagement, increase teamwork and improve student motivation in the disciplines. According to (Liuta et al., 2019). the introduction of game-based learning methods, increases the interest and understanding of engineering students and even improves their capacity of self-learning. Based on all these statements, the main objective of this paper is to answer the question: **In what degree are engineers' professional competences influenced by game-based learning methods?** This core question, and its related central objective were deployed in secondary ones. Table 1 shows the connection among them and the research steps addressed in the research, while Figure 1 summarizes the flow of the research steps mentioned in Table 1.

Table 1. Questions, objectives and research steps

Questions addressed	Objectives	Research steps
What are the professional competences developed by game-based methods?	Identify the competences developed by game-based methods within the literature.	Analysis of competences that are developed by game-based methods in the literature.
What are the competences expected of an engineering professional?	Identify what competences are expected of an engineering professional, in Brazil and worldwide.	Analysis of the competences expected of an engineering professional.
Based on the results of the literature, is there an expectation of an association between the competences worked by the game-based learning methods with those expected from an engineering professional?	Analyze what competences required of engineers are being developed by game-based methods.	Association of the results of the two previous questions.
What is the degree of influence perceived by users of game-based learning methods (teachers, students, and coordinators) on the expected competences?	Map users' and appliers perceptions in the degree of influence of game-based learning methods on the competences they develop.	Analyze from the perspective of users, both game developers and participants, to what degree these competences were developed within the scenario they participated in.

Figure 1. Flow of the research steps

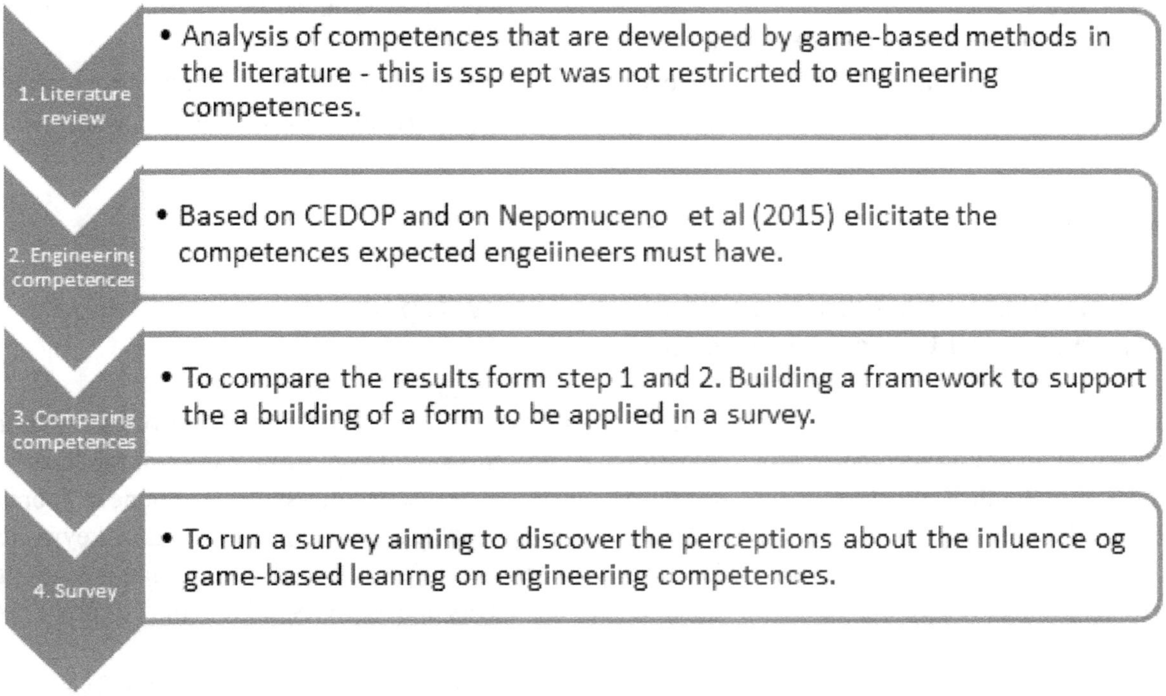

To clarify, the game-based methods that have been exposed here:

- Gamification: Gamification is the adaptation of game elements to environments and contexts that are not games.
- Serious Games: Serious games are games that were created without the goal of entertaining, but rather to train users (Zimmerling et al., 2016).
- Commercial games: Commercial games are games that were created to amuse those who play them, with no initial motivation to contribute to the training or development of any competence.
- Competence means an ability to perform a task or function. In the area of engineering and administration, the courses focus mainly on administrative competences, and the development of some of these competences is attainable with knowledge-based methods. However, competences such as analytical thinking, goal setting, teamwork, strategic thinking, analysis and problem solving, decision making, time management, communication, recognition of opportunities, creativity, conflict management, flexibility, courage and self-esteem are difficult to approach by classical methods (Mustata et al., 2017).

To facilitate the understanding of the topic addressed, it will be defined here which terms are related to which stages of the research and what are the meanings:

- Competences developed: These are the competences taken from articles in the Scopus database that have game-based methods.

- Expected competences: These are the competences selected to answer which competences are expected from an engineering professional, not only those determined by national curriculum guidelines, but also from other sources.
- Competences analyzed: These competences are the intersection between the competences developed and the expected competences. This junction was performed by the author's analysis.

BACKGROUND

Mapping in the Literature the Competences Developed by Game-Based Learning Methods

A search was carried out in the literature looking for articles that approached the influence of game-based teaching and learning methods on the professional competences of alumni - as a matter of the facts no restriction about the field of knowledge was introduced in this first step. So, a search was done in Scopus data base looking for articles that "game" AND ("teaching" OR "learning") appear their title, keywords or even in the abstracts. As a result, 327 articles were selected to pass through a screening. according to (Rodriguez et al., 2013) the use of a database such as Scopus or even WoS reduces the possibility of using no blinded review articles. "

Following a filtering process based in (Moher et al., 2009), (Siqueira & Aparecida, 2020)and (Da Silva et al., 2015), all 327 articles were screened looking for only those which listed the professional competences developed by game-based learning methods. This screening-based filtering process has resulted in selecting 44 articles for a deeper analysis.

Aiming to increase the confidence of the research, Table 2 reports the data related to the Scopus Cite Score of the articles' sources. One can note that most articles researched were published in journals ranked in quartile Q1 of Scopus.

Aiming to reduce the amount of space used naming each citation, this item brings a numbering of the 44 articles that are cited in this paper as it appears in Table 3.

Analysis of Articles' Content

These articles passed by a full reading and analysis of the texts, using the approach inspired in (Méxas et al., 2013) and (Oliveira et al., 2018). The data referring to the articles analyzed are shown in Tables 4 and 5. In these tables, the column with the title "Area" represents the classification taken from the Scopus website (this is related to the area where the source's highest percentile based on the cite score metric), while the column "Method" was written by the authors of this chapter taken into account the the analysis of the texts, and the column "Competences developed" shows the competences mentioned by the authors of the analyzed articles. The Table refers to those developed by gamification and commercial games, while Table 5 reports the ones mentioned as impacted by serious games.

Table 2. Position of the sources in Scopus cited score ranking

*Source	N. of articles	Highest percentile / Area
Technology, Knowledge and Learning	3	89% Mathematics
Simulation and Gaming	3	66% Business, Management and Accounting
Information Systems Frontiers	2	91% Theoretical Computer Science
Surgical Endoscopy and Other Interventional Techniques	2	96% Surgery
Nurse Education Today	2	93% General Nursing
Entertainment Computing	2	58% Software
Computer Applications in Engineering Education	2	80% Education
Revista Iberoamericana de Tecnologias Del Aprendizaje	2	60% General Engineering
Computers and Education	1	99% Education
Computers in Human Behavior	1	99% Arts and Humanities
Cognition	1	99% Language and Linguistics
Journal of the American Medical Directors Association	1	96% Health Policy
Virtual Reality	1	86% Computer Graphics and Computer-Aided Design
International Journal of Medical Informatics	1	82% Health Informatics
Liver Transplantation	1	92% Surgery
World Journal of Surgery	1	86% Surgery
Environmental Earth Sciences	1	74% Geology
American Journal of Surgery	1	84% Surgery
Journal of Surgical Education	1	81% Surgery
International Journal of Simulation Modelling	1	70% Modeling and Simulation
IEEE Computer Graphics and Applications	1	62% Computer Graphics and Computer-Aided Design
Clinical Simulation in Nursing	1	92% Nursing
SpringerPlus	1	82% Multidisciplinary
Journal of Laparoendoscopic and Advanced Surgical Techniques	1	62% Surgery
Journal of Business Strategy	1	46% Strategy and Management
International Journal of Emerging Technologies in Learning	1	57% General Engineering
Scientia Iranica	1	49% General Engineering
Journal of International Education in Business	1	41% General Business, Management and Accounting
The Clinical Teacher	1	16% Review and Exam Preparation
Revista Brasileira de Enfermagem	1	38% General Nursing
Electronic Journal of Research in Educational Psychology	1	32% Education
Industry and Higher Education	1	31% Business and International Management
Revista Facultad de Ingenieria	1	25% General Engineering
International Journal of Safety and Security	1	25% Safety, Risk, Reliability and Quality

source: Scopus (2020)

Table 3. Codes assigned to the articles

Article	Code	Article	Code
(Allal-Chérif et al., 2016)	A1	(Kriz & Auchter, 2016)	A23
(Barr, 2017)	A2	(Kumar & Raghavendran, 2015)	A24
(Bellotti et al., 2014)	A3	(Lagro et al., 2014)	A25
(Berns et al., 2016)	A4	(Lamb et al., 2017)	A26
(Butt et al., 2018)	A5	(Lin et al., 2015)	A27
(Castro & Gonçalves, 2018)	A6	(Liuta et al., 2019)	A28
(Cook et al., 2012)	A7	(Manrique-Losada et al., 2015)	A29
(Corrigan et al., 2015)	A8	(Mason & Loader, 2019)	A30
(Filella Guiu et al., 2016)	A9	(Meletiou-Mavrotheris & Prodromou, 2016)	A31
(García & Cano, 2018)	A10	(Mustata et al., 2017)	A32
(Graafland et al., 2015)	A11	(Nino & Evans, 2015)	A33
(Graafland et al., 2017)	A12	(Qin et al., 2010)	A34
(Graafland et al., 2014)	A13	(Seager et al., 2011)	A35
(Guenaga et al., 2013)	A14	(Sewilam et al., 2017)	A36
(Hernandez-Linares et al., 2017)	A15	(Shute et al., 2016)	A37
(Hummel et al., 2017)	A16	(Smith, 2017)	A38
(Johnsen et al., 2016)	A17	(Stone et al., 2009)	A39
(Kaczmarczyk et al., 2016)	A18	(Suki & Suki, 2019)	A40
(Kaliappen, 2019)	A19	(Verkuyl et al., 2016)	A41
(Katz et al., 2017)	A20	(Witte & Daly, 2014)	A42
(Khanzadi et al., 2019)	A21	(Zanasi et al., 2017)	A43
(Kowalewski et al., 2017)	A22	(Zimmerling et al., 2016)	A44

Table 4. Methods and competences developed by gamification and commercial games

Method	Competences	Code	Area (Scopus)
Gamification (9 articles)	Training of computer competences in nursing	A6	Nursing
	Construction design management knowledge	A21	Business, Management and accounting
	Drive worker engagement and transform organizational culture	A24	Business, Management and accounting
	Effective way to encourage learning in a geographically divided surgical residency	A26	Medicine
	Understanding hydraulics	A28	Business, Management and accounting
	Learning software quality concepts, teamwork, software process development	A29	**Engineering**
	Knowledge, competence and practical experience required to effectively integrate games into the mathematical curriculum	A31	Mathematics
	Student attitudes about learning and statistics	A38	Business, Management and accounting
	Management knowledge	A44	**Engineering**
Commercial games (3 articles)	Adaptability in resourcefulness and communication competences	A2	Social Sciences
	High-ranking thinking and decision-making, persistence, socialization, leadership competences, self-confidence, autonomy and self-regulation	A33	**Engineering**
	Proficiency in second language learning	A40	Multidisciplinary

Table 5. Competences developed by serious games

Competences	Code	Area (SCOPUS)
Socialization, "externalization"	A1	Mathematics
Complementing theory and practice, with basic concepts of entrepreneurship and business administration	A3	**Engineering**
Learning a second language	A4	Multidisciplinary
Procedural competence mastering and retention	A5	Nursing
Experience in life support, motivating continuous learning and engaging in the improvement of physical motor competences	A7	Nursing
Communication. Collaboration or negotiation and an overall improvement in collaborative learning	A8	**Engineering**
Emotional competences for conflict resolution	A9	Psychology
Experience and motivation in learning mathematical topics	A10	**Engineering**
Training surgical residents to deal with equipment problems	A11	Medicine
Improved response to equipment problems during surgery	A12	Medicine
Ability to teach and manage biliary tract disease	A13	Medicine
Entrepreneurial competences and problem solving	A14	**Engineering**
Competence in time management in the context of engineering training	A15	**Engineering**
Ability to solve complex problems	A16	Social Sciences
Decreased surgical trainees learning curve in clinical reasoning and problem solving, increasing efficiency in surgical training	A17	Medicine
Consolidation and review, training	A18	Medicine
Strategic administration	A19	Business, Management and accounting
Anesthesia administration in orthotopic liver transplantation	A20	Medicine
Learning cognitive aspects in laparoscopic cholecystectomy	A22	Medicine
Business knowledge and business planning competences	A23	Business, Management and accounting
Decision-making	A25	Medicine
Competence in surgical decision-making	A27	Medicine
Risk assessment and management competences	A30	Medicine
Analytical thinking, strategic thinking, teamwork, goal setting, opportunity recognition, problem recognition, problem solving, decision making, proactive thinking, time management, communication, intuitive thinking, responsibility, argumentation, creativity, diplomacy, conflict management, flexibility, courage and self-esteem	A32	Business, Management and accounting
Physical motor competences used in blood administration	A34	Medicine
Project management	A35	Computer Science
Dealing, having new ideas and managing floods	A36	Environmental Science
Problem solving	A37	Psychology
Transferring knowledge from the classroom to a real submarine environment	A39	Social Sciences
Pediatric nursing competences	A41	Nursing
Cultural awareness, critical thinking, leadership, teamwork ethics and interpersonal communication	A42	Business, Management and accounting
Fast and accurate analysis of information	A43	**Engineering**

Engineers Expected Competences

For this study, the competences researched are based in the curriculum guidelines of the (Ministério da Edcuação e Cutura [MEC], 2019), (European Centre for Development of Vocational Training [Cedefop], 2016), (Accreditation Board for Engineering and Technology [ABET], 2019) and in those selected by the article (Nepomuceno & Costa, 2015) as the expected ones for engineers profiles. The Cedefop, as the European Centre for Development of Vocational Training, states that there are five key competences for engineering professional: "problem solving", "job-specific competences", "learning", "moderate ICT competences" and "teamwork". The article by (Nepomuceno & Costa, 2015) is a compilation of literature competences and the opinion of experts about what competences an engineering course should provide to its students, serving as a basis for what is discussed in the scientific community. The Accreditation Board for Engineering and Technology (ABET) is a board that researches the engineering profiles need by industry and services, and provides a recognized accreditation system of engineer's careers,

Tables 6 shows the expected competences of engineering professionals, and it is a result from analyzing the competences brought in (Nepomuceno & Costa, 2015), (ABET, 2019), (Cedefop, 2016) and in MEC (2019). It takes into account similar meanings and intentions, beyond the use of different words for the same meaning.

Table 6. Summary of the engineers' expected competences

Expected competences	(Nepomuceno & Costa, 2015)	(MEC, 2019)	(ABET, 2019)	(Cedefop, 2016)
Base concepts and techniques	X	X	X	X
Ability to communicate in verbal, oral and writing way	X	X	X	X
Problem solving	X	X	X	X
Sociability	X	X	X	X
Continuous search for career improvement	X	X	X	X
Ability to manage and implement projects	X	X	X	X
Ability to manage, supervise and control team	X	X	X	X
Information and Communication Technology (ICT) knowledge		X	X	X

Association of Expected Competences with the Ones Developed by Games-Based Teaching

Applying the same reasoning that was applied in the previous topic to generate Table 6, the competences that appears in Tables 4 and 5, were merged into dimensions as it appears in Table 7. This action made possible to compare the engineers' expected competences (Table 6) against those mentioned in previous scientifical researches works that explored the effect of game-based learning over professional competences.

By comparing the data in Table 6 against those in Table 7, it is possible to identify that closer all the expected engineers' competences were already mentioned in the literature as having potential o be improved by game-based methods, the exception is "Continuous search for career improvement".

In the other hand, one can observe that almost competences mentioned in previous articles (Table 7) are covered by the expected engineers' competences that appear in Table 6, the exception is the ability to make decisions. Another point to be underlined is the competence to communicate in a second idiom. Despite it should be covered by "Ability to communicate in verbal, oral and writing way", it is especially important in countries that do not speak English as their first language and it should be considered one of the most expected competences an engineer not native to those countries should have. As a matter of the facts, the conclusion of this comparison is that Table 6 should be updated to Table 7, by the inclusion of two other competences: "Second language learning".

Table 7. Comparing competences mentioned in the literature as developing by game-based learning against those expected ones

Competences analyzed	Mentioned in the following articles	(Nepomuceno & Costa, 2015)	(MEC, 2019)	(Cedefop, 2016)	ABET (2020)
Base concepts and techniques	A3, A5, A7, A10, A11, A12, A13, A17, A18, A20, A21, A22, A24, A26, A28, A31, A34, A38, A39, A41	X	X	X	X
Sociability	A3, A8, A29, A32, A33, A42	X	X	X	X
Ability to communicate in verbal, oral and writing way	A1, A2, A8, A32, A33	X	X	X	X
Problem solving competences	A14, A16, A17, A32, A37	X	X	X	X
Ability to manage and implement projects	A4, A15, A19, A23, A30, A32, A33, A35, A36, A43	X	X	X	X
Ability to manage, supervise and control teams	A9, A23, A29, A30, A32, A33, A42, A44	X	X	X	X
Information and Communication Technology (ICT) knowledge	A3, A6, A29		X	X	X
Continuous search for career improvement	-------------------------	X	X		X
Decision making	A25, A27, A32, A33				X
Second language learning	A2, A40				

SURVEY

A survey regarding users of game-based learning methods was carried out looking to catch their perceptions about on the influence of such methods that appear in Table 5 on the competences shown in Table 7. Through a google form available in https://forms.gle/q8gpp8wB2igRff1Z9, the sample was asked to inform in what degree the game-based learning methods influences on the competences mentioned in Table 7. The sample could choose to answer the questions with one of the following options: "Influences very negatively", "Influences negatively", "Do not influence", "Influences positively", "Influences very positively", "I don't know how to answer". The form also included two open questions:

- "How much did your experience in the game-based learning method influence the (competence taken from the "Competences analyzed" column in Table 6, repeating this question for all competences) of your students or yours?"
- "In your experience, what were the barriers and advantages of applying or using these game-based learning methods?"

RESULTS

General Analysis

The questionnaire had 92 responses that were analyzed, of which came from people from 19 different countries, 69% from Brazil, 7% from the United States of America, 2% from each of the following countries, Germany, Spain, France, Malaysia and Mexico, and 14% joining the other 11 countries.

Respondents' ages range from 22 to 73, with 26% between 22 and 30 years old, 22% between 31 and 40 years old, 27% between 41 and 50 years old, 16% between 51 and 60 years old and 9% between 61 and 73 years old. In the questionnaire 35% of the respondents are women, the other 65% are men.

About the highest complete level of education, 37% are doctors or post-doctors, 39% are masters, 16% are graduates and 8% are graduated from high school. These data prove the respondents' high level of education, which was expected, since the research was aimed at possible applicators of learning methods.

Of those who have already participated or applied game-based learning methods, representing 57%, 20 participated as users, 8 participated as applicators of the method and 24 participated both as users and as applicators of the method. Of that same group individually, 35 participated using gamification, 10 participated with applied commercial games, 18 participated in serious games and 5 responded with not knowing how to answer the question, being possible to participate in more than one type of learning methods.

Response from Game-Based Learning Applicants

Figure 2 summarizes the answers collected in the survey, which got the opinion of the applicators of game-based learning methods. Of the 8 applicators, 6 used gamifications, 4 used serious games and 1 used commercial games. It is a need to notice that the number of respondents of this group is so low, despite the results have survived to statistical test as mentioned forward.

Within the group of applicators, it is said that the "second language learning" competence is not influenced. In the case of "Ability to manage and implement projects", although the positive influence was perceived in 3 of the answers, another 3 did not know how to evaluate, which leaves an uncertainty in this competence for the applicators.

Regarding their perceptions about barriers asked in the form, the applicators mentioned logistical barriers, such as obtaining computers, access to the network and support from the local site information technology sector. There are also barriers related to how games are perceived, how students perceive new learning methods and the time it takes to develop or find the right material for specific teaching goals.

Figure 2. Applicant responses

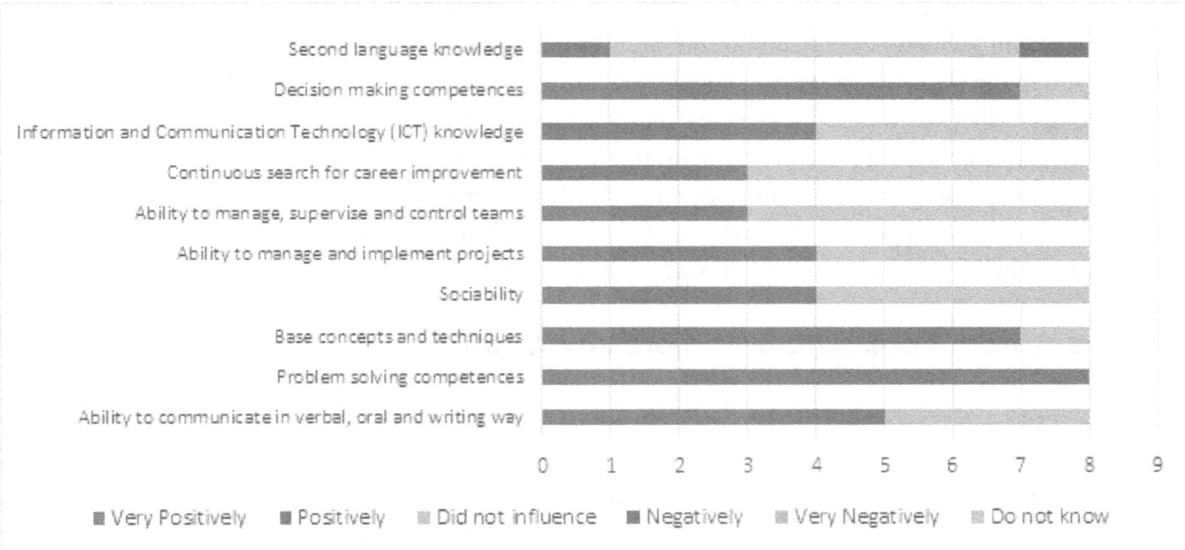

For the applicators, the advantages are the richness of the experience, the students' engagement, and the possibility of experiencing and practicing competences that are not normally worked on.

Users' Responses

Figure 3 shows the responses on the development of the competences of engineers in the opinion of users of game-based learning methods. Of the 20 users 13 used gamification, 4 used serious games, 1 used commercial games and 5 do not know the name of the one who used.

Figure 3. Users' responses

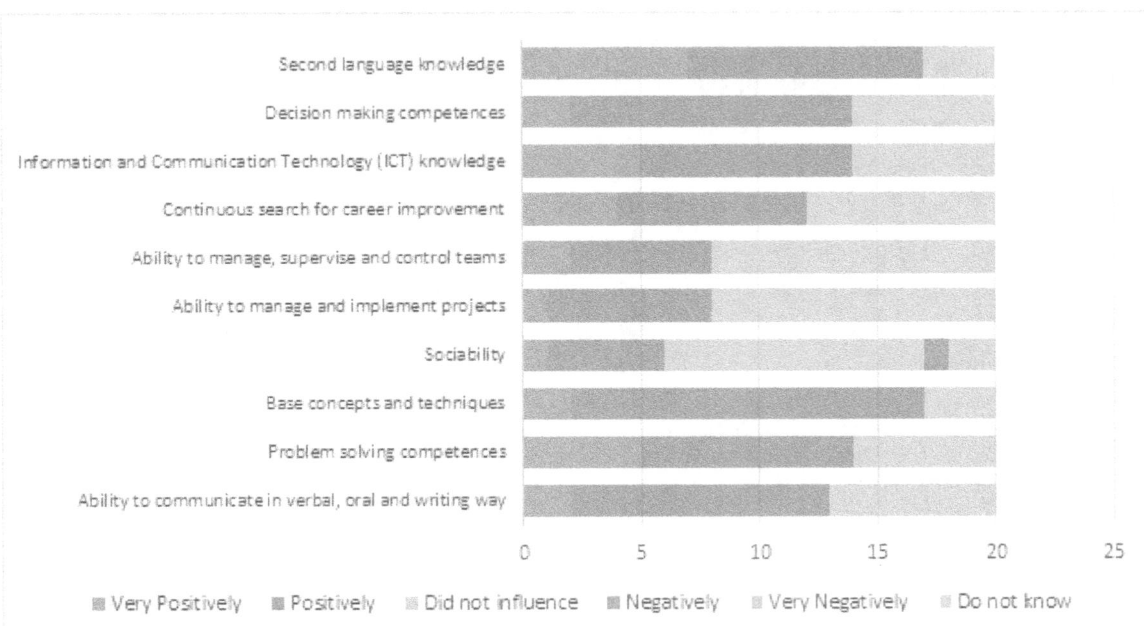

For users, the competences "Sociability", "Ability to manage and implement projects" and "Ability to manage, supervise and control teams" are not influenced by game-based methods. Highlighting "second language learning" which is the one with the highest numbers of "Very positive" influence responses, probably due to the popularization of language teaching applications using gamification.

Users believe that within the barriers are the misconceptions about how to improve these methods, the lack of game creators, the non-authorization to use these methods, the lack of interaction, the possibility of learning using these methods and the inability to provide adequate incentives to encourage participation. No advantage was pointed out by the users, even though most competences have many answers about the "Very positive" and "Positive" influence.

Responses from Applicants who were also Users

Figure 4 describes the responses on the development of the competences of the engineers in the opinion of the applicators who have also been users of game-based learning methods. Of the 24 users who were also applicators, 16 used gamifications, 8 used applied commercial games and 9 used serious games. One should take observe that in f the form, the respondent could inform that he/she could be enrolled as both: applicant or either a user of game-based learning method.

In the answers given by the applicators who have also been users, it is noticeable that all the competences analyzed are developed positively by the game-based methods. Of these responses, only two responses were negatively influencing, but they are in a large minority in relation to the other responses. Highlight for the competence "Assume the posture of permanent search for professional updating" which, despite the fact that the answer most frequently was "positively", adding the answers "Does not influence" and "Does not know if it influences", they are in the majority.

Figure 4. Responses from applicants who were also users

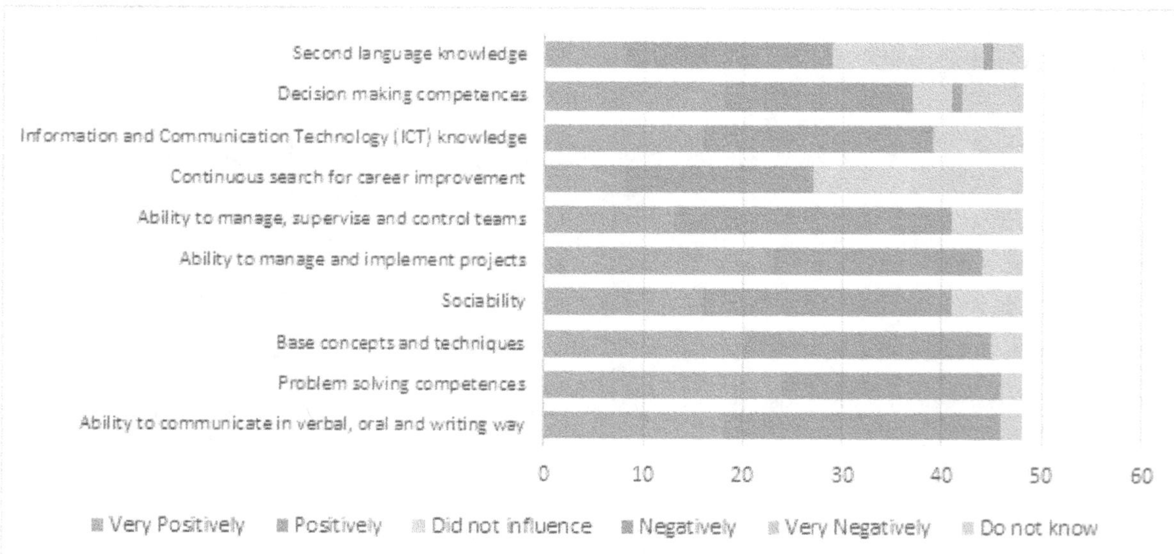

The barriers perceived by this group are the cost of the methods, the required level of English, the time that applying the method takes from the teacher and the student, the logistical problems, how to get computers, classrooms with specific equipment, extra materials, and the possibility of serving only small groups of students. The willingness of students to follow these methods and the lack of teacher training were also seen as barriers.

Applicants who have also been users see advantages such as self-knowledge created by the methods, students and teachers being on the same level, the student's continuous involvement in the method, the ability to serve students with physical and mental disabilities, the ability to see and to increase the evolution of students, the general increase in motivation and the possibility of working with something closer to reality.

Responses of Those who have Never Participated in Any Method

As one can see in Figure 5, of the respondents: 44% never participated in a game-based learning method, of these, 14 would like to participate as users, 8 would like to participate as applicators of the method and 18 would like to participate as users and applicators of the method. These respondents were asked how they believe that learning game-based methods would influence the analyzed competences It can be observed that the competence "Assume the posture of permanent search for professional updating" is the one with more answers than "Doesn't influence", tied with more answers than "I didn't know", and has the least number of answers "Very positive", however, it has the second highest number of "Positive" responses. This makes this competence the least defined by this study, and which would need another observation, more focused, even though it is credited as positively influenced.

The "Problem solving" competency is the one with the highest sum between "Very positive" and "Positive" responses, that is, it is the one that is credited by the respondents as the one most likely to be influenced by game-based learning methods.

Figure 5. Responses of those who have never participated in any method

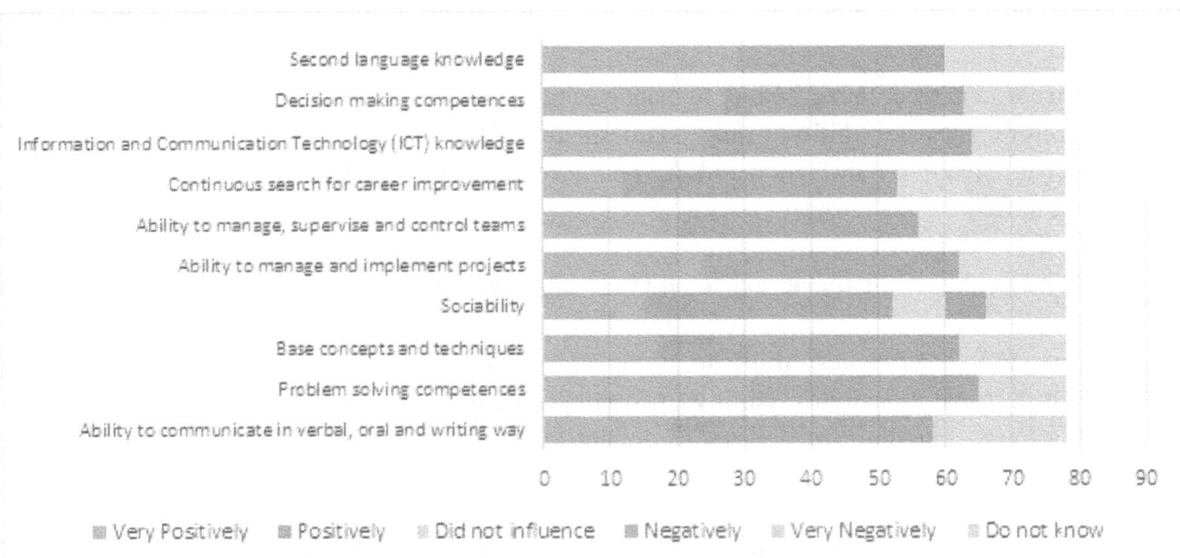

Validation and Data Consolidation

With the separation into groups, the need to validate the conclusions was noted, in view of the sample size, so the U test proposed by (Mann & Whitney, 1947) was used.

This test aims to express whether two samples have an expressive differentiation between them, used many times within medicine to check if there is a difference between a sample where a certain medication was applied and between a sample that was applied placebo. It was used in this research to verify if the samples discussed previously can be analyzed as one.

For this research, an acceptance rate of 5% was used, which results in a normal z of 1.96, therefore, applying the U test, all z results must be +1.96 and -1.96 to say that the samples do not have an expressive differentiation. The tests were done comparing all groups of respondents with all competences,.

All numbers were between +1.96 and - 1.96, which shows that the responses have a low rate of variation between the groups analyzed. Therefore, all responses can be analyzed together, which means that conclusions can be drawn from the total of responses.

CONCLUSIONS

Although it is not exhaustive, this study reached its core objective by contributing to discover in which degree game-based learning methods influences engineering competence, once it compiles and analyzes articles on game-based learning methods, and carries out unprecedented research on the influence of game-based learning methods on the professional competences of graduates from engineering.

Answers to the Research Questions

The questions that appear in a Table were answered in this paper as it follows:

1. **What are the professional competences that are developed on by game-based methods, according to the literature?**

Base concepts and techniques, Ability to manage and implement projects, Ability to manage, supervise and control teams, Sociability, Ability to communicate in verbal, oral and writing way, Problem solving, Decision making, Information and Communication Technology (ICT) knowledge, and Second language learning

2. **b) What are the competences expected of an engineering professional?**

Base concepts and techniques, Ability to manage and implement projects, Ability to manage, supervise and control teams, Sociability, Ability to communicate in verbal, oral and writing way, Problem solving, Decision making, Information and Communication Technology (ICT) knowledge, and Second language learning

3. **c) What are the results from comparing competences expected of an engineering against those mentioned in the scientifical literature as influenced by game-based learning?**

   a. The influence of game-based learning on the expected competence "Continuous search for career improvement", had not been previously researched or it was had not been detected in previous researches.
   b. Competences already detected as impacted by the adoption of game-based learning were not mentioned as an expected competence: "Decision making', and "Second language learning".
   c. Analyzing and associating the lists of competences mentioned above, a new list of competences was made, showing all the competences expected of an engineering professional.
4. **d) What is the influence perceived by users of game-based learning methods (teachers, students, and coordinators) on the expected competences?**
   a. It can be understood that game-based learning methods positively influence all the competences expected of engineering professionals. This positive influence is perceived by everyone who has used these methods and is expected by those who have not yet used them.
   b. The "Problem solving" competency is the one with the most "Very positive" influence responses, and is tied with "Basis of concepts and techniques", in relation to the sum of the "Very positive" and "Positive" responses, which makes the "Problem solving" competence, credited by the respondents, as the most positively influenced by game-based learning methods.
   c. "Continuous search for career improvement" is the one with the most answers "Doesn't influence" and "Didn't know", making this competence the least influential in studies, even with a sum of "Very positive" and "Positive" responses" greater than the sum "Doesn't influence" and "Didn't know".
   d. "Second language learning", "Decision making" and "Sociability" were the only competences assessed with a "Negative" influence, and even then, there are only 2 in" Second language learning "and 1 in the other two competences.
   e. As a whole there was an agreement among the answers provided by respondents that had not yet be part of a game-based learning and those that already participated in a such event. The only discordance was about "Sociability": it was the only one competence assessed with a "Negative" influence – even with only six negative responses, and the answers prevailing as "Very positively" or "Positively".

The research also elicited barriers for adopting adopt game-based learning methods: logistical barriers, materials lack for building the game, costs to develop and apply the game, unclear game explanation among others

Limitations and Future Works

The main limitation of this work is about its sample and the number of references . Despite the results pass by the statistical validation, the sample should be bigger and better spread around the world. So, as future works, it is suggested to the increase the sample and spread its geographical distribution, beyond the adoptions of other methods to analyze the collected data, such as clustering techniques and even multicriteria decision modeling to prioritize actions to face the barriers discovered in the research, looking to reduce cost and improve the results of adopting game-based learning techniques. It is interesting to notice that the evaluation of effects of adopting game-based learning under a multicriteria perspective, will make possible to compare the results against those reported in (Costa & Alves, 2021) that investigated the adoption of an outranking multicriteria decision method to evaluate the effects of Problem Based Learning (PBL) on engineering competences.

ACKNOWLEDGMENT

- Conselho Nacional de Desenvolvimento Científico e Tecnológico. Grant Number: 314352/2018-0
- Coordenação de Aperfeiçoamento de Pessoal de Ensino Superior. Grant Number: 001.

REFERENCES

ABET. (2019). *Criteria for Accrediting Engineering Programs, 2020 – 2021*. https://www.abet.org/accreditation/accreditation-criteria/criteria-for-accrediting-engineering-programs-2020-2021/

Allal-Chérif, O., Bidan, M., & Makhlouf, M. (2016). Using serious games to manage knowledge and competencies: The seven-step development process. *Information Systems Frontiers*, *18*(6), 1153–1163. doi:10.100710796-016-9649-7

Barr, M. (2017). Video games can develop graduate skills in higher education students: A randomised trial. *Computers & Education*, *113*, 86–97. doi:10.1016/j.compedu.2017.05.016

Bellotti, F., Berta, R., De Gloria, A., Lavagnino, E., Antonaci, A., Dagnino, F., Ott, M., Romero, M., Usart, M., & Mayer, I. S. (2014). Serious games and the development of an entrepreneurial mindset in higher education engineering students. *Entertainment Computing*, *5*(4), 357–366. doi:10.1016/j.entcom.2014.07.003

Berns, A., Isla-Montes, J.-L., Palomo-Duarte, M., & Dodero, J.-M. (2016). Motivation, students' needs and learning outcomes: A hybrid game-based app for enhanced language learning. *SpringerPlus*, *5*(1), 1305. Advance online publication. doi:10.118640064-016-2971-1 PMID:27547679

Butt, A. L., Kardong-Edgren, S., & Ellertson, A. (2018). Using Game-Based Virtual Reality with Haptics for Skill Acquisition. *Clinical Simulation in Nursing*, *16*, 25–32. doi:10.1016/j.ecns.2017.09.010

Castro, T. C., & Gonçalves, L. S. (2018). The use of gamification to teach in the nursing field. *Revista Brasileira de Enfermagem*, *71*(3), 1038–1045. doi:10.1590/0034-7167-2017-0023 PMID:29924165

Cedefop, E. (2016). Skills and jobs survey. *Researchers and engineers: skills opportunities and challenges.*

Cook, N. F., McAloon, T., O'Neill, P., & Beggs, R. (2012). Impact of a web based interactive simulation game (PULSE) on nursing students' experience and performance in life support training - A pilot study. *Nurse Education Today*, *32*(6), 714–720. doi:10.1016/j.nedt.2011.09.013 PMID:22082881

Corrigan, S., Zon, G. D. R., Maij, A., McDonald, N., & Mårtensson, L. (2015). An approach to collaborative learning and the serious game development. *Cognition Technology and Work*, *17*(2), 269–278. doi:10.100710111-014-0289-8

Costa, H. G., & Alves, A. C. (2021). *A Non-compensatory Multicriteria Model for Sorting the Influence of PBL Over Professional Skills*. Springer. doi:10.1007/978-3-030-71187-0_108

Da Silva, G. B., Costa, H. G., & De Barros, M. D. (2015). Entrepreneurship in engineering education: A literature review. International Journal of Engineering Education.

De Jesus, I. R. D. I. R. D., & Costa, H. G. G. H. G. (2013). Role of production engineers in the public sector. *22nd International Conference on Production Research, ICPR 2013*. https://www.scopus.com/inward/record.uri?eid=2-s2.0-84929359638&partnerID=40&md5=0754d9a5f95b4ee82c862b6c042c7b91

Filella Guiu, G., Cabello, E., Pérez-Escoda, N., & Ros-Morente, A. (2016). Evaluation of the Emotional Education program "Happy 8-12" for the assertive resolution of conflicts among peers. *Electronic Journal of Research in Educational Psychology*, *14*(3), 582–601. doi:10.14204/ejrep.40.15164

García, I., & Cano, E. (2018). A computer game for teaching and learning algebra topics at undergraduate level. *Computer Applications in Engineering Education*, *26*(2), 326–340. doi:10.1002/cae.21887

Graafland, M., Bemelman, W. A., & Schijven, M. P. (2015). Appraisal of face and content validity of a serious game improving situational awareness in surgical training. *Journal of Laparoendoscopic & Advanced Surgical Techniques. Part A.*, *25*(1), 43–49. doi:10.1089/lap.2014.0043 PMID:25607899

Graafland, M., Bemelman, W. A., & Schijven, M. P. (2017). Game-based training improves the surgeon's situational awareness in the operation room: A randomized controlled trial. *Surgical Endoscopy*, *31*(10), 4093–4101. doi:10.100700464-017-5456-6 PMID:28281117

Graafland, M., Vollebergh, M. F., Lagarde, S. M., Van Haperen, M., Bemelman, W. A., & Schijven, M. P. (2014). A serious game can be a valid method to train clinical decision-making in surgery. *World Journal of Surgery*, *38*(12), 3056–3062. doi:10.100700268-014-2743-4 PMID:25159118

Guenaga, M., Arranz, S., Florido, I. R., Aguilar, E., De Guinea, A. O., Rayón, A., Bezanilla, M. J., & Menchaca, I. (2013). Serious games for the development of employment oriented competences. *Revista Iberoamericana de Tecnologias Del Aprendizaje*, *8*(4), 176–183. doi:10.1109/RITA.2013.2285024

Hamari, J., Shernoff, D. J., Rowe, E., Coller, B., Asbell-Clarke, J., & Edwards, T. (2016). Challenging games help students learn: An empirical study on engagement, flow and immersion in game-based learning. *Computers in Human Behavior*, *54*, 170–179. doi:10.1016/j.chb.2015.07.045

Hernandez-Linares, R., Sánchez, H., Agudo, J. E., & Rico, M. (2017). Chronos: A tool to develop the time management competence among engineering students. *Computer Applications in Engineering Education*, *25*(1), 79–89. doi:10.1002/cae.21780

Hummel, H. G. K., Joosten-ten Brinke, D., Nadolski, R. J., & Baartman, L. K. J. (2017). Content validity of game-based assessment: Case study of a serious game for ICT managers in training. *Technology, Pedagogy and Education*, *26*(2), 225–240. doi:10.1080/1475939X.2016.1192060

Johnsen, H. M., Fossum, M., Vivekananda-Schmidt, P., Fruhling, A., & Slettebø, Å. (2016). Teaching clinical reasoning and decision-making skills to nursing students: Design, development, and usability evaluation of a serious game. *International Journal of Medical Informatics*, *94*, 39–48. doi:10.1016/j.ijmedinf.2016.06.014 PMID:27573310

Kaczmarczyk, J., Davidson, R., Bryden, D., Haselden, S., & Vivekananda-Schmidt, P. (2016). Learning decision making through serious games. *The Clinical Teacher*, *13*(4), 277–282. doi:10.1111/tct.12426 PMID:26213136

Kaliappen, N. (2019). Educational benefits of using Business Strategy Game (BSG) in teaching and learning strategic management. *International Journal of Emerging Technologies in Learning*, *14*(7), 209–215. doi:10.3991/ijet.v14i07.9792

Katz, D., Zerillo, J., Kim, S., Hill, B., Wang, R., Goldberg, A., & DeMaria, S. (2017). Serious gaming for orthotopic liver transplant anesthesiology: A randomized control trial. *Liver Transplantation*, *23*(4), 430–439. doi:10.1002/lt.24732 PMID:28133947

Khanzadi, M., Shahbazi, M. M., Arashpour, M., & Ghosh, S. (2019). Lean design management using a gamified system. *Scientia Iranica*, *26*(1), 15–25. doi:10.24200ci.2018.20325

Kowalewski, K.-F., Hendrie, J. D., Schmidt, M. W., Proctor, T., Paul, S., Garrow, C. R., Kenngott, H. G., Müller-Stich, B. P., & Nickel, F. (2017). Validation of the mobile serious game application Touch Surgery™ for cognitive training and assessment of laparoscopic cholecystectomy. *Surgical Endoscopy*, *31*(10), 4058–4066. doi:10.100700464-017-5452-x PMID:28281111

Kriz, W. C., & Auchter, E. (2016). 10 Years of Evaluation Research Into Gaming Simulation for German Entrepreneurship and a New Study on Its Long-Term Effects. *Simulation & Gaming*, *47*(2), 179–205. doi:10.1177/1046878116633972

Kumar, H., & Raghavendran, S. (2015). Gamification, the finer art: Fostering creativity and employee engagement. *The Journal of Business Strategy*, *36*(6), 3–12. doi:10.1108/JBS-10-2014-0119

Lagro, J., van de Pol, M. H. J., Laan, A., Huijbregts-Verheyden, F. J., Fluit, L. C. R., & Olde Rikkert, M. G. M. (2014). A Randomized Controlled Trial on Teaching Geriatric Medical Decision Making and Cost Consciousness With the Serious Game GeriatriX. *Journal of the American Medical Directors Association*, *15*(12), 957.e1–957.e6. doi:10.1016/j.jamda.2014.04.011 PMID:24913210

Lamb, L. C., DiFiori, M. M., Jayaraman, V., Shames, B. D., & Feeney, J. M. (2017). Gamified Twitter Microblogging to Support Resident Preparation for the American Board of Surgery In-Service Training Examination. *Journal of Surgical Education*, *74*(6), 986–991. doi:10.1016/j.jsurg.2017.05.010 PMID:28545826

Lin, D. T., Park, J., Liebert, C. A., & Lau, J. N. (2015). Validity evidence for Surgical Improvement of Clinical Knowledge Ops: A novel gaming platform to assess surgical decision making. *American Journal of Surgery*, *209*(1), 79–85. doi:10.1016/j.amjsurg.2014.08.033 PMID:25454955

Liuta, A. V., Perig, A. V., Afanasieva, M. A., & Skyrtach, V. M. (2019). Didactic games as student-friendly tools for learning hydraulics in a technical university's undergraduate curriculum. Industry and Higher Education. doi:10.1177/0950422218824507

Mann, H. B., & Whitney, D. R. (1947). On a Test of Whether one of Two Random Variables is Stochastically Larger than the Other. *Annals of Mathematical Statistics*, *18*(1), 50–60. doi:10.1214/aoms/1177730491

Manrique-Losada, B., Gasca-Hurtado, G. P., & Álvarez, M. C. G. (2015). Assessment proposal of teaching and learning strategies in software process improvement. *Revista de la Facultad de Ingeniería, 2015*(77), 105–114. doi:10.17533/udea.redin.n77a13

Mason, J., & Loader, K. (2019). Using a Serious Game to Train Violence Risk Assessment and Management Skills. *Simulation & Gaming, 50*(2), 124–135. Advance online publication. doi:10.1177/1046878119837314

MEC. (2019). *Diretrizes Curriculares Nacionais dos Cursos de Engenharia*. Author.

Meletiou-Mavrotheris, M., & Prodromou, T. (2016). Pre-Service Teacher Training on Game-Enhanced Mathematics Teaching and Learning. *Technology. Knowledge and Learning, 21*(3), 379–399. doi:10.100710758-016-9275-y

Méxas, M. P., Quelhas, O. L. G., Costa, H. G., & De Jesus Lameira, V. (2013). A set of criteria for selection of enterprise resource planning (ERP). *International Journal of Enterprise Information Systems, 9*(2), 44–69. doi:10.4018/jeis.2013040103

Moher, D., Liberati, A., Tetzlaff, J., & Altman, D. G. (2009). Preferred Reporting Items for Systematic Reviews and Meta-Analyses: The PRISMA Statement. *PLoS Medicine, 6*(7), e1000097. doi:10.1371/journal.pmed.1000097 PMID:19621072

Mustata, I. C., Alexe, C. G., & Alexe, C. M. (2017). Developing competencies with the general management II business simulation game. *International Journal of Simulation Modelling, 16*(3), 412–421. doi:10.2507/IJSIMM16(3)4.383

Nepomuceno, L. D. de O., & Costa, H. G. (2015). Analyzing perceptions about the influence of a master course over the professional skills of its alumni: A multicriteria approach. *Pesquisa Operacional*. Advance online publication. doi:10.1590/0101-7438.2015.035.01.0187

Nino, M., & Evans, M. A. (2015). Fostering 21st-Century Skills in Constructivist Engineering Classrooms with Digital Game-Based Learning. *Revista Iberoamericana de Tecnologias Del Aprendizaje, 10*(3), 143–149. doi:10.1109/RITA.2015.2452673

Oliveira, A. S. S., de Barros, M. D. D., de Carvalho Pereira, F., Gomes, C. F. S. F. S., & da Costa, H. G. G. (2018). Prospective scenarios: A literature review on the Scopus database. *Futures, 100*, 20–33. doi:10.1016/j.futures.2018.03.005

Qin, J., Chui, Y.-P., Pang, W.-M., Choi, K.-S., & Heng, P.-A. (2010). Learning blood management in orthopedic surgery through gameplay. *IEEE Computer Graphics and Applications, 30*(2), 45–57. doi:10.1109/MCG.2009.83 PMID:20650710

Rodriguez, D. S. S., Costa, H. G., & Do Carmo, L. F. R. R. S. (2013). Métodos de auxílio multicritério à decisão aplicados a problemas de PCP: Mapeamento da produção em periódicos publicados no Brasil. In Gestao e Producao. doi:10.1590/S0104-530X2013000100010

Seager, W., Ruskov, M., Sasse, M. A., & Oliveira, M. (2011). Eliciting and modelling expertise for serious games in project management. *Entertainment Computing, 2*(2), 75–80. doi:10.1016/j.entcom.2011.01.002

Sewilam, H., Nacken, H., Breuer, R., & Pyka, C. (2017). Competence-based and game-based capacity development for sustainable water management in Germany. *Environmental Earth Sciences*, *76*(3), 131. Advance online publication. doi:10.100712665-017-6416-0

Shute, V. J., Wang, L., Greiff, S., Zhao, W., & Moore, G. (2016). Measuring problem solving skills via stealth assessment in an engaging video game. *Computers in Human Behavior*, *63*, 106–117. doi:10.1016/j.chb.2016.05.047

Siqueira, B. J. S., & Aparecida, A. E. (2020). Socially responsible investments (SRIs) – mapping the research field. In Social Responsibility Journal. doi:10.1108/SRJ-06-2019-0204

Smith, T. (2017). Gamified Modules for an Introductory Statistics Course and Their Impact on Attitudes and Learning. *Simulation & Gaming*, *48*(6), 832–854. doi:10.1177/1046878117731888

Stone, R., Caird-Daley, A., & Bessell, K. (2009). SubSafe: A games-based training system for submarine safety and spatial awareness (Part 1). *Virtual Reality (Waltham Cross)*, *13*(1), 3–12. doi:10.100710055-008-0110-1

Suki, N. M., & Suki, N. M. (2019). Structural Relationships in the Embedding of Role-Play Games in a Class for Japanese Language Proficiency: Towards a Unified View. *Technology. Knowledge and Learning*, *24*(1), 65–87. doi:10.100710758-016-9294-8

SuperData Research, Inc. (2019). *Worldwide digital games market: September 2019*. SuperData Research.

Verkuyl, M., Atack, L., Mastrilli, P., & Romaniuk, D. (2016). Virtual gaming to develop students' pediatric nursing skills: A usability test. *Nurse Education Today*, *46*, 81–85. doi:10.1016/j.nedt.2016.08.024 PMID:27614548

Witte, A. E., & Daly, P. (2014). Proverbial wisdom - A "serious" international business game. *Journal of International Education in Business*, *7*(1), 2–13. doi:10.1108/JIEB-12-2012-0029

Zanasi, A., Ruini, F., & Bonzio, A. (2017). Intelligence analysts' training through serious games: The leila project. *International Journal of Safety and Security Engineering*, *7*(3), 380–389. doi:10.2495/SAFE-V7-N3-380-389

Zimmerling, E., Höflinger, P. J., Sandner, P. G., & Welpe, I. M. (2016). A system framework for gamified Cost Engineering. *Information Systems Frontiers*, *18*(6), 1063–1084. doi:10.100710796-016-9624-3

Chapter 5
A Scrum-Based Classroom Model for Learning Project Management

Erik Teixeira Lopes
University of Brasília, Brazil

André Luiz Aquere
University of Brasília, Brazil

ABSTRACT

Brazilian higher education uses traditional learning methods centered on the professor and lectures. However, active learning methodologies have recently been gaining ground, especially in courses in the health area, due to legal guidelines for their implementation in Brazil. At the same time, the use of active methodologies in engineering education to optimize learning results is already widespread in several countries. In this sense, this chapter aims to propose a structure that addresses the interface between the agile Scrum framework applied to education, known as EduScrum, and the active learning methodologies to develop a more applied and results-focused approach. Thus, the scope of this work includes a review of the literature and the structuring, application, and evaluation of a hybrid method adequate for training engineering students for modern technological advancements. Finally, the results obtained, as well as a roadmap, are presented to guide the application of the model in other learning contexts.

INTRODUCTION

After the publication of the book "Scrum, The Art of Doing Twice the Work in Half the Time" (Sutherland, 2014), Scrum became known worldwide as an exponent of agile methods of project management applied to the development of new technologies and knowledge. Initially proposed in the context of software development, the method became widely disseminated by the possibility of its application to other services and industries, such as education.

DOI: 10.4018/978-1-7998-8816-1.ch005

At the same time, in recent years, there has been a growing demand from the market and the students themselves for more dynamic, applied and participatory classes that would lead to the development of skills and competencies necessary for future engineering professionals. Thus, active learning has become the focus of discussions around the world (Aquere, 2017; Crawley et al., 2007; Filho et al., 2019; Kon, 2010; Lima et al., 2017; Prince & Felder, 2006; Vybornov et al., 2014).

Along this line, there is a synergy between the Scrum proposal applied to education, eduScrum, and active learning methodologies involving the concepts of transparency, inspection and adaptability and adapting its procedures and artifacts to the teaching context, thus supporting student-centered approaches and the development of cross-cutting technical skills focusing on the reality of current and future engineering needs.

This chapter aims to structure a learning model that acts on this agile-active interface, optimizing the strengths of both approaches in courses that develop both the technical and behavioral content of engineering students and preparing them for new technological and professional demands. It will discuss and present how to structure a course to operationalize this model and how to conduct classes and activities, as well as identify the main difficulties in working with this proposed hybrid approach.

To this end, a course structure has been proposed, centered on eduScrum, built collaboratively and frequently updated according to its development so that the course itself becomes an individual project of each student. Thus, the objective was to make possible the application and learning of the competencies developed in the course both in terms of evaluations and extra activities, as well as in projects that go beyond the academic environment, applying techniques of time, knowledge and risk management.

Finally, following the agile proposal of completing cycles of a better product, the course seeks to optimize the results obtained by reviewing theoretical models or the structuring of artifacts and tools for the facilitation of this structure.

Therefore, the present chapter is expected to contribute to the teaching of engineering, providing a model that will contribute to the training of students in the current context of technological advances in different areas through learning with agile methodologies used in companies that incorporate theoretical and behavioral dimensions.

This chapter applies the model proposed for the teaching of project management subjects.

METHODOLOGY

The methodological approach proposed for the development and implementation of this model is action research, characterized by presenting activities in both the practice and research areas, in addition to being innovative, participatory, interventionist, problematizing, documented and deliberate, aspects that differentiate it from other investigative research approaches such as routine practice and scientific research (Tripp, 2005). Given this participatory and continually evolving approach, it is understood that action research approaches scientific research in a similar way to eduScrum, using iterative cycles in education to obtain results. In this sense, the development of work using the Scrum framework and its artifacts, as well as the teaching of scrum in the developed course, introduces a scientific methodological character to students, contributing to their developing good research practices as complementary skills.

The model proposed here was applied to the Project Management and Multidisciplinary Teams (GEPEM) course offered by the Department of Civil and Environmental Engineering of the University of Brasília, with classes of up to 20 students with a tradition of using active methodologies to address the concepts of project management.

The complete process of the pilot project development consisted of five stages:

1. Exploration of the theme of the project: Exploratory research on the theme of the project, basing discussions and solutions on books, articles and other related works. According to a previous survey, the main topics to be addressed were eduScrum, active learning methodologies and project management.
2. Structure of the course: It consisted of the reformulation and formalization of the course based on the demands discussed with the teacher and the previous bibliographic review. In this respect, the following steps were contemplated, always accompanied in parallel by discussions with the advisor and the necessary adaptations:
    a. Requirements gathering and interface structuring.
    b. Structuring of the proposed course.
    c. The use of active learning methodologies.
    d. The project of the course.
    e. Adaptations to the initial proposal.
3. Application of the proposed model consisted of the application and monitoring of the Project Management and Multidisciplinary Teams course. Two researchers actively participate in this stage: one as a professor and the other as a student of the course, being present in the classes and performing the activities carried out by the other students. During the semester, notes were made about the perceptions of the student and the professor, always aligned at the end of each sprint, to review the main points and make the necessary adaptations. Additionally, at this stage, two questionnaires were applied, one at the beginning of the school semester and the other at the end, to compare the results expected and obtained by the students and the proposed model.
4. Analysis and consolidation of results: The study was carried out at the end of the school semester, consolidating the answers obtained from the questionnaires and the notes and minutes of meetings between the researchers, summarized in this study.
5. Diagnosis of improvements: This consisted of a second iteration of the proposed model, with an analysis of the main difficulties, theoretical reformulations and developed artifacts and tools. The objective was to use the knowledge acquired with the application to propose an updated version with greater ease of operationalization and possibly better results.

BACKGROUND

The model presented here approaches the teaching/learning process as a project (Project Management Institute [PMI], 2017). To this end, in this section, the authors present a brief review of the concepts of project management, agile management and Scrum. The review is completed based on the concepts of the flipped classroom and project-based learning, two active learning methodologies used in the model.

This section presents the conceptual bases that are fundamental for understanding the model. Thus, its understanding does not depend on the technical content or theoretical scope of the course in which it will be used.

Project Management

Projects are temporary efforts aimed at creating a single product or service (PMI, 2017). A project may contain repetitive activities, but its delivery necessarily needs to be unique. Projects occur at all levels within companies and in everyday life, for example, when we face personal issues as projects to be completed. A project is temporary in nature, understood as having a well-defined beginning and end and of being completed, for example, through the delivery of objectives, the exhaustion of resources or even by legal or strategic issues. During its execution, it acts as an agent of change in organizations, as it lists and organizes activities that will make a difference at some level and adds value to stakeholders because its purpose is to satisfy interests, meet legal requirements or even create or improve some processes.

Thus, a project needs to be executed efficiently using tools and techniques that will permit organizations to meet deadlines and achieve success, which is the essence of project management (Martin & Tate, 2001). On this basis, the results of the project can be linked with the objectives of stakeholders, compete more effectively in markets, sustain the organization, and respond in an agile way to local and global changes to the context.

Based on this initial understanding, one can frame several daily activities as projects, such as personal initiatives, approval in a course or the elaboration of a dissertation. Thus, project management becomes a way to optimize people's performance in various contexts and involving different initiatives.

Agile Management

Agile project management originated from the Agile Manifesto. The traditional management system was used from the 1970s, during the "software crisis", to the early 2000s, when it was found to be unfeasible. Instead of rejecting all concepts and techniques of this approach, a change in perspective and appreciation was suggested, such as a focus on the client, on regular deliveries and on the people who make up the team (Duarte, 2019).

Agile management addresses the phases of a project according to two recurring patterns in adaptive environments, both starting from activity prioritization lists, to reduce burdens by eliminating the starting and ending activities of the iteration. The first consists of sequential iterative phases, i.e., the decomposition of the project into iterations of predictable and agreed-upon durations, which facilitates the schedule. This model has advantages for complex projects with high risk and variability, even if constant repetitions may incur indirect costs. The second pattern consists of continuous overlapping phases during which it is understood that all groups of processes must be executed throughout the process because, with the passage of time and increased maturity of the project, new ideas and better solutions will emerge (PMI, 2017). Thus, the objective is to refine and aggressively improve all points of the work plan.

Within this context, many methodologies and variations arise to manage projects in a more agile way, some of which are already well established, such as Scrum, scaled agile framework (SAFe) and Six Sigma. In summary, two main points distinguish classical from agile methodologies: the exchange of the primary restriction of the scope for the time and the set of tools used. The main benefits of agile management are greater assertiveness, flexibility, collaboration, communication, and simplicity.

Scrum

Scrum is a flexible methodology that acts by defining an incremental process and that can be applied to projects and activities of different sizes, overcoming barriers and ensuring constant evolution based on adaptability (Bissi, 2007). In the book "Scrum the art of doing twice as much time" (Sutherland, 2014), Sutherland emphasizes three pillars for ensuring the success of the Scrum approach: transparency, inspection and adaptation. Each of these concepts allows the user to deliver creative solutions within a framework that allows the resolution of complex and adaptive problems with great value delivery. The importance of each of these is reflected in the division of roles in the procedures and artifacts adopted, addressed not only by Sutherland (2014) but also by Duarte (2019), Delhij et al. (2016) and Bissi (2007). The main terms associated with Scrum are:
Roles: Product owner, developers and Scrum Master.
Events: Sprint, Sprint Planning, Daily Scrum, Sprint Review, and Sprint retrospective.

Artifacts

The Scrum process begins with the definition of the team and the roles of the team members, with the Scrum Master being of paramount importance as being responsible for the management and transmission of project information to all team members. Next, the time boxes and artifacts used are defined. It is important to establish a fixed schedule for meetings and constancy in deadlines to ensure the focus on daily activities by all, in addition to the correct explanation of the steps to be completed and of the forms of teamwork, thus increasing the success of the sprints. The cycles are then repeated according to product development until final delivery (Bissi, 2007). A standard Scrum framework is presented in Figure 1:

Figure 1. Scrum framework
Source: Author

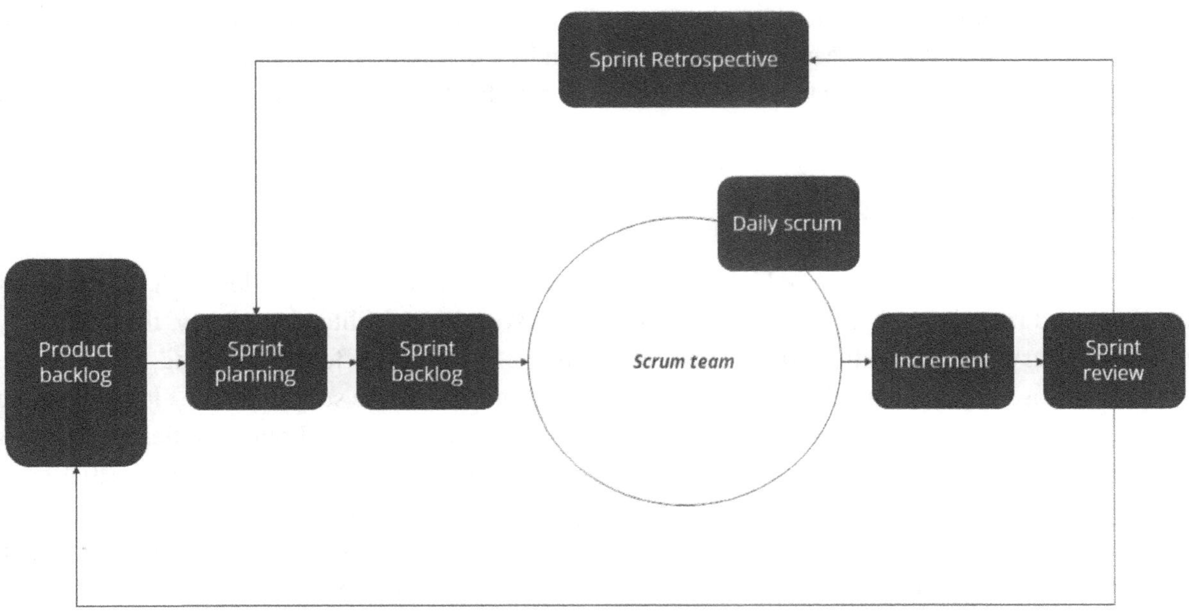

A scrum is an iterative approach, always starting with planning and ending with a review of the results and learning. During the process, daily alignments are made to monitor product development and identify possible problems and impediments.

EduScrum

The dynamism and adaptability of active methodologies such as project-based learning (PBL), in which the product is developed during execution, show similarities to scrum principles. In education, eduScrum has demonstrated improved results and motivations in case studies in higher education (Baldo et al., 2019; Ferreira & Martins, 2016; Souza et al., 2018), in addition to effectiveness in the development of other skills such as leadership, management, teamwork and receiving criticism (Borges et al., 2014). In experiences related to PBL, although there are reports of difficulties with some of the practices and artifacts, the students' evaluations have been predominantly positive (Dinis-Carvalho et al., 2017).

According to Borges et al. (2014), the Scrum methodology applied to education involves some main practices, such as Sprint Planning, Sprint Review, Retrospective, Daily Scrum, Preparation of the product backlog and sprint backlog, and use of Kanban.

Adaptations of the following events and actors are necessary when using eduScrum (Delhij et al., 2016):

Professor (Product Owner): determines training needs, monitors, enhances quality and evaluates educational outcomes.

EduScrum Development Team: forms self-organizing, multidisciplinary teams that track their own progress.

Team Leader (eduScrum Master): one of the members of the student team who assumes the leadership role and some of the responsibilities of the product owner as the project progresses.

As in Scrum, the same three concepts permeate the structure of an eduScrum process: transparency based on a language and definition of "Completed" shared by all participants; regular inspections of artifacts to detect undesirable deviations; and adaptation based on formal and practical events such as those described above (Sutherland, 2014).

It is also worth mentioning that in eduScrum, the product owner is responsible for achieving consensus to complete tasks and learning objectives (given by the completion of all activities of a Sprint), in addition to ensuring a pleasant work environment as a motivating factor for students in the context of higher education (Delhij et al., 2016).

Flipped Classroom

The Flipped Classroom is an active learning model used at different educational levels in various environments and models, such as the Thayer method at the West Point Military Academy, the case study method at Harvard University, and the 1998 publication of the book Effective Grading, in which the authors propose a model whereby the students first have contact with the subject before class to explore the content and then receive feedback during the face-to-face stage. Several other studies and articles also address this methodology (Bergmann and Sams, 2012; Filho et al., 2019; Valério et al., 2019).

In general, the flipped classroom is a methodology used by students to prepare theoretical content before class and to then develop activities such as exercises in person in an inverted sequence in relation to the traditional learning model. (Filho et al., 2019). The process can be classified into three stages: prelesson, class, and after-class.

In the prelesson, the teacher guides and provides materials so that students have a first contact with the topic to be addressed, using several tools. This allows the student to deal with doubts and questions, a process that plays an essential role in the development of their intellectual autonomy, encouraging contact with colleagues and the teacher, in addition to the search for other sources to resolve doubts. This helps students develop more confidence who participate less in exhibition classes, for example (Filho et al., 2019).

During the class period, the teacher then develops activities that aim at synthesizing and clarifying the content, such as discussions, analysis and debates, using different strategies for the optimization of learning. The content base previously obtained by the students allows the teacher to provide a moment of real learning in the class, in addition to the development of other students' skills in parallel to the technical content of the course, such as teamwork, argumentation and oratory. The evaluation process can also occur naturally during class time or not, using varied strategies such as seminars, papers or rapid tests (Filho et al., 2019).

Finally, the afterschool component consists of the student reviewing and extrapolating the content, for example, through applying what they learned to day-to-day experiences and other areas of knowledge. The teacher can guide these activities as the class moment progresses to ensure compliance with the program content. Thus, more advanced cognitive development is favored, such as analysis, synthesis and creation, because constructed knowledge has more meaning than knowledge passively received by the student (Filho et al., 2019). It is noteworthy that there is an intersection between the after-class of one theme and the prelesson of the next; at this transitional moment, the student initiates contact with the material provided for the next theme.

Project-Based Learning

Project-based learning is an active learning methodology based on real day-to-day situations (Bell, 2010; Pereira et al., 2017; Thomas, 2000). This methodology seeks to provide not only technical knowledge but also skills such as problem solving, investigation and decision-making abilities based on guidance by a teacher (Thomas, 2000). There are a wide variety of studies that address PBL, incorporating different definitions and the lack of conceptual and practical consensus (Thomas, 2000); thus, there are several models for the application of PBL.

PBL projects that utilize the methodology in question must follow standard criteria that should be applied. In his bibliographic review, Thomas (2000) and Pereira et al. (2017) propose five foundations that faithfully characterize an applicable project centered on autonomy, the orientation of the issue, constructive investigation, and its reflection of reality:

- PBL projects are central, not peripheral to the curriculum: they are the student's way of learning the concepts of the course and extrapolating their objectives, developing skills other than those expected for them.
- PBL projects are focused on questions that lead students to find and interact with concepts linked to the course: they should be constructed by guiding students to the objectives of the course, so they will absorb all the content that should be addressed.
- PBL projects involve students through constructive research: they should make students need to use an objective-focused investigation through design, problem framing, and resolution, among others.

- PBL projects are guided by students to a significant extent: certain deliveries or steps should not be determined by the teacher, but students should conduct the process in search of the solution and deliveries they deem relevant.
- PBL projects are realistic: they should involve real situations such as working with nonstudent stakeholders and delivering a product to some customer, among others.

COURSE DEVELOPMENT AND APPLICATION

This section discusses the structuring and application of the course, starting with the gathering of student requirements, the structuring of the interface and the curriculum, and the adaptations and learning during the application.

Gathering of Requirements and Interface Structuring

Prior to the beginning of the enrollment period at the University of Brasília, with the initial hypotheses of the model to be proposed for the course of Project Management and Multidisciplinary Teams, a virtual questionnaire published on social networks was used to identify the value perceived by the students of active methodologies, the targeted project management and behavioral skills, and the themes to be addressed in the course.

The questionnaire was structured into sections for rapid completion to obtain the greatest number of answers. The first section characterized the sample, identifying the course, semester, gender, and age of the research participants. Then, questions related to access inquired about the infrastructure for remote classes and the availability of the time needed to take the course. In addition, the questions were structured according to the students' previous experience with the active learning and investment methodologies they hoped to obtain when attending an optional course, in the case of extra class hours.

In total, 68 responses to the questionnaire were obtained, predominantly in the courses of civil engineering and environmental engineering. The main answers that led the research were the time of dedication expected by the students and the primary aspects that add value to the course, these with their respective scores on a Likert scale from 1 to 5: multidisciplinary (4.50), addressing team management (4.53), addressing real situations (4.79), addressing project management (4.57), and developing complementary competencies (4.51).

In parallel to the application of the questionnaire, the researchers identified the interface between eduScrum and active learning methodologies, mainly project-based learning and the flipped classroom, which were selected to be applied in the first restructured version of the course. The objective was to identify the similarities of the proposals to permit their optimization throughout the semester. Figure 2 illustrates the main considerations raised.

Figure 2. Interface between active methodologies and eduScrum
Source: Author

Course Structure

Initially, we chose to maintain the course load at 4 hours per week, divided into two classes during a total of 15 weeks, a model that had already been applied in other semesters. Based on this planning, a preliminary schedule and work plan were prepared. In addition, the class was limited to 20 students.

The classes were grouped into sprints to characterize eduScrum. All planning was carried out previously but later validated with the students in the first sprint. On this basis, we opted for a collaborative process using the first sprint to align expectations with those of the students, with adjustments made to the previous planning according to the students attending the course, provided that they met the minimum requirements provided by the professor.

The course was then divided into 5 sprints of equal duration to facilitate control of a first experience and establish a modular and replicable structure for courses of different durations. It is important that each of these sprints contain minimal planned events (planning, task division, stand up, sprint review, and retrospective) and a clear goal.

Sprint I (Introduction): The first cycle of the course is focused on welcoming the students and adapting to the methodology, in addition to aligning expectations and work agreements. The idea is to address some initial project management concepts to contextualize the students and to introduce the Scrum approach, which will be necessary for course development. The deliveries provided are the course plan, adjusted according to the agreement between students and the professor, and the project contract, with the rules of coexistence, deadlines and other agreements signed by the group.

Sprints II to IV (Development): The development cycles consisting of the main technical aspect of the curricular unit address the expected content of the course and structure learning with active methodologies. The order of presentation follows the logic of a project, using the conventional flow (start, planning, execution, monitoring, and closure), with some topics presented in advance to facilitate coordination and group work, such as communication management. The planned deliveries are related to the content addressed in the classroom and applied to the project of the course. Thus, there is a parallel between theory and practice, with knowledge being consolidated in smaller cycles that are closer to the reality of the students.

Sprint V (Closure): The last cycle aims at evaluating and finishing the course, in addition to addressing some additional content such as project management references. Both its delivery and its development are based on the closure of the project of the course and the final presentation to formalize the conclusion of the "contract" established in the first sprint. It is important to emphasize that this cycle addresses not only the professor's evaluation of the students but also the self-assessment of each student and the retrospective view of the entire project, in this case, of the academic semester.

Figure 3. Basic structure adopted for the course
Source: Author

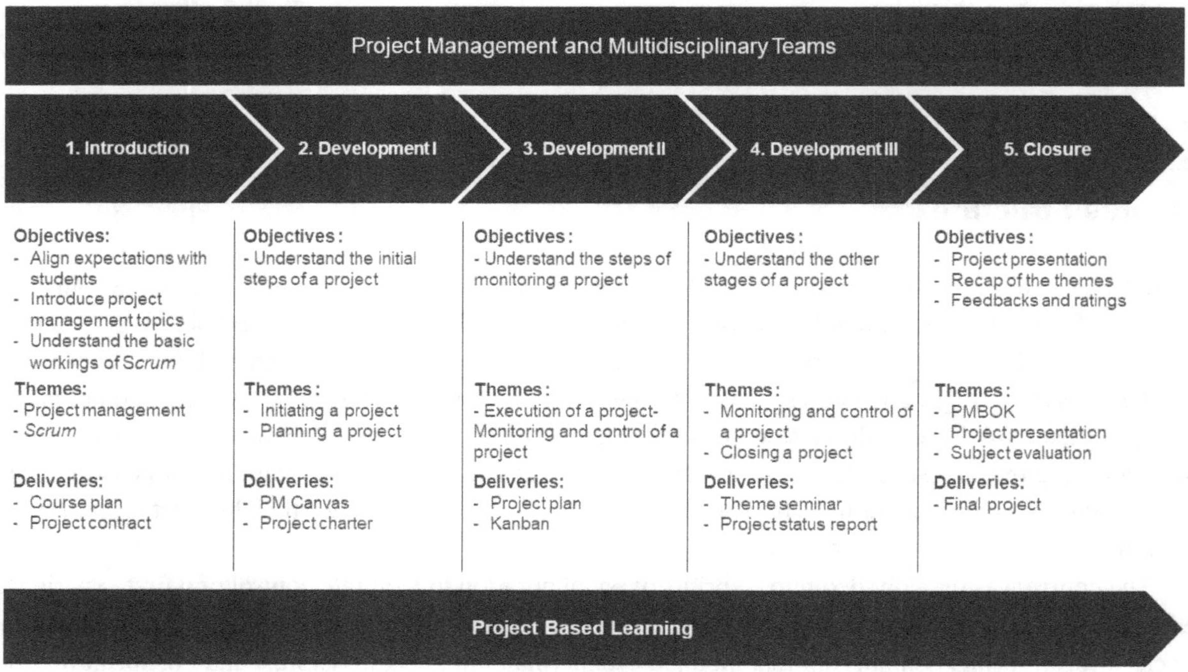

Figure 3 summarizes the structure adopted for the course.

A detailed approach to the 3-week cycles is presented to permit understanding of the sprints in a macro way. As already mentioned, the idea of fixed-size cycles facilitates both the organization and the planning as well as the execution of sprints, with the adoption of a simpler structure that is repeated throughout the semester. In total, there are 6 classes of approximately 2 hours:

Lesson 1: The first lesson of each sprint presents the objective of that cycle, proposes goals and expected results, and allows the students to plan. It also begins to address the concepts that will be developed.

Lessons 2, 3 and 4: These 3 classes cover the technical development of project management content, aiming to bring the robustness and concepts necessary both for the development of the project and for its application to day-to-day life.

Lesson 5: A class with two main functions in the schedule is planned: bringing an applied and complementary approach to project management, either with the application of workshops or cases based on real situations or using a discussion of complementary competencies for a project manager such as oratory and organization. This allows the schedule to be interrupted such that in case of unforeseen events or holidays, the primary content and events are not rushed, and this class may work in place of some of the others.

Lesson 6: The sprint closing class aims to recap and reinforce the main points addressed, in addition to allowing the evaluation of the cycle, both formal (specific activities) or informal (discussion and individual reflections of the students).

For the implementation of the model, we opted to use only the flipped classroom combined with project-based learning. The first presents a good proposal for classes that do not have significant experience with active learning methodologies. The second is essential because students must develop a project during the course. It is worth noting that the pre- and post-class moments were organized so as not to make excessive demands on the students, mainly because they were not in the habit of working with active methodologies and also because this is an optional course. On this basis, as the cycles are completed, the work required by the extra class is expected to gradually increase, with the students gradually consuming more content and becoming more active. This is one of the advantages of working with the Scrum methodology, i.e., a gradual increase in productivity.

Finally, lesson 5, which addresses applied issues and other complementary competencies of a project manager, also makes it possible for the students to develop innovative methodologies for validation. According to the project themes and interests of the students, this more informal moment permits the professor to apply methods to situations, such as problem-based learning, in a reduced context for validation and possible implementation in other semesters.

In parallel to the development of the sprints, the methodology of project-based learning is proposed for the execution of one or more projects by the students, divided into groups.

The project schedule is adjusted to the content addressed to apply the methods and concepts already discussed in practice, starting at the end of the first sprint and ending in the last sprint, representing the main part of the students' evaluation.

The themes of the projects can be freely chosen by students, the only requirement being that they should be real problems, should benefit society, and should not be based on financial objectives.

Adaptations to the Initial Proposal

The class plan was elaborated at the beginning of the semester, guided according to the results of the requirements to organize the sprints and the distribution of the classes. Given the context of the COVID-19 pandemic, some changes were made in relation to the initial proposal, including the use of remote teaching platforms.

While these restrictions may have imposed some limitations, they forced both the students who took the course and the professor to adapt to a new process that can be used to apply to the course in collaboration with national and international partner universities, with an even richer learning experience for those involved.

Application of the Proposed Model

The first sprint of the course exhibited a steady pace and fulfilled the objectives and schedule, with active participation by the students in general. It was noticed that the duration of the classes ended up limiting some discussions, possibly becoming a problem in the development sprints. In addition, the lack of more tangible deliveries and evaluations was pointed out, mainly due to the lack of clarity of the method used for evaluation.

The development sprints were able to approach the proposed themes with sufficient depth. This was due to the engagement of the students who participated more and who performed the readings, comparing them to the experiences they had outside of class. However, there was still a certain polarization among the more participatory students. The lectures with external guests proved to be one of the highlights, mainly because the themes were focused on market realities.

Another positive point was the structuring of the partial presentations of the work carried out. These "events" became regular and increased the professor's contact with the projects, in addition to allowing for a clearer evaluation. In addition, for those results that were unsatisfactory in sprint 2, the professor was able to take measures such as dividing the class for the preparation of the work and developing extra schedules with the groups that needed additional support.

On the other hand, the schedule occasionally proved to be unfeasible because it did not predict absences or unforeseen events, resulting in cumulative delays. In this case, the alternative was to assign the asynchronous reading of the material made available by the professor.

Another point of attention was the suitability of Scrum events. The way they were proposed, it was not feasible to perform the procedures without adapting them to a more informal model. This was identified as the greatest challenge for model rethinking, with the need to adapt both the time and dynamics of the events so that eduScrum would not be mischaracterized.

Since the projects became the central point of the course, the schedule was adapted so that the projects could be developed during class hours. The students were very receptive to this alternative; however, it is necessary to predict the time required for this synchronous development so that the schedule is not adversely affected.

The final sprint was fully directed at the closure of the projects. Most of the classes were made available for the students' work, and the professor was available for assistance. This was well received by the students and provided better access to the products. At the end of the course, each group presented its product and an account of the development process, addressing the difficulties encountered and associating the work with the concepts studied theoretically. In the end, the other groups and the professor asked questions and clarified doubts. This moment was important both to finalize the course and for the students to develop presentation skills. It also served as a tool for the professor to carry out evaluations. A later date was set for submitting the finished products to be evaluated asynchronously by the professor.

ANALYSIS AND CONSOLIDATION OF THE RESULTS

This section presents the results of the application obtained from the students who participated in the course, as well as a summary of the points for improvement identified by both the students and the researchers.

At the end of the course, the students responded to an anonymous questionnaire for a qualitative and quantitative evaluation of the academic semester. Regarding the general degree of satisfaction, all students scored the course with a 4 or 5 on a scale of 1 to 5, with an overall average of 4.47.

Next, the students evaluated the importance of some aspects of the course, represented in Figure 4. Only "Use of foreign literature" received "indifferent" grades, while the other items were classified as "important" or "very important" by the students. The use of active learning methodologies obtained the highest number of "very important" evaluations, showing considerable acceptance of the approach by the students.

Figure 4. Importance of the course scored by the students
Source: Author

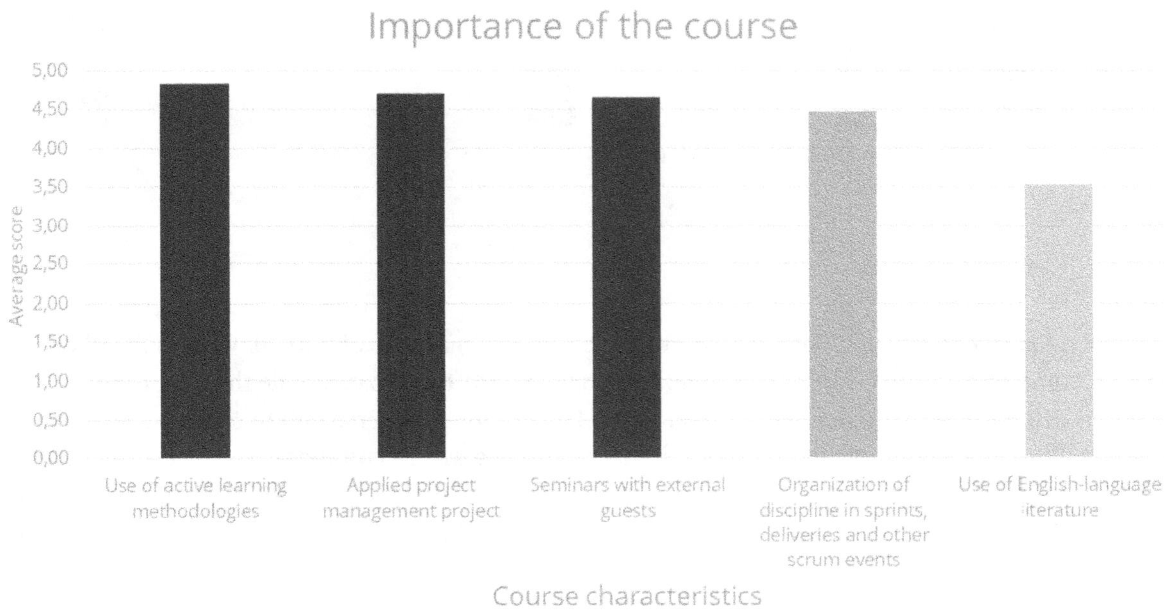

Regarding the operational aspects of the course, the content addressed, the project focus, the evaluative methods and the bibliography used received the best evaluations. The format of the classes received some "Indifferent" grades, with the students stating that remote teaching impaired communication and interaction. The results are shown in Figure 5.

Figure 5. Average satisfaction reported by the students
Source: Author

Finally, the students were asked to answer the question "On a scale of 0 to 10, what is the likelihood of you recommending the GEPEM course to someone else? ", used to calculate the net promoter score (NPS), as established by Frederick Reichheld (2003). On this scale, grades 0 to 6 characterize detractors, or individuals whose evaluation of the initiative are negative; neutral grades 7 and 8 characterize individuals who do not promote or are contrary to the course; and grades 9 and 10 are those promoters who actively recommend it. Based on these evaluations, the SPL obtained was 82, which is considered a score in the excellence ranking.

General Analysis of the First Application

Based on the development, the interface between active learning methodologies and an agile method of project management applied to education was satisfactory. The proposed model was applied within a limited context due to social distancing, but it was possible to validate and critique the model to be better understood. The rate of student satisfaction with the course was high, exceeding initial expectations. In addition, some key concepts proposed, such as active methodologies, the presence of guest speakers and the elaboration of real projects, were well received and approved by the students, and the main aspects were positively evaluated.

Regarding the active methodologies, the flipped classroom was predominant, mainly owing to the previous experience of the professor. Project-based learning proved to be the greatest engagement factor and an excellent evaluative instrument, with clearer deliveries and allowing the direct application of the project management concepts studied. Additionally, the unpaid character and the requirement that the projects contributed to the University of Brasília were relevant factors in the students' engagement and dedication, yielding significant results.

However, aspects to be improved were identified during the process of application and analysis of the results of the method. The greatest challenge was to define structures, artifacts, and tools that would allow eduScrum events to be performed within an even shorter period of time and with a larger number of people to avoid mischaracterizing the method. Another point of attention is the definition of a schedule with free days so that situations that may require classes to be canceled do not affect the deadlines. In addition, it is necessary to clearly define the methods of assessment and align them with the students to obtain greater transparency in the process, one of the defining characteristics of Scrum. Finally, in the case of the remote teaching scenario, it is worthwhile considering other collaborative platforms to better integrate less participatory students and lead to greater class homogeneity.

Diagnosis of Improvements

According to the methodology used, the main difficulties and points for improvement were identified based on the results and analyses of the first stage of the project. The objective was to prioritize performance with regard to the most relevant aspects and to add value to both the students and the professor. Table 1 summarizes the considerations made, synthesizing the diagnosis obtained from the first application.

Table 1. Diagnosis of application

Difficulty	Description
Small percentage of scrum events executed.	Many events were canceled during the semester and the procedures and time assigned were not clear. In general, eduScrum ended up being misrepresented, reducing its utility for, and understanding by, the students.
Duration of Scrum events.	The events proved to be unfeasible because of their duration, mainly due to the number of students. There was a lack of standardization and prior structuring of the procedures defining responsibilities and the model to be implemented.
Difficulty in following the schedule.	The course schedule had to be adjusted a few times during the semester.
Lack of clarity in the evaluation.	Some students did not evaluate the activities by assigning scores and weights in a clear manner, negatively affecting the aspect of transparency of eduScrum and reducing the credibility of the course.
Unequal participation of students in the class.	Depending on the student's profile and dedication, there was a discrepancy in their participation in synchronous activities, impairing the professor's ability to monitor the development of individual students.
Slow project development.	The students began to carry out the project from the middle to the end of the course, which caused them to lose time at the beginning of the semester. With a steadier delivery and diluted development during the sprints, better deliveries are expected at the end of the semester, with a reduced load at the end of the school period.
Difficulty in conducting dynamics in class.	The dynamics and participation of the students were below expectations, mainly due to the high number of students and the need for remote classes.

Source: (Author)

Figure 6. Lever map
Source: Author

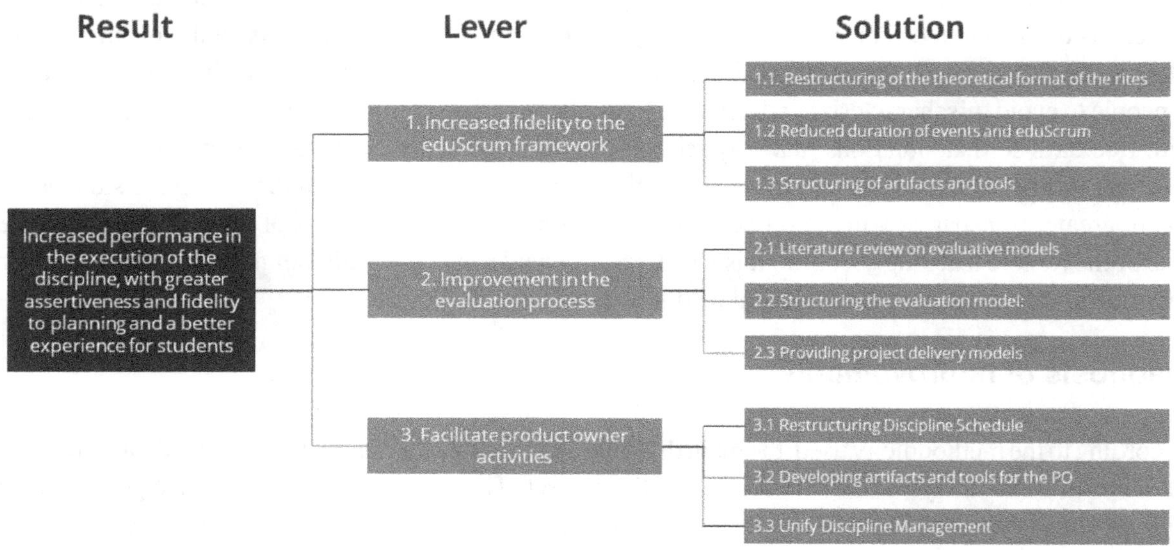

Figure 7. Prioritization matrix
Source: Author

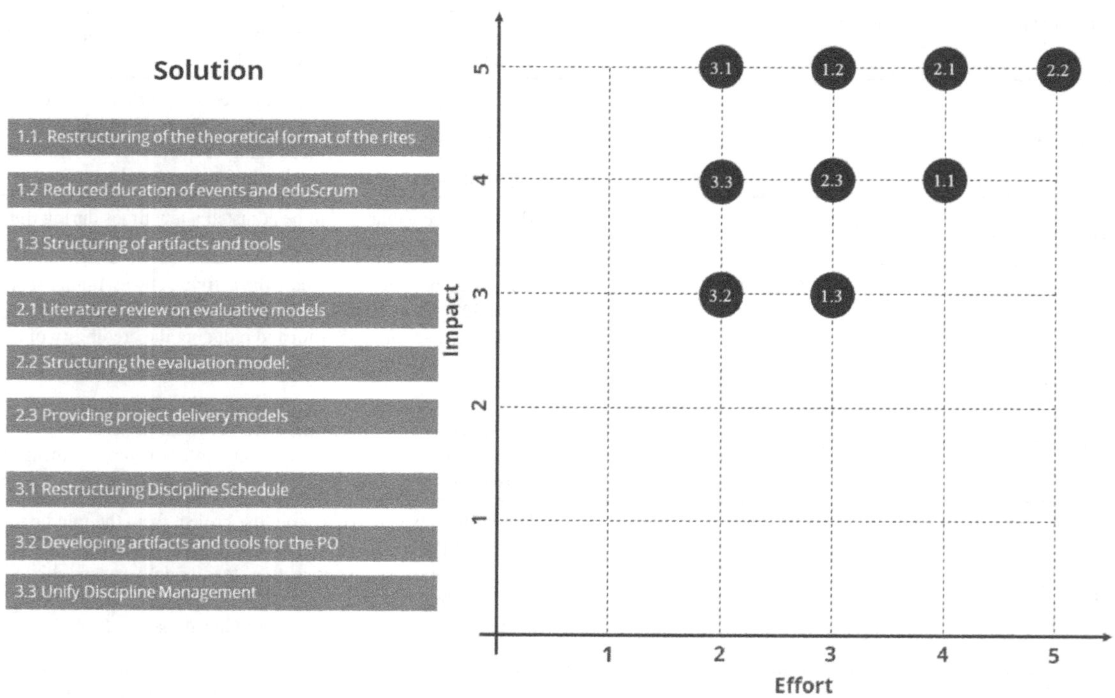

Based on the analysis of Table 1, a map of levers was structured, a tool used to associate the actions that must be taken to achieve a result. In this case, three main levers were identified, with a total of nine solutions for product success.

Given the number of solutions raised and the time available, a classification of the solutions was made according to their "impact" to improve the course and the "effort" necessary to develop it to prioritize those with the highest cost-benefit. This stage was managed jointly by the researchers to meet the interests and perceptions of both and was later positioned in a matrix of "effort x impact", presented in Figure 7.

Based on this prioritization, the solutions were developed and are presented and discussed in the next chapter, with preference given to those with the greatest impact, that is, in the upper positions of the "y" axis in the graph.

SOLUTIONS AND RECOMMENDATIONS

This section discusses the development of the solutions presented and prioritized and is divided according to the levers presented in Figure 6. Finally, a roadmap for applying the model in other contexts of engineering education is presented.

Lever 1: Increased Alignment with the eduScrum Framework

The first lever identified concerns the use of eduScrum in the developed teaching-learning model. EduScrum incorporates agile concepts and adapts them to the educational context, one of its main characteristics being the procedures or events held. The objectives of the procedures are to allow the professor to closely monitor the students and to develop a transparent relationship, which are essential for success in the educational context. During implementation, the procedures were found not to be well described or structured, without a defined procedure, with consequent inefficiencies and the need for additional time during the course, which was already scarce. In addition, no tools were available to help the professor, a fact that required a product owner who had already acquired several functions in an eduScrum. Thus, three improvements were addressed in this lever: a) restructuring of the theoretical format of the predicted procedures (solution 1.1); b) reducing the duration of events and of eduScrum (solution 1.2); and c) structuring the artifacts and tools for an assertive implementation of eduScrum (solution 1.3). New proposals were made for each of the procedures considered, as presented below:

1. **Sprint Planning:** Sprint planning is an event in which three objectives need to be met: forming the teams that will work together, defining and agreeing on the learning goals for the period and planning and creating a division of work to achieve the goals.
    a. Proposed procedure: At the beginning of the first sprint, the professor or product owner will divide the class into groups, which will be maintained in the other sprints. He will then present the course schedule, emphasizing the deadlines for deliveries, holidays and important milestones of the period. Next, the professor will present examples of deliveries that will be made in the sprint, showing work from previous semesters or from his personal collection. Each step is followed by clarification of doubts and collaborative validation with the students, during which changes can be made to align expectations and to obtain a commitment to the proposed schedule.

   b. Expected time: 30 minutes, approximately 10 for the professor's presentation and 20 for a discussion of doubts, concerns, and necessary adaptations.
   c. Artifacts and tools used: Course schedule and shared folder with examples of deliveries.
2. **Sprint Daily:** The daily meeting initially lasts 5 minutes and is held at the beginning of the class to inspect the produced material, to synchronize the activities to be performed according to the schedule and to identify the obstacles and problems detected.
   a. Proposed procedure: At the beginning of each class, the professor will open the class management tool, in this case, a frame model in the Trello application, and present the activities on the board. Together, the class will update the frame by moving to "completed" and "doing" the activities performed and planned for the day. Then, some randomly chosen students will answer the questions provided in the eduScrum framework: "What have I done to help the team achieve the goals of the previous Sprint?"; "What will I do in this class to help the team achieve Sprint goals? "; and "What are the obstacles that have blocked the team from achieving its goals?". At least one student should be selected per group to obtain alignment of the class as a whole and each of the groups separately.
   b. Expected time: 15 minutes, 5 of them for collectively updating the table and 10 for oral updating of the students. It can be seen that the time taken for this exercise is longer than predicted in the Scrum framework, but due to the difficulties of execution and the importance of this procedure, it was decided to schedule more time to ensure proper execution.
   c. Artifacts and tools used: Monitoring of the course using a framework in the Trello application. It is recommended that the groups also develop a similar framework to organize their activities in the course project, but this should be updated individually by the groups, not as part of the class procedure.
3. **Sprint Review:** The sprint review takes place at the end of each cycle and aims to verify and present what was learned by the teams during the period. It is an inspection process that aims to ensure iterative and agile learning, allowing us to identify possible failures to be worked on in the next sprint.
   a. Proposed procedure: At the beginning of the previous sprint class, the professor will provide an anonymous questionnaire for completion. The questionnaire poses questions about the understanding and execution of the work both in terms of the content of the period and the development of the course project, with objective questions of gradation. Finally, the descriptive questions that will be discussed in the retrospective are presented: "What went well in this sprint? "; "What can or should be done better in the next sprint? "; "What should we not do anymore?" and "What steps can we take in the next sprint? ".
   b. Expected time: 15 minutes, fully available to fill out the questionnaire by students and professors.
   c. Artifacts and tools used: Sprint review questionnaire and retrospective.
4. **Sprint Retrospective:** This was a retrospective event held after the review so that students could perform a self-assessment, creating an improvement plan. Its main procedures are the evaluation of the methodology and the way of working, a self-evaluation and evaluation of the team, and the discussion of improvements. In the case of the proposed course, the retrospective is held immediately following the review as a single event.
   a. Proposed procedure: After completing the sprint review questionnaire, the professor provides a self-assessment and peer assessment questionnaire to be completed by the students. Next,

the learning is discussed. At this time, the professor opens the answers to the first questionnaire (which is anonymous) and presents them to the students, discussing the results obtained. In addition, some students are asked to comment on the descriptive questions to discuss the improvement plan. Finally, there is a time for general comments on the course, the progress of the projects and whether expectations are met or not. All these points must be recorded by a monitor or by the professor to be used for the planning of the next sprint. The remaining time of the lesson can also be used to clarify doubts about the content of the course addressed in the sprint or to aid the development of the project.

b. Expected time: 30 minutes, approximately 10 of them for questionnaire responses and 20 for discussion of the completed sprint.

c. Artifacts and tools used: Self-assessment questionnaire and evaluation of the teams and minutes of the review/retrospective of the sprint.

Lever 2: Improvement in the Evaluation Process

During the application, it was noticed that the students did not clearly know how or when they would be evaluated; thus, failing to prioritize some activities that would be important in the researchers' view, such as partial deliveries of the project. In addition, since eduScrum values transparency and inspection, the evaluative model needs to be defined and clarified with the students.

To fill this gap, three solutions were developed in the project: a literature review of evaluative methods in active methodologies to incorporate good practices into the interface with eduScrum; the structuring of an evaluative model, incorporating learning and adapting to eduScrum events; and centralizing and developing examples of deliveries/evaluations to clarify what is expected from the students and to facilitate their understanding of the evaluation process.

1. Literature Review on Evaluative Models

When assessing active learning methodologies, it is important that both the professor and the students understand that they should not be restricted to traditional evaluative models and that diversification of the assessment instruments used is necessary (Filho et al., 2019). According to Mesquita (2015), written tests still constitute most evaluations in engineering courses. However, Filho et al. (2019) stated that students already recognize the importance of other forms of evaluation that promote the development of skills for professional practice and meaningful learning.

Evaluation during active learning is continuous, flexible and of paramount importance because it reveals the coherence between all the processes of curriculum development (Mesquita, 2015). According to the same author, the evaluation needs to be compatible with the approach used during classes; thus, it is not positive to use, for example, classes focused on day-to-day professional practice or evaluation tools that only require memorization and reproduction of studied models. In this respect, the evaluation needs to be articulated with the other curricular elements, promoting more participatory environments and more diverse, reflective processes (Mesquita, 2015) (Hoffmann, 2014).

Filho et al. (2019), apud Moretto (2017), approach evaluation not as an "end" in itself but as a way for the professor to collect data and direct the next steps of the course according to the feedback received. That is, evaluation during active learning is a time not only to assign grades but also to identify difficulties and solutions and to direct students toward scientific knowledge based on reflections and challenges, which is an essential aspect of the mediating process (Hoffmann, 2014).

In view of these understandings, it is necessary to establish a "didactic contract" with the students for evaluation in active methodologies, demonstrating their understanding of the evaluation process itself (Filho et al., 2019). To this end, using Perrenoud (1999), the cited authors mention some characteristics of this contract, with the professor being responsible for encouraging participation, accepting errors, valuing guesses and cooperation, directing the progress of the course and explaining the content addressed. Students, on the other hand, are responsible for actively participating, collaborating, and interacting with each other and with the professor. Thus, the evaluation process should not be standardized but should involve students in the assessment of their own competencies, favoring mutual evaluation and self-assessment by explaining the objectives and using debates (Perrenoud, 1999) and, thus, representing a new learning opportunity (Masetto, 2011).

Analyzing several authors, Filho et al. concluded that evaluating acquires the meaning of "mapping" or diagnosing the learning process to allow new interventions and help students overcome the difficulties encountered (Filho et al., 2019). To this end, questions should be answered in the evaluation process, such as "What are the difficulties? "; "What are the obstacles? "; "What are the advances? "; "What aspects need to be improved?" (Filho et al., 2019).

Finally, emphasizing evaluation as a continuous and ongoing process [REMOVED BIBLIOGRAPHY FIELD]during the learning period, several authors, such as Filho et al. (2019), Perrenoud (1999), Mesquita (2015), Masetto (2011) and Hoffmann (2014) discuss the importance of feedback for the process, providing security and collaboration between professors and students. To this end, in general, the principles of Hoffmann's mediating evaluation (2014) summarize the characteristics of an evaluative model: to provide students with many opportunities to express their ideas; to give students opportunities for discussion based on problem situations; to pay close attention to tasks, questions and comments; to offer students several opportunities to discover better solutions; and to complement the evaluation process.

2. Structuring the Evaluation Model

Based on the review performed, we started to structure the evaluation model to be adopted. First, we perceived synergy between the concepts presented by eduScrum and those found in the literature because both understand evaluation as a longitudinal and constant process that directs decision-making by identifying advances and difficulties, adding transparency to the teams. Table 2 explicitly demonstrates this interface, comparing the questions of the evaluation process of Filho et al. (2019) to those proposed by the eduScrum Guide (Delhij et al., 2016).

Table 2. Comparison of the evaluation in active methodologies and in eduScrum

Filho et al. (2019)	eduScrum guide (2016)
What progress?	• What did I do to help the team achieve previous Sprint goals?
What are the difficulties?	• What went well? • What can or should be done better?
What are the obstacles?	• What are the obstacles that blocked the team from achieving the goals?
What aspects need to be improved?	• What can or should be done better? • What should we not do anymore?

Source: (Author)

Thus, it can be seen that the questions and concepts of eduScrum events meet the requirements identified for evaluation in active learning methodologies and are, therefore, adaptable to teaching in this context. To this end, evaluation methods already used in active learning methodologies were identified and used as a starting point for the proposed model. Five were selected among the options studied: last-minute report, self-assessment, formative evaluation, peer evaluation, and physical or digital portfolio. The book "A new classroom is possible" (Filho et al., 2019) was used for the initial understanding of these aspects.

Each of the selected methods was classified as a "Content" or "Behavioral" evaluation to balance both the aspects of programmatic content and the other skills and competencies expected of a student. This classification was performed by the researchers, and although partially subjective, it provided a way to balance the evaluation to cover both approaches. Next, a relationship was established between these and scrum events to synchronize their use and periodicity. Finally, the model to be used was detailed, and a brief description of the understanding of how to conduct evaluations in the context of the course was elaborated. The suggestion was that these models be used in the evaluation of students, as agreed upon between the professor and the students in sprint planning.

3. Providing Project Delivery Models

The last solution related to the course evaluation lever was the centralization and consolidation of examples and models of project and course deliveries. The main purpose of this solution is to provide a shared folder with students so that they have references about deliveries, which were sometimes abstract in the first application. It is expected that, with this, they will be able to predict the effort required in each sprint, improving their organization, planning and distribution of tasks.

Figure 8. Shared folder architecture
Source: Author

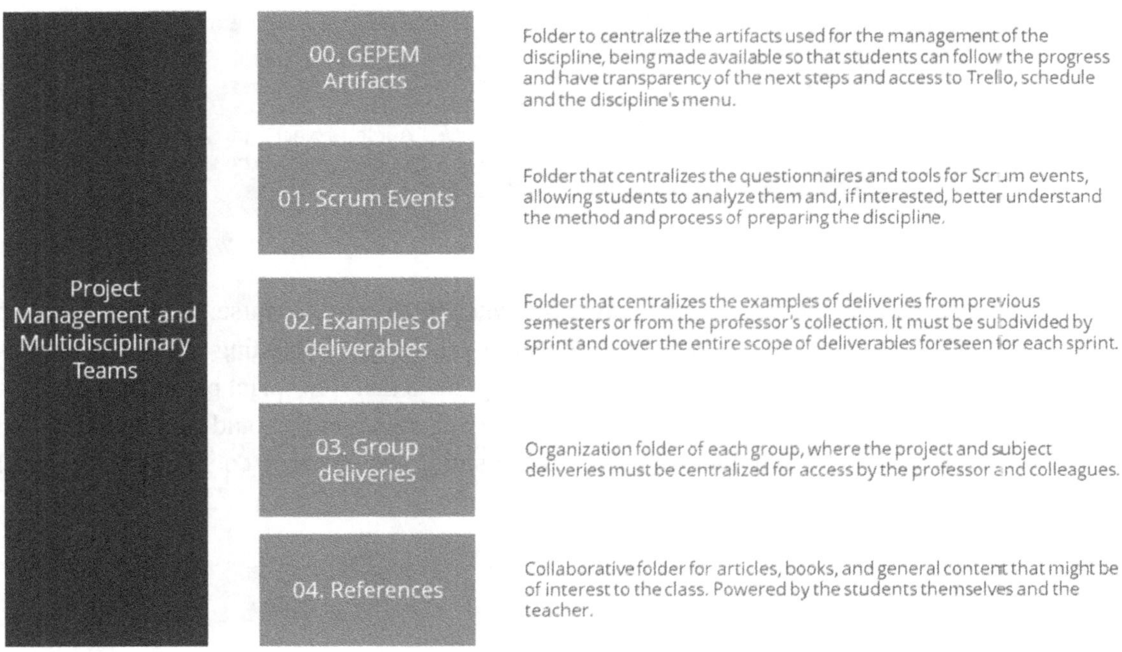

The shared folder can still be used to provide articles, references and the partial deliveries of the groups themselves, centralizing management of the knowledge of the class in a single environment, which can increase collaboration between groups. Figure 8 shows the architecture adopted in the folder organization, with a brief explanation of the contents of each subdivision.

The step of structuring the shared folder also serves as a conference if the documentation required for the course is ready. In this case, when structuring folder "02. Examples of Deliveries", for example, the researchers noted the need to define what would be minimally needed for the deliveries of the entire course to provide them with examples for the students. Thus, Figure 9 elaborates on the deliveries expected for the course.

Figure 9. Deliveries planned for the course
Source: Author

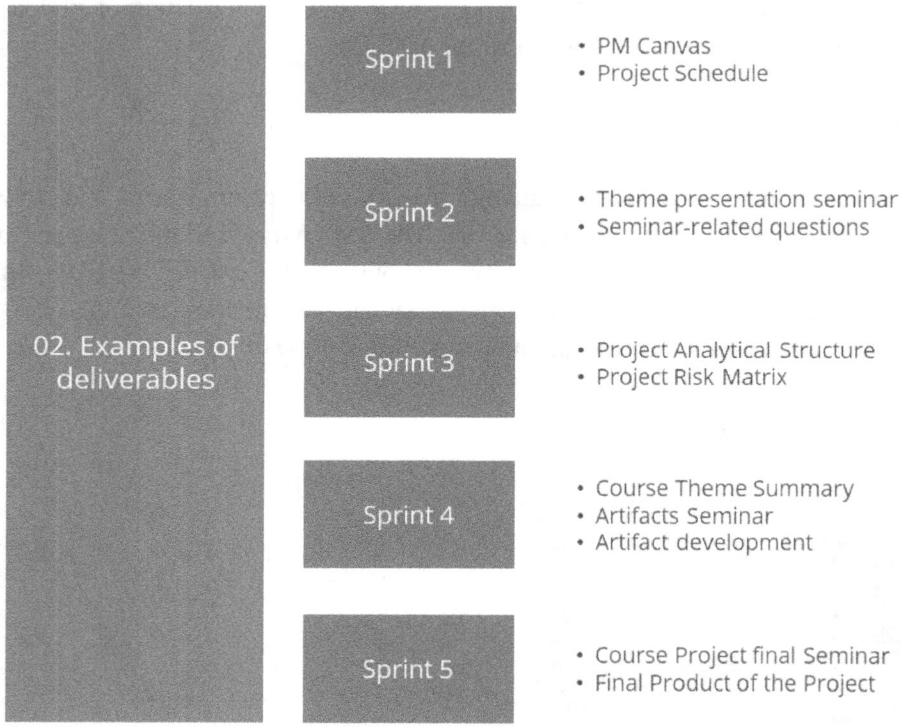

This stage represents good practice for the professor when planning the course, requiring him to prepare the documentation and make the necessary decisions prior to the beginning of the semester, with better security and clarity for the students. The important point is that if in sprint planning the professor and students agree to add or remove something from the scope, this folder should be updated with new examples so that it always reflects the full scope of the sprint and to avoid confusion or incomplete deliveries.

Lever 3: Facilitate Product Owner Activities

The third and final lever was to facilitate the activities of the product owner, represented by the professor. By developing the model and integrating different solutions, the degree of complexity of the course increased, with many artifacts, events and references that could make it very costly and render the model unfeasible.

The aim of this lever is to simplify and organize the activities of the professor and possible monitors or auxiliaries, permitting previous organization and monitoring of the course. The solutions proposed were the restructuring of the course schedule in terms of the demands of the students after the first application and the problems observed by the researchers. Artifacts and tools should be elaborated so that the professor would not need to design documents but only use them and adapt them, when necessary, with the unification of the management of the course in one place, avoiding different fonts and scattered files.

1. Restructuring Course Schedule

The first solution addressed was the restructuring of the course schedule. The students had previously reported that they would like to take more classes to elaborate the projects, a wish that was met during the first application. In addition, the researchers noted a lack of time off in the schedule and the definition of evaluation moments. Thus, a calendar was restructured with 30 classes, 2 per week (a model already adopted), with a preset schedule of the content addressed in each class. In addition, each class was defined according to three objectives: "Content", "Protected," and "Evaluation". "Content" refers to synchronous moments when some content of the technical scope of the course should be developed; "Protected" refers to the classes intended for either the development of the project or the visit of a guest speaker; and "Evaluation" refers to the time when students will be evaluated, either by deliveries and presentations or by performing the review and retrospective procedures of the sprint.

In the end, 16 meetings were held: 2 of the "Content" type, 7 "Protected" and 7 for "Evaluations". This distribution was expected to develop a schedule with flexibility for unforeseen events, in addition to ensuring a coherent and regular evaluation process.

2. Developing Artifacts and Tools for the Product Owner

The second solution of this stage was to elaborate artifacts and tools that could be used during the semester. Its objective was to standardize and enable the other solutions implemented to save time and effort by the professor responsible for the course. Table 3 briefly summarizes and explains each of the products developed.

Table 3. Artifacts and tools developed

Name	Description
Sprint review/retrospective Questionnaire	Online questionnaire to be filled out and the reflection on the sprint to be completed and used for oral discussion.
Self-assessment questionnaire/peer review	Evaluation questionnaire designed to inform the students of their performance and that of their team.
Course schedule	Schedule of activities with groupings and scheduled free days to avoid delays in the semester or a reduction of the course scope.
Shared knowledge management folder	Shared online folder for the centralization of all course information in an organized and transparent manner.
"Trello" Framework of Course Management	Online access board of the class for performing events and unifying the course management, redirected to the other important files.
Burndown Chart for course	Graph of the development of the course, proposed as an artifact by the Scrum methodology and adapted to the course
Students' delivery models	Centralization of delivery models used in other semesters to clarify to students what is expected and provide greater transparency of the scope and evaluation of the course.

Source: (Author)

Figure 10. Trello planning section
Source: Author

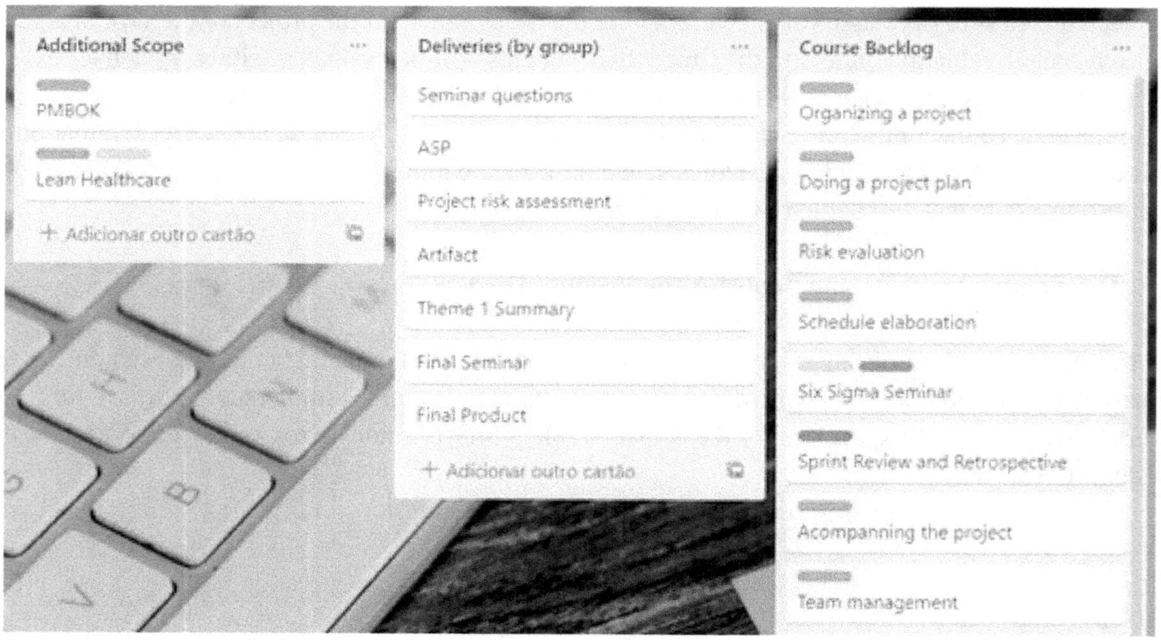

3. Unify Course Management

Finally, the last solution was related to the unification of management and monitoring of the course in a single platform. During the research, several axes and complements were explored, making it more complex and difficult for the professor to follow the model. Thus, it was necessary to centralize and simplify the monitoring to ensure the execution of all the procedures, events, evaluations, and scope provided, for which a "Trello Framework of course management" was developed. The idea was to use the Trello platform for conducting events, directing information to the knowledge management folder, providing links and artifacts and for managing the scope and schedule. The framework was then divided into three major groups: planning, execution and management.

Planning: This part of the board involves the one initially predicted by the professor and is used as a reference for what should be developed during the academic semester. Three lists make up this section: deliverables, concerning the products that the students must develop during the semester for later evaluation; backlog of the course, in which the entire schedule of activities is separated and detailed in chronological order; and additional scope, which refers to the activities that the class decides to carry out in conjunction with the professor in addition to what was previously planned.

Execution: The execution is essentially a Kanban frame of the sprint, used in sprint planning and updated at the beginning of each lesson, consisting of the following lists: "Backlog of Sprint", "Doing", "Review" and "Done".

Figure 11. Trello execution section
Source: Author

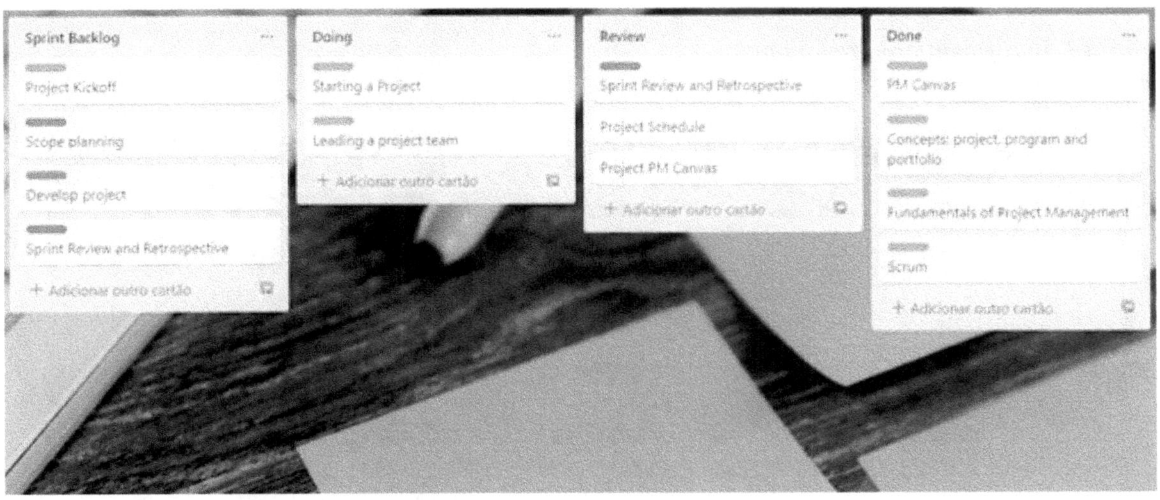

Management: The management part presents the "Canceled" list regarding what has been removed from the scope, the "Useful links" list, which contains directions for all artifacts developed and used for the course, and the "Burndown chart" covering the monitoring of the activities. Other management tools, such as Gantt charts, can be added later to other lists in the frame.

Figure 12. Trello management section
Source: Author

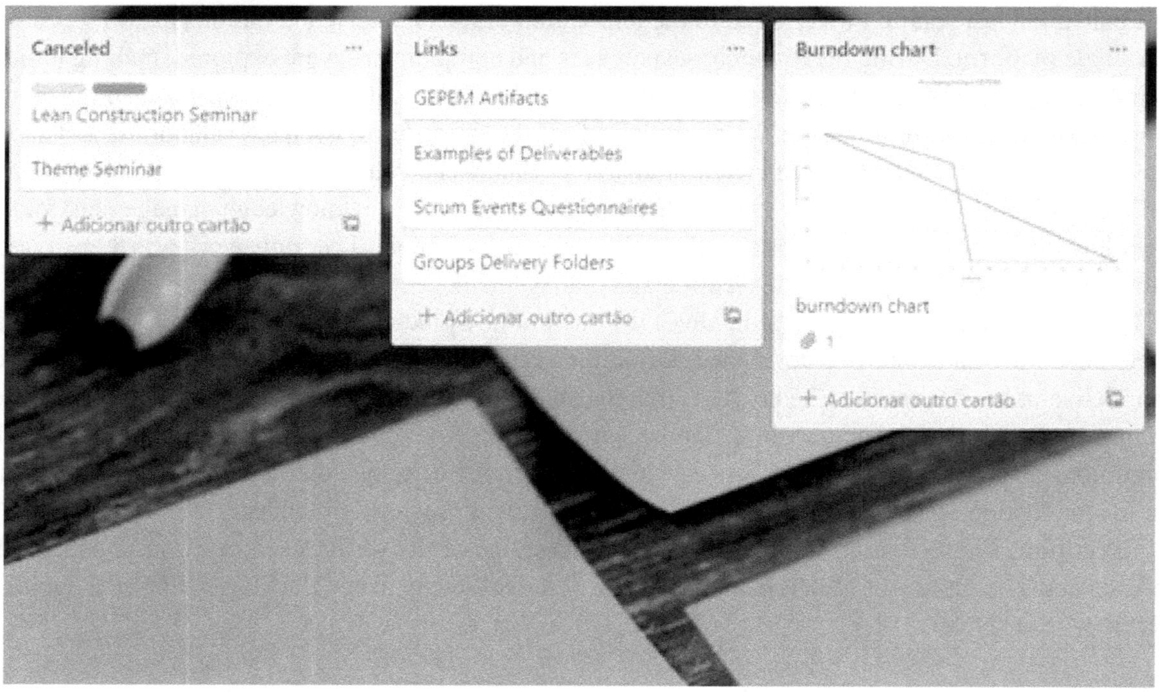

Figure 13. Implementation roadmap
Source: Author

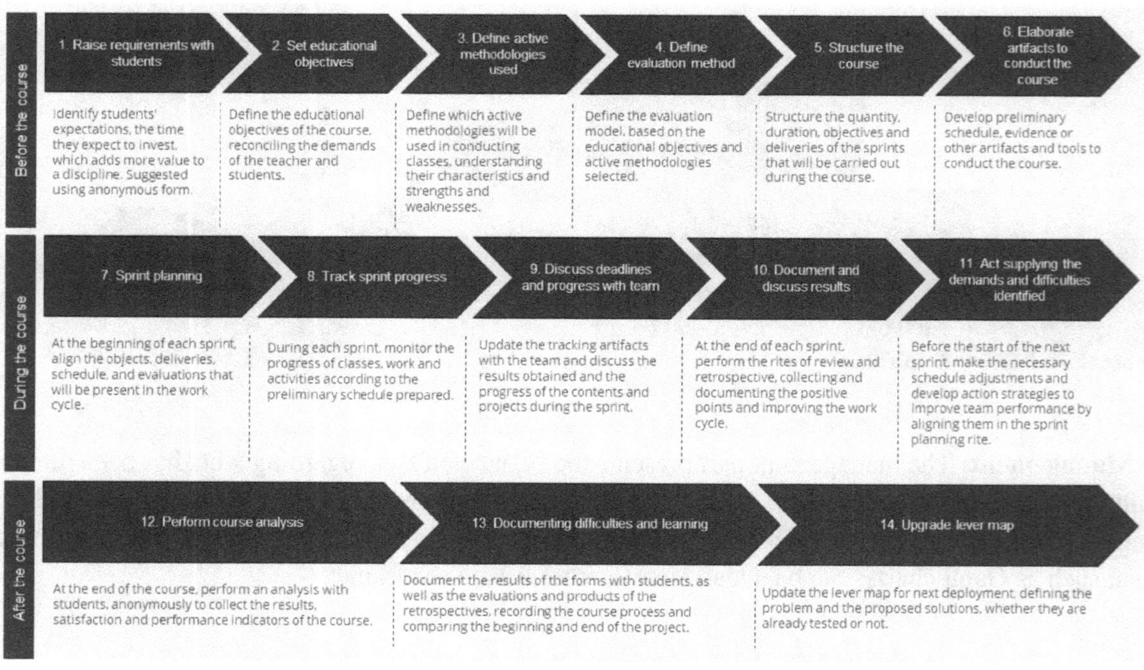

Implementation Roadmap

Finally, based on what was developed during the chapter, a roadmap was created that summarizes the professors' role in using this model in other applications, as shown in Figure 13. The roadmap includes the moments of preparation, execution and documentation of the course, defining the main activities and objectives. Details of each item were previously presented in the body of this work.

CONCLUSION

The constant changes that occur in engineering education, motivated by the development of new technologies and tools, attest to the importance of adaptive teaching. The model presented here aims to allow flexibility for use in other contexts and is feasible for implementation. It is understood that the research has limitations regarding the number of participants and validation time, but points to be considered are listed below for new applications:

- One should plan the scope of the course and divide it into activities that can be performed in a class so that the contents discussed start and end on the same day.
- The schedule needs to provide asynchronous and nonmandatory activities for use as a slack.
- The initial and final sprints feature more eduScrum events and features, reducing the technical load. Thus, it is important that the scope of the course is concentrated on the programmatic content of the intermediate sprints.
- It is important to provide a monitor or assistant for the professor, who can assist students in understanding the roles and organizing the project teams, working similar to a Scrum Master in the first sprint.
- It is essential to try to reduce the time of events so that the focus of the course is not to learn how to use the Scrum methodology. The explanations should be succinct and directed at the application so that they can be used as tools for the development of the course. The use of inexperienced teams and professors represents a high risk for a successful implementation.
- Even though the assessment is cross-cutting, students may feel insecure if it is not explicit. It is important to have transparency about which instruments will be used and their evaluation weights and to always use the sprint retrospective to validate the mood of the classroom.

In view of these developments, it is concluded that the pilot program proved to be a promising alternative for increasing student engagement and using active methodologies, in contrast to the expositional curriculum that predominates in current engineering education; it may also be applicable to other courses administered both face-to-face and remotely. The improvements proposed in the second stage, as well as the artifacts and tools developed, are expected to overcome some of the difficulties already identified. It is also expected that, following the iterative proposal of the agile methodology, other difficulties will be identified through new applications, permitting new improvements and the maturation of the model.

REFERENCES

Aquere, A. L. (2007). Projects Laboratory: building a university through PBL. *Proceedings of the PAEE'2017, 9th International Symposium on Project Approaches in Engineering Education (PAEE): Implementing Problem/Project-Based Learning (PBL) in different contexts of Brazilian Higher Education Institutions*, 72-81.

Baldo, R. F. G., Cerri, D. G. P. M., Silva, A. L., Zuim, J. B., Mergulhão, L. E. M., Adolpho, M. N., Boschi, T. F., Garcia, R. F. H., & Favalli, R. (2019). Aplicação da metodologia scrum em um estudo de caso de engenharia [Application of scrum methodology in a engineering case study]. *Revista Produção Online, Florianópolis, SC, 19*(3), 856–875. doi:10.14488/1676-1901.v19i3.3248

Bell, S. (2010). Project-Based Learning for the 21st Century: Skills for the Future. *The Clearing House: A Journal of Educational Strategies, Issues and Ideas, 83*(2), 39–43. doi:10.1080/00098650903505415

Bergmann, J., & Sams, A. (2012). Flip your classroom: reach every student in every class every day. International Society for Technology in Education.

Bissi, W. (2007, January/June). Scrum – Metodologia de desenvolvimento ágil [Scrum – Agile development methodology]. *Campo Dig. Campo Mourão, 2*(1), 3–6.

Borges, K. S., Schmitt, M. A. R., & Nakle, S. M. (2014). *EduScrum – projetos de aprendizagem colaborativa baseados em scrum* [EduScrum – collaborative learning Projects based on scrum]. Instituto Federal de Educação, Ciência e Tecnologia do Rio Grande do Sul, Campus Porto Alegre.

Crawley, E. F., Malmqvist, J., Ostlund, S., & Brodeur, D. (2007). *Rethinking Engineering Education: The CDIO Approach*. Springer.

Delhij, A., Solingen, R. V.., & Wijnands, W. (2016). *O Guia eduScrum* [eduScrum guide]. Time eduScrum.

Dinis-Carvalho, J., Fernandes, S., & Filho, J. C. R. (2017). Combining lean teaching and learning with eduScrum. *Int. J. Six Sigma and Competitive Advantage, 10*(3/4), 221–235. doi:10.1504/IJSSCA.2017.086599

Duarte, L. (2019). Scrum e Métodos Ágeis – Um Guia Prático [Scrum and agile methods – a practice guide]. Gravataí. RS.

Ferreira, E. P., & Martins, A. (2016). Educrum – The empowerment of students in engineering education? In *Proceedings of the 12th International CDIO Conference*. Turku University of Applied Sciences.

Filho, G. E., Sauer, L. Z., Almeida, N. N., & Villas-Boas, V. (2019). *Uma nova sala de aula é possível* [A new classroom is possible]. LTC.

Hoffman, J. (2014). *Avaliação mediadora: uma prática em construção da pré-escola à Universidade* [Mediator evaluation: a practice in construction from preschool to university]. Porto Alegre.

Lima, R. M., Andersson, P. H., & Saalman, E. (2017). Active Learning in Engineering Education: A (re)introduction. *European Journal of Engineering Education, 42*(1), 1–4. doi:10.1080/03043797.2016.1254161

Martin, P., & Tate, K. (2001). *Getting Started in Project Management*. John Wiley & Sons, Inc.

Masetto, M. T. (2011). *O professor na hora da verdade: a prática docente no Ensino Superior* [T he teacher at the moment of truth: teaching practice in Higher Education] (1st ed.). Avercamp.

Mesquita, D. I. A. (2015). O currículo da formação em engenharia no âmbito do processo de Bolonha: Desenvolvimento de competências e perfil profissional na perspectiva dos docentes, dos estudantes e dos profissionais [The engineering training curriculum under the Bologna process: Development of skills and professional profile from the perspective of teachers, students and professionals] (Doutorado em Ciências da Educação Especialidade em Desenvolvimento Curricular). Instituto de Educação, Universidade do Minho, Portugal.

Moretto, V. P. (2017). Planejamento: planejando a educação para o desenvolvimento de competências [Planning: planning education for skills development]. Petrópolis, RJ: Vozes.

Pereira, M. A. C., Barreto, M. A. M., & Pazeti, M. (2017). Application of Project-Based Learning in the first year of an Industrial Engineering Program: Lessons learned and challenges. *Production, 27.*

Perrenoud, P. (1999). Construir as competências desde a escola [Building skills from school]. Porto Alegre. *The Art of Medication.*

Prince, M., & Felder, R. (2006). Inductive Teaching and Learning Methods: Definitions, Comparisons and Research Bases. *J. Engr. Education, 95*(2), 123–138.

Project Management Institute (PMI). (2017). *A Guide to the Project Management Body of Knowledge (PMBOK Guide).* Project Management Institute.

Reichheld, F. F. (2003). The One Number You Need to Grow. *Harvard Business Review.* https://hbr.org/2003/12/the-one-number-you-need-to-grow

Souza, D. G., Vergottini, V. S., Bernini, D. S. D. (2018). Educação dos tempos modernos através da aprendizagem colaborativa: uma abordagem sobre EDUSCRUM [Modern time education through collaborative learning: an eduScrum approach]. *Anais do XXIX Simpósio Brasileiro de Informática na Educação (SBIE).*

Sutherland, J. (2014). *Scrum: The art of doing twice the work in half the time.* LeYa.

Thomas, J. W. (2000). *A review of research on Project-Based learning.* The Autodeks Foundation.

Tripp, D. (2005, September/December). Pesquisa-ação: Uma introdução metodológica [Action Research: a metodology introduction]. *Educação e Pesquisa, São Paulo, 31*(3), 443–466. doi:10.1590/S1517-97022005000300009

Valério M., Moreira A., Braz B. (2019). A sala de aula invertida na universidade pública Brasileira: evidências da prática em uma licenciatura em ciências exatas [The flipped classroom in the Brazilian public university: evidence of practice in a degree in exact sciences]. *Revista Thema, 16*(1).

Vybornov, N. A., Levchenko, A. A., Likhter, A. M., & Smirnov, V. V. (2014). Research practice as a constituent of the worldwide CDIO initiative. In *Proceedings of the 5th International methodological conference on "Changes in Education in the 21st century: Best International Practices and Russian Experience. How to Form Business Mindset and Innovative Thinking".* Astrakhan: Astrakhan State University Publishing House.

ADDITIONAL READING

Akçaylr, A., & Akçaylr, G. (2018). The flipped classroom: A review of its advantages and challenges. *Computers & Education*, *126*, 2018.

Barbosa, C., Lopes, C., & Macedo, H. (2019). Introducing scrum in a PBL context. Proceedings of the PAEE/ALE'2019, 11th International Symposium on Project Approaches in Engineering Education (PAEE) and 16th Active Learning in Engineering Education Workshop (ALE) (vol. 9, pp. 276-283). University of Minho.

FEPECS. (2018). *Projeto Pedagógico do Curso de Graduação em Medicina da Escs – Escola Superior de Ciências da Saúde* [Pedagogical Project of the Undergraduate Course in Medicine at Escs - School of Health Sciences]. Fundação de Ensino e Pesquisa em Ciências da Saúde.

Mendes, N. L., Vinhal, N. R., Albuquerque, A. V. D., Barcelos, I. S. M., & Aquere, A. L. (2019). PBL in teaching Project Management: Mais que Já Civil. Proceedings of the PAEE/ALE'2019, 11th International Symposium on Project Approaches in Engineering Education (PAEE) and 16th Active Learning in Engineering Education Workshop (ALE) (vol. 9, pp. 560-566). University of Minho.

Prince, M. (2004). Does active learning work? A review of the research. *Jounal of Engineering Education*, *93*(3), 223–231. doi:10.1002/j.2168-9830.2004.tb00809.x

Chapter 6
Contribution of Project-Based Learning on Social Skills Development:
An Industrial Engineer Perspective

Ariana Araujo
https://orcid.org/0000-0002-3155-9603
ALGORITMI Center, University of Minho, Portugal

Heidi Manninen
University of Minho, Portugal

ABSTRACT

The scope of this chapter is to describe and share experiences of two industrial engineers that had practiced project-based learning (PBL) during their engineering degree. Currently, authors look backward with a different perspective related to PBL as they are working as industrial engineers in different areas for 10 years in a multinational environment. Such experiences provide to the students the opportunity of developing soft skills that would be difficult to obtain following a traditional expositive lecture, more focused on individual work. Several challenges and advantages of learning by doing with PBL prepare students and contribute for their professional life because this kind of learning is closer to the professional daily life. In this chapter, four main experiences faced by the authors as engineering students are reported. Furthermore, the importance of experience like that and its contribution for the professional life is explained from the authors' point of view.

INTRODUCTION

Bologna Process led some changes in the European University learning methodologies and it also brought transparency and standards creation to ensure mobility and comparability between European countries. The process implied changes such as curricula, structures, education paradigm and also the

DOI: 10.4018/978-1-7998-8816-1.ch006

teaching culture, adopting different learning approaches student-centered (Alves et al., 2016; Reinalda & Kulesza, 2006). The university to which the authors belong also drove this transformational process, including PBL approaches in some courses. Authors experienced PBL in the first editions (2006-2007 academic year) in an Integrated Master degree in Industrial Engineering and Management (IEM) of the approach pioneer in the engineering degree (Lima et al., 2007). At that time, Project-Led Education (PLE) presented by Powell & Weenk (2003) was one of the approaches in engineering courses that promoted active learning by doing a opened project, presenting a team solution for it, being able to discuss the topic, fulfilling deadlines and reporting outcomes.

PBL is an active learning methodology that aims to develop technical and transversal skills in student-centered environment involving stakeholders cooperation and collaboration (Alves et al., 2018; Knoll M., 1997). That approach provides to the students useful tools and methods to deal with onsite challenges and situations, developing argumentation and social interaction skills in a professional environment, being prepared for conflicts management and also having a proactive approach (Alves et al., 2018; Frank et al., 2003; Guerra et al., 2017). Some authors (Dewey, 1996; Kilpatrick, 1918, 1921) also referred the contribution of PBL on the autonomous, independent, and responsible individuals' development as students are creating their own knowledge combining practice and theory in a project environment.

The main contribution of this chapter is to describe and present an overview of the PBL experiences of two Industrial Engineers and provide their perspective about this learning approach. Authors participated in four main PBL experiences for five years: in both semesters of the first year and in both semesters of the fourth year. Authors have faced some challenges during the implementation of this approach in their university once they were enrolled in the first Bologna curricula. The approach development led some experiences and adjustments from teachers and organizational committee during several years (Lima et al., 2017). Therefore, authors experienced two PBL projects in the first-degree year and the integration of different engineering courses and architecture in the same project, a framework that was not repeated until now. There are many studies (Alves et al., 2018; Lima et al., 2007; Mills & Treagust, 2003; Soares et al., 2013) performed by researchers and teachers regarding PBL approach and its impact on students' competences acquisition, however this chapter presents the personal opinion of two alumni with ten years into their engineering careers. Each experience had different levels of importance and it was felt in different way. These projects demanded a group of students working for the same project, building up ideas, discussing problems and managing conflicts between them. Additionally, skills such as communication, teamwork, time management and many others had to be developed. The ability of doing public presentations and defend an idea was a big advantage in this kind of learning that was not demanded before. Furthermore, teachers' role was not the same as authors were used to see. Its role was more focused on guiding students to find their own answers. Teachers were not anymore at the center of teaching process and this is also part of the teaching paradigm change in a PBL approach (Alves et al., 2016; Angelva et al., 2017; Powell & Weenk, 2003).

According to the Council of the European Union (2018) there are three elements that define competences in a learning process such as knowledge, skills, and attitudes. Knowledge is the combination of different existent concepts that allows to understand a given subject. Skills are the ability and capacity of achieving results applying acquired knowledge. Attitude is related to the mindset to deal with people, ideas, and situations. Key competences are determinant for a personal, professional, and social development, and they might be acquired throughout life. Critical spirit, teamwork, initiative, communication, intercultural relationship, creativity, analytics skills, problem solving, and negotiation are example of key competences that might be acquired in an education, training and learning process. That kind of competences are demanded and highlighted in current labor market that faces a technological transformation, so critical thinking and innovative spirit are more and more appreciated (Henshaw, 1991; Lang et al., 1999; Trevelyan, 2019). Inductive learning methods based on project approach, collaborative learning, traineeships in companies, laboratories, practical experiences among others are beneficial and contribute for competences development. Furthermore, this learning approach can increase motivation and engagement of the involved ones (Council of the European Union, 2018; World Economic Forum, 2015).

The importance of experience in the learning process is also in line with model of organizational knowledge creation presented by Nonaka & Takeuchi (1995) that referred the importance of Tacit knowledge in the knowledge creation. Sharing experience, practical exercise, physical proximity and also learning-by-doing are methods that contribute for Tacit knowledge (Nonaka & Takeuchi, 1995). These methods and development of competences mentioned above are clearly visible and presented in this chapter on the PBL framework. Using this approach, the authors intend to answer to the questions from their perspective: is it possible to combine technical and social also called soft skills in the same learning process? Can PBL work as soft skills promoter in an education process?

In this sense, from the methodological point of view, this chapter follows part of a narrative approach once that authors are going to structure and explain their experiences (Moen, 2006). Although other types of narratives exist (Williams & Carvalho, 2020), the authors choose this approach as it is intended to present this like a story. "*A narrative is a story that tells a sequence of events that is significant for the narrator or her or his audience*" (Moen, 2006). That story should be part of the individual' social, cultural and institutional context. From the social constructivism point of view, social activities contribute for the individuals learning and development (Moen, 2006; Wertsch, 1991). Authors structured their work describing their PBL experiences during the engineering degree. Then, authors explain from their point of view the main positive aspects they have faced in such experiences and main challenges they have overcome concerning organizational topics and transversal competences development answering the research questions previously mentioned. Additionally, authors describe their professional careers and contribution of PBL in different contexts of their engineer workplaces. Although it seems a simple storytelling about the experiences and perceptions, according to Moen et al. (2003) and Gudmundsdottir (2001) this is a natural method to structure ideas and recount experiences that might be valuable contribution.

This chapter is organized in six sections. The current one introduces the topic and the main objective of this chapter. The second one presents a brief literature about PBL approach and its perspective implementation in the university that authors belong. The third section describes the main PBL experiences lived by authors. The fourth one reports the positive aspects and main challenges of PBL from authors point of view. The contribution of this kind of experiences for the professional career is discussed in the fifth section. The last section presents the final remarks of this chapter, including the work limitations and also some proposals for future work.

PBL OVERVIEW AND UNIVERSITY PERSPECTIVE

PBL is an active learning methodology that has its roots in the original work of Dewey (1996) and Kilpatrick (1918, 1921). The active learning contrasts with traditional methods where students usually have a passive role in the learning process. So, active learning methodologies requires a strong student commitment and engagement throughout learning activities, doing by themselves, which it does not mean that traditional activities could not be included in the process (Freeman et al., 2014; Prince, 2004). Students must be capable to analyze, evaluate and criticize activities they are doing to understand the content purpose and meaning (Bonwell & Eison, 1991; Prince, 2004). According to Bonwell & Eison (1991) there some characteristics that define an active learning method in the classroom such as: students are more involved than listening, teaching process is focus on students skills development, students are requested to think about topics, they are engaged in the activities and their own actitudes and values are also considered. Literature reports different definitions for active learning depending of authors interpretation however methods such collaborative learning, cooperative learning and problem-based learning have common goals where students are learning process centered (Prince, 2004). Collaborative learning is focus on the student interactions instead of solitary activity. Cooperative learning includes the following assumptions according to Johnson et al. (1998a, 1998b): "*individual accountability, mutual interdependence, face-to-face promotive interaction, appropriate practice of interpersonal skills, and regular self-assessment of team functioning*". Problem-based learning is always active, includes the development of a relevant problem mainly self-directed learning by the students where collaborative and cooperative approach are usually used (Prince, 2004).

This approach has been implemented in several education institutions around the world, starting from pre-school till university level. PBL has been used in professional training in medicine and health related professions since 1960´s. PBL has been suggested as a solution to engineering education issues and therefore it has been deployed, since many years, in engineering schools, in a globally increasing trend .

In the context of engineering education, Powell & Weenk (2003) designated PBL as Project-Led Engineering Education and defined it as "*Project-led engineering education focuses on team-based student activity relating to learning and to solving large-scale open-ended projects. Each project is usually supported by several theory-based lecture courses linked by a theme that labels the curriculum unit. A team of students tackles the projects, provides a solution, and delivers by and agreed delivery time (deadline) a "team product", such as a prototype and a team report. Students show what they have learned by discussing with staff the "team product" and reflecting on how they achieved it*". This definition shares key features with different PBL models that are described in literature, mainly learning by doing, real world problems, role of the tutor, interdisciplinary, collaboration and group work and a product.

The development of these projects allows students to develop technical and transversal competences, as students build their own knowledge by combining theory with practice. The social skills are related with the ability to engage with others, allowing the knowledge sharing and growth. Therefore, the social skills are of utmost importance to establish the network of people and deal with the complex multidisciplinary systems created by the Fourth Industrial Revolution (Alves et al., 2018). The National Academy of Science and Engineering (Kagermann et al., 2013) report addressed the need to train future workforce to be able to manage complexity, abstraction and problem-solving. It is expected that the engineers can have autonomy to organize their work and great communication skills (Alves et al., 2018).

The implementation of PBL approach is a challenging task, as it involves the changes in curricula, students grading and assessment, school architecture and the need to educate the teachers in this method. The new role of teacher in this approach is one of the major differences when comparing with the traditional learning methodologies, where teachers are in the center of the process. As the role of the teacher in project-led education changes to interaction oriented, the students become more responsible for their own learning process (Alves et al., 2016, 2018; Angelva et al., 2017; Frank et al., 2003; Guerra et al., 2017; Powell & Weenk, 2003).

The implementation of PBL in engineering programs has been reported by literature. Considering the current challenges in industry and the skills demanded for future engineers, these demands will not be satisfied by more traditional engineering curriculum and learning methodologies. A mixed-mode model has been adopted by several institutions, with some traditionally taught courses, mixed with project-based education. This model has been quite successful in addressing the needs of industry, without sacrificing knowledge of engineering fundamentals (Mills & Treagust, 2003; Trevelyan, 2019).

The university to which the authors belong has promoted PBL education initiatives since academic year 2004/2005 and the research has highlighted the success of this approach in engineering education in this institution. Several reasons have motivated the implementation of this approach in this institution, mainly the Bologna process, the institutional support from Rectory team, as they have challenged the teachers to introduce new learning methodologies, due to the demotivation from some students and also teachers with the traditional approaches. After running a training session with the professor Peter Powell, a group of teachers from this institution decided to implement PBL approach. The opportunity to learn the principles of Project-led Engineering Education with Professor Peter Powell represented a huge importance for IEM, as this contact provided the necessary elements to support previous ideas and experiences with project approaches in the context of engineering education. The first PBL experience was implemented, with the support from Rectory of the University, in 2005, involving several teachers from different departments. This first experience and the further experiences thought the following years were supported by educational researchers (Lima et al., 2017).

The interdisciplinary PBL approach has been implemented during several academic years of IEM (Industrial Engineering and Management) program, in different semesters and in different academic phases. Since the first edition, teachers have been working in order to plan, prepare, execute and monitor PBL, in a cycle of action-research of a continuous improvement effort. The general approach has been quite similar in all the PBL editions, but due to the difficulties and challenges faced in every edition, new solutions have been developed through the years, by the teachers and the educational researchers. Most of the changes implemented along the years have been introduced after the assessment of students and all stakeholders feedback, collected during and at the end of each project edition. This feedback has represented an opportunity to adjust the teaching strategy to better adapt to students needs and project requirements (Fernandes et al., 2020). One of the major changes suggested by the students was the creation of a course called project to be integrated in the program and this change was introduced in the academic year of 2012/13.

The PBL approach in this institution in general has consisted in splitting the class in several teams and challenge them to develop a project during one semester, by applying different disciplinary fields, to propose a solution for one open-ended problem. So, students should present solutions to open-ended problems from real world, by developing technical and transversal competences. The class is divided in teams with several students, not only to increase the social interaction between them, but also due to human and physical resource limitation, related with scarcity of project rooms and material for prototype development, as well limited number of tutors to follow-up and support the teams. Every year, the institution has tried to carefully choose the project theme. The teachers aim is to choose an innovative topic to increase the motivation and bring awareness and engage the students in current real problems of society. Therefore, most of the projects developed have involved environmental and sustainability issues (Alves et al., 2016). In some of the PBL editions, the project has also integrated external partners, mostly industrial companies.

PBL requires huge interaction between the team members during the semester, but also closer and very frequent contact with lecturers and tutors. As students must perform several presentations about project status during semester, it represents a great opportunity to develop presentation skills to big audience. In some of the PBL editions, presentations involved also external companies. University has also provided to the teams coaching in the process of peer assessment, teamwork and conducting presentations. The project assessment methodology consists in the evaluation of multiple items, each of them having different weights and graded by several lecturers. The monitoring of project progress consists in several milestones and most of them are associated with deliverables, usually in form of presentations, reports, or prototypes.

Several studies done on PBL approach in IEM have shown that this methodology has ensured the development of technical and transversal competence. Students themselves realize that they developed communication and teamwork skills, besides conflict management competence. These findings also report an increase in enthusiasm, cooperation, and competition between the students (Alves et al., 2018; Fernandes et al., 2014). The university which authors attended has conducted several studies (Alves et al., 2016, 2018; Fernandes et al., 2020; Lima et al., 2007, 2017; Soares et al., 2013) to understand student's perception about this methodology. Students have identified several outcomes, as engagement in teamwork and collaboration with the colleagues. Conflict management has also been pointed as one of the most significant learning outcomes of the process. Another relevant outcome that has been highlighted is the ability to communicate knowledge in writing and oral presentations. The motivation and the set-up of own learning goals has been also addressed as one of the most positive results. Students also refer as a key feature the increasing relevance of course content, as they could articulate the theory and practice. PBL has also ensured a clearer and more realistic vision of professional world.

As the biggest challenges in the PBL approach, students highlighted the time and tasks management, once that leadership and responsibility management represented a problem in most of the teams. Other major challenges identified have been the assessment methodology and the balance between the demands of the project and subject's individual evaluation. Despite all the challenges, the general opinion from students has been very positive. The PBL has been described as a good way to prepare students for professional life, even by the less engaged students (Fernandes, 2014; Lima et al., 2007).

Teachers involved in the implementation of PBL approach have noticed that IEM students have increased their motivation and engagement in the courses, as the students have more possibility to actively participate in the learning process. This approach has also ensured a closer teacher-student relationship and higher collaborative work between teachers from different departments and schools (Alves et al., 2016; Fernandes et al., 2014).

According to Lima et al. (2017), the experience of PBL approach in IEM has shown evidence of the advantages of this methodology at different levels and highlights that there is still further room for improvement and development of the concept: there should be opportunity to more projects, more interaction with companies and society, more projects with interaction with other cultures and students from other programs, further increase the interaction with other departments, engineering education research recognized by peers and more opportunities for staff development (Lima et al., 2017).

DESCRIPTION OF PBL EXPERIENCES

In this section, it is presented the main PBL experiences that authors have been challenged during its engineering degree. Four experiences are reported that occurred in different phases of the degree: two in the first year and more two in the fourth year. So, from the student point of view, these experiences reported different student maturity stages once in the first year was the beginning of a long journey, but in the fourth year, students already had enough maturity to understand main PBL purpose and the main outcomes of having such learning method. Therefore, this chapter mentions different perspectives depending on the phase it occurred. Figure 1 represents some examples of work developed by authors. In Table 1, it is identified and characterized the four experiences. For each experience is described the project theme, the period it occurred, number of members of the team, course subjects included on the project and if project involved a real industrial environment (a company). Additionally, it is presented the assessment included in each experience as an input for the course final grade.

Figure 1. Examples of project results a) Product developed; b) Production system developed; c) Failure tree analysis of a company machine

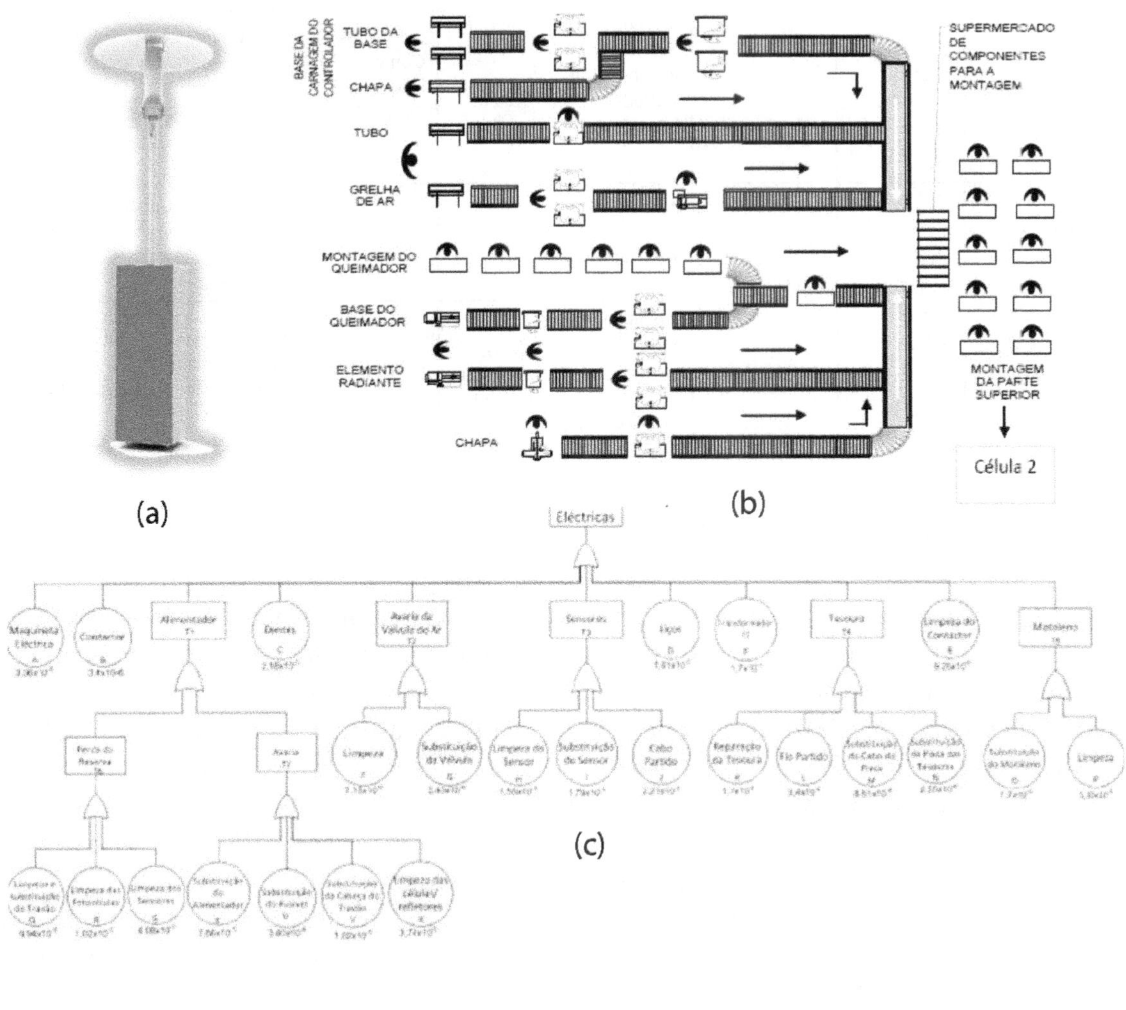

Table 1. Summary of project-based learning experiences (#)

#	Project Theme	Period	Number of members	Course Subject	Real case (on site)	Assessment
1	Fuel Cells Production System	First year - first Semester	6-7	• Calculus; • Computer Programming I; • Introduction to Industrial Engineering and Management; • General Chemistry.	No	• Peer Evaluation; • Preliminary reports and presentations; • Final Report and presentation; • Individual assessment on subject; • Project individual written test; • LEGO Prototype.
2	Space tourism	First year- second semester	6-8	• Linear Algebra and Differential Calculus; • Computer Programming II; • Cost Analysis I; • Physics.	No	• Peer Evaluation; • Preliminary reports and presentations; • Final Report and presentation; • Individual assessment on subject; • Project individual written test.
3	Gas outdoor heater	Fourth Year - first Semester	10	• Production Information Systems; • Simulation; • Production Integrated Management.	No – National Competition "Hotspot Design"	• Peer Evaluation; • Preliminary reports and presentations; • Final Report and presentation; • Individual assessment on subject; • Poster.
4	Evaluation of a workstation	Fourth Year - second Semester	7	• Workstations Ergonomic Study; • Reliability and Maintenance;	Yes	• Final Report and presentation; • Individual assessment on subject;
	Technology and Corresponding Systems			• Computer-Aided Manufacturing (CAM); • Computer-Aided Design (CAD)/ Computer-aided Process Planning.	No	• Final Report and presentation; • Individual assessment on subject.

Experience 1 – First Year, First Semester

The university which the authors belong introduced PBL concept in an Industrial Engineering and Management degree in the academic year 2004/2005 and has run several editions, with overall very positive feedback from students and teachers (Fernandes et al., 2007; Lima et al., 2017).

The academic year 2006/2007 was the third edition that PBL was introduced in an Industrial Engineering and Management of the university which the authors belong. In this particular year, it involved six groups of six to seven students from the first year. The students had the responsibility to create the groups by themselves but had to follow two criteria: each group need to have at least 50% of students of each gender and, at least, two students with knowledge in Chemistry.

The project theme was Fuel Cells Production System and the coordination team defined it before the semester started. Every year the project theme for PBL changes, but teachers always choose an appealing and challenging subject related with sustainability topics, in order to motivate the students and increase awareness for environment and social problems (Alves et al., 2019; Moreira et al., 2011).

Each group had a tutor to guide the teamwork and the project management. Tutor role was particularly important; since for majority of the students involved, this was their first academic experience and many of them lack the maturity to cope with project and time management, as well as the capacity to handle interpersonal conflict during the teamwork. Therefore, tutors had an important role in promoting the teamwork and conflict resolution(Alves et al., 2007; Alves et al., 2017; van Hattum-Janssen & Vasconcelos, 2008). In addition to the tutors and teachers, the experience also involved the participation of educational specialists and a recently graduated assistant developing research in project management (Fernandes et al., 2007).

This project had several deliverables along semester (Table 1), including reports, public presentations and a production system prototype in LEGO Mindstorms (Sousa et al., 2013). The milestones were of extreme importance in order to teachers monitor the status of the project, support and guide the students in solving blocking points and problems in the development of the project, including adjustments in case needed. These milestones were usually done as public presentation and represented a great opportunity to students develop public presentation skills and creativity.

Every deliverable previously mentioned were considered on the students' evaluation. The individual assessment was performed by peer assessment. Peer assessment of teamwork consisted in the evaluation of performance of the other team members, according criteria previously defined and negotiated with students. These criteria included topics as presence in team meetings, level of effort in work, original contributions, interpersonal relationship, and deadline fulfilment. In the end of semester, there was an individual written test to check the knowledge from the students in the project and work developed during semester.

The technical competences developed in this project were related with the four courses from 1st year first semester, described above in the Table 1.

In this case, the students had to design a production system to produce fuel cells, as well as simulate a behavior of this production system in a prototype in LEGO Mindstorms. Students developed also computer program in Basic programming language. As the topic of fuel cells is related with chemistry, it was also easy and interesting to integrate General Chemistry in the project. So, PBL offered the possibility to learn with practical and concrete examples, providing the chance to apply the technical skills learned in the courses in the real world. This process also made the students to study continuously along the semester and not only before the individual exams. However, in the case of Calculus, it was not so easy to implement this particular subject into the PBL process, due to the nature of the topics involved in this course.

This project was technically quite challenging for students, since fuel cells was a new technology and there were not many companies producing this technology in Portugal. There was very little technical information available about production systems related with fuel cells. However, students managed to find a small company that offered them not only the possibility to a visit a real production system of fuel cells, but also provided them some technological information about fuel cell production system. After this blocking point was solved, it was easier to students start developing the design of the production system.

This PBL edition was very intensive and remarkable experience to students, as the workload was very high, due to several deliverables and milestones. This experience demanded a big effort not only from students, but also from teachers. Besides the PBL, students had to study to several individual exams along semester so, sometimes it was challenging to conciliate both approaches. Also, the evaluation of

PBL project had a high influence in the final grades and, sometimes, PBL contributed to decrease the grades obtained in the individual assessment of courses. Therefore, in some situations this contributed to decrease the motivation from the students. The teamwork was in general rewarding and motivating, as usually the team spirit in the teams was good, despite some interpersonal conflicts that arise during the project development, especially because the students did not have the same level of commitment to teamwork. During the semester, there were some situation of lack of engagement of some of less motivated students – this situation created some interpersonal conflicts in some of the teams.

Apart from the huge workload and preparation needed from teachers and students, the other challenge involved in PBL are the resources needed, as students needed a room to meet each other and work together. The department building was at that time quite new and offered great conditions to students develop their work, with large and new and well-decorated rooms. Besides the dedicated rooms, also laptops were offered to students (one per group) as well as LEGO Mindstorms equipment to develop the prototype.

Experience 2 – First Year, Second Semester

In the academic year of 2006/2007, PBL was carried out, for the first time, in two consecutive semesters. It involved again six groups of six to eight students from the 1st year of Industrial Engineering and Management. Most of the groups remained the same as the ones created in the first semester (Experience 1).

The project theme was also quite innovative, but this time it was not related with sustainability, as usual, it was about Space Tourism.

The project ran in quite similar way as in the first semester, but in this edition, there was no prototype in LEGO Mindstorms involved.

Also, a major difference was the fact that the students could create, in the first week of the project, their planning of the project management for their group and present it to teachers. This was a big step to student's autonomy and gave them more confidence to develop the teamwork during semester.

The technical competences developed in this project were related with the four courses from first year second semester described above in the Table 1.

In this case, the students had to design an agency to explore Space Tourism. Similar to previous semester, students developed also a computer program in Basic programming language, but this time also a prototype of a website in HyperText Markup Language (HTML). The topics learned in Physics and Linear Algebra and Differential Calculus could also very easily be integrated in the PBL. Students performed the economic evaluation of the project with the skills learned in Cost Analysis I course.

In general, the second experience of PBL ran in a quite similar way as in the first semester, although this time students had more maturity and confidence in project management and interpersonal conflict resolution. Therefore, the skills acquired in the first semester were without any doubt unquestionable.

Experience 3 – Fourth Year, First Semester

In the fourth year first semester, Industrial Engineering and Management students had the chance to choose between two different approaches to develop PBL project: participation in a project with the colleagues from Industrial Engineering and Management or participate in a voluntary multidisciplinary project called Innovation and Entrepreneurship Integrated Project (IEIP). The students could participate in the IEIP in a voluntary basis and they had to apply in advance already in third year.

IEIP is a concept of PBL involving students from different engineering integrated master courses from the same university, that work together in a project during one semester. This concept was introduced for the first time in the academic year 2007/2008 and ran during several years. In each edition of IEIP, there were groups of six to ten students, competing with each other, to develop a new product, improve an existing one or proposing changes to production systems or products from real companies.

In the edition of academic year 2009/2010, IEIP was in its third edition and it involved four groups of ten students from Mechanical Engineering (ME), Industrial Electronics and Computer Engineering (IECE), Polymer Engineering (PE), Industrial Engineering and Management (IEM) and a special participation from students from Architecture (AR) as shown in the Figure 2. In the academic year 2009/2010, the project consisted in participating in a national contest from the Portuguese company GalpEnergia, to develop a Hotspot (gas outdoor heater). Therefore, as design was a very important topic in this project, Architecture was involved for the first time. Each group consisted from two students from the five integrated master courses.

Figure 2. Courses involved in IEIP

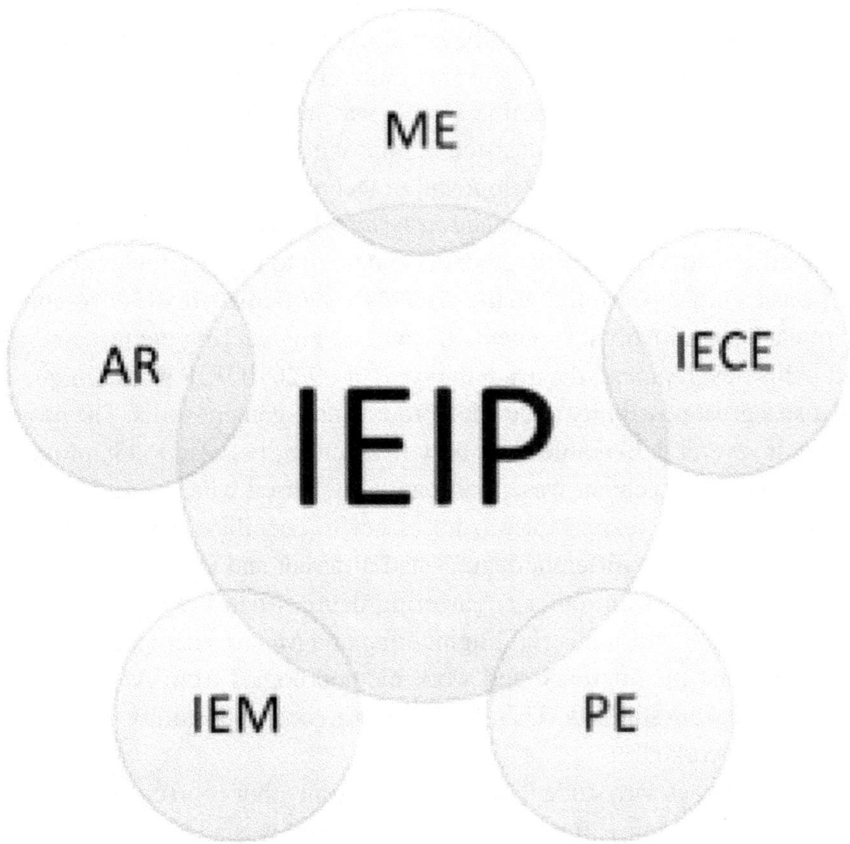

Each group had a tutor to guide and support during the project. Students also received training from TecMinho, an interface institution that is part of the university in the area of Entrepreneurship and Industrial property. This was particularly important, as students had to analyze several patents to research the products already existing in the market of gas outdoor heater.

The project had several milestones and deliverables in order to control the progress of the work. Industrial Engineering and Management department also had available a dedicated room to students meet and work with each other during the semester. During the project, besides the regular public presentations to present the status of the project, several meetings with all the teachers involved in the project were organized. These meetings were important to guide the students and provide an overall feedback from the teachers involved.

The IEIP evaluation considered several components: reports assessment, preliminary report and final report, public presentations, and the prototype evaluation. The students that were selected by GalpEnergia to the second phase of the contest got 5% extra bonus in the evaluation. The individual assessment was performed by peer assessment.

In case of Industrial Engineering and Management, the technical competences developed in this project were related with the three courses from fourth year first semester involved in IEIP, as described in the Table 1.

In this case, IEM students had to design the production system to produce gas outdoor heaters, relating organizational functions and methods of manufacturing, production management and planning and production control, as well as simulating a behavior of the production system in software ARENA. In the academic year of 2009/2010, the project was quite challenging for Industrial Engineering and Management students, since no real production system was involved, as in the usual IEIP editions. In the previous editions of IEIP, the project consisted in improving a real production system in a company. Therefore, in this particular edition, the development of technical skills was compromised because students did not have the chance to see a real production system. Due to this limitation, the work developed by Industrial Engineering and Management students ended up to be very theoretical. Nevertheless, on the other side, this was a unique possibility to the students to participate in the development of a product with a multidisciplinary team involving students from other engineering degrees and architecture. So, regarding the soft skills development, the academic year 2009/2010 IEIP was a unique opportunity.

IEIP provided also a great possibility to develop project management skills. The project timeline was very challenging, with several deliverables, not only to teachers, but also to GalpEnergia contest. The complexity of the project management was significantly increased when compared with the previous PBL experiences from previous years, as it was necessary to conciliate very different time schedules, since the students involved in five different degrees had different and very often conflicting schedules. On the other side, the students from other engineering degrees had very little experience in project management when compared with Industrial Engineering and Management students. Other interesting experience was the fact that the approach and work methodologies from Architecture students were very different from engineering students. Therefore, it was a good opportunity to see how other courses outside engineering areas worked.

IEIP was a chance to interact with students and teachers from other courses. The network created with some of the students remained long after the project ended and was a chance to keep a good network with people working in very different engineering fields, even after graduation from university (Soares et al., 2013).

At the end of the project, students could compare the experience of having such project with different courses involved and the common project, only with IEM students. These experiences were described in an article presented in an international symposium (Araújo et al., 2010). Additionally, the participation in an international symposium in a different country was also an important experience promoted by teachers that encouraged students to share their experience and get the opportunity of knowing colleagues from different universities.

Experience 4 – Fourth Year, Second Semester

In the second semester of fourth year, PBL was developed in groups of six to seven students, but this time the PBL ran in two separate projects, each of them involving two different courses. The students were allowed to create the groups in a voluntary basis and this time there were no restrictions imposed.

The four courses involved in the two separate projects were the ones as described in the Table 1.

One of the projects consisted in a practical work developed in a manufacturing company, in order to perform an ergonomic study of a real workstation and analyze the reliability and maintenance of equipment.

Students had the opportunity to visit the company several times during the semester and to apply the concepts learned in courses of Workstations Ergonomic Study and Reliability and Maintenance in real world. This PBL experience represented the chance to some of the students (the ones that participated in IEIP edition) to visit a real production system and work in suggestions for improvements for the first time during their academic experience. In this project, the cooperation with the company stakeholders was crucial to ensure the success of the project.

The second project involved the application of the concepts learned in the courses of Computer-aided design (CAD)/ Computer-aided Process Planning and Computer-Aided Manufacturing (CAM). Students had to project the creation of a company based in the concepts learned in the courses.

Similar to the PBL editions from previous years and first semester of fourth year, students had deliverables and milestones during the project, and they had a dedicated room to work together.

In this PBL experience, students had already developed project management skills, during the previous PBL editions and also had acquired more maturity to deal with the demands and challenges involved in the teamwork. It is also important to mention that students could choose their work group without any restrictions and in general they ended up by creating teams with colleagues with whom they have already worked for several years, either in previous PBL editions, or in academic teamwork done in the previous three and half academic years. So, the project management and interpersonal relationship management in general ran quite smoothly, when compared with previous experiences. However, as this time all the students had to develop a project with a real manufacturing company, it was also necessary to cope with other new challenges, in this case, communication in context of a real work and corporate environment. Students had to perform a presentation to company stakeholders and also create a written report. Furthermore, students wrote an article and presented the work developed in a conference (Araújo et al., 2011).

POSITIVE ASPECTS AND CHALLENGES OF PBL

Engineers need to design, operate and manage systems, by applying mathematic and scientific concepts, in order to solve real life problems in the society. Besides the technical skills, engineers need to have a strong sense of human interaction, not only when they design the solution, but also when planning, executing and managing the deployment of these solutions. Therefore, PBL offers to student's new ways of tackling problems, promoting the creativity, critical thinking and collaboration. PBL promotes soft skills development when compared with more traditional learning methodologies, as these are focused in technical knowledge and academic skills.

PBL represents a big endeavor to students but also to teachers. The work involved in the preparation, monitoring and follow-up the PBL for teachers is much higher when compared with traditional learning methodologies. Therefore, the teachers that promote these initiatives need to be highly motivated. One of the big advantages of this approach is the connection created between the students and teachers involved. This connection reveals as very important not only during the academic experience, but even after in the work life, when some of the students need to work in projects involving cooperation with university. Also, a challenge is to choose topics that are in the forefront of innovation and technology and that allow to students to explore innovating and creative topics during their academic experience. In addition, PBL demands good working conditions, by giving to students dedicated rooms and laptops and, also, e-learning-based portals, where students can submit their works digitally. All these mentioned topics involved a tremendous amount of human effort and, also, physical and material resources, and represent an extra challenge to universities when implementing PBL. The traditional learning methodologies are much less demanding and easier to implement from authors point of view.

Looking back after ten years of working experience in industrial engineering and management in a multinational manufacturing company, authors realize the clear advantages and the contributions of PBL in the university education and the preparation to face the real-world problems and complexity.

Each of the four editions of PBL offered unique experiences, that contributed in different ways to develop and prepare the students to face the real working environment.

Experience 1 – Managing the First Academic Impact

The first edition was the most difficult to manage regarding interpersonal interaction with other students. In general, for most of the students involved, this was the first academic experience involving PBL. It was sometimes very difficult to manage and deal with less motivated students that were not very much engaged in the teamwork. However, some of the students, despite young age and first academic experience, revealed great leadership skills and developed very quickly maturity and capacity to manage the challenges of the project and the teamwork. Nevertheless, both authors highlight as one of the biggest challenges of PBL the difficulty to get all the students to work and commit to teamwork in the same level. PBL is not very efficient in less motivated students as they tend to be left behind during the teamwork and end up by contributing very poorly in the activities. Therefore, these students do not develop themselves and do not take advantage of all the benefits from PBL. The PBL evaluation had a big impact in the final grades and contributed to lower the grades obtained in individual assessment. During the PBL, this felt as unfair and was sometimes very demotivating to some of the students.

Since the concept was introduced immediately in the start of the first semester of the first year, the students ended up working in separate groups, with the groups being split in separated rooms. Therefore, the interpersonal relationship and team spirit and connection in the class as a whole team was poor – and this situation did not improve significantly in the next academic years. It was interesting to note that some of the groups created in the first year, first semester last until the end of the all five years of the integrated master degree. So, part of the students ended up in working with the same teammates during all their academic experience, with advantages but also some disadvantages. However, some groups got the chance to share the same room with teams from other course years, so they had the possibility to know and share experience with them.

Other of the big challenges was the workload, as students had to study in parallel to several individual exams and tasks. Nevertheless, it was also very positive as it demanded that students had to develop time management skills and study continuously and not only before the final exams, as in more traditional learning methodologies. The summary of all positive aspects and challenges from authors point of view are described in the Figure 3.

Figure 3- Positive aspects of PBL and challenges of experience 1

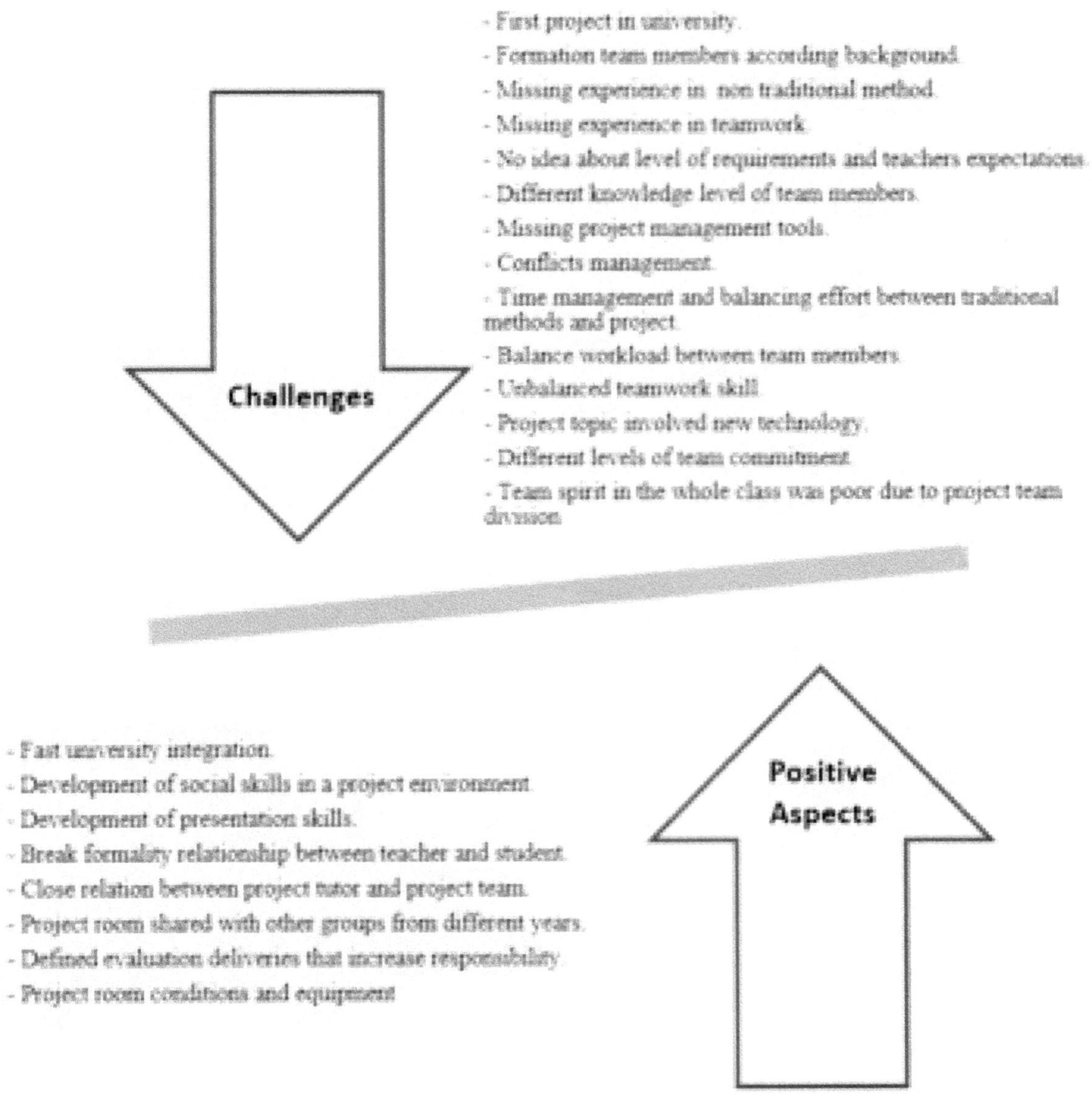

Experience 2 – Consolidation of Project Management Skills

The second experience ran more smoothly, as the students already had developed much more maturity and were familiar with the learning methodology and project management. As a challenge, authors highlight the fact that students felt overwhelmed and tired from the previous semester experience. The project topic choose was very futuristic therefore it was sometimes challenging to integrate the subjects learned in the courses. The most important advantages and challenges of this experience are presented in Figure 4.

Figure 4 - Positive aspects of PBL and challenges of experience 2

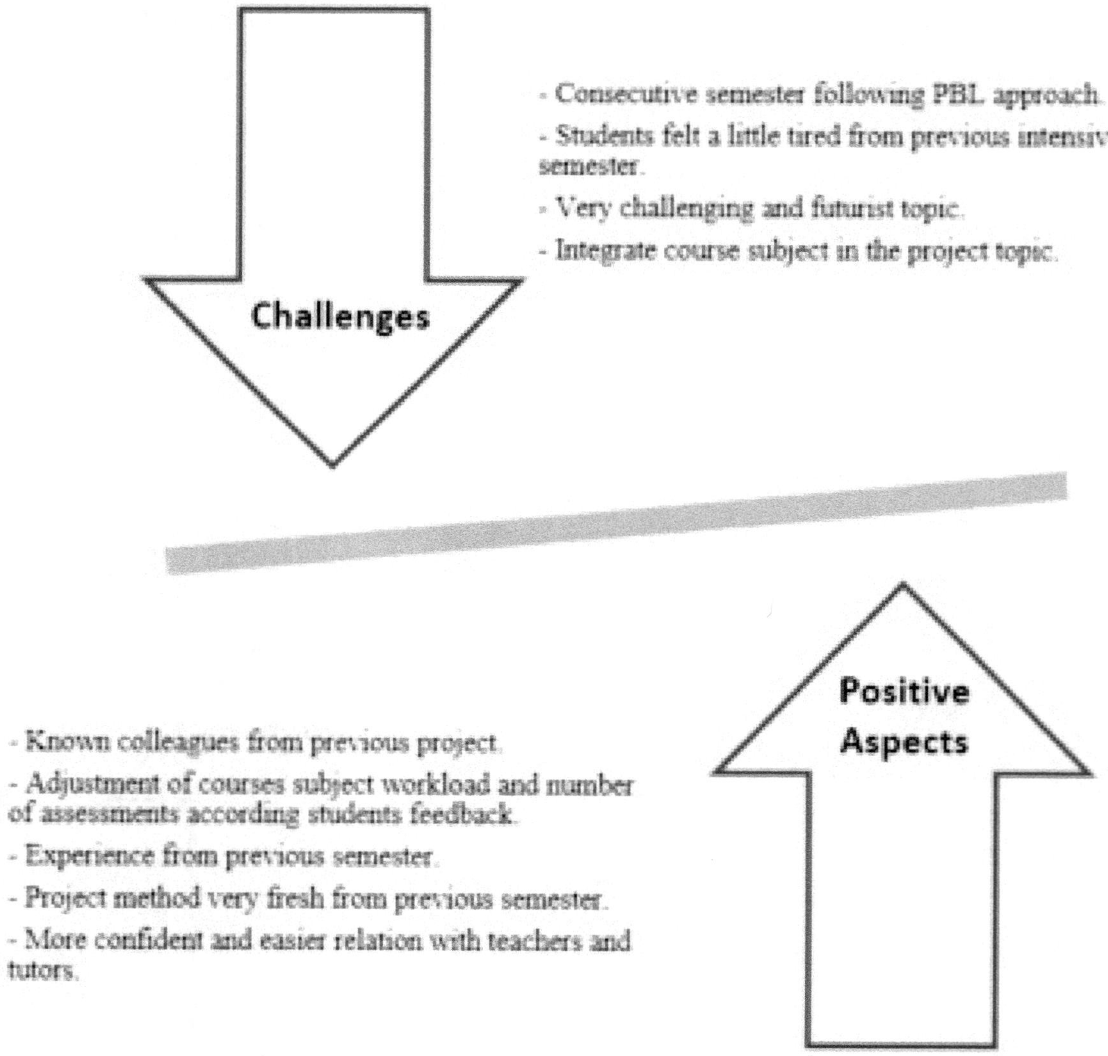

Experience 3 – Challenged by the Multidisciplinary Voluntary Project

Figure 5. Positive aspects of PBL and challenges of experience 3

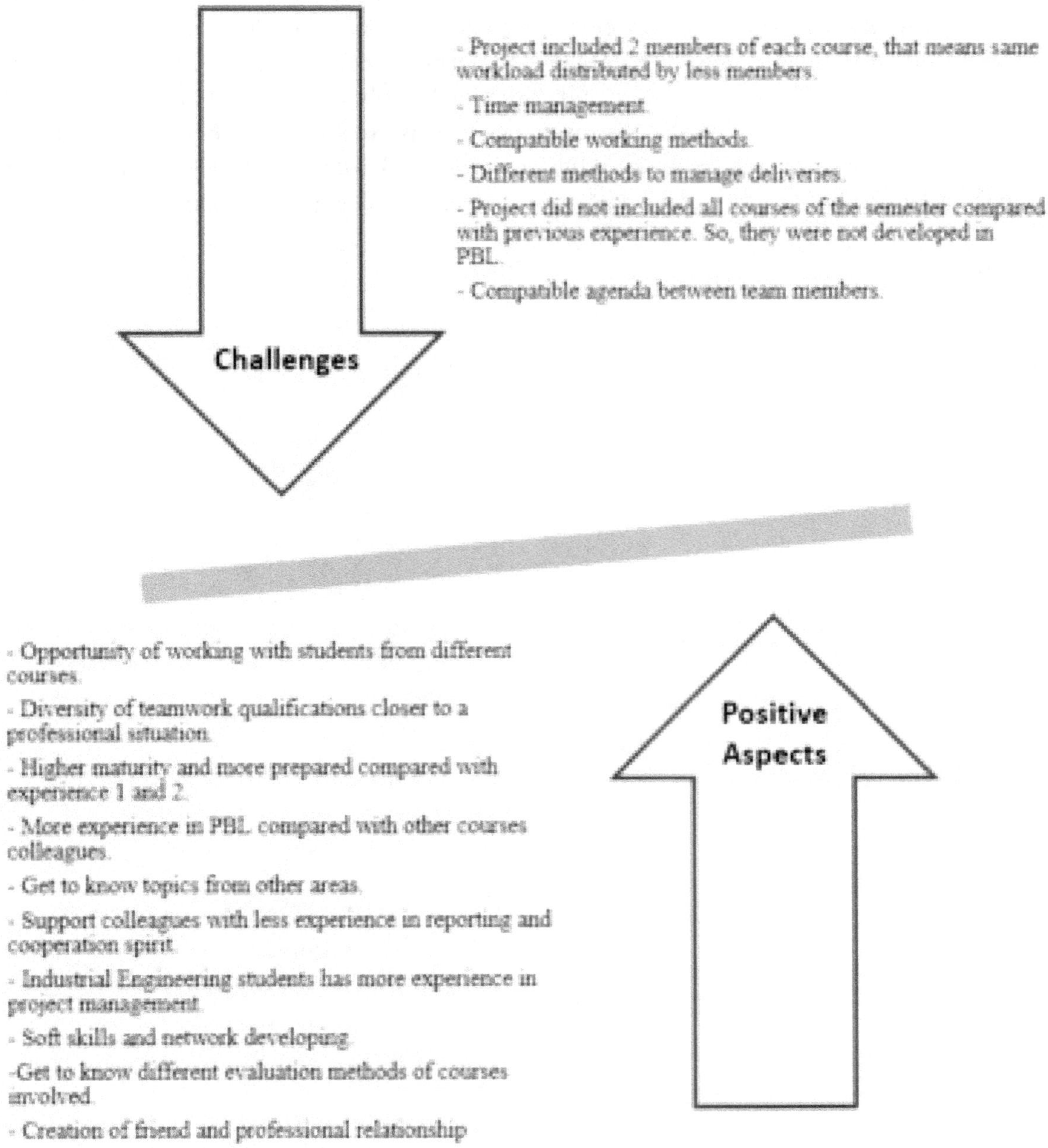

The IEIP was a great experience to the authors. The teamwork with other students was really a great chance to know students from other engineering degrees and understand the particularities and the methodologies applied in the learning process of other degrees, especially architecture. Despite the other students did not have much experience in project management, the project ran quite smoothly and there were no significant interpersonal conflicts in the teams. The workload did not seem as intense as in the first PBL experiences, as all the students that participated in the project were in general highly motivated and contributed significantly to project development. The deadlines were fulfilled in time without any major issues, despite the difficulties in conciliating conflicting time schedules between ten students from five different degrees. The network created in this project was also an opportunity to understand better the mindset from the students from other engineering degrees and architecture, as it was very different from the industrial engineering and management students. In overall, this type of multidisciplinary projects represents a huge advantage, despite the workload involved to students and teachers, as they contribute significantly to transversal skills development. This experience was close to the real world-working environment, as in their professional life, engineers work with many people from very different academic backgrounds. Summary of main positive topics and challenges are presented in the Figure 5.

Experience 4 – Dealing with Two Projects

Figure 6. Positive aspects of PBL and challenges of experience 4

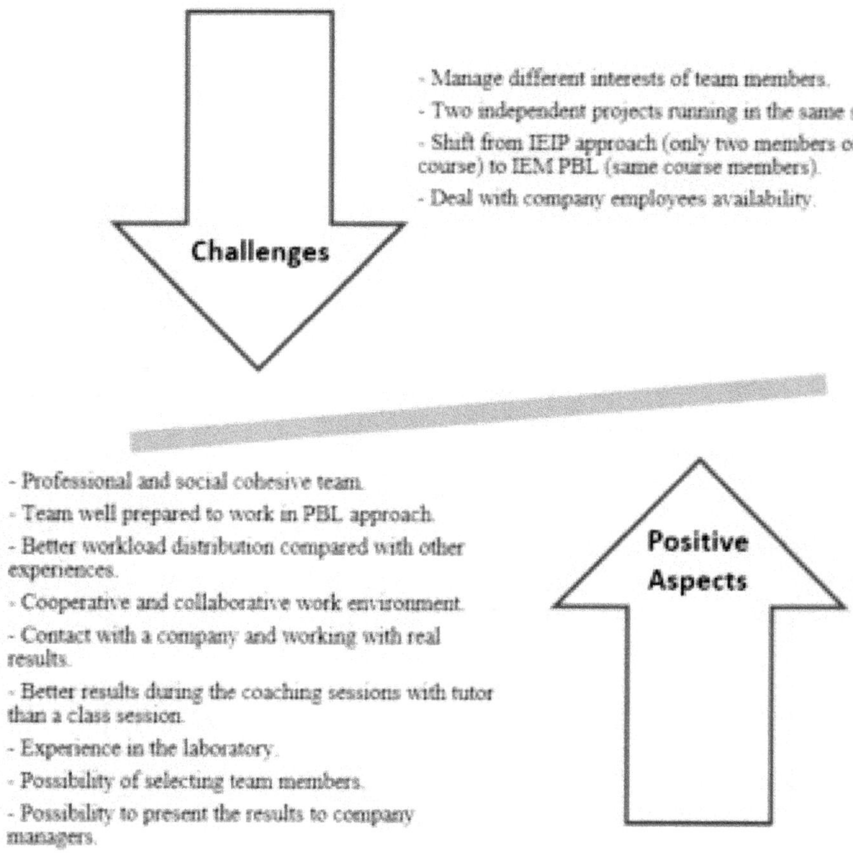

The last experience of PBL was in the fourth year second semester. In this experience, the project was developed with the team of students that could choose by themselves. Therefore, the project management was very easy this time, as the team worked very well together, and all students were equally committed and motivated to achieve good results. So, it was quite different from the two first PBL experiences. However, this edition of PBL brought other new opportunities for technical and personal development (presented in Figure 6), as it was done, for the first time to some of the students, in a context of a real manufacturing environment. Therefore, it was very interesting to visit a company, provide improvement suggestions and apply the concepts learned in courses in the real world. This PBL experience provided also an opportunity to develop even further presentation and communication skills, as students had to meet and present result to the senior management level during the visits to the company and also had the opportunity to present the work in an international conference (Araújo et al., 2011).

CONTRIBUTION OF PBL FOR THE PROFESSIONAL LIFE

In this chapter is described the contribution of PBL experiences in the authors' professional life and hopefully answer to the question put in the introduction: "Is it possible to combine technical and social also called soft skills in the same learning process? Can PBL work as soft skills promoter in an education process?" Authors are both Industrial Engineers however they have different professional career and experiences. Next section describes the professional career of each author and the following one describes the main contribution of PBL from Authors 'point of view.

Description of Professional Career – A.Araujo

A.Araujo started its professional career in a multinational company in August 2010 for one-year internship. During this period, author developed its master thesis related to Lean Production and Pull systems in an industrial environment with a real problem (Araújo & Alves, 2012).

This internship occurred in the Continuous Improvement department that included ten members. Although it was not the first industrial contact, it was the first time that author was by its own, without colleagues from projects and responsible by developing its work. Integration was relatively fast because department elements welcomed very well, they used to have internships before, and their mentoring was crucial to easily understand company behavior. During that internship, author also had opportunity to participate in many activities that provide her get to know the entire value stream chain, continuous improvement practices and also contact with several departments and understand their interaction.

After internship, author change role and department as there was no chance to keep in the same one due to resources restrictions. Then, author changed to Production area as production team leader in the night shift. The internship was crucial to start that function once author has already known the production area and people involved, so far, she was integrated in the company. Team leader is responsible to manage and coordinate production lines including human resources and machines, quality issues, machine breakdowns and assure production planning and its performance to fulfil the target Key Performance Indicators (KPI). During that time, author was responsible for two production lines and two cells that involved roughly twenty-five operators. It is a very demanding and stressful role in terms of personal relationship because sometimes was not easy the direct contact with operators, manage conflicts and different motivations.

Eventually, author had the opportunity of changing for Industrial Engineering section. Her main responsibilities were time data management, workstation definition according ergonomic guidelines and target cycle time, line capacity calculation, operators standardized work, calculation of manufacturing cost, participation in process and product development phase, participation in new products quotation phase, collaboration with process engineers for line and station improvement and also participation in some multicultural projects in Malaysia and Hungary. This function allowed to have a closed contact with production, different areas included in industrialization process such project management, product and process development concerning new products, customer audits, controlling related product cost and yearly business plan that represent a very stressful phase. In this function, author also participated in an innovation project with university, a great opportunity to contact and work again with teachers. That function represents the main author background once it represents roughly five years of her professional carrier.

Currently, author has a different role in a different area: Continuous Improvement section. So far, author had the opportunity to come back to the internship section. Author is responsible for promoting simultaneous engineering activities during the industrialization phase with different teams inside and outside company such as development, engineering, manufacturing, logistics, purchasing, sales, among others. For that, author is responsible to organize and moderate workshops like Design for manufacturing and assembly (DFMA), Value Stream Mapping (VSM), Line Design, Material Flow Design and others. Additionally, training and coaching concerning Lean Production tools are also part of her function. Author also participates in several continuous improvement activities in the company. To perform such activities, author had to be certificated by central department in Germany that lead several international activities with different colleagues. This function demands several interactions and communication channels between different teams. The communication and management skills are crucial to get teams committed and find solutions for many situations. Presentation skills are very important to present workshop and project results to the managers. Furthermore, that skills also are important on training and coaching conditions.

The relation between author and university is still very close as author is getting her PhD about Lean Production and Pull systems (Araújo et al., 2021). Furthermore, author had a small experience as Invited Assistant in the same department and university where she took her degree.

Description of Professional Career – H.Manninen

H.Manninen finished the graduation as Master in Industrial Engineering and Management in November 2011 and has worked since then in multinational company. Author started in this company in March 2011 as a trainee in the Logistic department responsible for logistic projects, costs and process management. The internship was a chance to develop the master thesis, under theme *End of life planning process review*, and to have a good integration in the company logistic process. The colleagues responsible for mentoring the project during internship were senior engineers and project managers with huge experience in logistics and supply chain process management. They ensured a great background to start and develop a career in logistic and supply chain, giving to author the big picture of logistic and supply chain in the company. The knowledge and the mindset acquired during the internship were crucial to the author future career development. After the internship, author started the professional activity as a procurement planner in the logistic department. The tasks and working environment were very different from the one experienced during the internship, since it was more close to the daily business operations. This task was very demanding regarding stress and time management and also interpersonal relationship, since it involved working with many people from different areas and backgrounds, not only logistics, but from other departments, like manufacturing, engineering and also external suppliers and service providers. During the four years in this position, author was also the SAP (main integrated software for business management used in the company) key-user in the area and this task represented a great opportunity to learn more deeply about material planning process in SAP, as well as to participate in small projects related with SAP. After four years in this position, author started a new job in the company, as inbound process engineer. By this time author had the chance to return back to the department where she did her internship, working again with senior logistic process and project managers, in more strategic topics. In this new position, author developed several projects related with supplier development, mainly Electronic Data Interchange (EDI), Vendor Managed Inventory (VMI) and consignment program development and implementation with several international automotive suppliers. This task demanded, besides the technical knowledge in logistic process, good project management skills, but above all, great communication and negotiation skills, since author had to interact with several stakeholders from the suppliers, from technical EDI experts to senior key account managers, lawyers, daily business planners, among others. During these years author was involved in other topics, as for example, leading a joint project between the company and the university where she took her degree, under innovation at procurement. Back in 2018, author changed to a new position inside the logistic department, joining to the logistic crisis management department. The position consisted in the implementation of logistic key account concept in the divisional logistic department. This task was extremely demanding as it involved several interaction and communication with high management level, not only inside the company but also with external suppliers. The communication skills were crucial to ensure a fast and smooth resolution of critical situations in the supply chain, during times of uncertain and unpredictable situations. During crisis management, author had the chance to work four months in Germany and one month in Japan and had to deal in a daily basis with multicultural teams, involving colleagues and contact people from suppliers from all around the world.

After nine and half years working in logistics, author decided that it was time to start a new position in the company, moving into a different department inside the organization. Author is working currently as an overhead cost controller and controlling business partner in the Controlling department. The work is very different from logistic and represents an opportunity to develop new technical skills and knowledge in cost controlling, as well as to interact with several departments inside the company.

The author feels that the skills acquired in the PBL have been crucial to perform the tasks so far and is looking forward to new opportunities to perform other positions in the company in the future, in a world that is changing so fast and needs a constant adaptation and flexibility. This is the greatest advantage and the reason why author decided to study Industrial Engineering and Management: it provided the needed skills and competences, both technical and transversal, to adapt to several different tasks in multicultural and demanding industrial environments.

Importance of PBL in the Professional Life and Development of Social Skills

Since the beginning of professional life as internship that soft skills revealed a huge importance for both authors. In an industrial daily business, mainly in a multicultural environment, people work with many different ones with diverse characteristics. Engineers are always in contact with different departments and external partners, discussing several topics, interacting with managers, present and defend ideas. Engineering practice demands working in multicultural environments and a global context (Williams et al., 2018)

Among many technical tasks to be performed in a daily work, expectations, and conflicts have to be managed. For instance, in a project team, work packages have to be completed by different people according precedence and work packages, which only are finalized when everyone perform their work. This is very similar to the PBL experiences already mentioned in the previous sections. Additionally, in an industrial environment, no one works individually. Every day, engineers need to contact with someone in many situations such as requiring support, working in project team, existence of precedence tasks, due to a supplier process, bureaucracy, a small call or an email, in a training session or even in home office approach, mainly during the actual COVID pandemic. PBL provides tools and experience to get more prepared for this kind of cooperation. In fact, in a traditional learning approach, students work individually to achieve a grade and he/ she does not depend of a group work. PBL shows real situations to the students that will also happen in the professional life.

Furthermore, the close relation created with teachers in a PBL approach was maintained even after degree finalized. That kind of networking allowed to have a cooperation in different situations such as conferences and seminars, experience exchanging with students. Additionally, there are a strong cooperation in projects and other activities with university and company that authors are working currently. That kind of interaction is very important in order to feedback market requirements and needs, discuss potential competences, receive new internships and cooperate with students and teachers. Authors believe that the relation between students and teacher become more closer and promote this kind of networking that would be difficult in a traditional learning where there is a barrier between students and teachers. In fact, contribution of PBL has already been mentioned in the whole chapter.

Figure 7 summarizes the most important contribution of having PBL approach as learning methodology point out by authors.

Figure 7. Summary of main soft skills acquired by authors in PBL

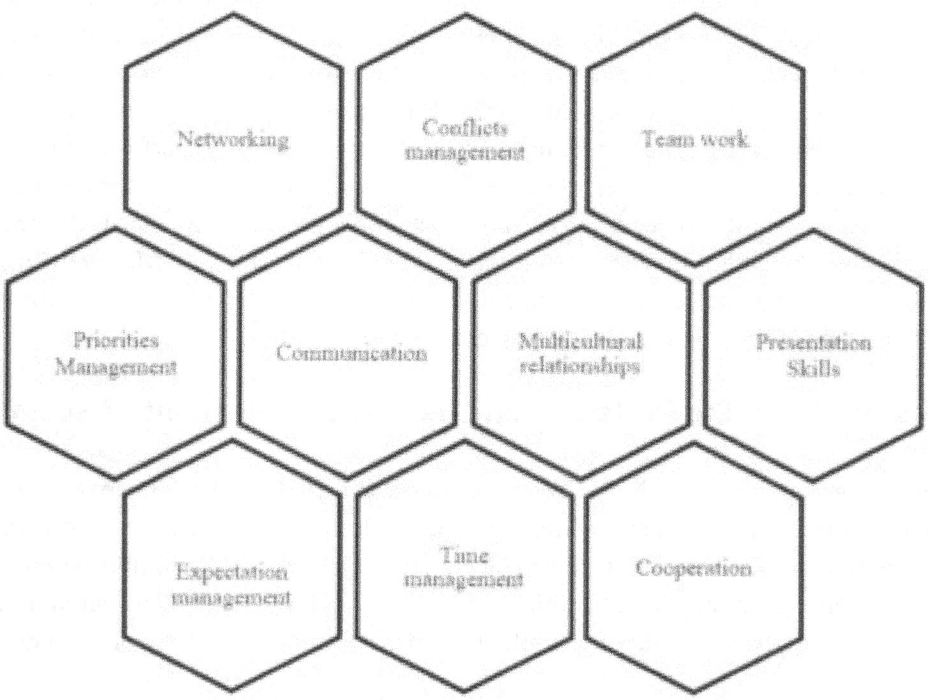

CONCLUSION

In this chapter, four PBL experiences were described by two Industrial Engineers that lived such experiences during their IEM degree. Each experience was carefully described focusing on organizational topics and team operation once that main contribution of this chapter is the influence of PBL on authors 'social skills development.

This chapter addresses important questions such as: "Is it possible to combine technical and social also called soft skills in the same learning process?" "Can PBL work as soft skills promoter in an education process?" In fact, many studies mentioned during the chapter proved that PBL has been demonstrated strength learning and educational competences to develop more prepared engineers for the industrial environment. Authors perceptions about the PBL approach are very consistent with the findings related in the extensive research done thought the years by the teachers and educational researchers from the University where they studied. Students develop active participation competences that would be useful in a professional life. Additionally, teacher's role might also change, they encourage students to learn by themselves and to be creative, guide them to achieve targets and apply technical concepts in project environment. That conditions are challenging for both: students and teachers, however this kind of learning approach demonstrates being closer to the business conditions, means an advantage for the future.

Both authors faced similar PBL experiences as students, their professional career are related different areas however authors have the same opinion about PBL contribution. From authors' point of view, PBL provides methods and tools to develop important soft skills demanded in industrial daily basis environment. Authors are also convinced that with traditional education approach, they would not have a chance to develop transversal competences that revealed being an important part of the engineer's professional skills. The key success of many projects is the great combination of technical and transversal competences of team members mentioned also by Trevelyan (2019), and PBL might work as booster of these competences in higher education. According to authors' perspective, students can be better prepared for real business environment, acquiring pro-active and agile behaviors suitable for market requirements.

As a suggestion for future improvement of PBL concept, authors highlight that it would have been interesting to have a chance to experience more multicultural and international experiences in context of PBL, besides the participation in the international conference. Maybe the integration of Erasmus students from other universities in the PBL approach projects or the creation of international projects or more frequent international conferences between IEM students in order to share their experiences would be a suggestion to challenge and improve even further the education of future engineers.

The limitations of this work are related to the sampling. The perspectives described in this chapter are based only on authors' experiences, that cannot be representative. In this way, as a future work, authors propose a deep and detailed research regarding the impact and relation between engineering career practices and PBL experiences during engineering degree.

Furthermore, it might be also interesting to understand the market perspective. Is it a value contribution for the market point of view hiring engineers with this kind of skills? It would also be interesting to understand the market performance impact of transition from traditional to non-traditional engineering education. Maybe, with the recent pandemic situation due to COVID 19 where students and teachers have been adapted to new technologies and different learning models represents another challenge for active learning approaches. This proposal represents a very challenging research due to the several factors involved and also the long period that such research might request. The work presented might be the research starting point and a valuable contribution to empirical knowledge of engineering practice.

REFERENCES

Alves, A., Moreira, F., & Sousa, R. (2007). *O papel dos tutores na aprendizagem baseada em projectos: três anos de experiência na Escola de Engenharia da Universidade do Minho*. Libr. Actas Do Congr. Int. Galego-Portugués Psicopedag.

Alves, A. C., Leão, C. P., Moreira, F., & Teixeira, S. (2018). Project-Based Learning and its Effects on Freshmen Social Skills in an Engineering Program. In Human Capital and Competences in Project Management. InTechOpen. doi:10.5772/intechopen.72054

Alves, A. C., Moreira, F., Carvalho, M. A., Oliveira, S., Malheiro, M. T., Brito, I., Leão, C. P., & Teixeira, S. (2019). Integrating Science, Technology, Engineering and Mathematics contents through PBL in an Industrial Engineering and Management first year program. *Production*, *29*(e20180111), e20180111. doi:10.1590/0103-6513.20180111

Alves, A. C., Moreira, F., Leão, C. P., & Teixeira, S. (2017). Tutoring Experiences in PBL of Industrial Engineering and Management Program: Teachers vs Students. *ASME 2017 International Mechanical Engineering Congress and Exposition Volume 5: Education and Globalization*. doi:10.1115/IMECE2017-71306

Alves, A. C., Sousa, R. M., Fernandes, S., Cardoso, E., Carvalho, M. A., Figueiredo, J., & Pereira, R. M. S. (2016). Teacher's experiences in PBL: Implications for practice. *European Journal of Engineering Education*, *41*(2), 123–141. doi:10.1080/03043797.2015.1023782

Angelva, J., Tepsa, T., & Mielikäinen, M. (2017). Team teaching experiences in engineering education a project-based learning approach. *Proceedings of the 45th SEFI Annual Conference 2017 - Education Excellence for Sustainability, SEFI 2017*, 1182–1189.

Araújo, A., & Alves, A. C. (2012). Pull System implementation through FIFO lane to achieve synchronism between lines and assembly cells. *International Conference in Industrial Engineering and Operations Management (ICIOEM2012)*, 351.

Araújo, A., Alves, A. C., & Romero, F. (2021). Barriers to Lean and Pull System implementation: a case study. *9th Manufacturing Engineering International Conference MESIC2021*.

Araújo, A., Manninen, H., & Peixoto, B. (2011). Estudo Ergonómico de um Posto de Trabalho numa Empresa de Tecelagem. *Livro de Actas Do Encontro Nacional de Engenharia e Gestão Industrial 2011 (ENEGI 2011)*, 219–220.

Araújo, A., Manninen, H., Peixoto, B., & Sarmento, A. L. (2010). Análise do Contributo de um Projecto Interdisciplinar na Aquisição de Competências em Processos de Gestão. In R. M. Lima & D. Carvalho (Eds.), *Natascha van Hattum-Janssen* (pp. 195–198). Ibero-american Simposium on Projectt Approaches in Engineering Education. https://repositorium.sdum.uminho.pt/bitstream/1822/19191/1/2010_book_ed PAEE2010_proceedings.pdf#page=139

Bonwell, C. C., & Eison, J. A. (1991). *Active Learning: Creating Excitement in the Classroom*. ASHE-ERIC Higher Education Report No 1. The George Washington University, School of Education and Human Development.

Council of the European Union. (2018). *Proposal for a Council Recommendation on Key Competences for Lifelong Learning*. Author.

Dewey, J. (1996). *Democracy and Education. An Introduction to the Philosophy of Education*. Free Press.

Fernandes, S., Flores, M. A., & Lima, R. M. (2007). Project-led education in engineering: Monitoring and assessing the learning process. *Proceedings of SEFI and IGIP Joint Annual Conference 2007: Joining Forces in Engineering Educations Towards Excellence*, 345–346.

Fernandes, S., Alves, A. C., & Uebe-Mansur, A. F. (2020). Student-Centered Assessment Practices: An Integrated Approach With Project-Based Learning (PBL). In Determining the Reliability of Online Assessment and Distance Learning (p. 340). IGI Global.

Fernandes, S., Mesquita, D., Flores, M. A., & Lima, R. M. (2014). Engaging students in learning: Findings from a study of project-led education. *European Journal of Engineering Education*, *39*(1), 55–67. doi:10.1080/03043797.2013.833170

Fernandes, S. R. G. (2014). Preparing Graduates for Professional Practice: Findings from a Case Study of Project-based Learning (PBL). *Procedia: Social and Behavioral Sciences*, *139*, 219–226. doi:10.1016/j.sbspro.2014.08.064

Frank, M., Lavy, I., & Elata, D. (2003). Implementing the project-based learning approach in an academic engineering course. *International Journal of Technology and Design Education*, *13*(3), 273–288. doi:10.1023/A:1026192113732

Freeman, S., Eddy, S. L., McDonough, M., Smith, M. K., Okoroafor, N., Jordt, H., & Wenderoth, M. P. (2014). Active learning increases student performance in science, engineering, and mathematics. *Proceedings of the National Academy of Sciences of the United States of America*, *111*(23), 8410–8415. doi:10.1073/pnas.1319030111 PMID:24821756

Gudmundsdottir, S. (2001). Narrative research on school practice. In V. Richardson (Ed.), *Fourth handbook for research on teaching* (pp. 26–240). MacMillan.

Guerra, A., Ulseth, R., & Kolmos, A. (2017). *PBL in Engineering Education* (A. Guerra, R. Ulseth, & A. Kolmos, Eds.). SensePublishers., doi:10.1007/978-94-6300-905-8

Henshaw, R. (1991). Desirable attributes for professional engineers. In J. B. Agnew & C. Creswell (Eds.), *Broadening Horizons of Engineering Education, 3rd Annual conference of Australasian Association for Engineering Education* (pp. 199–204). Academic Press.

Johnson, D. W., Johnson, R. T., & Smith, K. A. (1998a). Active Learning: Cooperation in the College Classroom (2nd ed.). Interaction Book Co.

Johnson, D. W., Johnson, R. T., & Smith, K. A. (1998b). Cooperative Learning Returns to College: What Evidence is There That it Works? *Change: The Magazine of Higher Learning*, 26–35.

Kagermann, H., Wahister, W., & Helbig, J. (2013). *Recommendations for implementing the strategic initiative INDUSTRIE 4.0: Securing the Future of German Manufacturing Industry*. Academic Press.

Kilpatrick, W. H. (1918). The project method. *Teachers College Record*, *19*(4), 319–335.

Kilpatrick, W. H. (1921). Dangers and difficulties of the project method and how to overcome them. *Teachers College Record*, *22*(4), 283–287.

Knoll, M. (1997). The project method: Its vocational education origin and international development. *Journal of Industrial Teacher Education*, *34*(3), 59–80.

Lang, J. D., Cruise, S., McVey, F. D., & McMasters, J. (1999). Industry expectations of new engineers: A survey to assist curriculum designers. *Journal of Engineering Education*, *88*(1), 43–51. doi:10.1002/j.2168-9830.1999.tb00410.x

Lima, R. M., Carvalho, D., Flores, M. A., & Van, N. (2007). A case study on project led education in engineering : Students ' and teachers ' perceptions. *European Journal of Engineering Education*, *32*(3), 337–347. doi:10.1080/03043790701278599

Lima, R. M., Dinis-Carvalho, J., Sousa, R. M., Alves, A. C., Moreira, F., Fernandes, S., & Mesquita, D. (2017). Ten years of project-based learning (PBL) in industrial engineering and management at the University of Minho. In PBL in Engineering Education: International Perspectives on Curriculum Change (pp. 33–52). Sense Publishers. doi:10.1007/978-94-6300-905-8

Mills, J. E., & Treagust, D. F. (2003). Engineering Education – Is Problem-based or Project-based Learning the Answer? *Australasian Journal of Engineering Education*. http://www.aaee.com.au/journal/2003/mills_treagust03.pdf

Moen, T. (2006). Reflections on the Narrative Research Approach. *International Journal of Qualitative Methods*, *5*(4), 56–69. doi:10.1177/160940690600500405

Moen, T., Gudmundsdottir, S., & Flem, A. (2003). Inclusive practice: A biographical approach. *Teaching and Teacher Education*, *19*(3), 359–370. doi:10.1016/S0742-051X(03)00020-9

Moreira, F., Mesquita, D., & van Hattum-Janssen, N. (2011). The importance of the project theme in Project-Based Learning: a study of student and teacher perceptions. *Proceedings of the 2011 Project Approaches in Engineering Education, 53*(9). doi:10.1017/CBO9781107415324.004

Nonaka, I., & Takeuchi, H. (1995). *The knowledge-creating Company: How Japanese companies create the dynamics of innovation*. Oxford University Press.

Powell, P., & Weenk, W. (2003). *Project-led Engineering Education*. Lemma.

Prince, M. (2004). Does active learning work? A review of the research. *Journal of Engineering Education*, *93*(3), 223–231. doi:10.1002/j.2168-9830.2004.tb00809.x

Reinalda, B., & Kulesza, E. (2006). *The Bologna Process – Harmonizing Europe's Higher Education*. Verlag Barbara Budrich. doi:10.2307/j.ctvbkk39v

Soares, F. O., Sepúlveda, M. J., Monteiro, S., Lima, R. M., & Dinis-Carvalho, J. (2013). An integrated project of entrepreneurship and innovation in engineering education. *Mechatronics*, *23*(8), 987–996. doi:10.1016/j.mechatronics.2012.08.005

Sousa, R. M., Moreira, F., & Alves, A. C. (2013). Active Learning using Physical Prototypes and Serious Games. *Proceedings of the Fifth International Symposium on Project Approaches (PAEE2013)*, 1-9.

Trevelyan, J. (2019). Transitioning to engineering practice. *European Journal of Engineering Education*, *44*(6), 821–837. doi:10.1080/03043797.2019.1681631

van Hattum-Janssen, N., & Vasconcelos, R. M. (2008). The Role of the Tutor in Project-Led Education: the development of an evaluation instrument. *International Conference on Engineering and Technology Education*, 27–31.

Wertsch, J. V. (1991). *Voices of the mind: A sociocultural approach to mediated action*. Harvard University Press.

Williams, B., Figueiredo, J., & Trevelyan, J. (2018). Engineering Practice in a Global Context. CRC Press.

Williams, B., & Carvalho, L. C. (2020). Trajectory of a fourth-generation female entrepreneur as seen through a monomyth lens – Casa Ermelinda Freitas. *Journal of Small Business and Entrepreneurship*, 1–21. doi:10.1080/08276331.2020.1789827

World Economic Forum. (2015). *The skills needed in the 21st century. New Vision for Education - Unlocking the Potential of Technology*. https://widgets.weforum.org/nve-2015/chapter1.html

Chapter 7
Project-Based Learning Application in Higher Education:
Student Experiences and Perspectives

João Eduardo Teixeira Marinho
Universidade do Minho, Portugal

Inês Rafaela Martins Freitas
Universidade do Minho, Portugal

Isabelle Batista dos Santos Leão
Universidade do Minho, Portugal

Leonor Oliveira Carvalho Sousa Pacheco
Universidade do Minho, Portugal

Margarida Pires Gonçalves
Universidade do Minho, Portugal

Maria João Carvalho Castro
Universidade do Minho, Portugal

Pedro Duarte Marinho Silva
Universidade do Minho, Portugal

Rafael José Sousa Moreira
Universidade do Minho, Portugal

ABSTRACT

Learning methodologies that are active and centred on the student are concepts that accentuate the learning process instead of the teaching process. In this way, the following chapter aims to present the application of project-based learning in higher education and the different impact it might have on the students, as well as the experience and perspective from the students' point of view on this kind of teaching approach. To be able to collect data, the authors used initially a qualitative approach to comprehend and understand the research subjects' perspectives and points of view, followed by the focus group method as a method for collecting qualitative data. Throughout the students' experience, the development of transversal competences is mentioned several times as a great aspect related to this kind of methodology. However, the need for effective management of conflicts between members and the mandatory integration of some subjects' contents are also mentioned as some less-positive aspects.

DOI: 10.4018/978-1-7998-8816-1.ch007

INTRODUCTION

The constant changes in technology and the globalization of markets requires skills, attitudes, and different learning behaviours from all the students in higher education. To prepare the students for the challenges that lie ahead of them in today's world, universities confront them with real-life problems and assessments during their studies. When the students are confronted with real-life problems, they can seek help from their teachers to develop their knowledge or skills, but mostly, students only need to believe that they can solve the proposed assessments, even the most complicated ones.

Furthermore, with this approach to better prepare the students, they can also learn how to split their time between the task that they are supposed to perform and their personal life. However, it should be noted that this kind of task should be aligned with the knowledge and ability of the workgroup as a very difficult task could be impossible to solve due to the lack of specific knowledge, even when the will and effort, from the workgroup, to solve the problem is remarkable.

In this way, this chapter reports the experience of a group of fourth-year students of the Industrial Engineering and Management Master's Degree, who experienced the application of Project-Based Learning (PBL) methodology in the execution of projects throughout their learning process.

The chapter is structured in five main sections. After a brief introduction, it is presented the active methodology PBL through a literature review. Then, it is explained the Focus Group methodology used for collecting the data. Finally, the students' point of view related to the PBL experiences, including the difficulties, feelings experienced, and skills learned. In the discussion section, it is discussed how these aspects influenced personal development and will impact their professional role as future engineers. Finally, the conclusion section presents some final concluding remarks.

LITERATURE REVIEW

Learning methodologies that are active and centered on the student are concepts that accentuate the learning process instead of the teaching process. Active learning is defined by Bonwell & Eison (1991) as "instructional activities involving students in doing things and thinking about what they are doing". These types of activities create excitement in the classroom to make learning something natural for the students (Alves *et al.*, 2018). Several authors, in particular Vygotsky (1986), defend the theory of social constructivism, where the ground idea is that concepts and the construction of meaning are learned by students interacting together. On the other hand, Kolmos (1996) also argues that learning is an active process of investigation, and is an outcome of the interest, curiosity, and experience of the learner (Alves *et al.*, 2019).

PBL is an example of this kind of methodology. PBL is a methodology that has been used for years. It has its roots in learning-by-doing ideas that were developed in the first half of the 20th century, initially by Kilpatrick in 1918 and Dewey in 1996 (Lima *et al.*, 2012). According to Dewey, learn from experience means "to make backward and forward connections between what we do to things and what we enjoy or suffer from things in consequence. Under such conditions, doing becomes trying; and experiment with the world to find out what it is like; the undergoing becomes instruction-discovery of the connection of things" (Dewey, 1916).

Another definition for PBL was given by Adderley in (1975). According to them, PBL has five different phases/aspects: (i) resolution of a problem that can be proposed by the students themselves, though it is not required to be; (ii) initiative to solve the problem comes from the students and requires integration of a range of educational activities; (iii) delivery of a final product, coherent with the initial problem; (iv) the solution for the problem will usually be handled as a project and, most of the time, it is time-consuming; (v) changing the role of the instructor from an authoritarian position to a consultant position (Pereira *et al.*, 2017).

More recently, Thomas (2000) described what was essential for a project to be considered an example of PBL and gave five different criteria that must be respected (i) centralization – fundamental strategy to the development of activities; (ii) existence of a "driving" question; (iii) the necessity of a constructive investigation – by the students; (iv) incentive to autonomy; (v) realism – focusing on problems related to students' reality (Pereira *et al.*, 2017).

This type of educational project was first applied in an educational context in the sixteenth century as reported by Knoll (1997). However, it was first in the 20th century that this methodology established its roots in the educational community, especially in higher education, in particular in universities in Denmark. Specifically, in engineering courses, PBL projects normally include Science, Technology, Engineering and Mathematics (STEM) courses. When students solve a problem in a PBL context, they become more aware of how the real world connects with the STEM thematic, through learning how to solve problems with the STEM knowledge. Approaching the problem with this strategy, the students can apply theories, formulas and principles, and therefore understand the contexts in which they can be applied (Alves *et al.*, 2019).

PBL is a student-centred active methodology that allows the development of aptitudes, skills and mindsets. For the teachers, this methodology represents leaving behind their specific knowledge in their disciplines as well as traditional teaching methods, in favour of an open mind and promotion of team-work. This skill is essential, for both teachers and students, since collaborative work is a competence that potentially allows easier insertion into the labour market (Cargnin-Stieler *et al.*, 2019).

PBL aims to engage students in acquiring knowledge and skills through real-world experiences and well-planned activities. Usually, students are distributed, forming teams of about ten elements, and work together to solve case studies or even concrete problems. The main purpose is to apply theoretical concepts to practice, placing the student as the main actor of the teaching-learning process.

A PBL project is built to be an integrated project and, to be well-accomplished, students must use and relate the content learned throughout the course and the various subjects. Besides the importance of integrating different expertise, like any other project, the organization is the key to success. The students need to delegate tasks for each group member, define deadlines and assign responsibilities so that in the end, it is possible to integrate the different parts into the final project (Pereira *et al.*, 2017).

Nowadays, knowledge and skills are being replaced by the concept of competencies. This concept is much more than pure knowledge and skills as it also includes attitudes (Rychen & Salganik, 2000). In this concept, knowledge is described as facts, concepts, ideas, and theories previously recognized, reinforced and comprehended by a certain area or subject. As for the skills, they are described as the ability and competence to carry out procedures and use the prevailing knowledge to accomplish the desired results. Lastly, the attitudes characterize the disposition and mindsets to act or react to ideas, individuals, or circumstances. In the 21st century, the most desired competencies are critical thinking/problem solving, creativity, communication and collaboration (World Economic Forum, 2015). On this note, to develop this kind of competency, it is suggested to use learning methodologies such as Project-Based Learning (Alves *et al.*, 2020).

According to Dewey (1996) and Kilpatrick (1918), the PBL methodology is a suitable method that prepares independent, autonomous and trustworthy citizens for their future social behaviour. For these authors, the morals are learned through the students' self-made path to their knowledge, through a combination of theoretical learning and its application leading to a meaningful education (Alves *et al.*, 2018).

Through PBL, students acquire skills of various types, such as problem-solving ability, oral communication, written communication, teamwork, self-directed learning skills and critical thinking. While developing their projects, students are also exposed to many aspects that don't happen in theoretical classes or appear in books, which brings them closer to real-life expectations. It is also important to enhance that this type of learning environment allows students to improve their knowledge at their own pace (Duch *et al.*, 2001; Frank *et al.*, 2003; Prince & Felder, 2006).

For some authors such as Costa (2007) and Mills & Treagust (2003), the labour market is becoming increasingly competitive and demanding extraordinary professional skills, so knowledge alone is not enough and differentiating. Thus, teaching through PBL provides many benefits for students and improves their academic development. These authors emphasize the following benefits for students: "they do not only gain knowledge, but they learn to do a project; they practice their skills and acquire others; they know how to behave in a group; they gain as practical activity, as it approaches those of their profession". In addition, the authors propose that "projects, whenever possible, should involve the university and the communities in surroundings; should evaluate students based on the reality that they will find in the labour market, and should increase communication and unity within the classrooms" (Pereira *et al.*, 2017).

Although there are many advantages of using PBL, when this type of teaching is compared to traditional teaching exclusively based on lectures, there are some disadvantages pointed out by students, as reported by Alves *et al.* (2020). These disadvantages are associated with the intricacy of the assessment and how it is understood by the students. The workload is also pointed out as a problem, especially how unbalanced it becomes to the members of the team, as well as the knowledge acquired between all the members. Finally, the students also mention how difficult it is to integrate all the different contents of the different subjects to develop the proposed project (Alves *et al.*, 2020).

MATERIALS AND METHODS

This chapter aims to collect and analyze students' experiences with PBL methodology. Thus, it was decided to resort to qualitative research for data collection. The qualitative approach allows the interpretation and comprehension of reality as it is since it essentially privileges the understanding of a given subject from the research subjects' perspective and points of view (Ribeiro, 2008).

It was then decided to resort to the Focus Group method, which is an investigative method for qualitatively collecting data and interpret the nature of the speech. This method can be applied at any stage of the investigation and can even be combined with other techniques.

Focus Group consists of a group discussion, concerning the theme covered by the research, to obtain information that could scarcely be obtained through interviews with only one subject (Sá *et al.*, 2021).

Due to its group nature, Focus Group may create both complimentary and contradictory ideas among the various subjects, resulting in a collective opinion or divergent opinions. It stands out from the interview survey due to the interaction between different participants.

This method was chosen as the basis for this study due to its qualitative nature and, in addition, because it helps to attain results that go beyond words, that can be translated into ideas, developed around the topics addressed. However, considering the present type of study and the results that were intended to obtain, it was decided to hold a single session, between thirty minutes and one hour, rather than multiple longer sessions.

Like other data collection techniques, Focus Group is also governed by a sequential set of steps, that include planning, preparation, moderation, data analysis and presentation of results, which will be described briefly in the following subsections (Sá *et al.*, 2021).

1. Planning

In this phase, crucial aspects leading up to the preparation and moderation of the session were defined. Here the goals of the study were defined, the main one being to analyze and discuss the experiences of students in PBL and how it affects learning in higher education.

Moreover, in this phase, the script to be used in the moderation phase was prepared, whose issues narrowed through the progression of the moderation, gradually deepening the theme. The questions were drafted in such a way as to allow the collection of the desired data, considering that the following research questions were intended to be answered:

a. What are students' perceptions of the advantages of implementing PBL?
b. How do students see PBL as a teaching methodology, and what is its contribution to student learning?
c. What are students' perceptions on the challenges of implementing PBL for higher education, and how could it be improved in the future?

It should be noted that the topics covered in the script intended to collect students' opinions about their collective and individual experience with PBL methodology, hence not following a specific theoretical framework.

In addition, at this stage was also defined the number of participants in the session, which would be eight. To create a group with a certain degree of homogeneity and in congruence with the goals of the Focus Group, it was defined that the participants should be current students of the Industrial Engineering and Management Master's Degree at the university where the study for this chapter was conducted and that they had to have experienced at least two projects in PBL context, between September 2016 and June 2021. It should be noted that, when choosing the eight students, it was considered the fact that they all belonged to the same workgroup in at least one of the projects, having therefore lived the same experiences. In this way, the Focus Group session would become more casual and pleasant, with positive interactions between members. Furthermore, the classifications in the developed projects were also considered in the students' selection.

2. Preparation

In the preparation phase, the eight participants were selected. These met the criteria established in the planning phase. In addition, the location and date of the session were defined.

Moreover, the objectives, specificities and rules of the session were communicated and explained to the participants by the moderator.

Since it was an informal session, the moderator chosen to conduct it was another student of the same master's degree and was not one of the participants selected.

3. Moderation

The moderation phase is where the meeting between the moderator and the participants takes place to streamline the session. A particularity of the Focus Group is the presence of a moderator and an observer. However, considering the informal nature of the meeting, there was no need for an observer. Though, audio recordings and a collection of notes by the moderator were used to proceed with the collection and conservation of the meeting.

Thus, the moderator was responsible for conducting the conversation according to the previously prepared script to avoid possible deviations from the central topic of the conversation. His role was to promote and facilitate dialogue and interaction between participants, to collect data for the study, through questioning the students. It should be noted that the moderator was prepared with relevant tips, particularly regarding impartiality, which was followed throughout the session.

This phase was essentially crucial to gather opinions from the group members regarding their experience with PBL methodology, through a healthy and uninhibited interaction between them. However, and as mentioned above, Focus Group can be combined with other methods, for which an interview survey was also used. This method aimed to collect individual testimonies from students anonymously and without interference from the opinions of the other colleagues. It should be noted that the individual interview moments were brief, as the goal was to obtain a summary of what the experience with PBL was and not go into details, as this was achieved with Focus Group.

4. Data analysis

The next phase was data analysis. At this stage, a transcription of the recording was prepared, from which all relevant information for the study was extracted. The information collected originated an aggregate of the aspects raised by the students when in groups. Also, individual testimonies were obtained from the interviews.

It was concluded that the opinions of the elements resulted in a collective opinion, as they complimented each other, presenting limited divergence. This may have been due to the positive dynamics of the group, since all elements knew each other and all assumed an equally active role, with no dominant or insecure members.

5. Presentation of results

The presentation of results was the last phase of the implementation of the Focus Group method. At this stage, the results of the moderation were transmitted to the participants, through an informal report.

It should be noted that, with the results provided by the application of this method, it was intended to get to know the profile of the students who experienced PBL, their views and experiences about this type of learning, as well as their difficulties and perspectives of future developments in this context.

PBL EXPERIENCES

The following section reports the experience of the group of eight students that participated in the Focus Group mentioned in the previous section. Therefore, it is important to note that all the conclusions that follow were drawn from the application of the Focus Group.

The confrontation with PBL methodology occurred in three different moments of the academic path of these students: in the first semester of the first year and, consecutively, in two semesters of a more advanced phase of the programme. Although the method used in the three projects developed was the same, all of them required and provided very different soft skills and feelings. Considering the duration of the work done, these quickly became the moments of greatest tension in the students' lives, mixed with the companionship, dynamics, and fun that all the meetings provided. Although the tension was constant throughout the months of the development of the projects, in the end, the pleasure of delivering an intense, laborious, tiring and extremely demanding project, both intellectually and emotionally, overcame any negative moments along the way.

Analysing the three moments mentioned above, it is clear how one is influenced by different experiences and how the management of the various aspects and stages evolves from project to project.

1. First year, first semester project – first impact with PBL methodology (2017)

The first project was developed in the context of Project Integrated of Industrial Engineering and Management course of the first year, first semester and tested, above all, the creative and organisational capacity of the whole group, while at the same time demanding the integration of the theory learned in all the subjects of this particular year: Linear Algebra, Algorithms and Programming, Mathematical Calculation, Industrial Engineering and Management Introduction and General Chemistry.

On a sentimental level, it is impossible not to mention that it was the most turbulent. The lack of experience, the fear of failure and rejection, or the fear of not meeting expectations lived side by side with the enthusiasm of the first contact with the world of engineering and the first simulation of the profession to exercise.

Additionally, the initial lack of knowledge of the working group and of each members' way of working contributed to the slow progress that needed to be adapted according to the perception of each one's potentialities. In this project, the theme was "Valorisation of Forest Biomass". The goal was to use this energy/renewable resource and transform it into a useful product, designing the production system to produce the product. In this way, the aim was to make new aspiring engineers aware of the need to look into the future of the career, taking into account the current environmental panorama, while at the same time building a business from the start.

Since it required the creation of a fictitious company and product, from the choice of the name to the sale of the final product, it forced the students to reconcile all their imaginative capacity with the speed of execution, allowing them to adapt to the frenetic pace at which the technological and industrial world currently lives. On the other hand, the lack of technical and specific knowledge in important areas took some of the rigour out of the final result, constituting, however, in an enormous gain of experience for the projects that followed.

Although the technical skills suffered a huge buff during this path, the student's best enhancement occurred at the level of interpersonal relationships. Examples are the management of conflict situations, caused, for example, by the divergence of opinions and ideas, confrontation of postures and attitudes, divergence of individual goals, or lack of communication within the team. This was also the first impact with peer assessment, which promoted the evaluation within the members of the team.

The following images portrait one of the most important deliverables of this first project – the layout of the fictitious factory. The first image (see Figure 1) refers to the idealized layout and the second one (see Figure 2) refers to its prototype representation.

Figure 1. Idealized layout of the fictitious factory

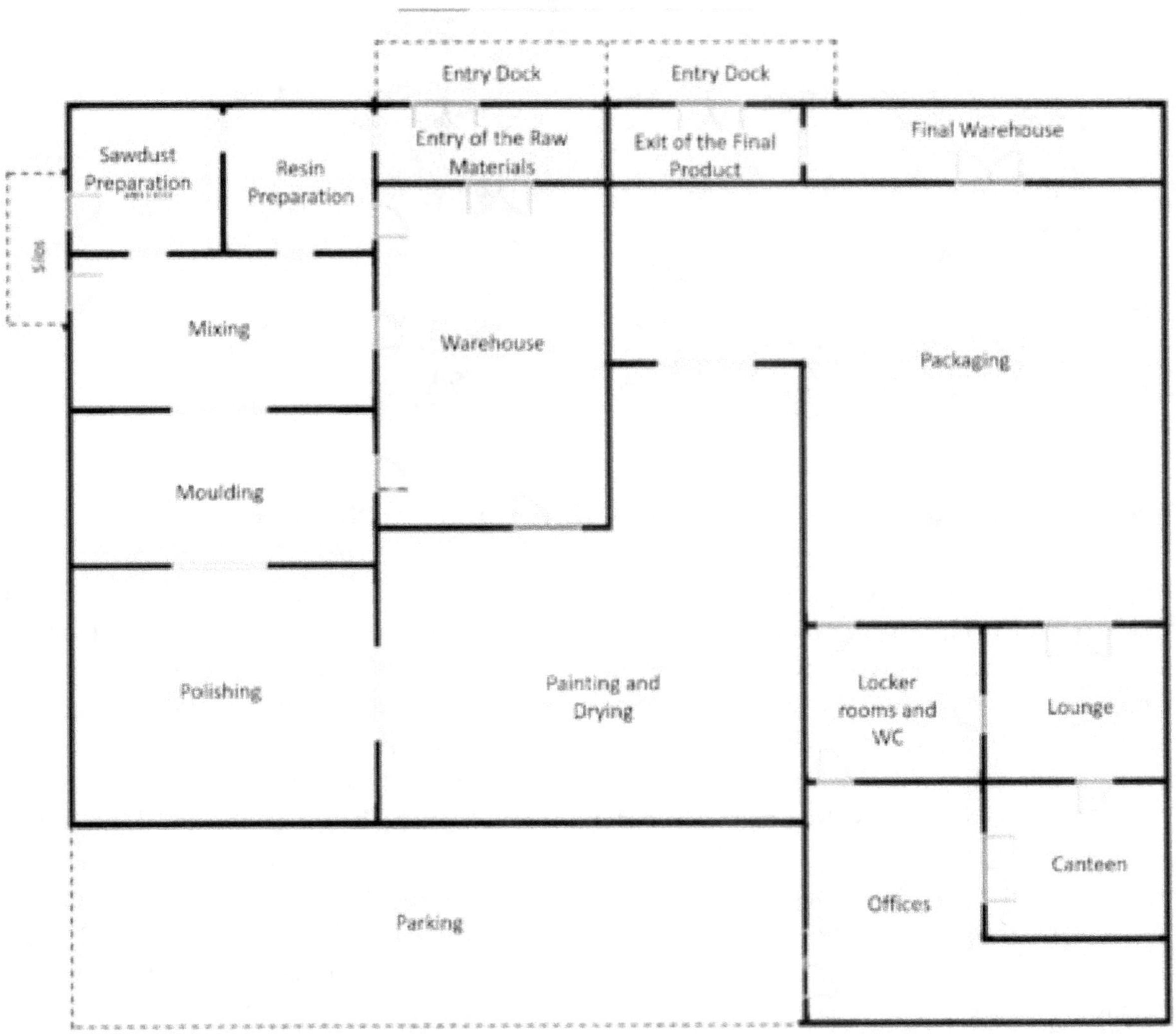

Figure 2. Prototype representation of the layout

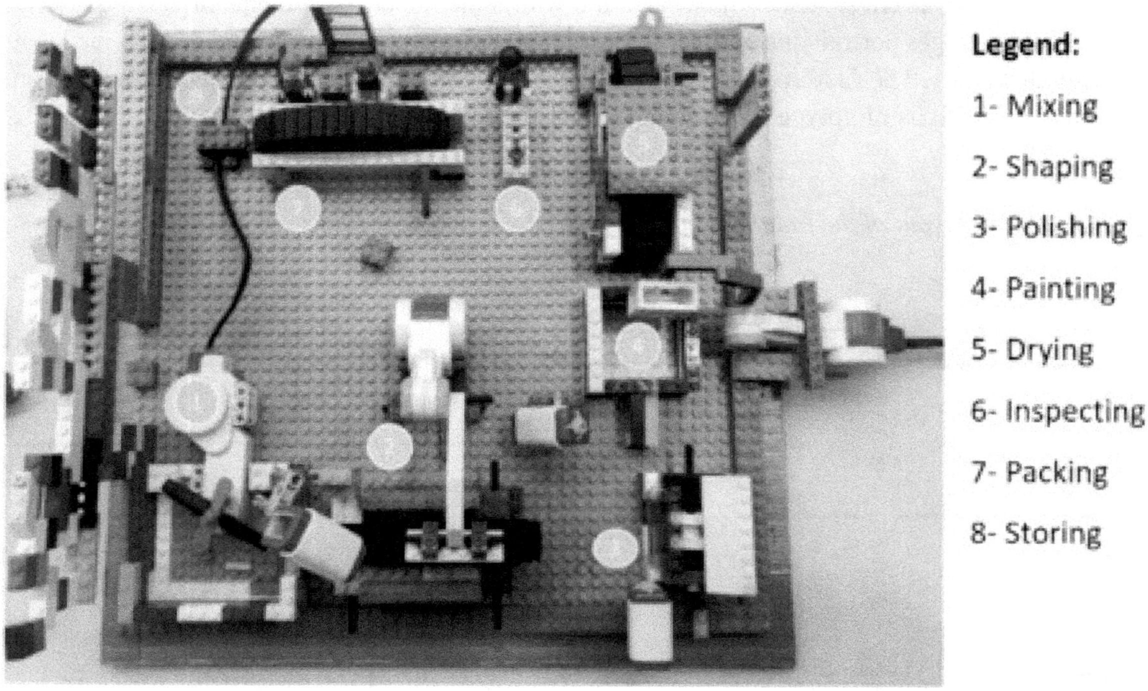

2. Fourth year, first semester project – first contact with a real company (2020)

The second moment of the PBL method emerged at an advanced stage of the programme, also in the context of Project Integrated of Industrial Engineering and Management, specifically in the first semester of the fourth year and proved to be totally different from the first project. It involved the subjects of Ergonomic Study of Workstations, Integrated Production Management, Simulation, Information Systems for Production and Organization of Production Systems II. Instead of creating a fictitious company and all the stages of creating a product, the group of students was integrated into already established and well-solidified company, that was also a market leader and a reference at both national and international level.

Given the need to show concrete and reliable results, the organisation of the whole project was very meticulous and required maximum commitment from all members of the group. For the students this was, perhaps, the greatest difficulty of the project. The ability to keep all elements of the team focused for a long time on a common goal and synchronized in the execution of tasks and data collection, was something that the technical knowledge that education provides does not promote.

Thus, some of the subjects learned throughout the year, although valuable in solving the problems encountered, were not so useful in interpersonal relationships. Still, many of them did not prove to be adequate to the problem encountered, even if their integration in the project was mandatory. Besides this personal resources' management, which proved to be difficult, the approaches and contacts with company collaborators of the most various hierarchies proved to be another complicated aspect, given the group's lack of experience in the labour market. Thus, this was a project that demanded a lot of technical knowledge, but where the decisive and most enhanced factors along the way were the soft skills such as discipline, self-management, leadership, and creativity. Despite this, the fact that it was carried out on the Gemba overrides any negative issues that may have occurred, as it allowed contact with the evolutions that modern engineering has undergone in recent times and confrontation with the later challenges that the profession of Engineer will require.

This project culminated in the realization of an article published and presented at the IFIP International Conference on Advances in Production Management Systems (Freitas I. *et al.*, 2021). In this paper, the company's difficulties are mentioned, and solutions to possibly overcome them are proposed. One of these suggestions is a milk-run where different boxes are left at each workstation with the components that will be applied there (see Figure 3).

Figure 3. Prototype of the proposed milk-run

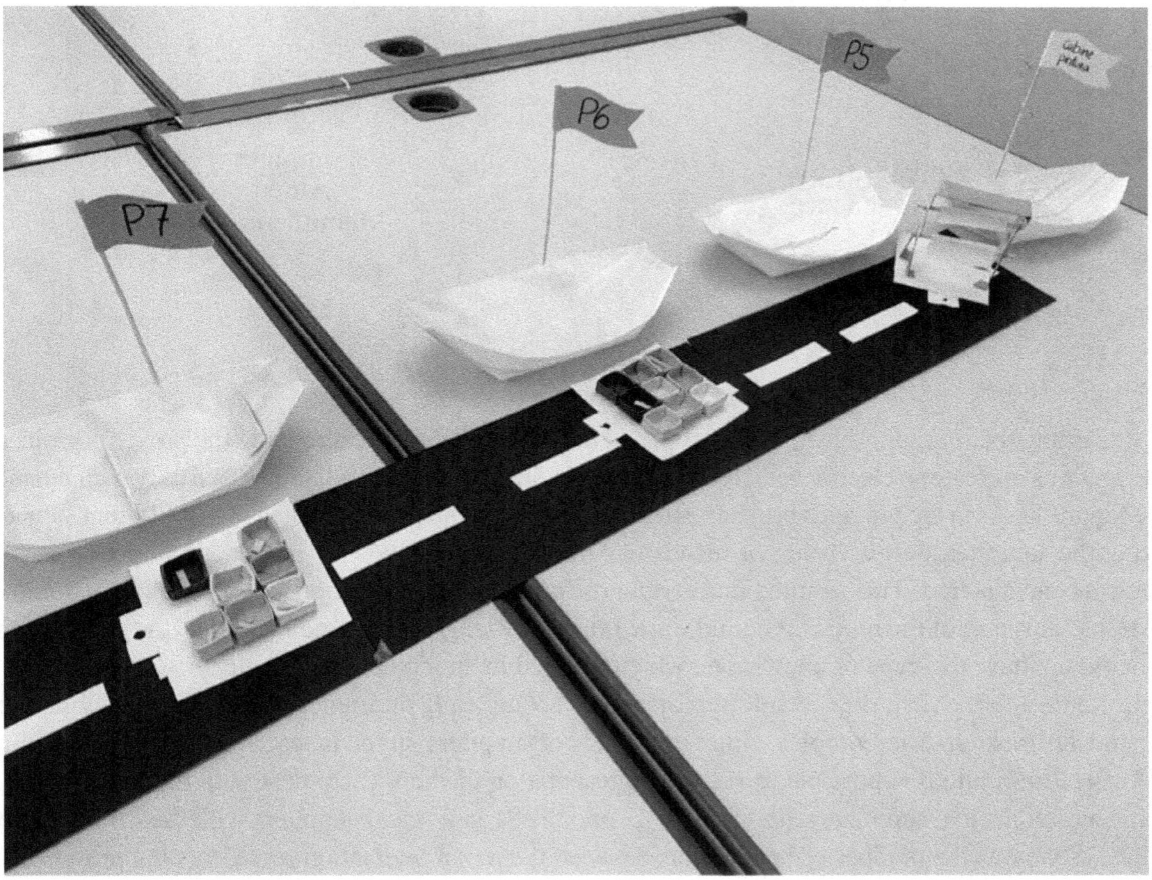

3. Fourth year, second semester project – advanced concepts (2021)

Finally, the third and last project executed was quite different from the previous ones, as it was divided and approached by the students in two distinct subprojects, which were performed simultaneously. Instead of one project that included all courses of the particular year, in this case two subprojects were developed involving two courses in each (Graphic 1).

Figure 4. Courses involved in each subproject

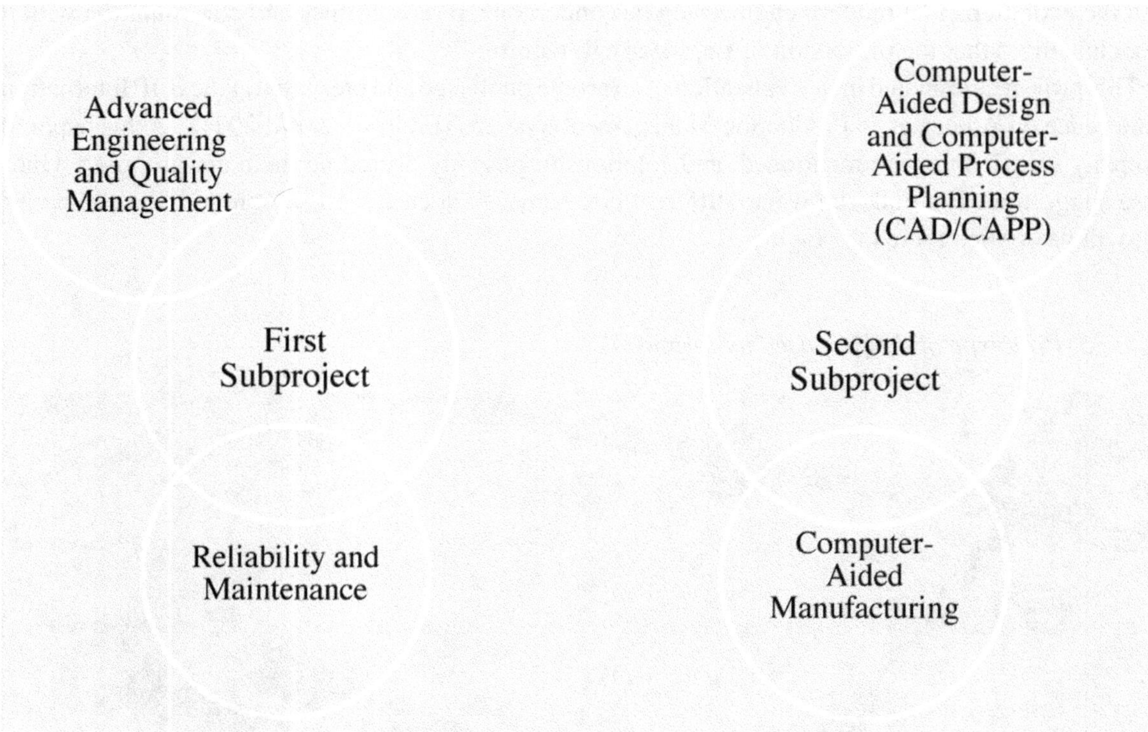

Regarding the first subproject, the concept and the organisation were similar to the previous projects. The goal was to create an electric vehicle from scratch, with high quality, reliability, and easy maintenance (see Figure 4). As in the previous projects, sometimes some of the theory learned in the different subjects during the semester, despite being of mandatory integration, was not the most appropriate to certain problems encountered. This resulted in a certain difficulty in the realisation and perception of some of the tasks. This part of the project required some prior knowledge of the product design. As students, the team did not have this type of knowledge, which resulted in an increased difficulty.

In contrast to the first, the second subproject had a completely different dynamic. In this subproject, the students took on different roles, either customers or suppliers, in the latter case, both individual and collective. In Figure 5 is possible to see the representation of the 9 customers with the letter C, the 9 companies/collective suppliers with the letter E, and the 91 individual suppliers with the denomination of FIS, as well as the number of interactions between them. An interaction represents the provision of a service to a customer by a supplier, either individual or collective.

The differentiating factor about this project was the demonstration of market competitiveness in the success of a company. Although it required, similarly to the two previously mentioned, technical knowledge in very specific areas, the goal was autonomous learning. Thus, the expectation was that each of the students would acquire knowledge and search for information in various areas of engineering and, later, sell their knowledge in the form of a product to a client that would request a particular service.

Although it had a slow start, given the confrontation with a teaching methodology with which the students were not familiar, it quickly gained a dynamic that came very close to the movement of the services market. This subproject contributed in a very positive way to personal growth, as it required students to negotiate the value of their knowledge. This situation required self-awareness of the students that were part of the project, in terms of their learning process and degree of knowledge required for the requested services. In other words, it confronted the students with the famous question "What is my monetary value?" which is one of the great paradigms of a new graduate entering the labour market.

Figure 5. Prototype of the electric vehicle designed by the team for the first subproject

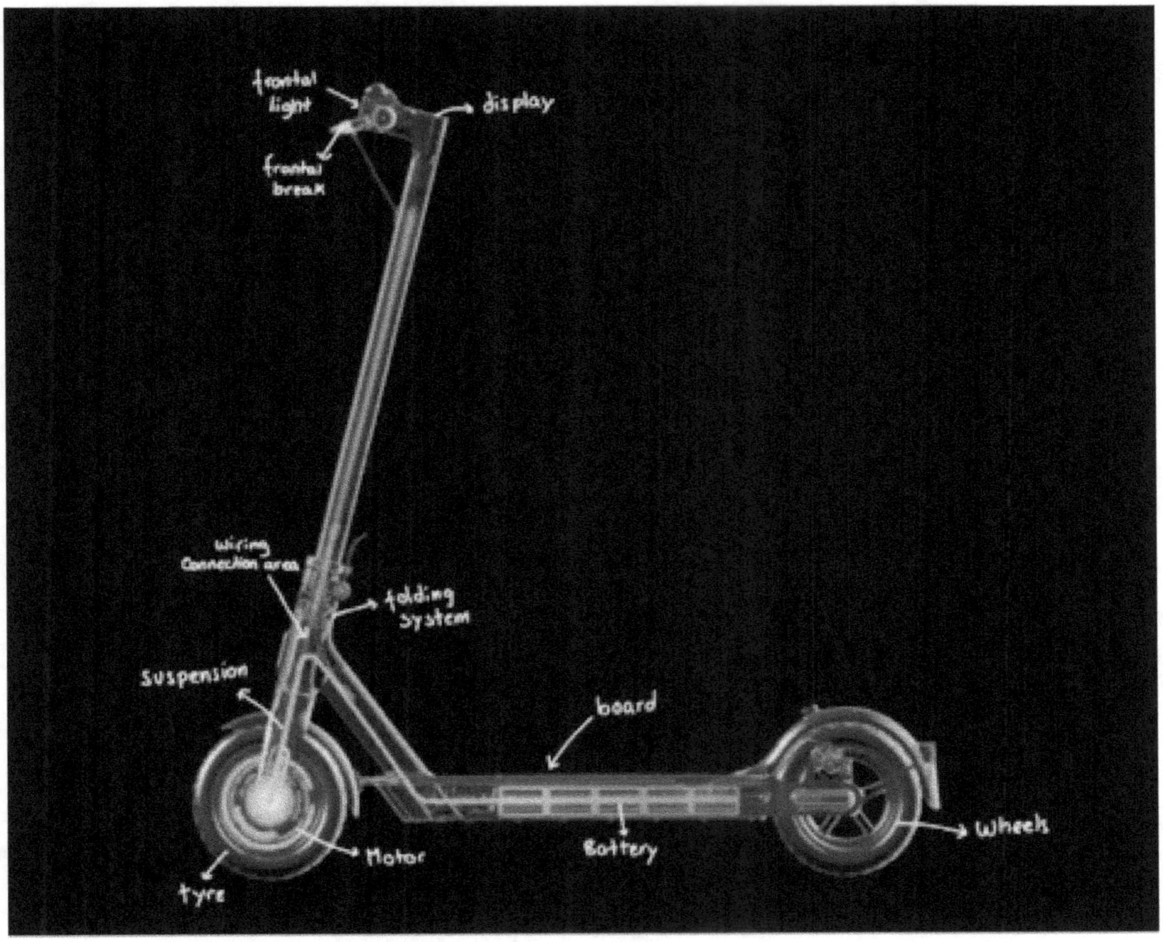

The fast dynamic and the pace at which services were requested led to the tiredness of the students, who needed to organise their discipline and self-management to achieve the intended goals. The big difference to the other projects was that the work was carried out mostly individually and not in groups, which, although demanded more from each student, took away the interaction and the socialisation aspects which were a constant in the other projects.

An important point to mention is the fact that these two subprojects were being executed at the same time. Therefore, it was at times difficult to organise the team and all its elements, in terms of tasks' allocation for each student, and consequently, to keep everyone focused and aligned towards a common goal.

Figure 6. Representation of the crowdsourcing context of the second subproject

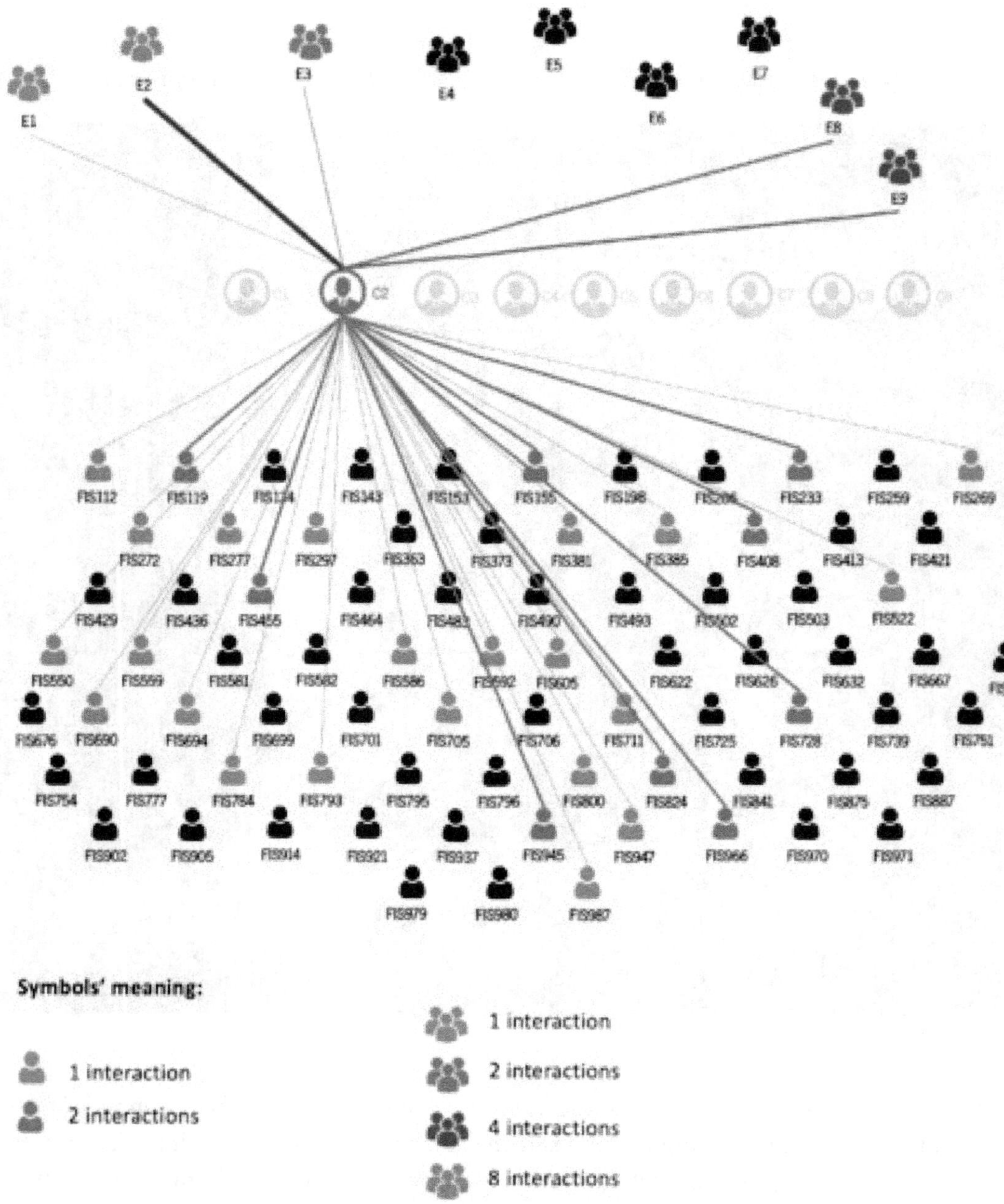

The evaluation of each one of the three projects was carried out both by teachers and students, who used the peer assessment method. While the teachers' evaluation was done traditionally, i.e., evaluation of all the reports made and the respective presentations, the peer evaluation consisted of a ready to fill form. This form should be filled by each student, who evaluated the commitment, dynamics, originality, autonomy, or even the punctuality of the other members of the team, assigning values from 0 to 1. However, in the second subproject of the third project (fourth year, second semester project), a traditional evaluation was adopted by the teachers, thus evaluating both the tasks developed by each group that represented a company and by each supplier, as well as a presentation made by each client.

The importance of the introduction of the peer evaluation was not the differentiation of the students, but the self-awareness of each of the members. Since the peer was completed more than once, it allowed everyone to correct their behaviour after being confronted with their peers' evaluations. This is one of the most important elements of human development that is accomplished through this type of project. Usually entering the academic and university world translates into an enormous inner growth, but moments like this allow the students to smooth any edges that may prove to be more persistent and stubborn.

RESULTS

For a better understanding of the impact of this methodology, eight testimonies of the students who participated in the Focus Group and experienced PBL methodology, are presented below. The aim is to describe the experience from the students' point of view, explaining the difficulties, the feelings experienced, the skills learned, and how these aspects influenced personal development and will impact the professional role of the students as future engineers. In other words, it will be discussed the need for the implementation of methodologies such as PBL and how this will bring competitive advantages in the labour market to the students when adopted by educational establishments. The testimonies are:

- **Testimony 1:** "For me, the highlight of my academic student journey coincided with the development of the projects where the PBL methodology was implemented. The elaboration of each of the projects allowed an enormous growth, both professionally and socially, enhancing my strengths, such as conflict management and responsibility, and developing new characteristics, such as autonomy and search for solutions, which were being demanded as the projects went on."
- **Testimony 2:** "Unlike all the other members of the group, my experience in carrying out this type of project was almost null since I had done my first year at a different university and, therefore, the project carried out by me had nothing to do with the rest of my colleagues. However, although there were some doubts and difficulties at the beginning of each project, they allowed me to grow as an engineer both technically and socially, as it allowed me to understand the daily life in the real world of a company and, at the same time, to work in a team in which the supervisor was included."
- **Testimony 3:** "PBL was a very valuable experience because it allowed me to apply theoretical concepts to a real context, which will certainly be very useful for my professional future. This type of learning environment helped me to develop soft skills and brought me closer to real-life expectations. I think that despite being a challenging project, this type of methodology brings truly rewarding results."
- **Testimony 4:** "The PBL methodology, present in three different moments of my academic career, was of great importance since it allowed me to acquire transversal skills, such as critical thinking and group work, that I believe will be an asset in my professional future. Although these were moments of great demand, where challenges and tension were constantly present, the final results were undoubtedly rewarding, not only in terms of deliverables related to the project but also in what concerns the gained skills and friends that will unquestionably continue to accompany me."
- **Testimony 5:** "As a student who experienced three different applications of the PBL methodology, I can say that it helped me develop a wide variety of skills, such as critical thinking and communication.

Regarding more technical aspects, I can say that the application of this methodology motivated me to learn the contents taught in the different curricular units since I knew that they would later be applied in the project, and it also helped me consolidating those same contents. Thus, I consider this methodology to be very advantageous not only for academic formation but also for future similar experiences at work."

- **Testimony 6:** "For me, PBL was the most immersive experience of this course and takes learning to a higher level. Learning the contents and applying them in parallel, in a practical context, is the greatest asset of this learning methodology. Also, through teamwork, high-value transversal skills are obtained from each member of the team and each of them can offer a different set of skills to the work being done."
- **Testimony 7:** "As a fourth-year student and future Industrial and Management Engineer, I consider that the PBL played an essential role, not only about the learning and consolidation of the content taught in the different curricular units but also at a personal level, namely in the promotion of teamwork, in the ability to resolve conflicts and face unexpected situations. In this sense, I believe that this methodology is fundamental to prepare and train better professionals, able to integrate the demanding markets today."
- **Testimony 8: "**For me, as a student, I experienced the active learning methodology PBL in three different times of my academic course. All experiences culminated in the acquisition of transversal skills, such critical thinking, group work and complex information processing, amongst others. Through this methodology I had the opportunity to apply more theoretically aspects into a real industrial problem in a company, making me realize how useful the theory will be in my future professional life. Even though the three different projects had a different impact in me as person given the different years of the course that they were developed, I can firmly say that they have helped me to develop my professional and personal side."

It is important to mention how the human side of each student changes. The students have already discussed the soft qualities and technical skills and how important these aspects will be in their future, but there is no greater appreciation than that which occurs in the emotional aspect of the student. The projects described above accounted for 18 months of work with people whose ages are between 18 and 22. The development of these projects involved many hours of work, socialisation with hundreds of people and many off-topic conversations throughout this time frame. The impact of all these factors on the socialisation process is enormous for teachers and students. Everyone becomes a better version of themselves, with even more enhanced awareness and the ability to realise that the profession, solutions, and Engineering cannot be performed without people.

After presenting the different gathered opinions, it is time to discuss and answer to the questions raised in the materials and methods section. The following sections present such discussion.

- **What are students' perceptions of the advantages of implementing PBL?**

The great advantage of having projects following the PBL methodology lies in two fundamental aspects. The first is related to the fact that it brings a broad perspective, since the students encounter the most varied areas, acquiring technical knowledge that will influence the professionals that they will become. In addition to this specific and technical component, the development of soft skills for the industrial world, personal growth, or the capacity to integrate into multidisciplinary teams are highly potentiated. The second aspect is how different it is from the traditional teaching methodology, based on memorization, which in no way resembles, for example, noble engineering.

Looking specifically at the professional world of Engineering, companies and organizations nowadays are increasingly aware of the need to form multidisciplinary teams in internal projects to solve problems. That is achieved by integrating elements from the most varied areas to obtain consistent and solid solutions that take into account all the aspects inherent to the problem, from the most obvious to the most hidden. Additionally, and even though the wheel has already been invented, even if the problem has clear similarities with other problems that have already occurred internally or externally, the context of the problem's occurrence demands that the team outs their creativity to use and build an appropriate solution to the context, considering all the conditioning factors of the situation. These characteristics of the current Engineering panorama are related to what PBL tries to develop: enhancement of transversal qualities and a marked development of critical thinking, and the search for innovative solutions.

In addition to the aforementioned, it is essential to note that Engineering has always been seen as a synonym for development and it is one of the most sought-after careers by students who are starting their academic life. As such, it is necessary to look critically at the way of teaching and training Engineers, because while it is true that students are required to be increasingly innovative, creative, and alert to the threats of the current world, it is also necessary that teaching practices are directed at promoting these qualities. Moreover, as in any profession, the study phase is everlasting, as in all areas closely related to technology. Furthermore, it is important to ingrain in students from an early age the ability to search relentlessly in all possible information sources, without any room for sluggishness and procrastination.

- **How do students see PBL as a teaching methodology and what is its contribution to student learning?**

As can be seen from the different testimonies, the opinion is unanimous regarding the impact of this methodology: there is a direct relationship between PBL, the enhancement of various transversal skills and the technical knowledge required to solve problems. Although access to the internet is currently an enormous competitive advantage compared to the previous years' education, critical thinking, creativity, and thinking outside the box have stagnated. The PBL methodology is the key that will allow aligning the high accessibility currently available with the constant promotion of student creativity. This fact is achieved through the presentation of current problems, aimed at the issues that are of most concern nowadays and that are expected to be of most concern in the future, such as the environmental aspect.

Thus, by requiring autonomous learning and a relentless search for what they want, it makes students focus on perception rather than memorization, allowing them to think outside the box and be able to better solve unexpected problems.

In other words, PBL methodology grants very tenuous guidelines. However, there is a "controlled freedom" since it requires that the search for solutions is done autonomously, in a team and in a self-taught manner, but also takes into account the technical knowledge learned. These two antitheses perfectly represent the amplitude that a methodology of this nature has in the promotion of varied skills that will play a significant role in the success of the student, engineer, company, and future world.

- **What are students' perceptions on the challenges of implementing PBL for higher education, and how could it be improved in the future?**

To guarantee the success of this methodology and, consequently, the student's learning, it is also important to mention the importance of similar experiences and models being implemented in preuniversity study cycles. That exercise would prepare and empower students to integrate more complex academic projects and experiences, and to face the adversities and obstacles of the labour market.

Furthermore, for the adequate functioning of the project, it is fundamental that the institutions can provide working spaces and materials to the project groups. Those spaces are crucial so that the groups can organize themselves, discuss ideas and work.

Finally, it is also of utmost importance that, in the case of projects developed in companies, that these are aligned with the project's goals. This way students can do a good job that benefits them as well as the institutions involved.

CONCLUSION

Given the feedback gathered from the students, reflected in the testimonies presented above, it can be stated that all the students that participated in the mentioned projects felt that the course of these projects brought several advantages. On the one hand, it enabled the development of transversal competencies, which otherwise would not have been stimulated. An example of this is the increased ability to work in a team, by accepting and valuing the opinions of all members of the group. Another competence acquired was the achievement of correct time management, both due to the great overload of tasks to be carried out for the different subjects and to the need for coordinating the personal schedules of each of the members of the group.

Furthermore, due to the existence of students with distinct personalities and ways of working, it was fundamental to have an effective management of the conflicts that arose, in order to maintain a healthy working environment, appropriate to the achievement of the outlined goals. Finally, the last two years proved to be challenging for the students who took part in these projects, as they found themselves hindered or limited in the implementation of the projects. For example, not being able to meet face-to-face or having their visits to organizations restricted as they used to, meaning that everyone had to adapt to a new context, emphasizing the more persevering side of each one.

On the other hand, the implementation of these projects allowed a better consolidation of the contents taught in the different subjects since they had to be integrated throughout the development of these works. Moreover, the fact that one of the projects was developed in an organization allowed the students to get closer to a real context, having a better perception of the future that awaits them as future Industrial Management Engineers.

However, as highlighted above, the application of this methodology still has some disadvantages associated. The mandatory integration of some subjects' contents was sometimes a little demotivating for the students because they were not the most suitable for that specific context. Additionally, concerning the project carried out in a company, the commitment and availability of the company and its managers in providing the necessary support to the work team was fundamental. This support included, for example, the clarification of doubts, the provision of data or the clarification of how the processes worked. If this did not happen, the team would feel great difficulties in the development of the project, and its success would be compromised.

However, the overall assessment of the application of the PBL methodology in the development of the projects is positive. Therefore, its continuity in the training of future engineers is considered extremely important, as it prepares them for the adversities that may arise, not only throughout their professional future but also in their personal life.

REFERENCES

Adderley, K., Ashwin, C., Bradbury, P., Freeman, J., Goodlad, J., Greene, D., Jenkins, J. R., & Uren, O. (1975). *Project methods in Higher Education*. Society for Research into Higher Education.

Alves, A. C., Leão, C. P., Moreira, F., & Teixeira, S. (2018). *Project-Based Learning and its Effects on Freshmen Social Skills in an Engineering Program*. Human Capital and Competences in Project Management. doi:10.5772/intechopen.72054

Alves, A. C., Moreira, F., Carvalho, M. A., Oliveira, S., Malheiro, M. T., Brito, I., Leão, C. P., & Teixeira, S. (2019). Integrating Science, Technology, Engineering and Mathematics contents through PBL in an Industrial Engineering and Management first year program. *Production*, *29*, e20180111. doi:10.1590/0103-6513.20180111

Alves, A. C., Moreira, F., Leão, C. P., & Fernandes, S. (2020). Ten Years Of Positive Feedback On Project-Based Learning From First-Year Engineering Students' Prespective. *IMECE2020-23212*, 1–10.

Bonwell, C., & Eison, J. (1991). *Active Learning: Creating Excitement in the Classroom*. Wiley.

Cargnin-Stieler, M., Malheiro, M. T., Alves, A. C., Lima, R. M., & Teixeira, M. C. M. (2019). Learning calculus through PBL in an industrial engineering and management program-a seven-year study. *Advances in Engineering Education*, *7*(3), 1–28.

Costa, L. R. J., Honkala, M., & Lehtovuori, A. (2007). Applying the problem-based learning approach to teach elementary circuit analysis. *IEEE Transactions on Education*, *50*(1), 41–48. doi:10.1109/TE.2006.886455

Dewey, J. (1916). *Democracy and Education. An Introduction to the Philosophy of Education*. Free Press.

Duch, G., Duch, B. J., & Allen, D. E. (2001). *The power of problem-based learning: a practical "how to" for teaching undergraduate courses in any discipline*. Sterling: Stylus Publishing.

Frank, M., Lavy, I., & Elata, D. (2003). Implementing the project-based learning approach in an academic engineering course. *International Journal of Technology and Design Education*, *13*(3), 273–288. doi:10.1023/A:1026192113732

Freitas, I. (2021). More Checks for Less Waste in the Lamination Process of a Shipbuilding Company Pursuing Lean Thinking. In A. Dolgui, A. Bernard, D. Lemoine, G. von Cieminski, & D. Romero (Eds.), *Advances in Production Management Systems. Artificial Intelligence for Sustainable and Resilient Production Systems. APMS 2021. IFIP Advances in Information and Communication Technology* (Vol. 630). Springer. doi:10.1007/978-3-030-85874-2_69

Kilpatrick, W. (1918). The project method. *Teachers College Record*, *19*(4), 319–335.

Knoll, M. (1997). The project method: Its vocational education origin and international development. *Journal of Industrial Teacher Education*, *34*(3), 59–80.

Kolmos, A. (1996). Reflections on project work and problem-based learning. *European Journal of Engineering Education*, *21*(2), 141–148. doi:10.1080/03043799608923397

Lima, R. M., da Silva, J. M., van Hattum-Janssen, N., Monteiro, S. B. S., & de Souza, J. C. (2012). *Project-based learning course design : A service design approach.* Academic Press.

Mills, J. E., & Treagust, D. F. (2003). Engineering Education: Is Problem-Based or Project-Based Learning the answer? *Australasian Journal of Engineering Education, 3*(2), 2–16. https://www.researchgate.net/publication/246069451_

Pereira, M. A. C., Barreto, M. A. M., & Pazeti, M. (2017). Application of Project-Based Learning in the first year of an Industrial Engineering Program: Lessons learned and challenges. *Production, 27*, 1–13. doi:10.1590/0103-6513.223816

Prince, M. J., & Felder, R. M. (2006). Inductive teaching and learning methods: Definitions, comparisons, and research bases. *Journal of Engineering Education, 95*(2), 123–138. doi:10.1002/j.2168-9830.2006.tb00884.x

Ribeiro, E. A. (2008). *A perspectiva da entrevista na investigação qualitativa.* Academic Press.

Rychen, D. S., & Salganik, L. H. (2000). *Definition and selection of Key competencies.* https://www.oecd.org/edu/skills-beyond-school/41529556.pdf

Sá, P., Costa, A. P., & Moreira, A. (2021). *Reflexões em torno de Metodologias de Investigação: recolha de dados* (Vol. 2). UA Editora.

Thomas, J. (2000). *A review of research on project-based learning.* Autodesk Foundation. http://www.ydae.purdue.edu/LCT/HBCU/documents/AReviewofResearchofProject-BasedLearning.pdf

Vygotsky, L. S. (1986). *Thought and language.* MIT Press.

World Economic Forum. (2015). *"The skills needed in the 21st century," New Vision for Education - Unlocking the Potential of Technology.* https://widgets.weforum.org/nve-2015/chapter1.html

Chapter 8
Training Graduated Students in an Industrial Context of a Retail Company

Rui Mota
Sonae SGPS, Portugal

Carolina Mesquita
Sonae SGPS, Portugal

ABSTRACT

Graduate students are a source of knowledge to companies. Their youth, readiness to show recently acquired abilities, and high levels of motivation to "change the world" are appreciated by human resources hiring teams to complete their purpose: to identify talent that can enhance business areas accomplishing relevant goals. However, "competences" do not always come along with the "full package" of a recent graduate. This chapter describes how a Portuguese retail company developed and implemented a Lean School to (1) upskill internal knowledge, skills, and behaviors about Lean in the existing work force and to (2) prepare the newcomers to use Lean in such a good way as if they had been part of the company for years. The authors will also describe some of the active learning methods used in the Lean School programs and report the evolution on some performance indicators like number of students in attendance and satisfaction levels.

INTRODUCTION

One of the most perishable assets for a company is knowledge. Continuous innovation pushes the state-of-the-art knowledge boundary in all fields of work. Innovation will never cease; therefore, companies need to constantly groom new knowledge in people; in other words, companies need to continuously develop competences in the collaborators.

Graduate students are a source of knowledge to companies. Their youth, readiness to show recently acquired abilities, and high levels of motivation to "change the world" are appreciated by human resources

DOI: 10.4018/978-1-7998-8816-1.ch008

hiring teams to complete their purpose: to identify talent that can enhance business areas accomplishing relevant goals. However, "competences" not always come along with the "full package" of a recent graduate. In a recent paper of Alves AC et al. (2021), the integration of interdisciplinary content in an engineering course raised the students competences. This key milestone demonstrated the benefit of an active learning approach, rather than of a traditional siloed one. Indeed, active learning methods can generate more prepared professionals, as concluded in the article of Fernandes (2014). In this study, students and professors both recognized that a PBL-Project Based Learning approach generated more content retention by connecting theoretical knowledge with practical application and strengthened teamwork and problem-solving skills also perceived as very useful for ease the access to professional life. Since the active learning approach is not yet standardized in Universities, companies need to fill this gap as discussed by several authors (Alves et al., 2012; Alves et al., 2017; Flumerfelt et al. 2016).

The Lean word was used for the first time in an article written by John F. Krafcik "Triumph Of The Lean Production System", Krafcik (1988), to translate to the Occident the spirit of Toyota's way of work, or the TPS – Toyota Production System. Later, in the book "The Machine That Changed the world", Womack et al. (1990), James Womack compared automotive industry's performance and the term Lean became part of the occidental management jargon. Furthermore, in a subsequent book, "Lean Thinking: Banish Waste and Create Wealth in Your Corporation", Womack & Jones (1996), the authors define the five principles of the Lean philosophy: (i) Define Value: what a customer is willing to pay for a certain product or service. On the contrary, the non-value activities are called waste, and must be eliminated; (ii) Value Stream: is the set of processes in place to deliver the value customers want to receive. Making the Value Stream visual is key to understand possible break points and constraints that affect value delivery to customers; (iii) Flow: means that all activities must be delivered with no interruptions. Events cutting the flow are considered waste and, therefore, must be eliminated; (iv) Pull-Production: activate the production when required by the customer, avoiding inventory with all costs related and (v) Pursuit of Perfection: means that the company is always looking for improving performance upon deviations or improvement opportunities: Continuous Improvement.

Lean Production, started to be applied to production areas, but rapidly jumped the shop floor boundaries to all kind of areas within a company and to all type of businesses and not for profit organizations. Whenever there are people and processes, there is room for improvement or, room for Lean, whatever the type: Lean Office, Lean Services, Lean Startup, Lean Six Sigma, Lean Education and so on, as explained in the book "Lean Education: An Overview of Current Issues", Alves et al. (2017).

In the searching activity to find training programs to comply to a legally imposed minimum of training hours, a Portuguese retail company stepped in a Lean consultancy firm that challenged not only to deliver the necessary number of training hours, but also to use them as a way to increase productivity and global performance level. This chapter describes how the Portuguese retail company developed and implemented a Lean School to (i) upskill internal knowledge, skills and behaviors about Lean in the existing work force and to (ii) prepare the newcomers to use Lean in such a good way as if they had been part of the company for years. The authors will also describe some of the active learning methods used in the Lean School programs and report the evolution on some performance indicators like the number of student's attendance and satisfaction levels.

THE NEED FOR A LEAN SCHOOL

At the end of 2000, a governmental rule obliged all the companies to give at least 35 training hours to all employees. As a result, the retail company under study started the implementation of a Lean/Kaizen program in the stores on its food retail business unit. During the first 3 years, with external support, more than 25.000 employees have received the required number of training hours. Therefore, all internal training needs were successfully suppressed, solving the compliance issue. The process to deliver such a training effort contained a bundle of different technics in a very structured standard process called Teams Development Program (TDP). This consisted of a Train-the-Trainer's approach where team leaders were encouraged to train their own team members. TDP emphasized the use of simulations, gamification and real case practice around Lean Tools, although it also included basic theoretical concepts. The company started to have a first glimpse of what a future Lean School would look like…

Due to this training effort, a Lean Culture started to flourish across the business areas like logistics, finance, stores, and to become part of the culture of the company. A new challenge was then ahead of all top managers: how to preserve and reinforce this Lean Culture in a long-term perspective and how to embed it into the newcomers? One of the things that soon became clear was the need to have internal Lean Expert to support Lean implementation, training and other supporting activities. To increase the number of Lean Experts, the company started to hire graduates from Engineering courses, specially from Industrial Engineering and Management courses. But, despite the profiles matched with what the company thought it needed, managers soon realized that full matching was only achieved after internal practical training programs, like the TDP has previously deployed to all internal employees. In other words, full competence was only achieved after the pragmatic "how to use" a certain tool complemented the academic theoretical background, with practice and experience. This need was the driver for the creation of the Lean School: a "learning by doing" school where students learn the concepts and practice the tools, to deliver faster and better results.

THE SCHOOL: CONTEXT, SCOPE, ORGANIZATION AND RESULTS

The Context: The Lean System

The retail company is one of the largest businesses conglomerate in Portugal, being the food and non-food retail the most relevant business units it owns. The company turnover for 2020 was above 6 billion euros and it has a workforce of more than 50.000 employees. Besides the retail business units, there are also finance, telco, tourism, industry, energy, services, and other businesses included in the conglomerate.

Lean was already somehow embedded in the way the company thinks and works, for decades. It fits with the values of the company and it highlights the best the company has, its own people with their own ideas and a powerful motivation to change for the better. However, at that time, there was no structured process to make Lean durable and sustainable. Therefore, in 2012, the President of the company launched a program to create the Company's Lean System.

The Lean System is composed by (i) a **Model**, the guideline for good practices and (ii) a **Governance Model**, the way the Lean System is managed and developed.

1. **The Model:**

The Model, presented in Figure 1, represents the set of working good practices that can be seen in use by world class companies. It is divided into 4 main blocks, containing Methods and Tools for the purpose of each block.

Figure 1. The lean model framework

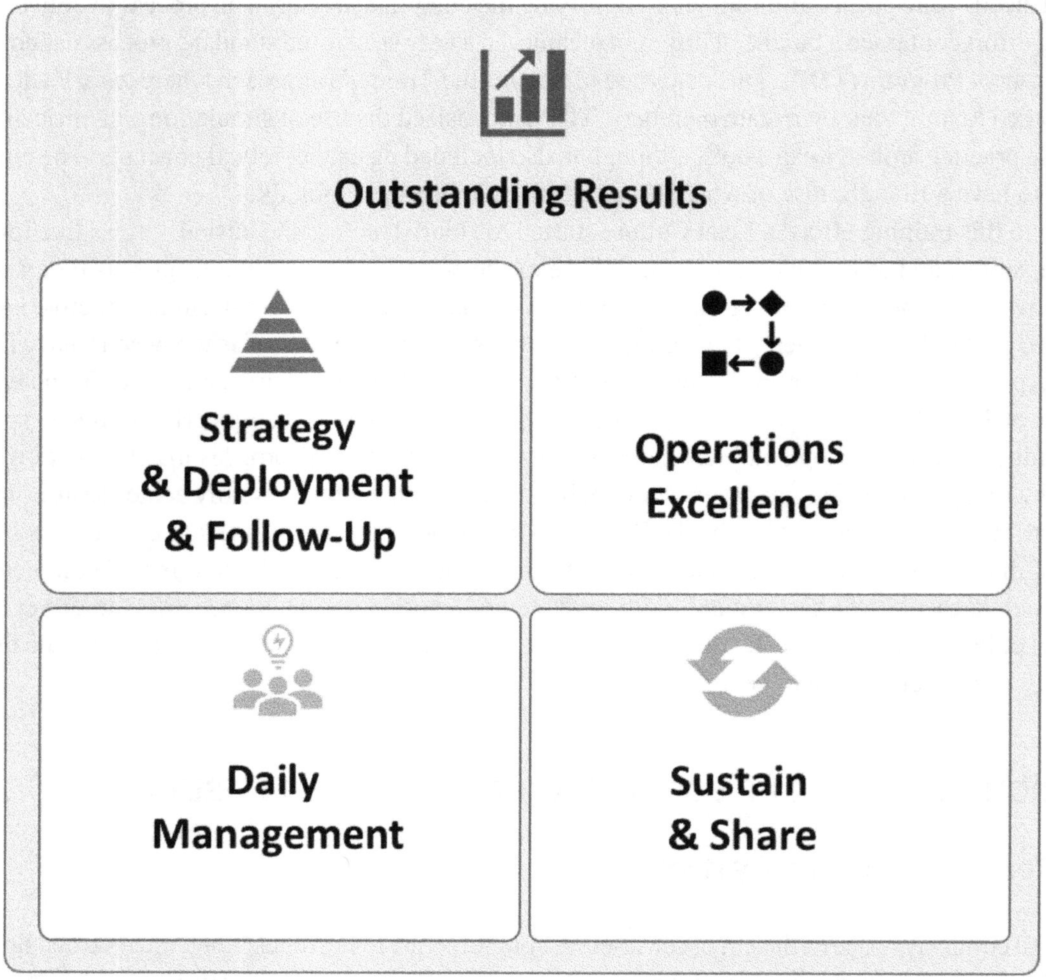

a. Strategy, Deployment & Follow-Up: Ensures company alignment throughout all team levels, from Top Management to the lowest team level. It contains methodologies and good practice to support the (i) business strategy design, (ii) strategy deployment: the cascade of key initiatives and key objectives to all teams to full alignment and (iii) a systematic follow up standard to track and react to deviations.
b. Operations Excellence: Creates or streamlines Excellency Processes in any area of Business (for example: Offices, Services, Industry, Operations, Sales & Commercial). It contains methods and good practices to support project management and problem-solving initiatives, organized by type of problem and by area of business.

c. Daily Management: Is the way each team measures, improves its performance every day and delivers results aligned with its mission and business strategy. It contains methodologies and good practices to improve team's performance and sustains continuous improvement mindset.
d. Sustain & Share: Ensures Lean culture maturity and its sustainability as part of the Company's culture. Sharing learnings from improvement initiatives accelerates the knowledge exchange among teams. It also measures team's Lean maturity levels according to the Model and provides guidance for evolution. It contains methodologies and good practices to benchmark, to share and recognize teams' achievements, reinforcing culture.
e. Outstanding Results: are the goal all 4 blocks want to contribute to. There are five types of results: financial, quality, operational efficiency, service levels (flow) and people.

2. **The Governance Model:**

To implement the guidelines and scope each business unit decides to use from the Model, it is necessary to have a strategy with clear goals, resources and a management cycle to keep the pace and act upon deviations.

The retail company created a multi-business team composed by different business units' top executives, that decides the best Lean Model and Lean System to have, as well as the resources to be allocated. This team is also composed by the Lean Managers, the ones responsible for the implementation of the Lean Strategy decided by each executive business team.

Each year, this team evaluates businesses' maturity levels and learnings from each business unit experience and decides proper actions. The pace and scope for Lean adoption is decided by each executive team taking in consideration the main challenges from individual business strategies.

The Lean School is one of the resources created by the multi-business team to foster the Lean culture implementation across all businesses, delivering training contents organized (i) by formats: from on-line, blended to fully presential sessions; (ii) by levels of responsibility: from natural teams to Top Management; (iii) by needs: from standard contents on basic concepts, also called "Lean foundations" to training-on-demand (ToD) for specific or experts content. ToD courses are developed by request, according to specific needs and purposes to be achieved.

The Lean School Manager is also a multi-business team member.

The Scope

The Lean School main purpose is to develop people's skills to a point where everyone can be autonomous by recognizing improvement opportunities in their work area and being able to implement action to improve them, delivering high performance results.

1. **The content:**

The training portfolio, in the current year of 2021, is composed by 33 courses that cover the full range of knowledge required by the company's Lean Model. Not being exhaustive, here are some contents that can be found in the training catalogue:

a. Daily Management: Lean principles, definitions and concepts like: Visual Management, Daily Meetings Dynamics, Waste, Gemba, 5S, Standardization, A3 Problem Solving, Process Mapping, Process Confirmation.
b. Strategy & Deployment & Follow-Up: Policy Deployment, Portfolio Management, KPI – Key Performance Indicator Tree, VSA – Visual Stream Analysis, SIPOC – Supplier Input Process Output Customer.
c. Operational Excellence: Project Management, Structured Problem Solving, SMED – Sigle Minute Exchange of Die, VSM – Visual Stream Mapping, SIPOC, Kaizen Event, as examples. (Lean tools like VSA, SIPOC, SMED presented above are part of the Lean Toolkit that can be found in "Lean Manufacturing - Tools, Techniques, and How to Use Them", Feld W.M. (2000) and in "Lean Six Sigma Pocket", George M. L. et al (2005)).
d. Sustain & Share: Lean Maturity Level Assessment tool.

Since the beginning it was clear for the Lean School team that contents must address all types of teams and levels of responsibility needs. For this reason, we can see in Figure 2 a variety of training programs organized by levels of responsibility, where "Top" means Top Management teams, who are responsible for the definition, the communication and the follow-up of the execution of the company's strategy. "Middle" stands for Middle Management teams, or the teams responsible for the coordination and execution of the strategy, using projects or coordinating Base teams. "Base" are the operating teams, usually the lowest hierarchical level of the organization.

Figure 2. The lean school training catalogue

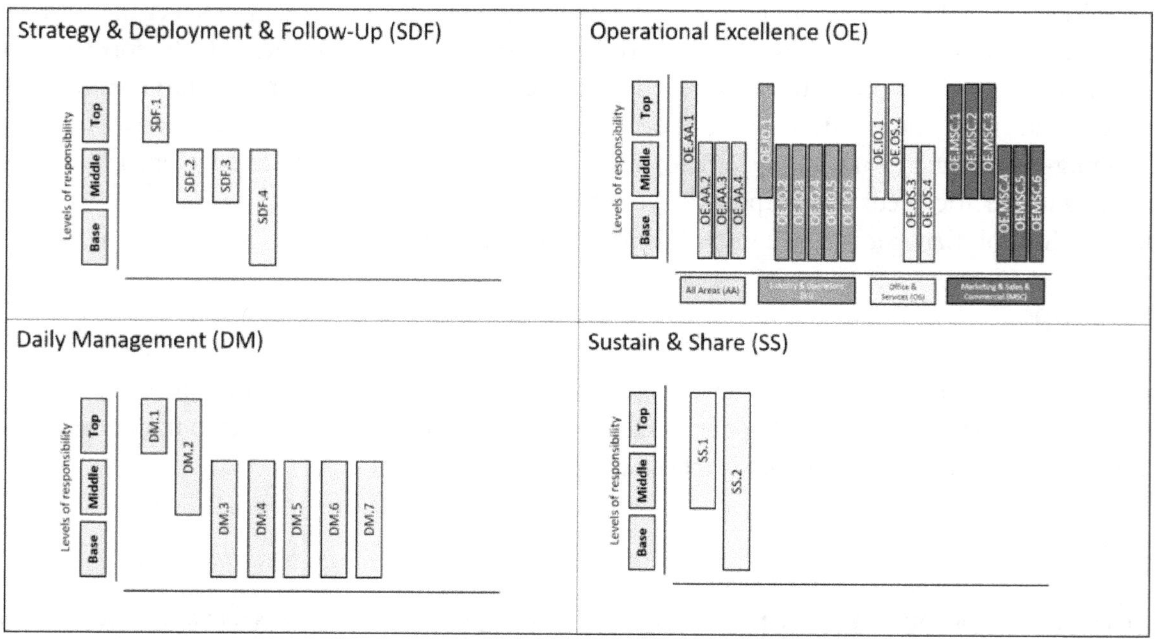

In the Operations Excellence pillar, the training programs are subdivided into macro-processes or Value Streams, once different problems require differentiated approaches. There is a common block of methods and tools, equally applied to all areas, regarding project management skills. Then, the organization by Value Streams allows specialized training programs according to specific needs or problems to address.

Some courses are designed for more than one level of responsibility because the content is the same. The same principles, methods and tools are explained to different team types, nevertheless it is also explained the differentiation on content to use with the same tool. For example, the daily Management course "DM.3" explains the relevance of 5S, Standards and visual management to both team type, Base and Middle, because the tool is the same, only the content to be used will differ in practice. And this is also a way to align a common language throughout all company, reinforcing the culture of continuous improvement. Another example can be the "SS.2". It refers to the Maturity Assessment tool. It is the same to all levels, common language, common training programs.

2. **The students:**

This company owns several business units, mainly retail ones related but not exclusively. So, as a cultural trait, Lean is being adopted across all the companies of the group. The Lean School delivers training to all the group companies, mixing in the same class students from different business units, contributing to accelerate knowledge sharing and learning from others' experience.

All newcomers, regardless the responsibility level, must attend the Daily Management training programs at the Lean School, considered to be the "Lean foundations". This training is not mandatory but highly recommended for the first 12 months of work because to sustain and reinforce continuous improvement mindset is critical that newcomers comprehend and adopt from the beginning of the work journey the culture of the company in ways of work.

The Organization

Nowadays, the Lean School is part of the Corporate Academy of the company and it serves the entire population of all business units. Each business unit has a Head of Lean, responsible to deploy, reinforce and sustain the Lean strategy defined by the correspondent Executive Team. The Head of Lean is also responsible for identifying the profiles of the employees, new or current ones, that should attend the Lean School to deliver Lean Strategy of the business unit, beyond newcomers.

Figure 3. The lean school organization chart

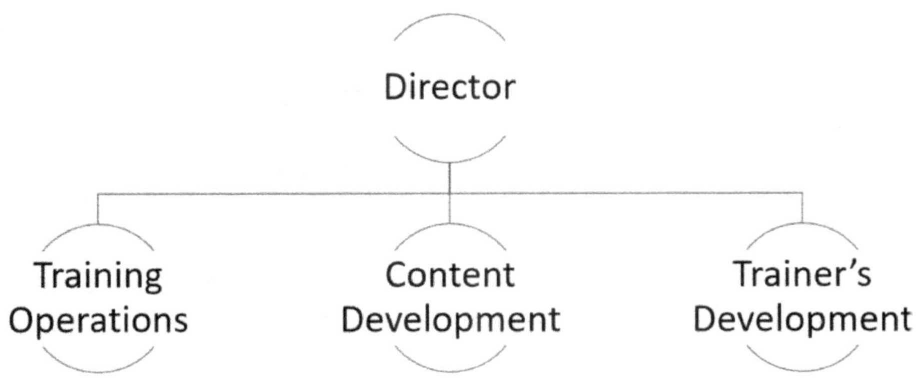

As presented in Figure 3, the Lean School is composed by a core team of 4 people. The (i) Training Operations person is responsible for coordinating with all business units the number of students to be trained. (The company explained the option for the usage of the word students instead of trainees because they consider students to be the trainees (newcomers) and current employees reinforcing knowledge in current position). Then, she/he must produce a quarter calendar considering training format, places where the training will occur and trainer's availability. (ii) The Content Development person produces all the training content according to the needs and goals of each business unit. She/he can also coordinate content development proposed and or developed by trainers. Content development means the incorporation of a new method or tool for the first time in the Lean School or an adaptation of an existing one. It also includes the production of the final design of all the training material, from the manuals to the games and dynamics used for training purposes, most of the time, subcontracted to a graphic designer. (iii) Trainer's Development person is the responsible to identify, certify and continuously develop the Lean Experts, acting as Trainers. Trainers are selected among the Lean Experts available in the business units. They belong to the business and are selected considering past Lean training background, years of Lean experience as implementation leaders or users. Communication skills are also an important ingredient to be a successful Trainer, inspiring others for the Lean Culture. (iv) The Director is responsible for providing the resources the Lean School needs to fulfill its mission, for coordinating the team and for establishing the cooperation needed with all Heads of Lean and respective Executive teams in order to guide the Lean School towards the future.

Every year, the net incorporation average of new employees is around 2.000 people, half of them being graduates from universities, of several courses and diverse locations. In the current year of 2021, since the foundation of the Lean School in 2013, more than 10.000 employees were already trained in the various courses along the way.

Results

Two indicators express the size and the impact of the Lean School. (i) Number of students gives the size and (ii) satisfaction level gives the perceived quality of the training delivered.

The (i) size reflects part of the success the Lean School is having in its journey. As it can be observed in Figure 4, the number of students has been growing continuously since 2017. The sudden growth in 2018 is due to a specific Training on Demand (ToD) program requested by the retail company, not repeatable in the following years. This growth is consistent with the introduction in 2018 of a new training program using gamification methods to deliver a more engaged student's participation: The Lean Lab Store Experience. This game will be detailed next.

Before this game, the training sessions already included practical exercises and other type of active learning methods, like Lego® Serious Play, Project Based Learning (PBL) for small projects, simulations, role-plays and sharing of internal cases where Lean is being producing good results. The training sessions were classified as good, but some written quality feedback from students also tells that the sessions were more theoretical than practical and too long. Some Human Resources (HR) training pivots did not recommend the Lean Training programs at all because they were comprehended as "boring" and somehow theoretical. Since the foundation of the Lean School, the training programs evolved from six-day training sessions, with lots of lecture style to two-day programs with more active learning methods. However, it was not enough for students needs because they wanted an even more practical approach and shorter time frames. The Lean Lab Store Experience started with a 2 days format in 2018 and evolved for a 1-day format in 2019.

Figure 4. Number of attendant students

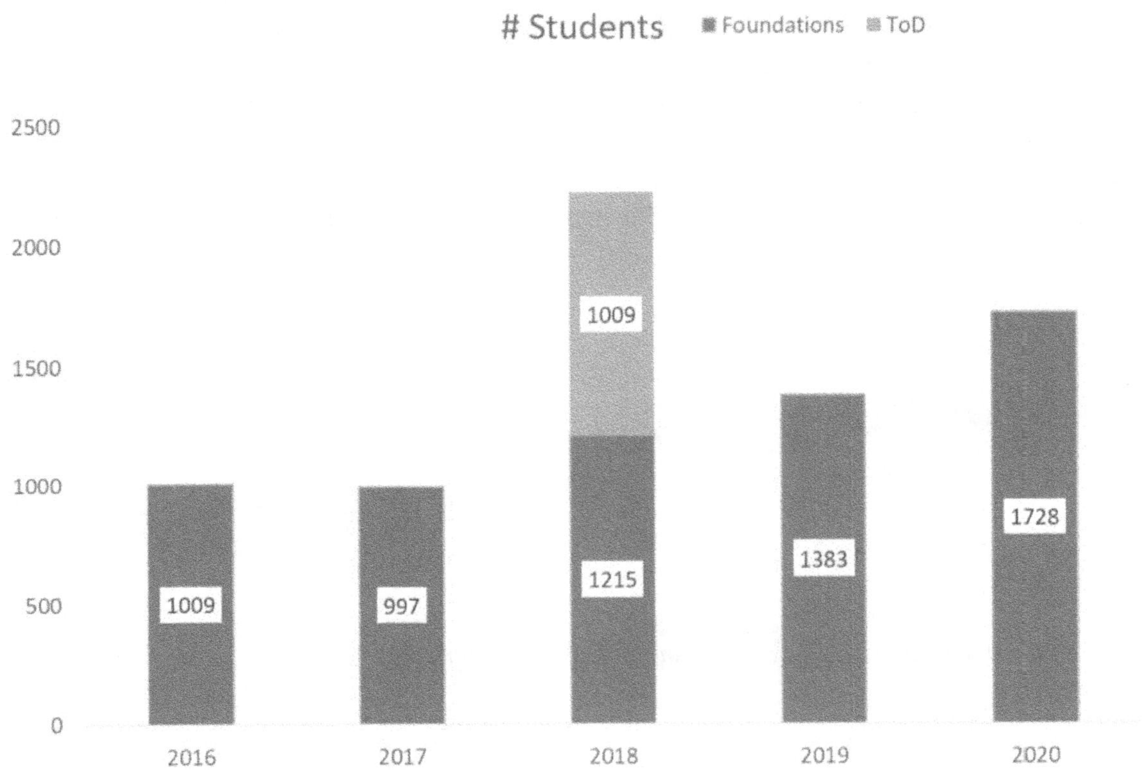

The number of students in 2020 is due to the special year for training activity: the lockdown period imposed by COVID-19 in March. All the Lean Lab Store Experience presential sessions were suspended, bringing to the Lean School team a huge challenge to solve. On-line training sessions came as the solution for it. But, presential sessions are not the same as on-line ones; they tend to be more boring and low engaging ones. How did the team solve this issue? Will the School satisfaction level decrease with this new format? These questions were all mixed and answered in the Training Strategy section below.

The usage of the number of students' metric as a performance metric one can generate some discussion. Legal number of training hours are completely overcome with a wide variety of training programs beyond Lean's. Although some business units have decided to make the Daily Management training programs mandatory to newcomers, it does not happen in full. So, this metric is used as a sign of the attraction provoked by the word-of-mouth's effect.

The (ii) quality indicator of the Lean School is the student's satisfaction level, that recently evolved to the Net Promoter Score®, (NPS®), Reichheld (2006). Fred Reichheld from Bain & Company Consultancy company is the founders of the NPS®. He explains all the details of the NPS® in his book "The Ultimate Question: Driving Good Profits and True Growth" Reichheld (2006). This metric can be implemented to all types of industries and sectors and, when applied to training schools indicates the student's loyalty and brand advocacy levels to the training sessions. It consists in answering to the following question:

"On a scale of 0 to 10, how likely do you recommend this training session for your company?"

The students rating zero will very unlikely recommend the training session. On the other hand, the ones rating ten will extremely likely recommend it to their colleagues. The results are classified in three categories: (i) promotors (rating 9 or 10): will be loyal to training and recommend it in public and to colleagues; (ii) neutral (rating 7 or 8): will not criticize training but will not promote it either; or (iii) detractors (rating 0 to 6): will criticize training in public and will not be present again unless it is mandatory.

The NPS® is the difference between the weight of promotor and the weight of detractors, as explained in Bain & Company website, co-authored with the authors.

NPS® = % Promoters - % Detractors

The result can go from -100 to +100 points. The higher the result, the better.

Figure 5. Scale in use by the lean school

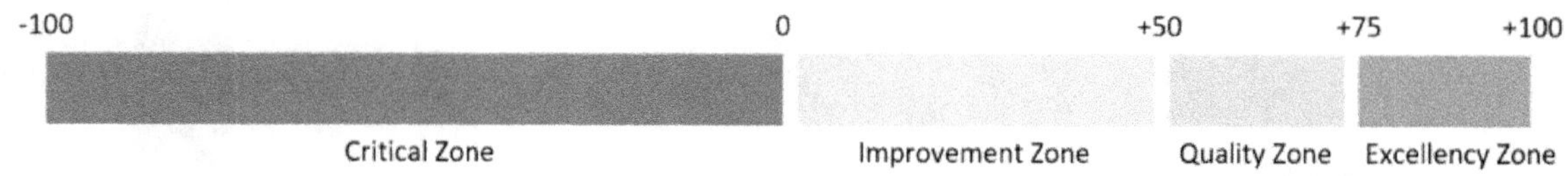

Figure 6. Student's evaluation of classes using the net promoter score

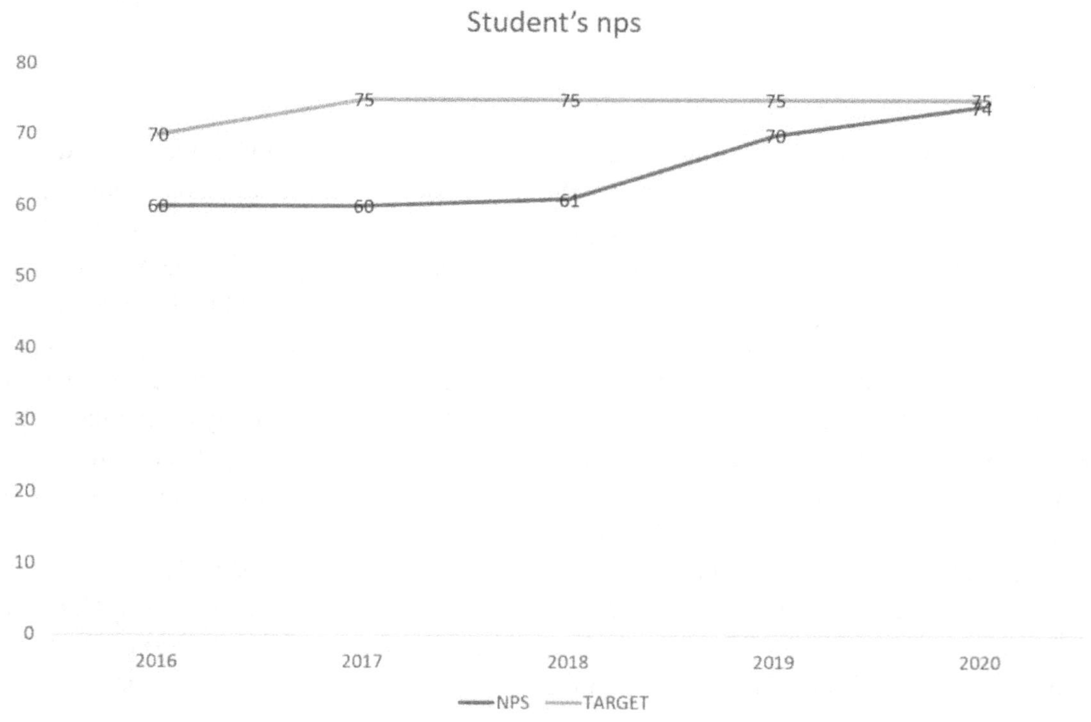

Figure 5 represents the interpretation of the NPS® scoring. The result to be conclusive needs at least 50 respondents. This indicator is being used worldwide because it allows some comparison between industries and guides companies on how to improve service levels to customers. On the other hand, customers can also decide from which company to buy a product or service based on the NPS® performance. For some comparison we mean that it is not perfect. Scales, scope and processes to collect customers' feedback is not standard and the results are not totally comparable. However, one thing is clear: the higher, the better! Due to the lack of an available and reliable NPS scoring for an inhouse executive training service, the Lean School produced its own NPS Scoring Scale, presented in Figure 6. There are some common patterns generally accepted in the interpretation of this metric. Below 0 (zero) is a bad performance; it means that people will say bad things about your service or product, in this case, training programs. Positive scoring is good and there are 3 clusters for this type of results: (i) improvement zone; (ii) quality zone; and (iii) excellency zone. The size of each positive cluster varies from industry to industry, but the most important part to the NPS® usage is the single evolution of this metric throughout the years, as we can see in Figure 6, below.

For 2016 and 2017, the NPS® of 60 points is considered a very good score, the training sessions are recognized as having quality. The introduction of the Lean Lab Store Experience in the second semester of 2018 with a two-day format, raised the NPS to 61 points. People loved the radical change to a training session based on a game played during 2 days in 4 rounds of simulation with some theoretical concepts explaining what was going on with the game results. But the change for 1-day format in the beginning of 2019, boosted +9 points to the NPS® to a total of 70 points, still quality, but closer to excellency zone, the 75 points level. For 2020, the on-line training sessions disrupted the internal paradigm that on-line sessions are worse than presential ones. After 2 months waiting to comprehend what was going to happened with the COVID-19 evolution, the Lean School team launched an internal workshop on how to deliver excellency level training session in the on-line format, considering that stopping training sessions was not an option. If COVID-19 lasts for the next 2 or 3 years, how will the Lean School continue its mission? And the answer was the implementation of an on-line synchronized training program of 5 modules of 90 minutes each. The content was divided into on-line exercises and delivering each module as a story and not as a lecture. The Daily Management contents, considered to be the foundations for Lean concepts and tools, that were in the Lean Lab Store Experience training program, were organized in a story with 5 episodes and in each episode (or module) there are hands on exercises like: (i) 5S numbers game from (http//Leantools.info/5sgame), (ii) standards with the fish drawing game using paper, pencil and exhibiting the result in the computer camera; (iii) on-line surveys to class discussion using Mentimeter platform (www.mentimeter.com) and Forms from Microsoft Office 365 and (iv) visual management using Miro (www.miro.com) white board platform to deliver collaborative exercises on process mapping, structured problem solving with fishbone tool and five whys. The Lean School team learned that smaller practical session and more frequent ones, are perceived as easier to attend and the interaction during the 90-minute session increased the satisfaction level of the students. When the Lean School launched the on-line program of 5 modules in July, the demand blew through the roof. In one week, the team received more than 400 subscriptions for July's sessions. Until the end of the year the numbers continued to grow. Since the sessions were shorter, the number of hours were fewer than previously, but the impact on the real usage of the learnings from the training sessions increased. The Heads of Lean from the businesses started to have more requests to help and support Lean implementations from people that recently attend and on-line training. If the Lean School could find a way to measure what people do with the training they receive, this could be the real indicator of the effectiveness and impact on the businesses. It does not exist, yet.

The number of student's (the size metric) can be interpreted as a good attractiveness sign, NPS® (the quality metric), is the one where the Lean School has more influence. The number of students depends more on the number of available people to be trained from current work force to newcomer's entry pace. This availability can also depend on business dynamics, workload peaks and management decisions on who is going to attend training session and so on. So, the number of students is a good metric, but its evolution or growth is highly dependent on businesses side. Quality is fully dependent on the Lean School team's action. It must articulate several success factors that contribute to this metric such as (i) Training Method, (ii) Session Format and (iii) Trainers. Next block will describe the Training Strategy implemented to answer all this question marks around the new training paradigm, the on-line format.

THE TRAINING STRATEGY: TRAINING METHOD, SESSION FORMAT AND TRAINERS

As previously described, the content for the training sessions is fully aligned with the Lean Model of the company for the standard sessions and adapted whenever a Head of Lean requires a Training On Demand session.

Once the content is clear, the paradigm shift from presential to on-line sessions in 2020 highlighted the need for clarification on what contents must be delivered to whom, in what format and for how long. The following Table 1 describes the solution produced to answer to all those doubts around the training strategy.

Table 1. Training strategy matrix

Training Purpose	Population	Format	Methods	Duration p/ training unit
Know the basics (Foundations)	All employees	On-line	Lectures, simulations	Less than 2 hours
Do by yourself	Managing roles	On-line + Coaching	Gamification, Hands-on	Less than 1 day
Be Expert	Project Managers	Presential + Coaching	PBL, Role-Plays, Coaching	3 to 6 months
Become a Master	Lean Managers	Presential + Coaching	PBL, Role-Plays, Coaching	6 to 12 months

The training strategy starts with the training purpose to be reached. Afterwards, target population, format, methods to use and duration are easier to identify.

"Know the basics" means that everybody needs to receive the foundations of the Lean Model, at all levels and it must be delivered in the most democratic way possible. Short sessions in an on-line format, synchronized or asynchronized is the way to go.

"Do by yourself" are training programs designed to increase Lean knowledge autonomy. There are some methods and tools that any leader in the organization must do by himself. Those sessions can start in an on-line format and students are invited to present extra autonomous work to validate the competences achieved.

"Be Expert" programs are dedicated to the people that want to be recognized as Lean Expert in a subject of the Lean Model. Training sessions are presential, since the methods used are more Project Based Learning (PBL) with coaching sessions to support the progression, up to a final presentation.

"Become a Master" is for those experts that want to become Head of Lean. They are fully autonomous; they master the methods and tools and contribute with new knowledge to the Lean Model.

Training Methods (The Lean Lab Store Experience Case)

As explained above, the Lean School has been using active learning methods to increase the value of training time. The Lean School wants to be recognized as a School that teaches how to work better every day and that is only possible when the students use the knowledge they have learned from the training. The 70-20-10 approach means that 70% of the session time is hands on, 20% good practices sharing and 10% theoretical knowledge. The Lean School is committed with this approach and the Lean Lab Store Experience is a good example to share here.

The purpose of this training session is to promote autonomy in basic Lean tools usage. 5S, visual management and standards in a gamification format, especially for newcomers.

Figure 7 represents the flow of the game: flow of materials and flow of information. The context: the game is played in a room with 9 persons: 2 customers, 1 store manager, 1 store stock keeper, 1 warehouse operator, 1 administrative clerk, 1 cafeteria operator, 1 pizza maker, 1 plant manager.

Figure 7. The lean lab store experience flow of information and material

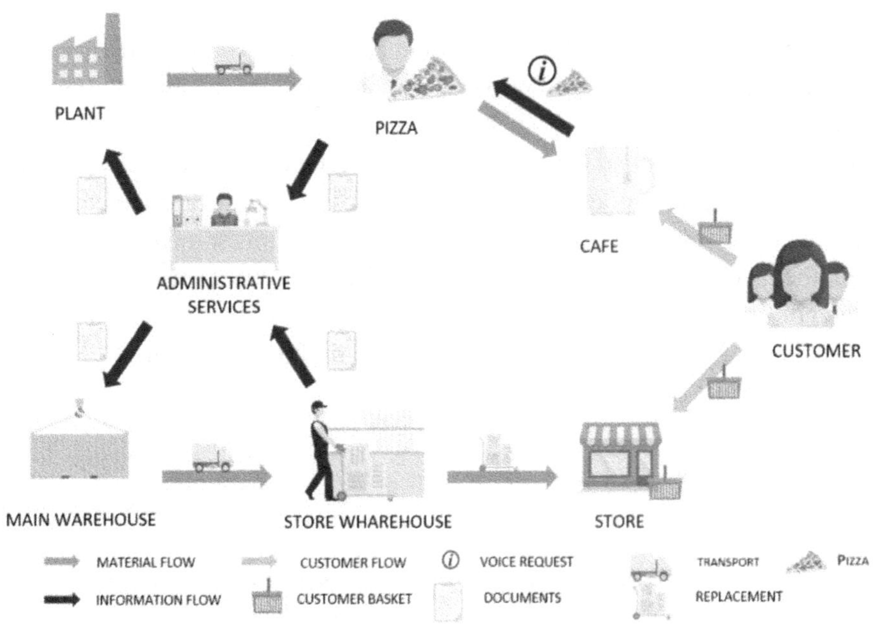

Each position has an organized workplace with all the pieces needed to play the game, like the ones presented in Figure 8:

Figure 8. Workstation examples

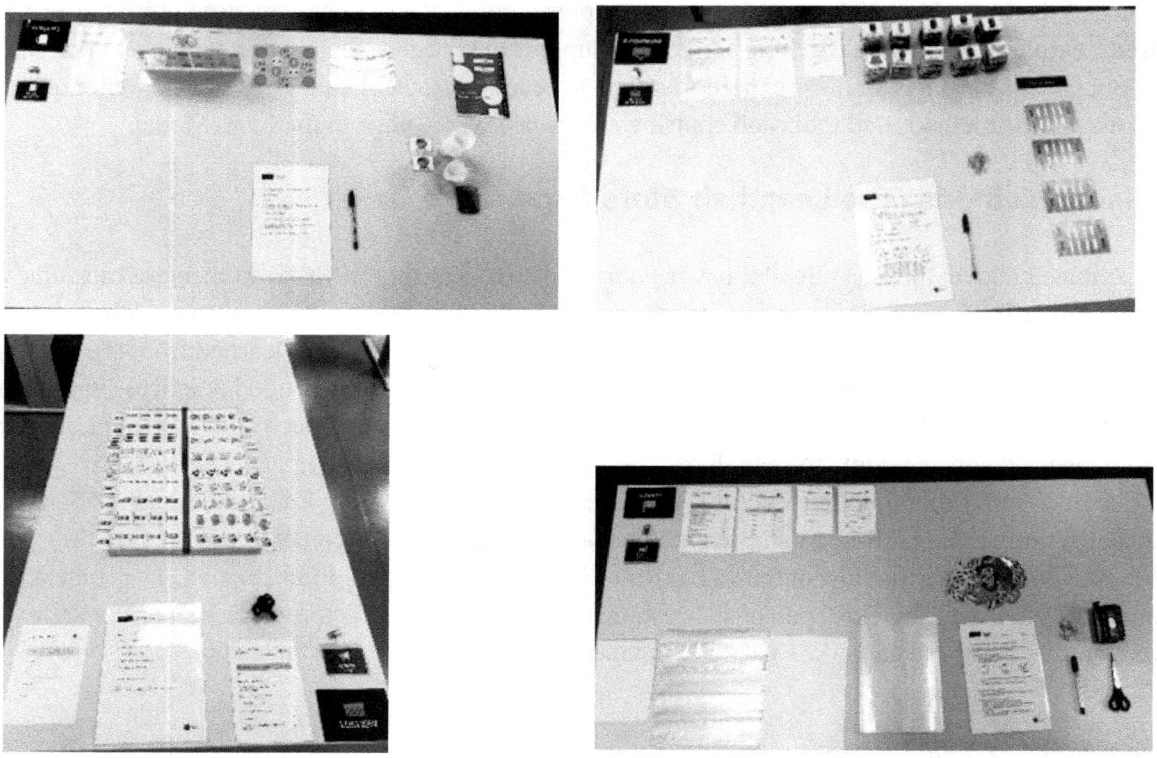

The objective to be reached by the team of 9 students is to increase the store global sales with high customer satisfaction levels. At the beginning very little information is shared about the game. Only some rules and the explanations of each position to each player.

This game is played in 3 rounds and typically the first round presents the lowest performance level of all. After each round the team role-plays a team's meeting using visual management to identify deviations on performance metrics and possible causes. The trainer explains some Lean Tools and share some examples and then, the team goes back to the role-play and decides what action to implement before the following round and defines the new goals to be reached.

This game is a 1-day training event and the students rate high scores on the NPS® because they are so fully engaged in all the playing that do not realize the time passing. The NPS® for this training session was 80 points before the lockdown period in the early 2020.

Session Format

As mentioned above, the lockdown period brought to the team a new format of training delivery: On-line Microsoft Teams (MS Teams) training sessions. After the launching success in July 2020, the program continued to be very appreciated with an NPS score around 74 points. Quality rated, but not Excellency rated, yet. The feedback from students and from Trainers was that the sessions needed to be even more practical and with hands on. But how to grab student's attention when they are much of their time working from home through a computer screen all the time? What are they telling us about this experience? And the Lean School started to realize that some students wanted more support and follow-up after the sessions. And then came the coaching part of the format for some types of training programs and people. It is important to support the students in the first steps to encourage them to continue.

Trainers

Trainers are a key pillar of the Lean School. With their ability to inspire the students, they strongly influence the next step a student will take to use what they have learned from the time invested in training. The same content can be delivered with a totally different experience and therefore, totally different satisfaction level and impact.

Trainers are selected from business units and with experienced background in the Lean Model, as implementers and as users. Having received training is not enough to be a trainer, experience is a key requirement. There is a process to select and certify each trainer according to the knowledge. And there is another process to develop already active trainers to evolve maturity levels and expertise. The more experienced in Lean usage and as Lean Trainer, the more impactful a trainer becomes.

The role of the Lean School is to develop peoples' competences in Lean ways of work. The stronger the Trainers' team is in size and quality the better this mission is accomplished. So, the Lean School recently launched a Trainers Development Program to upskill Trainers competences, as represented in Figure 9.

When Trainers are accepted as future Lean trainers, they go through a certification process, before delivering the first training hour in the Lean School. They need to master the contents, the methods, the dynamics, the rules, and the desired behaviors to adopt as Trainers. This requires learning time. So, they are "Learners".

Figure 9. Trainer's Development Program, maturity levels

After the first training session, usually as rookies shadowing an experienced Trainer, they are graded as "Advanced" Trainer.

This Trainer's Development Program is based on the train the Trainer's approach so that the more experienced Trainer develops the less experienced ones.

To be a "Lead" a Trainer needs to suggest and implement new contents or new methods to the training material in one program, with positive impact in the NPS® score afterwards. This content development is supervised by the Lean School Development responsible or by a Master of that knowledge area, if existing.

A good example of this case were the improvements incorporated in the on-line training programs (the 5 modules) at the end of 2020. The NPS was around 74 points, as mentioned above, and 5 Trainers accepted the challenge to upgrade them. It was the first time in the Lean School that contents were updated by people outside the Lean School Team. The results were much better than expected. The NPS of these modules rose from 74 to 84 NPS® points, a growth of +10 points or +13% in satisfaction level. Not less relevant is also the increasing number of students attending these on-line sessions. Up to the current month of June 2021, there are already 1.250 students with the training completed. Something is attracting more people and the Lean School teams believe it's the result of the increased quality that the new 5 "Lead" Trainers added to the modules. More exercises, even less talk, and more support afterwards.

To become a "Master" a "Lead" needs to develop 2 "Advanced" trainers to the "Lead" position. Currently, the Lean School has 1 Master, 5 Leads and 27 Advanced active trainers.

DISCUSSION

The usage of active learning methods like gaming, simulations, role-plays, project based-learning, have the final purpose to simulate, as closer as possible, real life situations in a safe environment where a student can fail and learn from failures, without damaging the business or the company. According to the experience of this retail company, a learning-by-doing approach proved to increase the speed of competences development in the newcomers, specially engineers. People learn faster when they experience a certain task or activity. This trait also justifies the usage of internal Lean Experts, with a wide experience of real-life situations to be the Trainers of the Lean School, using the methods, games, simulations, and so on.

The impact of the training delivered by the Lean School is not yet well quantified. The metric "number of students" only measures if a student attended or not a training session; it does not evaluate the value of that specific training, nor if the student really knows how to apply a certain method or tool. However, the authors recognize that the new strategy of using 70% of the training time applied to an "hand-on" approach can be a successful way to evaluate the real impact of a given training program. The mentorship provided by the Trainer to a real-life case usage, with a final presentation by the student of the work developed, is a more accurate way to evaluate success. For example, number of certified students (with a validated real case delivered).

The migration of the training sessions to the on-line format carries some challenges in terms of metrics comparison, like the number of training hours. It seems that the Lean School management team is more concerned with the spread of "lean foundation" contents through the wider range of business units as possible rather than having perfect metrics comparison. The process to access to the on-line training sessions proved to be very soft and simple and showed a new way to increase for the number of students. The Training Strategy described above is a reflex of this new opportunity that will require more experienced Trainers to be fully implemented. Growth brings several challenges. Increasing the number of students requires more capacity from the Trainers and requires a strong certification process to deliver training session of excellency.

CONCLUSION

Notwithstanding the existence of research on the positive impact of the usage of active learning methods in universities, significatively increasing student's motivation and content retention, those methods are not yet a common reality in the Universities, at least in the Engineering courses. Of course, the role of Universities is not to provide a practical learning experience only, but they could benefit from a more balanced combination of this active learning methodologies, specially Engineering courses from where the company of this study hires regularly. The effort to setup and develop a Lean School like the one described in this document could be placed to other challenges and the newcomers from Engineering schools would deliver faster and better performance.

The usage of active learning methods in the training programs is critical to deliver impact in the organization, increasing the effectiveness of the training investment. There is no Return on Investment metric implemented yet, but the Lean School would benefit from it, whenever it could implement it. The increasing number of students and the positive evolution of the NPS® scoring, demonstrates the value of the active learning methods.

The lockdown period opened new opportunities for the training activity. Delivering training to more than 1.250 students in six months with a satisfaction NPS® of 84 points in an on-line format is the demonstration that there is still a lot to do to reach even more people with quality standards, combining on-line with presential. The presential format gained a lot more perceived value. In a near future, to call a student for a presential session requires a higher level of perceived value, than an on-line session and this is a new challenge to be solved by the Lean School team.

There are other opportunities to be developed soon by the Lean School organization as the digital era is bringing to all areas of life. Training will be affected too, and a study group could be launched to foresee the next step for training industry.

REFERENCES

Alves, A. C., Dinis-Carvalho, J., & Sousa, R. M. (2012). Lean production as promoter of thinkers to achieve companies' agility. *The Learning Organization*, *19*(3), 219–237. doi:10.1108/09696471211219930

Alves, A. C., Flumerfelt, S., & Kahlen, F.-J. (Eds.). (2017). *Lean Education: An Overview of Current Issues*. Springer. doi:10.1007/978-3-319-45830-4

Alves, A. C., Flumerfelt, S., Moreira, F., & Leão, C. P. (2017). *Effective Tools to Learn Lean Thinking and Gather Together Academic and Practice Communities.* Paper presented at the ASME 2017 International Mechanical Engineering Congress and Exposition. 10.1115/IMECE2017-71339

Alves, A. C., Leão, C. P., Uebe-Mansur, A. F., & Kury, M. I. R. A. (2021). The knowledge and importance of Lean Education based on academics' perspectives: An exploratory study. *Production Planning and Control*, *32*(6), 497–510. doi:10.1080/09537287.2020.1742371

Feld, W. M. (2000). Lean Manufacturing: Tools, Techniques, and How to Use Them (1st ed.). St. Lucie Press.

Fernandes, S. R. G. (2014). Project-based Learning (PBL) Preparing Graduates for Professional Practice: Findings from a Case Study of Project-based Learning (PBL). *Procedia: Social and Behavioral Sciences, 139*, 219–226. doi:10.1016/j.sbspro.2014.08.064

Flumerfelt, S., Alves, A. C., Leão, C. P., & Wade, D. L. (2016). What do organizational leaders need from Lean graduate programming. *European Journal of Training and Development, 40*(5), 302–320. doi:10.1108/EJTD-01-2015-0005

George, M. L., Maxey, J., Rowlands, D., & Price, M. (2004). The Lean Six Sigma Pocket Toolbook: A Quick Reference Guide to 100 Tools for Improving Quality and Speed (1st ed.). McGraw-Hill.

Krafcik, J. F. (1988). Triumph Of The Lean Production System. *Sloan Management Review, 30*(1), 41–52.

Reichheld, F. (2006). *The Ultimate Question: Driving Good Profits and True Growth.* Harvard Business School Publishing Corporation.

Womack, J. P., & Jones, D. T. (1996). *Lean Thinking: Banish Waste and Create Wealth in Your Corporation.* Free Press.

Womack, J. P., Jones, D. T., & Ross, D. (1990). The Machine That Changed the World. New York: Free Press.

Chapter 9
Teaching Circular Economy and Lean Management in a Learning Factory

Angel M. Gento
https://orcid.org/0000-0002-8741-5780
Universidad de Valladolid, Spain

Carina Pimentel
Universidade de Aveiro, Portugal

Jose A. Pascual
Universidad de Valladolid, Spain

ABSTRACT

Traditionally, industries followed a linear process of resources consumption: taking raw materials from nature, transforming them into products, and selling them to consumers (who discarded them when they were no longer useful). Nowadays, due to the sustainable development concerns, there is an increasing awareness on the society for reuse, repair, recycling, and remanufacturing to avoid resource depletion and achieve waste reduction. Following this idea, with the aim to train students and practitioners in lean manufacturing and circular economy concepts and tools, a learning process organized in three sequential phases was developed, starting with the manufacture of a toy car (25 kg and over 100 pieces) using a traditional push system, then reengineering the process to implement pull system and lean manufacturing concepts, and finally, considering a circular economy pull system through the reuse and recycling of parts and components. In this way, the importance of reducing waste in manufacturing and the reduction in the use of raw materials by considering the 3Rs is highlighted.

DOI: 10.4018/978-1-7998-8816-1.ch009

INTRODUCTION

The rapid and continuous changes that are taking place in the business world, due to the Volatile, Uncertain, Complex and Ambiguous (VUCA) environment currently in place in the world demands learning throughout life, becoming essential not only to learn, but to learn how to learn. Thus, the development and maintenance of learning systems becomes a requirement. Furthermore, the emergence of new teaching approaches, putting emphasis on learning by doing and experiential learning, led to the increased use of learning factories, as a mean to simulate authentic factory environments. Although this topic is not new, it gained major importance in academia and industry in the last decade (Abele et al. 2015).

It is in this context that a collaboration between the University of Valladolid and the car manufacturer Renault started, aimed to develop a learning factory for students and workers to learn lean manufacturing concepts and how to apply them in different situations similar to those found in factories (Gento et al., 2016).

But the training does not end there. Nowadays, the European Commission (2011) has identified several critical raw materials and, moreover, some of those materials are widely used in batteries and catalytic converters, and their consumption is expected to increase in the coming years. Therefore, recycling and reusing components and materials is fundamental in the automotive manufacturing industry and has gone from being an environmental consideration to a necessity to minimize manufacturing costs.

Therefore, the initial training in this learning factory has evolved to introduce the concept of circular economy into it, taking into account the awareness of society towards the development of a more sustainable world, and the European directives about end-of-life vehicles and recycling of batteries (European Union, 2000, 2005). In this context, Circular Economy offers resource efficient solutions to keep the planetary boundaries while still creating economic growth (Kirchherr et al., 2017).

Within production operations, lean manufacturing is currently the most successful operations management paradigm, implying high impact on the sustainability of operations. However, Circular Economy concepts are not so emphasized (Kurdve & Bellgran, 2021). Thus, it becomes important to raise the awareness of future engineers on this issue and equip them with the required knowledge.

The objectives of this chapter are threefold. Firstly, it aims to show the importance of the university-industry collaboration as a mean for learning and continuous improvement of engineering students. Secondly, it is intended to present a "learning by doing" approach, embedded in a Learning Factory. This approach combines Lean Manufacturing with Circular Economy concepts, incorporating different levels of difficulty, using e.g. different management tools, and different workers and students. Furthermore, to facilitate concepts assimilation, the "learning by doing" approach is focused not only on learning concepts but also on their incorporation into everyday tasks. Finally, this chapter aims to show how the introduction of the circular economy thinking in a production process helps to reduce waste and to make better use of resources, reducing costs and increasing the efficiency of the overall production process.

This chapter is structured into seven sections. After this introduction, section two summarizes the theoretical background. The third section explores the research methodology, while section four is dedicated to the presentation of the Lean School developed jointly by Renault and the University of Valladolid, while in section five the solutions and recommendations are explored. Then, in section six some future developments are discussed, followed by a conclusion presented in section seven.

BACKGROUND

This section presents a brief literature review related to the topics explored in this chapter, namely the University-Industry collaboration, the circular economy, and the learning factories.

University-Industry Collaboration

University-industry collaboration (UIC), with a long tradition in several countries around the world (Ankrah & Al-Tabbaa, 2015), refers to the relationship between universities and companies with the main objective of promoting the exchange of knowledge and technology between them through students, professors, researchers, workers and managers at different levels.

Although there are many forms of university-industry collaboration, Bonarccorsi and Piccaluga (1994) classified them into six main categories:

- Personal Informal Relationships,
- Personal Relationships,
- Third Party,
- Formal Targeted Agreements,
- Formal Non-targeted Agreements,
- Creation of Focused Structures.

In recent years, these collaborations between the two parties have increased notably (Elsevier, 2021) because, although the objectives of both are different, there are synergies that can be exploited. According to Rybnicek and Königsgruber (2019), in universities, several objectives can be identified from the academic point of view, such as to be able to offer students training adapted to the needs of their future professional environment and thus improve their employability. Also, from the research point of view, objectives such as to be able to test applications to different theories, develop and exploit patents and access government grants and funds for applied research and laboratory equipment can be identified. On the other hand, companies need highly qualified personnel, access to research networks and the latest technologies as they are immersed in a VUCA environment where competitors are located all over the world, not forgetting tax benefits and social responsibility policies to promote innovation and regional and national development.

Circular Economy

Throughout its evolution, the industrial economy followed a take-make-dispose model. According to Ellen MacArthur Foundation (2012), approximately 65 million tonnes of raw materials entered the economic system in 2010 and are expected to reach 82 tonnes by 2020. However, recently many companies have realized that this linear manufacturing and consumption model increases their exposure to risks, mainly due to rising resource prices caused by shortages or extraction difficulties, and supply disruptions (World Economic Forum, 2014). To avoid this, some industries use recycled materials, which guarantees a constant flow of materials by adopting a circular economy model.

A circular economy is considered an economy "*where the value of products, materials and resources is maintained in the economy for as long as possible, and the generation of waste is minimized*" (European

Commission, 2015) (see Figure 1). Moreover, the European Commission (2008, 2011) has identified critical raw materials and recycling as key aspects for improved resource efficiency and the competitiveness of European industries

Figure 1. Conceptual diagram illustrating the Circular Economy in a simplified way (European Commission, 2018)

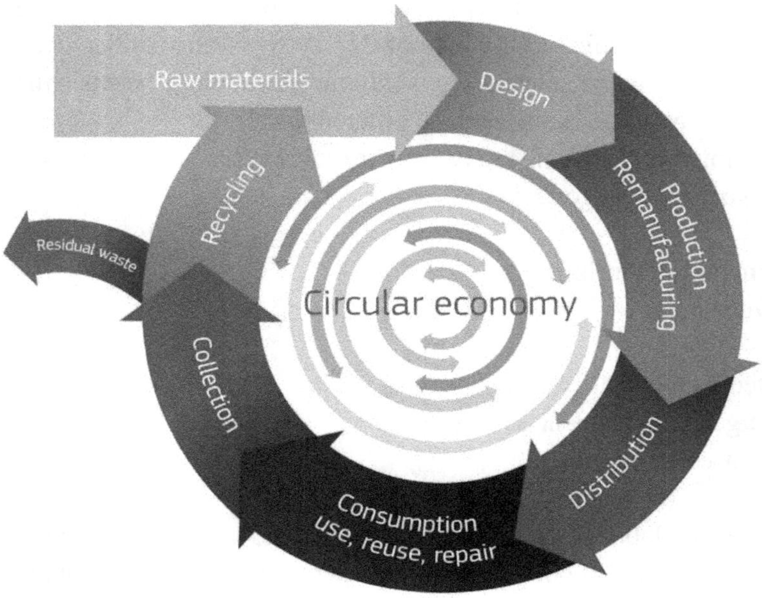

In the car industry, the sector to which our industrial partner (Renault) belongs, the European Commission (2018) has identified several critical raw materials. These materials range from very common in nature materials, such as graphite, to some less common metals such as palladium, platinum and rhodium that account for up to 70% of total consumption. In addition, these materials are widely used in batteries and catalysts so it is foreseeable that their consumption will increase in the coming years. Therefore, recycling and reusing components and materials is fundamental in the car manufacturing industry, so Renault and the University of Valladolid seek to sensitize their students and employees through training that helps to reduce the consumption of raw materials.

Learning Factories

For some years now, universities have been concerned with achieving a better balance between theory and practice in engineering (Lamancusa et al., 2008). This has led to the emergence of new learning paradigms, with an emphasis on action-oriented learning and experiential learning.

Action-oriented learning is tied to specific working situations (Cachay et al., 2012) being focused on the learner's active integration, through their own actions, leading them to deal with complex problems independently, while the teachers assume a moderating role (Abele et al., 2017). On the other hand, experiential learning is an instructional approach in which students learn through direct concrete experience (either spontaneous or designed and organized by the teacher), observation and reflection (McComas, 2014, Kolb, 1984), abstract conceptualization (Kolb, 1984), and active experimentation (Kolb, 1984). Thus, experiential education first immerses learners in an experience and then encourages reflection about the experience to develop new skills, new attitudes, or new ways of thinking (Lewis & Williams, 1994).

The Learning Factory concept is not new, dating back to the First World War, when an Instruction Factory was established at Loughborough Institute in 1916. This factory was aimed to train in the manufacture of war materials (Foden, 1963; Osborne & Case, 2013), but has returned strongly to academia and industry in the last decade (Abele et al., 2015) (Figure 2), from the project led in 1994 by three universities (Penn State, University of Puerto Rico-Mayagüez, University of Washington), a government laboratory (Sandia National Laboratories) and 36 industrial partners (Lamancusa et al., 2008). In it the concept of Learning Factory (LF) was developed. Learning Factories pursue an action-oriented approach (Cachay et al., 2012, Tisch et al., 2013) with participants acquiring competencies through structured self-learning processes in a production-technological environment (Tisch et al., 2013), being a major benefit of them the possibility of Experiential Learning (Abele et al., 2017).

Figure 2. Nº of yearly indexed documents on Scopus for the term "learning factory"

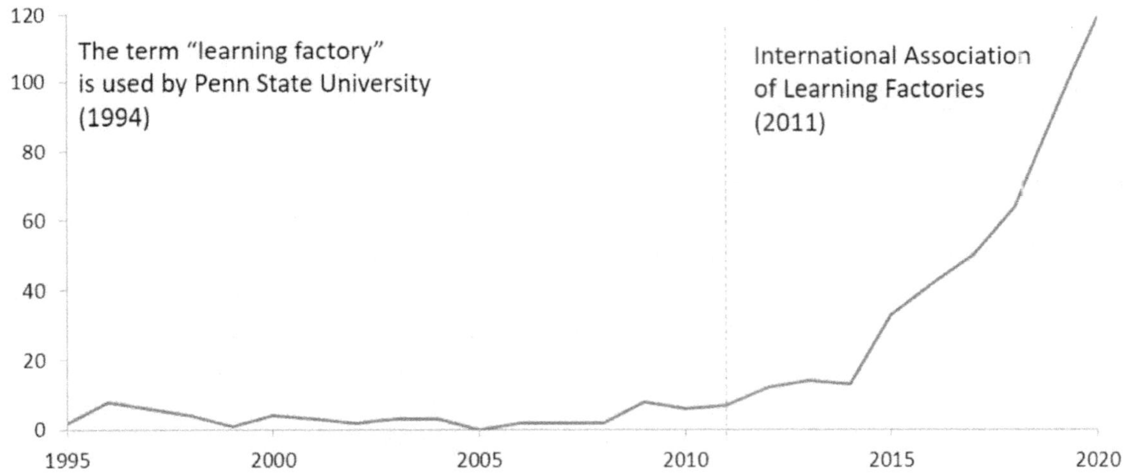

Although many definitions of "learning factories" can be found, the Collaborative Working Group on "Learning Factories for Future-oriented Research and Education in Manufacturing" within CIRP (also abbreviated: CIRP CWG on learning factories) has adopted the following one (Abele, 2019): "*A learning factory in a narrow sense is a learning environment specified by processes that are authentic, include multiple stations, and comprise technical as well as organizational aspects, a setting that is changeable and resembles a real value chain, a physical product being manufactured, and a didactical concept that comprises formal, informal and nonformal learning, enabled by own actions of the trainees in an on-site learning approach. In a broader sense, learning environments meeting the definition above but with a setting that resembles a virtual instead of a physical value chain, or a service product instead of a physical product, or a didactical concept based on remote learning instead of on-site learning can also be considered as learning factories*".

In engineering science, learning factories are not only the most consistent but also the most widespread approach towards education and training (Pittich et al., 2020). Typically, they integrate a wide variety of teaching methods with the objective of moving the teaching–learning processes closer to real industrial problems (Tisch et al., 2013), covering a wide range of different topics, such as automation, process improvement, energy efficiency, logistics and many more (Tisch et al., 2016).

But the greatest added value of learning factories is the integration. The integration of equipment and materials into manufacturing systems, the integration of people from several engineering and business disciplines into effective teams that design and produce products and processes, the integration of undergraduate students and industry workers of different levels and of course the integration of theoretical teachings with their practical application. So, students will work at university or in factories in interdisciplinary teams on real-world problems; and industrial workers will acquire key concepts that they will be able to apply to their jobs.

Despite the several benefits reported in literature about learning factories, such as effective learning, student motivation, and the higher degree of applied knowledge that is achieved, they also have some disadvantages among which it can be mentioned: the amount of resources needed and the associated costs, the lack of generalisation (they are usually focused on a specific topic), the low number of participants in each training process, and the difficulty to move the learning factory to different locations (Tisch et al. 2016; Abele, Metternich and Tisch 2019).

RESEARCH METODOLOGY

In this chapter the research methodology is described through the presentation of the Learning Factory learning process. The learning process is evolutive and incremental, being organized in different phases and stages. When the Lean School was created, the learning process consisted of two major phases during which a reengineering was applied, from a traditional manufacturing system, organised according to push system, to a lean manufacturing system, organised according to pull system (with several stages associated to each phase depending on the training objectives). More recently, a third large phase was added to the learning process, aimed to integrate lean manufacturing concepts with circular economy concepts, by incorporating in the system the recycling of finished products process, once they have been used by the consumer. In this way, the recycled parts are reused/reintroduced in the production of new cars, reducing waste generation. These three phases can be carried out independently or sequentially according to the objectives of the training and of the initial knowledge and competencies of each group. Also, at the beginning of the training an elementary version of a car model with 4 options is used, to which options are added along the several phases and stages, as concepts are introduced into the training.

The methodology followed in each phase (and stage) is the Plan-Do-Check-Act (PDCA) cycle (Figure 3). The training always starts with a short theory session where the concepts and tools that are required to the practical work are taught, followed by the manufacture of 20-30 cars on a designed line. Based on the results obtained, a small reflection session is held (10-15 minutes) where the participants present their vision and the problems found in order to then analyse possible solutions (individually and in groups) and make decisions on the changes to be implemented in the production line.

Figure 3. Model for game-based learning (adapted from De Vin et al. 2017)

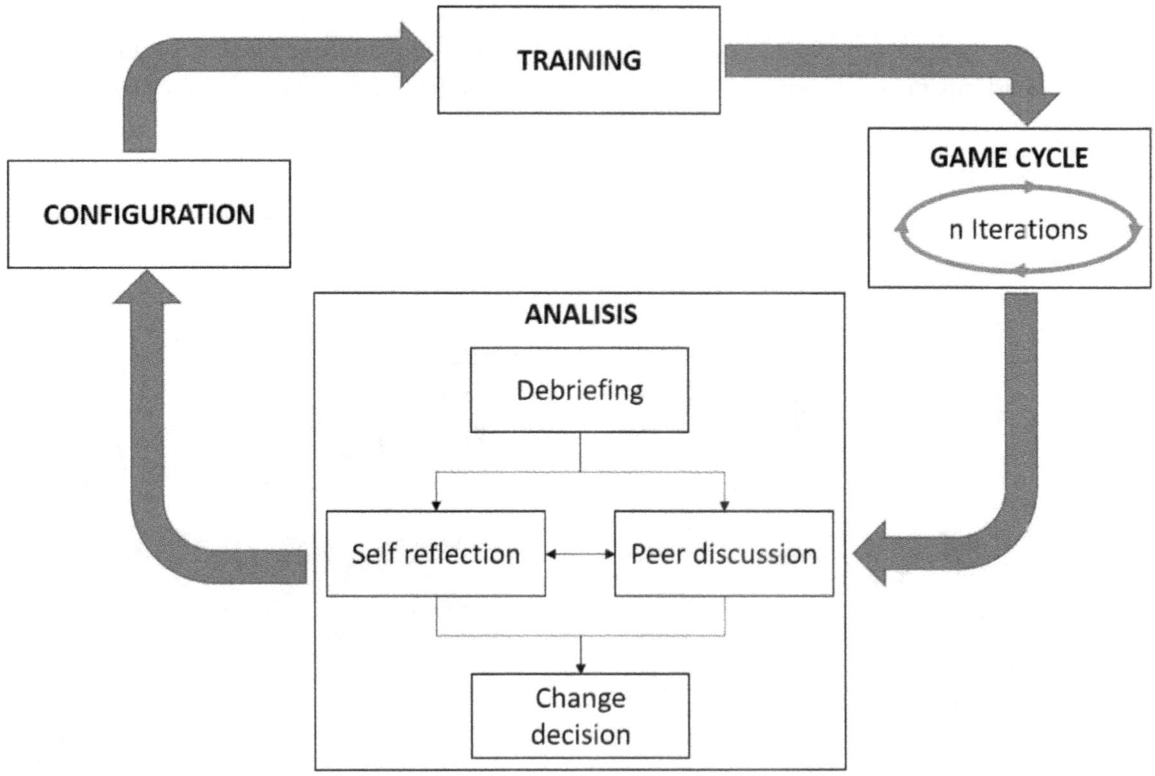

LEAN SCHOOL

In this section the learning factory that resulted from the University-Industry collaboration is presented, since its beginning till the integration of the circular economy concepts. Furthermore, some details of the learning process and of the experience are also provided.

Layout and Product

In 2014, Renault-Nissan Consulting (RNC) (a consultancy company of the Renault-Nissan group, one of the world's top automotive manufacturers) after more than 20 years collaborating in different ways with the University of Valladolid, created a laboratory in the Faculty of Engineering for teaching and development of lean tools for the university students and for employees of Renault and auxiliary companies: the Lean School (Pascual et al., 2019). For the past six years, this space has been used jointly for the training of students from the University of Valladolid (more than 700), Renault employees (around 500) and other employees from surrounding companies (approximately 250).

The Lean School is organized in several areas (Figure *4*): machining and stamping (green continuous rectangle), assembly (blue dashed rectangle) and warehouse (red dotted rectangles). Each of these areas contains different industrial and educational resources necessary for the manufacture and assembly of different products: mainly process equipment and tools (screwdrivers and wrenches of several types, hammer, pliers, shop knives, wrecking bar, rules, carpenter squares and slide calipers), handling and transport of equipment (trolleys and forklifts), storage and packaging equipment, etc., and consumables for participants to adapt equipment to each specific situation. As safety equipment (as in a factory), all participants must wear safety gloves and safety shoes. The modular configuration of the different positions allows participants to change the layout configuration quickly from one to another (based on the training progress).

Figure 4. Initial layout of the lean school to manufacture educational cars (L34N)

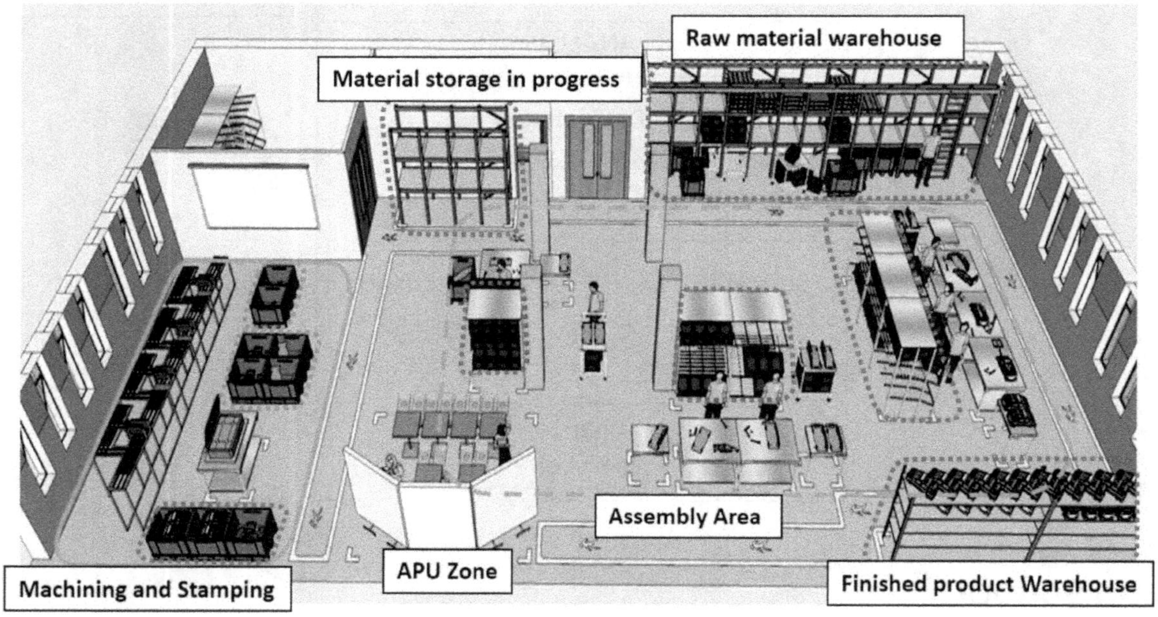

As the main partner of the Lean School is Renault-Nissan Consulting, the initial product manufactured in the learning factory was the L34N car (Figure 5). It is a toy car weighing about 25 kilos with two main models (minivan and pick-up), constituted by more than a hundred parts that result in many car options: colour, headlights, dashboard, wheels, sunroof, seats, etc.). So, more than 1500 different car versions can be manufactured (Gento et al., 2021).

The participants (students or employees depending on the training) receive a written manual that contains a list of the parts needed to manufacture the car, as well as the procedures for manufacturing and assembling the different parts and the quality controls required by customers (Figure 6). Moreover, grounded on that information, participants must develop flowcharts, workstation standards and process diagrams as part of their learning and it serves the teachers to check where reinforcement is needed.

Figure 5. Lean school car model and its main parts

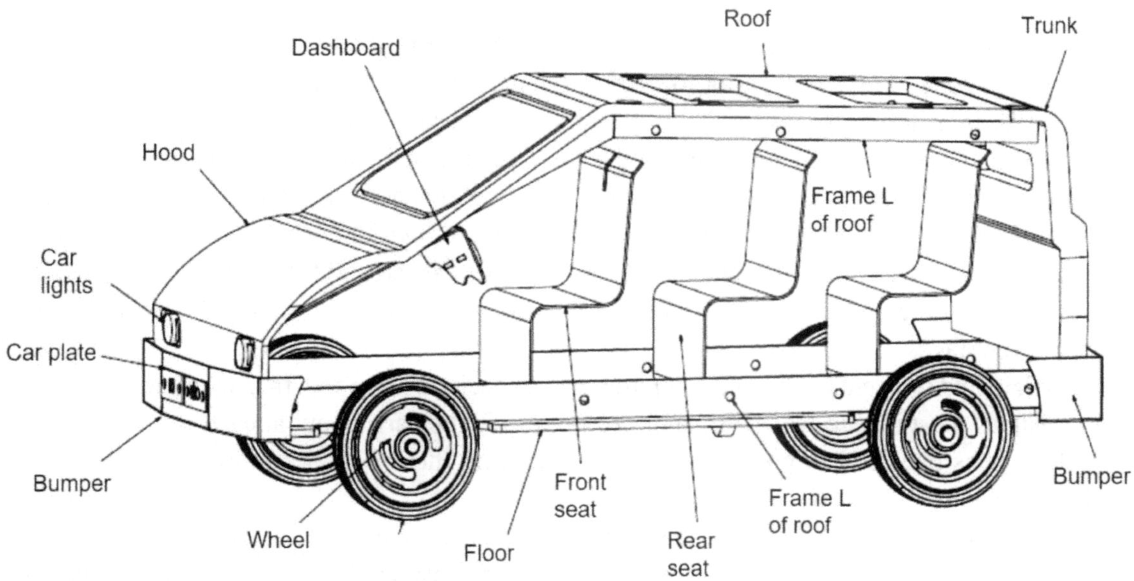

Figure 6. Typical operation process sheet por assembly

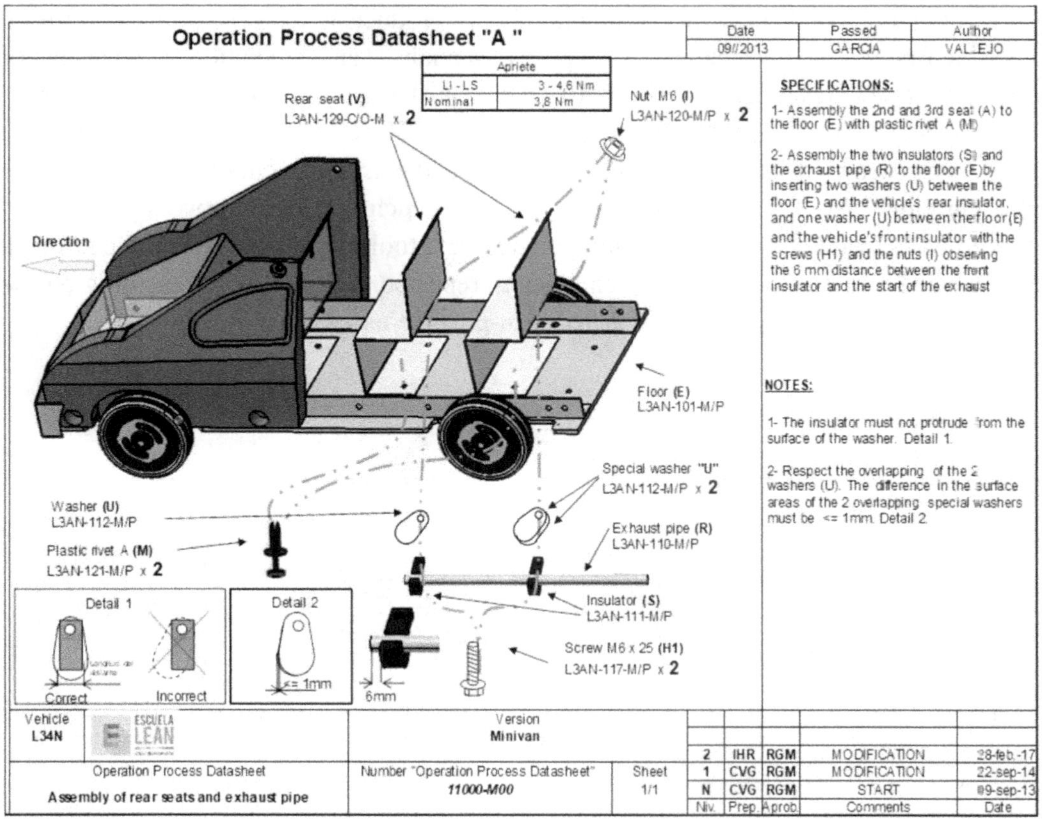

From Traditional Manufacturing to Lean Manufacturing

After a short theoretical training session in the area of the meeting area (or Autonomous Production Unit zone (APU zone in Figure 4), prior to starting the production, the participants are divided into six groups to work in different places by the instructors: a first group composed by a coordinator and an assistant is responsible for the management of the "learning factory", another similar group is in charge for machining operations, a third group, composed by one element, for stamping operations and six elements in the assembly stations is also constituted; a group of two/three people that is responsible for logistics activities between the different workstations is also created and finally a group of participants are assigned as method analysts and timekeepers in order to be able to have data for decision making in future phases/stages.

The participants (students or employees) assigned to each position together with the manager (and his/her assistant) must decide which assembly activities will be carried out at each workstation so that the workload is as balanced as possible taking into account the different size of the pieces and their difficulty to assemble.

Once this has been decided, the participants in each workstation must write down the standard of their position: e.g. operations to be carried out, key points to be considered, time of each operation, quality controls, etc, and test the standard during several cycles to guarantee the uniformity of the production. Logistics operators must also establish their procurement standards, ordering points, routes, and so on, in order to prevent breakdowns at any workstation. Furthermore, the director of the factory (assisted by his/her assistant) must control the production at each workstation and resolve any difficulties that may arise during the manufacture of the L34N car.

The production is carried out until a certain number of products is achieved or until a preestablished duration for the first round 1 is achieved. Afterwards, an analysis and evaluation of the results obtained is performed, after which participants will try to create a push organized flow, by balancing workloads, improving work standards, and producing in smaller batches. Thus, a first step towards lean manufacturing system is approached. Then, participants will work on the achievement of a pull flow by planning a new configuration (layout, workstations, flows, etc.) in which various improvement goals will have to be achieved and several lean manufacturing practices and tools will have to be applied. This training process is repeated a minimum of 4 times and can be repeated up to 9 times depending on the initial training of the participants and the lean tools to teach-practise-use (Figure 7).

Figure 7. Training process

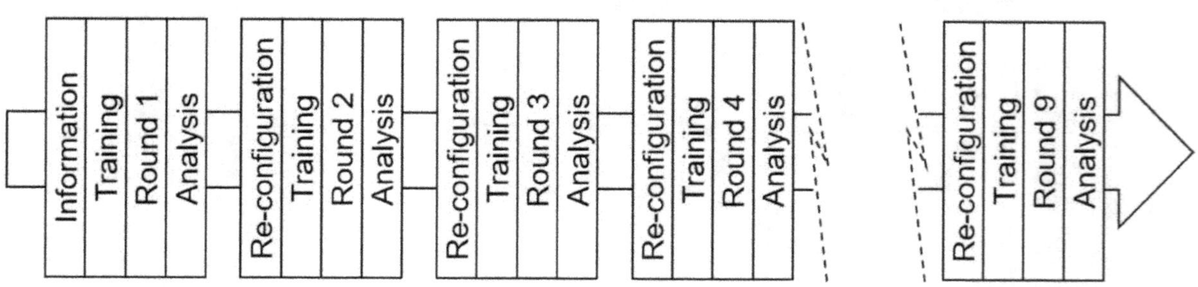

Typically, during these steps, participants manage to reduce space used in the factory by half (Figure 8a). In the last step, to achieve an improved pull flow, they must manage to eliminate the stocks of parts in the workstations considering the wide variety of L34N car versions (more than 1500) that can be manufactured to meet the requirements of customization by clients. To this end, the concept of working with kits that warehouse operators prepare is introduced and explained as well as its introduction on the assembly line (Figure 8b).

Figure 8. (a) Lean School layout ("pull" flow optimized). (b) Final manufacturing line with kits

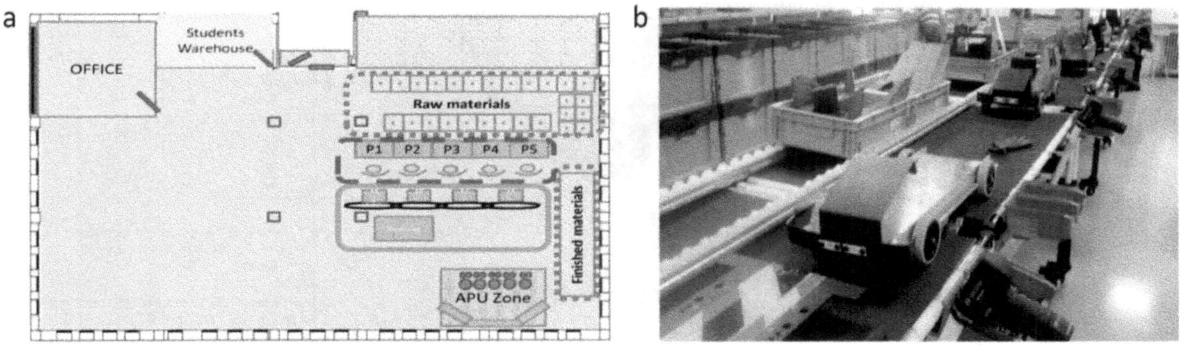

Circular Manufacturing

This last phase has been designed to meet a growing demand from society such as awareness of the need to reduce waste at the end of product life. The Lean School partner (Renault) manufactures cars that can be recycled in more than 85% and with recoverable materials in more than 95%, being manufactured with more than 35% from recycled materials. It is therefore essential that all employees (and students) are aware not only of the environmental gains but also of the cost improvements that can be achieved.

Considering the space gains that have been achieved in the previous phases of the learning process (more than 50% of the laboratory is empty), in this phase a recycling line constituted by four workstations is integrated in the production system, with the most modern concepts of lean manufacturing applied (learned in the previous phase).

Participants must identify vehicle parts and which ones can be reused (plastic parts, unpainted metal components), which ones should be refurbished (painted metal parts, screws, nuts and washers) and which ones should be recycled into raw materials to make new parts (plastic rivets, wheels). Similar to the assembly process, the participants have the operation sheets (Figure 9), but now they must define the disassembly/recycling process and assign them to each workstation so that they are balanced taking into account the space available at each workstation and the time required to complete the disassembly operations.

Figure 9. Operation process sheet for disassembling/recycling

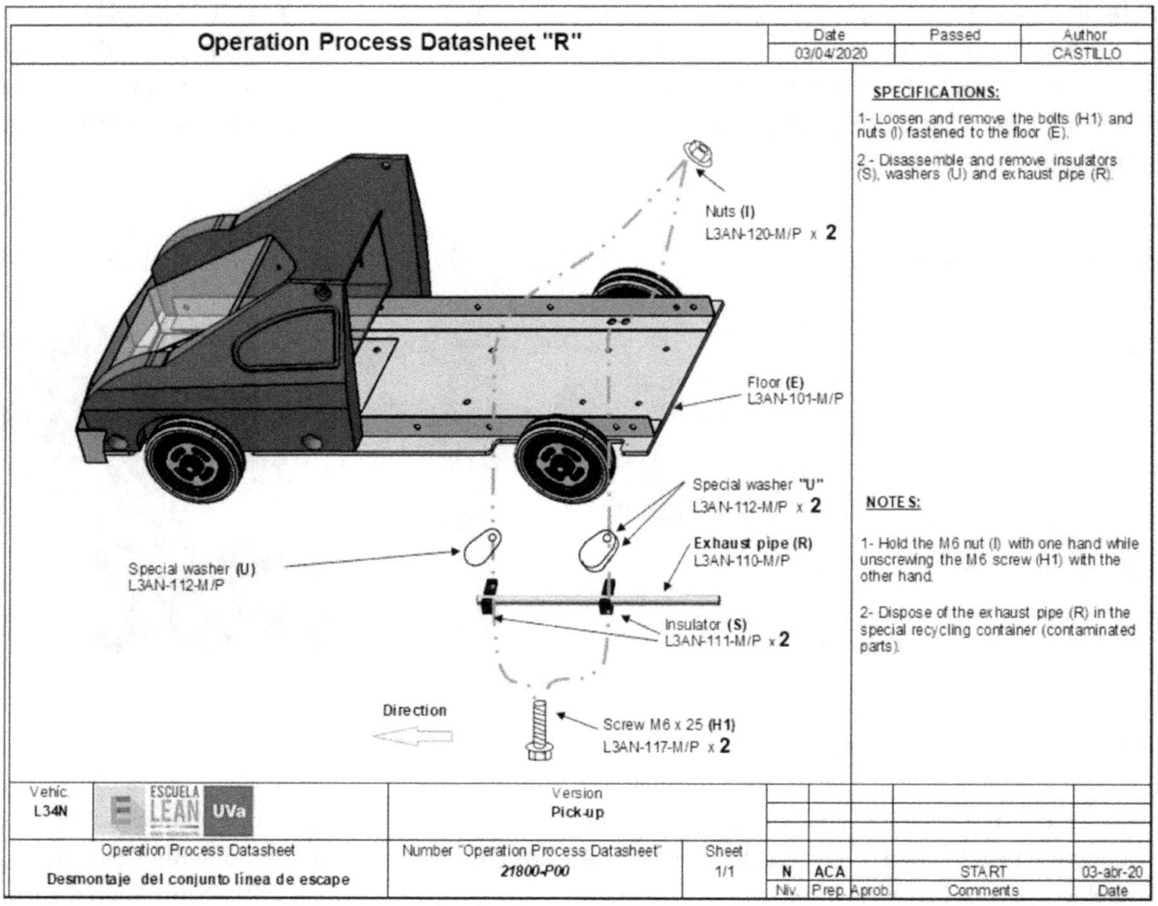

SOLUTIONS AND RECOMMENDATIONS

In this section the solutions and recommendations associated with the learning results and with the learning process are presented.

Learning Results

One of the solutions proposed by the participants in different training courses is the one shown in Figure 10. In this solution the layout is organized in 4 workstations, although on some occasions participants have proposed solutions with 3 or 5 workstations. The distribution of operations and parts to be dismantled/recycled in each of the four workstations is presented in Table 1.

Figure 10. Lean school layout (lean and circular manufacturing)

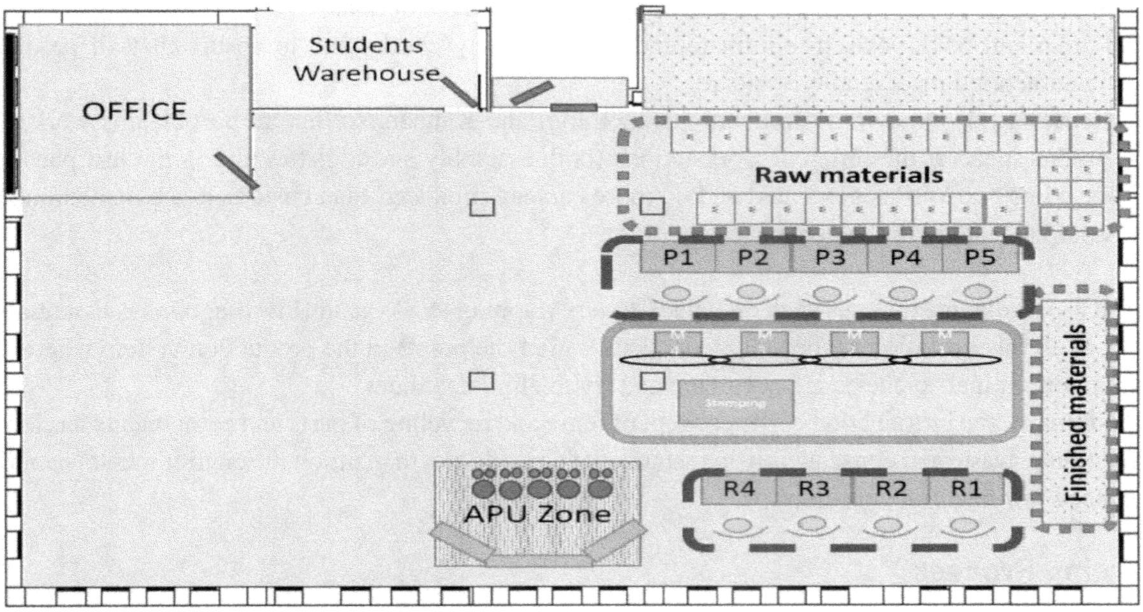

Table 1. Operations and parts to be dismantled/recycled by car version and workstation

	Workstation Recycling 1	Workstation Recycling 2	Workstation Recycling 3	Workstation Recycling 4
Version Pick-up	• Roof. • Hood • Car plate	• Rear bumper • Rear doors	• Seat • Exhaust pipe • Front doors	• Front bumper • Dashboard • Wheels
Version Minivan	• Roof. • Hood • Trunk	• Rear bumper • Rear doors	• Seats • Exhaust pipe • Front doors	• Front bumper • Dashboard • Wheels • Car plate
Boxes	5 Big, 7 Small.	12 Big, 4 Small.	8 Big, 3 Small.	3 Big, 5 Small
Shelves	2	2	2	3

During the whole training process, the participants learn to use different lean manufacturing tools:

- 5S and standardization, to organise and standardise each workstation, as the participants rotate between different workstations during the training process and must be able to perform the assigned operations after a short training by the previous user of the workstation.
- Single Minute Exchange of Die (SMED), because although at the beginning the participants could work in batches, in the successive steps the customer orders are car by car, and in addition the number of versions increases as new improvements are made to the initial models, making it impossible (or at least highly inadvisable) to work in batches, since storage space is limited and

costs increase. By using SMED changeover time and costs of producing very small batches is minimized. Thus, small batch production becomes a viable option.

- Jidoka and poka-yoke, to eliminate end-of-line quality checks and to ensure that all products manufactured meet quality standards.
- Heijunka (also known as production smoothing) and Kanban, so that all participants work at a similar speed at the different workstations (both assembly and logistics and in the last phase in recycling) no WIP is generated and no more cars are produced than customers are demanding or versions that are not in demand.

The use of diagnostic tools such as Value Stream Mapping (VSM) and flow diagrams is essential to carry out all this evolution as they allow the detection of the points in the production system where the most important inefficiencies are occurring and try to eliminate them.

And finally, the introduction of the concept of reuse and recycling of parts and components has led to a reduction in waste and consequently in manufacturing costs, not to mention the environmental benefits of reduced raw material consumption.

Learning Process

The Lean School follows an approach supported in action-oriented learning and experiential learning, promoting students' development and acquisition of learning skills, through their direct involvement and autonomy in the learning process, as well as through their responsibility with the knowledge they acquire during the training. The use of written form is privileged to ensure students will have some elements to support their remembering and memory of the new knowledge. Furthermore, the use of facilities with a good degree of reconfigurability is a prerequisite of the Lean School, to be able to respond to the student's improvement ideas.

An intangible and difficult to quantify result is the empowerment of teamwork, where participants see that the work of a team is more important than the improvement of a single workstation. Without forgetting that when there is a problem it is important to give quick solutions without having to obtain the best solution at the moment: it is better to improve 60% now than 80% tomorrow, since improvement is continuous (kaizen) and tomorrow that 80% improvement can be also reached.

Regarding the instructor role, in the Lean School they mostly assume the role of facilitators and coaches, not forgetting the need to be objective and results-oriented and setting performance objectives to be acquired in the training process. Typically, the instructors are 2 or 3 persons (university professors and Renault consultants/workers) working simultaneously and in a coordinated manner.

FUTURE RESEARCH DIRECTIONS

The economic and business world continues to evolve and with it must continue to evolve the training of today's students and of those who were students and are now workers but who cannot remain stuck in the past.

With this objective in mind, several improvements to the Lean School as future research steps are suggested:

- Incorporating position and pressure sensors towards workstation digitalization, so as to calculate production times and identify the accurate inventories levels at each assembly line point.
- Incorporating active Radio-frequency identification (RFID) tags on the most important parts (mainly plastic and metal components), so that their position in the plant layout is known on real-time, which will allow not only to know the inventory of parts at any given time, but also when they are scheduled to arrive at each workstation.
- Introducing AGVs (automated trucks and tractor units) to eliminate unnecessary (and non-value-added) travel by participants.

With this step towards Industry 4.0, without forgetting the elimination of waste and the recovery of products, components and materials (key aspects in today's world), new goals could be achieved: 1) it would be possible to focus the participants on value creation activities, leaving data collection for automatic systems; 2) the professors would be able to compare the results of many training courses and the participants themselves will know their level of performance compared to previous training editions; and 3) the teams would be able to anticipate maintenance work when abnormal operating values of some machines are detected, etc. In short, the professors team involved in the Lean School would be able to continue advancing at the pace at which companies are advancing and training the workers of the future.

CONCLUSION

In this chapter three clearly distinguished conclusions can be identified. Firstly, the importance of university-business collaboration, with numerous advantages for both: mainly, the opportunity for companies to identify students with high growth potential in their factories using modern management tools; and the university provides students with updated training adapted to the needs of companies, which facilitates their integration into the labour market. Furthermore, through the collaboration the requirements associated to the teaching and learning process/maintenance of a learning factory, such as space, cost and time, become more affordable, if a mutual aid scheme is followed.

Secondly, all the training courses developed incorporate incremental levels of difficulty to adapt to the different needs of students and workers, focusing not only on learning concepts but also on their application in a real environment, facilitating assimilation by "learning by doing". This is important because participants are more likely to believe the results when they experience them than when a teacher or expert explains a concept or gives a master class.

And finally, the introduction of the circular economy concept in the last phase of the learning process, after the elimination of many sources of waste using lean manufacturing, call the attention of participants about the importance of reusing components and recycling materials to continue reducing costs and manufacturing in a sustainable way.

ACKNOWLEDGMENT

We are grateful for the contributions of the Renault workers and the students of the University of Valladolid who have participated in the different trainings because together we continue to develop new cases that respond to their training needs. Special thanks to Renault-Nissan Consulting for its commitment to the University of Valladolid which facilitates the training of students in the first learning factory in Spain.

Carina Pimentel would like to acknowledge financial support from Fundação para a Ciência e Tecnologia grants: UIDB/04058/2020.

REFERENCES

Abele, E., Chryssolouris, G., Sihn, W., Metternich, J., ElMaraghy, H., Seliger, G., Sivard, G., ElMaraghy, W., Hummel, V., Tisch, M., & Seifermann, S. (2017). Learning Factories for Future Oriented Research and Education in Manufacturing. *CIRP Annals*, *66*(2), 803–826. doi:10.1016/j.cirp.2017.05.005

Abele, E., Metternich, J., & Tisch, M. (2019). *Learning Factories. Concepts, Guidelines, Best-Practice Examples*. Springer. doi:10.1007/978-3-319-92261-4

Abele, E., Metternich, J., Tisch, M., Chryssolouris, G., Sihn, W., ElMaraghy, H., Hummel, V., & Ranz, F. (2015). Learning Factories for Research, Education, and Training. *Procedia CIRP*, *32*, 1–6. doi:10.1016/j.procir.2015.02.187

Ankrah, S., & Al-Tabbaa, O. (2015). Universities–Industry Collaboration: A systematic Review. *Scandinavian Journal of Management*, *31*(3), 387–408. doi:10.1016/j.scaman.2015.02.003

Bonarccorsi, A., & Piccaluga, A. (1994). A Theoretical Framework for the Evaluation of University—Industry Relationships. *R & D Management*, *24*(3), 229–247. doi:10.1111/j.1467-9310.1994.tb00876.x

Cachay, J., Wennemer, J., Abele, E., & Tenberg, R. (2012). Study on Action-Oriented Learning with a Learning Factory Approach. *Procedia: Social and Behavioral Sciences*, *55*, 1144–1153. doi:10.1016/j.sbspro.2012.09.608

De Vin, L. J., Jacobsson, L., Odhe, J. E., & Wickberg, A. (2017). Lean Production Training for the Manufacturing Industry: Experiences from Karlstad Lean Factory. *Procedia Manufacturing*, *11*, 1019–1026. doi:10.1016/j.promfg.2017.07.208

Ellen MacArthur Foundation. (2012). *Towards the Circular Economy - Economic and Business Rationale for an Accelerated Transition*. Ellen MacArthur Foundation.

Elsevier. (2021). *University-industry collaboration*. https://www.elsevier.com/research-intelligence/university-industry-collaboration

European Commission. (2008). *Communication from the Commission to the European Parliament and the Council — The raw materials initiative: meeting our critical needs for growth and jobs in Europe*. COM (2008) 699.

European Commission. (2011). *Communication from the Commission to the European Parliament, the Council, the European Economic and Social Committee and the Committee of the Regions, Tackling The Challenges in Commodity Markets and on Raw Materials*. COM (2011) 25.

European Commission. (2015). *Communication from the Commission to the European Parliament, the Council, the European Economic and Social Committee and the Committee of the Regions, Closing the Loop - An EU Action Plan for the Circular Economy*. COM(2015) 614.

European Commission. (2018). *Report on Critical Raw Materials and the Circular Economy*. Commission Staff Working Document.

European Union (2000). *Directive 2000/53/EC of 18 September 2000 on end-of life vehicles*. Official Journal of the European Communities, (21.10.2000) 269/34–269/43.

European Union (2005). *Directive 2005/64/EC of 26 October 2005 on the type-approval of motor vehicles with regard to their reusability, recyclability and recoverability and amending*. Official Journal of the European Communities, (25.11.2005) 310/10–310/27.

Foden, F. E. (1963). Herbert Schofield and Loughborough College. *The Vocational Aspect of Secondary and Further Education*, *15*(32), 231–246. doi:10.1080/03057876380000271

Gento, A. M., de Benito, J. J., Sanz, P., & Pascual, J. A. (2016). Lean School: A Practical Space of Cooperative Learning from the Factory to the University. In C. Hernández (Ed.), *Advances in Management Engineering. Lecture Notes in Management and Industrial Engineering* (pp. 209–220). Springer International Publishing.

Gento, A. M., Pimentel, C., & Pascual, J. A. (2021). Lean school: An example of industry-university collaboration. *Production Planning and Control*, *32*(6), 473–488. doi:10.1080/09537287.2020.1742373

Kirchherr, J., Reike, D., & Hekkert, M. (2017). Conceptualizing the circular economy: An analysis of 114 definitions. *Resources, Conservation and Recycling*, *127*, 221–232. doi:10.1016/j.resconrec.2017.09.005

Kolb, D. A. (1984). *Experiential Learning: Experience as the Source of Learning and Development*. Prentice Hall.

Kurdve, M., & Bellgran, M. (2021). Green lean operationalisation of the circular economy concept on production shop floor level. *Journal of Cleaner Production*, *278*, 1–11. doi:10.1016/j.jclepro.2020.123223

Lamancusa, J. S., Zayas, J. L., Soyster, A. L., Morell, L., & Jorgensen, J. (2008). The Learning Factory: Industry-Partnered Active Learning. *Journal of Engineering Education*, *97*(1), 5–11. doi:10.1002/j.2168-9830.2008.tb00949.x

Lewis, L. H., & Williams, C. J. (1994). Experiential Learning: Past and Present. *New Directions for Adult and Continuing Education*, *62*(62), 5–16. doi:10.1002/ace.36719946203

McComas, W. F. (2014). *The Language of Science Education: An Expanded Glossary of Key Terms and Concepts in Science Teaching and Learning*. Springer., doi:10.1007/978-94-6209-497-0

Osborne, W., & Case, K. (2013). Instructional Factories and Training Courses: The Inception and Implementation of Training Courses for Un-Skilled and Semi-Skilled Munitions Workers during the Great War. *Proceedings of the 11th International Conference on Manufacturing Research (ICMR2013)*, 331–336.

Pascual, J. A., Hoyuelos, I., Mateo, M., & Gento, A. M. (2019). Lean school: A learning factory for training lean manufacturing in a physical simulation environment. *Management and Production Engineering Review*, *10*(1), 4–13.

Pittich, D., Tenberg, R., & Lensing, K. (2020). Learning Factories for Complex Competence Acquisition. *European Journal of Engineering Education*, *45*(2), 196–213. doi:10.1080/03043797.2019.1567691

Rybnicek, R., & Königsgruber, R. (2019). What Makes Industry–University Collaboration Succeed? A Systematic Review of the Literature. *Journal of Business Economics*, *89*(2), 221–250. doi:10.100711573-018-0916-6

Tisch, M., Hertle, C., Abele, E., Metternich, J., & Tenberg, R. (2016). Learning Factory Design: A Competency-Oriented Approach Integrating Three Design Levels. *International Journal of Computer Integrated Manufacturing*, *29*(12), 1355–1375. doi:10.1080/0951192X.2015.1033017

Tisch, M., Hertle, C., Cachay, J., Abele, E., Metternich, J., & Tenberg, R. (2013). A Systematic Approach on Developing Action-Oriented, Competency-Based Learning Factories. *Procedia CIRP*, *7*, 580–585. doi:10.1016/j.procir.2013.06.036

World Economic Forum. (2014). *Towards the Circular Economy: Accelerating the scale-up across global supply chains*. https://www3.weforum.org/docs/WEF_ENV_TowardsCircularEconomy_Report_2014.pdf

ADDITIONAL READING

Alves, A. C., Flumerfelt, S., & Kahlenva, F. J. (Eds.). (2017). *Lean Education: An Overview of Current Issues*. Springer International Publishing. doi:10.1007/978-3-319-45830-4

Ellen MacArthur Foundation. (2014). *Towards the Circular Economy - Accelerating the scale-up across global supply chains*. Ellen MacArthur Foundation.

Kolb, D. A. (1984). *Experiential Learning: Experience as the Source of Learning and Development*. Prentice Hall.

Kolb, D. A., & Kolb, A. Y. (2017). *The Experiential Educator: Principles and Practices of Experiential Learning*. Experience Based Learning Systems, Inc.

Mavropoulos, A., & Nilsen, A. W. (2020). *Industry 4.0 and Circular Economy: Towards a Wasteless Future or a Wasteful Planet?* Wiley.

Rother, M., & Shook, J. (1999). Learning to See: Value Stream Mapping to Add Value and Eliminate Muda. In The Lean Enterprise Institute. Productivity Press.

Shingo, S., & Dillon, A. P. (1989). *A Study of the Toyota Production System: From an Industrial Engineering Viewpoint (Produce What Is Needed, When It's Needed)*. CRC Press.

Womack, J., & Jones, D. (2003). *Lean Thinking: Banish Waste and Create Wealth in Your Corporation*. Simon and Schuster. Free Press.

Womack, J., Jones, D., & Roos, D. (1990). *The Machine That Changed the World: How Lean Production Revolutionized the Global Car Wars*. Simon and Schuster. Free Press.

KEY TERMS AND DEFINITIONS

3Rs/9Rs: In circular economy it stands for Reduce, Reuse, Recycle. Nowadays, the concept has been extended by adding more R-words to reach 9R covering the whole life cycle of a product from the initial design: Refuse, Rethink, Reduce, Reuse, Repair, Refurbish, Remanufacture, Repurpose, Recycle, and Recover.

Circular Economy: An economic system aimed at eliminating waste and the continuous use of resources. It uses "R" to define the different activities that can be done with a product at the end of its useful life: reuse, repair, refurbishment, recovery, remanufacturing and recycling to make a closed loop system, and minimise the use of resources and the creation of waste. Even adding more "R's" in the design phase: rethink and reduce.

Lean Manufacturing: (Also known as lean production) is a production system whose main objective is to reduce lead times by eliminating waste and facilitating the flow of materials within the production system, thus improving quality and response times to customers.

Learning by Doing: It is a methodology based on the idea that students perform actions and learn by observing their results. It is based on Edgar Dale's concept of the learning cone where against the 10% of concepts that are retained when heard, up to 90% are retained when people practice them.

Learning Factory: Is a laboratory that simulates a simplified factory where participants learn different concepts using the same tools as in a real factory. Its main characteristic is the active learning of the participants doing tasks in a similar way as they do in a real workplace and the modularity to adapt to the learning process.

Lifelong Learning: Refers to continuous learning throughout a person's life. It is nowadays considered necessary given the need for continuous updating due to the enormous amount of information and changes that occur.

Plan-Do-Check-Act (PDCA) Cycle: It is a management method used for the continuous improvement of products and processes. It consists of four phases: Plan (definition of the goals and processes required to achieve them), Do (implementation of the processes), Check (review of the results compared to the proposed goals), and Act (improvement of the processes if necessary). This cycle can be repeated indefinitely and is also known as the Deming cycle.

University-Industry Collaboration: Is the relationship that exists between universities and industry to promote the exchange of knowledge and technology. It can take different forms, but the most important are collaboration and research agreements, talks and conferences and student internships.

VUCA Environment: VUCA is an acronym, first used in 1987 by the U.S. Army War College to describe the more volatile, uncertain, complex, and ambiguous world that has emerged since the end of the Cold War. This concept has been extended to business and education as a result of the evolution of technology and the continuous changes in which we live.

Waste (Muda in Japanese): Is anything that does not add value to the product or service and for which the consumer is not willing to pay. According to the Toyota production system, 7 types of waste have traditionally been identified: Transportation, Inventory, Motion, Waiting, Overproduction, Over Processing, and Defects, but nowadays one or two more are usually added: Resistance to Change and Inaction.

Chapter 10
Graduate Lean Leadership Education:
A Case Study of a Program

Shannon Flumerfelt
Oakland University, USA

Calandra Green
Oakland University, USA

ABSTRACT

A midwestern university in the USA implemented a Lean Leadership Graduate Certification Program in the 2018-2019 academic year for current and emerging leaders seeking to extend, enrich, or establish leadership knowledge, skills, abilities in the workplace. The purpose of this chapter is to share the results of an evaluation on the effectiveness of this Lean Leadership Graduate Certificate Program. The results from this case study on the Lean Leadership Graduate Certification Program indicated a need to market to a larger group of emerging leaders. Leadership development findings suggest the need to further advance knowledge development in Lean students and consideration for program goals that include strategies having a significant impact on Lean student's emotional well-being in meeting leadership challenges. A continuous need to reinforce Lean Leadership competencies as a core dimension of the program resulted in the largest impact of the program with the Lean Leadership students.

INTRODUCTION

In the fall of 2018, an American midwestern university introduced a unique model of leadership development called the Lean Leadership Graduate Programs, offered as a dual option transdisciplinary graduate program. One program option, the Lean Leadership Graduate Certificate, was designed for practicing or emerging leaders from any sector or any workplace who desire to extend, enrich or establish leadership knowledge, skills and abilities through mastery of Lean tenets, tools and competencies. The second program option, the Lean Leadership Cognate for the Ed.D. in Leadership or Ph.D. in Educational

DOI: 10.4018/978-1-7998-8816-1.ch010

Leadership, offered an additional semester of study for those with master's degrees pursuing doctoral study. Both offerings intended to enable students to identify strengths and weaknesses as leaders and provide clear pathways for their future roles.

A series of two employer studies were conducted as the program was designed, developed, improved and adopted (Flumerfelt, Alves, Leao & Wade, 2016). For those two studies a pre- and post-design assessment was delivered to 37 American employers using Lean. The program was examined for workplace relevance in program design and delivery, student outcomes and organizational impacts. Determining the effectiveness, impact and outcomes of the Lean Leadership Graduate Certificate were the focus areas of this case study. Employers did endorse the final program design and feedback was positive as to the need for transdisciplinary Lean Leadership graduate programming, the program design and delivery, the content and competencies masteries included in student outcomes and anticipated organizational outcomes.

This chapter is focused on student outcomes of the program from its first year of deployment. To complete the study, current leaders across different organizations (finance, healthcare, higher education, and manufacturing) who were currently enrolled in the first Lean Leadership Certification course were surveyed. The survey was conducted to determine motivation for enrollment, an assessment of leadership competencies before and after the course competition, and a query of the most value-added components of the program. This case study was designed to provide insight into the question as to whether this unique type of lean training added value to the profession of leadership. This chapter explores lean theories, philosophies, and tools and examines the results of the program's case study as a qualitative measure of program effectiveness for student outcomes.

THE CASE FOR LEAN LEADERSHIP

Many leaders today share stories of silent suffering in their roles as leaders (Dahl, 2020). Job titles are matched with unrelenting deliverables to meet hierarchical needs. The constant pressures of process improvement, evolving technological advances, and overall organization change, creates an uneasiness for leaders who are trying to maintain confidence in their own individual abilities. When leaders have overwhelming feelings of underdevelopment, they may seek opportunities outside of the organization. Some have chosen to pursue advanced degrees aimed at improving their leadership skills. However, leaders are also seeking programs to help inform them of a better understanding of today's leadership challenges. Lean Leadership has become a favorable option for many who struggle with balancing rapid change in unsteady organizational climates.

The challenge of leadership today is viewed as a multifaceted approach to simultaneously working to improve finance, growth, customer service, quality, and technology (Longenecker & Yonker, 2013). Further, organizations are experiencing tumultuous change at unprecedented levels, due to societal change, Industry 4.0, and tough competition, meaning that complexity is increasing dramatically (Flumerfelt, Schwartz, Mavris & Briceno, 2019). Research also makes it clear that without effective leadership practices, at all levels of an enterprise, performance improvement and achieving better results will be difficult or even nonexistent (Longenecker et al., 2009). Authors Higgs and Rowland (2005) suggested that the inability of organizations to compete with economies of scale results in a condition that "runs the risk of being put out of business" (p. 159). Leadership sustainability requires a skill level beyond what currently exists to compete in today's market.

Leaders today are experiencing rapid changes in organizational climate. As work is compounded, leaders lose focus on what takes precedence. For example, how do leaders determine whether to focus on cost or quality? Neither benchmark should be sacrificed, however, leaders must develop skills to meet the bifurcated needs of their organizations (Deming, 1993).

Researchers view present day leadership development needs to be inclusive of leadership development increasingly occurring within the context of work, including critical reflection about the role of competencies in leadership development (SHRM, 2008) and revisiting the issue of work/life balance (Riordan, 2013; Hernez-Broome & Hughes, 2004). Leaders are expected to comprehend how the work is being delivered to the customer and determining its value is key to the work environment (Association for Manufacturing Excellence, 2007).

Personal family relationships and responsibilities combined with a constantly changing work environment add additional stress to work/life balance and drive significant impacts on leadership development (Hernez-Broome & Hughes, 2004). Researchers introduced the need for the development of resilience in leaders to balance these challenging work/life situations. Moxley and Pulley (2004) are cited who stated that resilience "is an active process of self-righting and growth that helps people deal with hardship in a manner that is conducive to development" (p. 29). It was found that having resilience allowed leaders to face future hardships successfully by taking difficult experiences and using them as opportunities (Hernez-Broome & Hughes, 2004). In other words, leaders must leverage their strengths and identify strategies to improve upon weaknesses.

It is also important to consider what elements could prove lacking in a leader. According to Longenecker and Yonker (2013), "leadership performance deficiencies can result in significant performance and career advancement problems for individuals" (p.160). The authors described two scenarios where leaders felt blindsided and ill-prepared for rapid changes within an organization. First, leaders must rise to any challenge when faced with new tasks in which they are unconsciously unskilled. Second, leaders must adapt to an environment where minimal feedback and direction is offered by senior leaders. Longencker and Yonker sampled 153 middle and front-line managers in the manufacturing and service industries to participate in a multi-source feedback (MSF) exercise using these two scenarios. As a result, Longenecker and Yonker (2013) outlined the top five leadership deficiencies based on the MSF group feedback as:

1. Ineffective communication practices
2. Poor time and priority management skills
3. Failing to clarify direction and performance expectations
4. Ineffective interpersonal/teaming skills
5. Inability to handle stress and staying poised (p. 162)

Leaders are often given uncharted tasks with expectations that are unclear or poorly defined. The lack of understanding could play a major role in the leader's ability to be successful.

In turn, leadership development is a key component to successful leadership practice and it serves as a countermeasure to confounding organizational life (Bolman & Deal, 2013). While classroom training plays a major role in leadership development, life experiences and storytelling has a significant impact in leadership development. Bolman and Deal (2013) present four frames for leadership and one frame, symbolism, is an extremely powerful frame for connecting oneself to leadership via meaning and humor. In fact, developmental experiences have the greatest impact when they can be linked to or embedded in a person's ongoing work and when they are an integrated set of experiences (Hernez-Broome & Hughes, 2004). For example, determining key elements to previous successes and failures on difficult tasks could lay the foundation on how to move forward with new tasks. These researchers maintained that, "Activities like coaching and mentoring are increasingly key elements of leadership development initiatives" (p. 24). We have known what to do for many years in terms of adult leadership learning. For example, Hernez-Broome and Hughes cited Hall et al., (1999) who stated that when coaching is typically done one-on-one, then insights into behavioral changes are even possible. Ting and Hart (2004) also stated that "the most effective coaching allows for collaboration to assess and understand the developmental task to challenge current constraints while exploring new possibilities, and to ensure accountability and support for reaching goals and sustaining development" (p. 24). In other words, leadership development is a continuous process of life as a leader that relies heavily on personal and shared experiences.

Therefore, leadership development today has shifted from knowledge acquisition to action learning whereby training is more substantial when it occurs within the context of work (Hernez-Broome & Hughes, 2004). To create relevancy with critical real time business matters, developmental leadership is combining instruction with a real business setting (Hernez-Broome & Hughes, 2004). Hernez-Broome and Hughes (2004) argued that the "development of leaders today means providing people opportunities to learn from their work rather than taking them away from their work to learn" (p. 27). Leaders can learn a great deal from visually overseeing workflows, providing real-time feedback, and interacting with internal and external customers and employees.

Leadership attributes in the past relied on a person's skills or traits. Competencies in leadership development are at the center of most organizations today (Hernez-Broome & Hughes, 2004) and are in high demand. Hernez-Broome and Hughes (2004) cited Barrett and Beeson (2002) who argued that "a recent benchmarking study found that leading-edge companies define leadership by a set of competencies that guide leadership development at all levels" (p.29). Additionally, the authors cited APQC (2000) who maintained that "leadership competencies of a best-practice organization uniquely fit the organization, its particular strategy, and its business model" (p. 29). Leaders cannot solely rely on job performance as a key indicator of success. Understanding what core competencies best align with the vision, mission, and values of an organization could be beneficial.

After coaching, mentoring, leading within the context of work, and balancing of stress/peace is accomplished, leaders continue to need a framework for continuous improvement in an ever-changing environment. This is where the impact of Lean Leadership becomes critical. Liker and Convis (2012) stated, "We believe that the biggest gap in capabilities in the lean movement, and the root cause of failure of many lean programs, is in leadership" (p.xiii).

Lean evolved in the manufacturing industry as an improvement measure for vehicle production and line efficiency (Dombrowki & Mielke, 2013). Lean in manufacturing decentralizes positioning between managers and employees to optimize operational issues. Researchers suggested that to achieve and sustain results, a change in behavior and mindsets of leaders must occur. For example, Dahl (2020) stated, ". . . the older and more established an organization becomes, the harder it is to ensure it continues to change and evolve. This is where Lean leaders must step in and ensure the spirit of evolution. . ." (p. 22). Lean is aimed to improve every process every day and research argued that front-line staff are often the first to recognize a deviation from the standard (Dombrowki & Mielke, 2013), however, the need for leaders to ensure that unacceptable variances and errors are removed and new methods are sustained through leader standard work (Ptacek, Coats, Ptacek, 2012).

Lean is not limited to manufacturing as its philosophy can be used for any application whose goal is a team effort to add value by meeting the needs of the customer first by eliminating waste and reducing costs (Sawhney & Chason, 2005). Sawhney and Chason (2005) suggested the transition to Lean is about organizational culture change and confronting work force problems like "resistance to change, lack of necessary job skills, low morale, and decision to recruit new employees or retain current workforce" (p. 77). Employees are the key factor to successful Lean implementation (Sawhney & Chason, 2005), but there must exist some philosophy and implementation of leadership. Leaders can benefit from understanding employee satisfaction and continuously improving organizational culture through increased leadership awareness (Flumerfelt & Soma, 2012).

Lean as a solution is being implemented across multiple industries today (Alvaro, Alves, Sousa, 2019). A study in Canada reported Lean being largely embraced in healthcare showing 73% of Canadian healthcare regions implementing Lean thinking into its practice (Goodridge et al., 2015). The investment was a sizable aim to "thinking and acting as one to create better health, better value, better care, and better team" (p. 2). Their goals included a formal infrastructure with the entire workforce participating in daily continuous quality improvement. While this study is in its early stages, tensions between decentralizing blue- and white-collar worker authority were noted. Additionally, attitudes of both leaders and employees created tension and need to change to create a better organizational culture.

Lean does not come without failure or limitation. Industry Week conducted a survey in 2007 and reported only 2% of companies who implemented Lean achieved their objectives and less than 1% reported achieving significant results (Goodridge et al., 2015). Contributing factors are attributed to lack of commitment by senior management and the unwillingness to "accept the cultural change required for Lean to be a success" (p. 4). However, where Lean is implemented with ongoing vigilance by leaders and regression to customary ways of doing business can be avoided, results of Lean are positive (Goodridge et al., 2015). It has also been found that a key contributor to the success of Lean implementation for leaders is continually reinforcing communication with key stakeholders (Emiliani, 2008).

THE CASE STUDY TO EVALUATE A LEAN LEADERSHIP GRADUATE PROGRAM

The need for an outcome evaluation on the effectiveness of the Lean Leadership Graduate Certificate program at an American midwestern university was based on the inaugural launch of the program so that faculty could better understand the program's impact or lack thereof. This assessment was chosen to learn more about the vulnerabilities of the Lean Leadership program and how well it informs the practice of its students in leadership today. Research was conducted to measure where the program has added value to leadership competencies and to learn where to reduce barriers to sustaining leadership skills.

Researchers Glick and Kulbok (2001) argued that "The purposes of program evaluation include providing timely data to inform judgments and decisions about improving, continuing, or terminating a program" (p.37). These authors suggested that program evaluation is critical to understand the need to adjust and align goals. They maintained that "The outcome of a program evaluation, a process that may be grounded in objective evaluation data but driven by a political process of negotiating and balancing different perspectives and interests" (p.37). The motivation for exploring an outcome evaluation of the Lean Leadership program is to research student perspectives of program outcomes.

The Context of the Case Study

A modest case study took place at an American midwestern university, established in 1957 and located in Oakland County amidst its partners in hospitals, educational institutions, Fortune 500 companies, cities, and government agencies. An American midwestern university supports over 19,000 enrolled students, including 810 international students, on and off campus living and has 300 student organizations. Programs include 143 bachelor's degree programs, 130 doctoral, master's degree and certificate programs. An American midwestern university benefits from the Pawley Lean Institute, an endowment established to promote student learning in Lean. The James Womack Foundation endorses the Pawley Lean Institute and has established a scholarship fund for students (Flumerfelt, Ross and Wade, 2017). The Department of Organizational Leadership within the School of Education and Human Services offers both undergraduate and graduate programs in Lean Leadership.

The Lean Leadership Graduate Programs at an American midwestern university offer two post graduate level curricula for enrollment: the Graduate Certificate program, compatible with any undergraduate degree, and the Cognate for Ed.D in Leadership, compatible with any Master's Degree. The Lean Leadership Graduate Programs has three or four separate semesters of study, ranging from six credits in the fall and winter semesters to two four-credit summer semester courses that require specified Graduate Certificate or Ed.D plan enrollments. The program included critical learning methodologies, such as problem-based, work-embedded approaches to discovery learning. Content and competency mastery was achieved through flipped learning methods with asynchronous online and in-person intensive Friday-Saturday sessions, content delivery via online lectures, work-embedded improvement projects, and mentorship and menteeship experiences.

Flumerfelt, an Endowed Professor of Lean and the Lean Leadership Program Coordinator, defined Lean as "The highest value that you can deliver to a customer using the fewest resources in the most efficient way and also making sure that all people in an enterprise itself whether it's an external customer or employees internally are always treated with respect" (Flumerfelt, 2017). The Lean Leadership Program at Oakland University was developed to respond to workplace needs based on reflective feedback from stakeholders across several different industries.

Lean Leadership is defined as the knowledge, beliefs, and actions that demonstrate respect for people and continuous improvement. Mr. Sperl, a Lean Leader Faculty and Lean Consultant, asserted the importance of Lean within the scope of organizational growth corresponding to leader growth. This Leader described Lean as helping workers become problem solvers. By determining the root cause of a problem, leaders can help guide workers improve work processes that align with organizational goals.

In 2018, the inception of the Lean Leadership Graduate Programs enrolled seven students, two students dropped out of the program in the first two semesters, two students graduated and received a Graduate Certificate in Lean Leadership, and three students completed the fourth semester cognate requirement for acceptance into the Ed.D in Organizational Leadership Program. In 2019, the second year of the Lean Leadership Programs enrolled seven students, five students graduated from the program and received a Graduate Certificate in Lean Leadership, and two students completed the fourth semester cognate requirement and were accepted into the Ed.D. in Organizational Leadership Program. All students were from a variety of public and private sectors, including manufacturing, healthcare, finance, and higher education, representing engineering and management.

Design of the Study

The program was evaluated as a case study using a basic qualitative research design with a qualitative survey. The survey focused on three main leadership topics that included leadership development, leadership challenges, and the impact of the Lean Leadership program on the students. A survey was administered to seven students, representing those enrolled in the program at the time or during that academic year. Their feedback was used to gain insight regarding the effectiveness of the program.

The design approach used to conduct the Lean Leadership certificate program evaluation was a basic qualitative research design. Merriam and Tisdell (2015) described a basic qualitative research study as having philosophically been derived from constructionism, phenomenology, and symbolic interaction and is used by researchers who are "interested in "(1) how people interpret their experiences, (2) how they construct their worlds, and (3) what meaning they attribute to their experiences. The overall purpose is to understand how people make sense of their lives and their experiences" (p. 38).

To determine the effectiveness of the Lean Leadership certificate program, a qualitative survey was developed using Qualtrics, a qualitative survey platform. Statistical data were obtained about the characteristics of the population being surveyed. Leadership development questions were defined in the survey as an assessment of current leadership knowledge, application, values as it applies to leadership skills. Leadership challenge questions assessed trust and respect, management communication and effectiveness, career development and work-life balance. The impact of Lean questions was an assessment as to the level of leadership mastery prior to and after participation in the Lean Leadership Programs. Key factors to program success, individual growth through program implementation, and identifiable strategies for continuous improvement. Merriam and Tisdell (2015) argued that this type of qualitative approach can uncover and illuminate the details of an educational or leadership process from the experiences of the individuals who created and or applied the process.

Prior to the survey being administered, feedback on the draft construct to items was requested from a select Jury Review Committee. The feedback provided was reviewed and additional edits made to the draft constructs to items considered. The first part of the survey collected background information about the participants, including age, ethnicity, gender, marital status, level of education, employment status, and income level. The remaining parts of the survey addressed three main areas – Leadership Development, Leadership Challenges, and Impact of Lean.

Final approval of the Qualtrics survey was obtained by the client to ensure the scope of research requested by the client was met. Once full permissions were acquired, a single reusable link was generated and forwarded to the client. The survey link was uploaded to Moodle, forwarded by the client, for students to access who are currently enrolled in the program and emailed to former students of the program by the researcher.

A mixed approach of survey type questions was used to elicit responses ranging from 6 text entry, 18 multiple choice, 3 net promoter score, and 7 Likert agreement scale questions for a total of 34 total survey questions. Participants were asked to rank from 1 (lowest) to 10 (highest) on the net promoter score questions. For the Likert agreement scale questions, participants could select one of the following responses: "strongly disagree," "disagree," "somewhat agree," "agree," "strongly agree" for many of the questions and "very unlikely," not likely," "somewhat likely," "very likely," for one of the questions.

Table 1shows the survey items that corresponded with the construct of Leadership Development.

Table 1. Correspondence of survey items to leadership development

Material Being Assessed	Survey Items
Leadership Context – position, level, amount of responsibility	What is your current level of leadership What are your overall years as a leader in an organization How many direct reports do you lead How many departments do you oversee What percentage of your time is spent with front-line staff on the work floor overseeing work as it is being delivered to customers
Leadership Self-Assessed Level	How would you rate yourself as a leader and How would others rate you as a leader
Knowledge Development	I talk with others about leadership and seek support, criticism I seek to understand what others know about leadership I examine leadership mistakes and how to improve I use new language, strategies, ideas to solve, meditate, manage, change or connect theory to general settings I learn from a mentor I read, research, talk, observe, if leadership theories have personal relevance and relate to my prior knowledge I lead in new ways based on theory, research or documentation I discuss potential results of new concepts with family, peers, allies, leaders I understand leadership constructs to influence others, share information, where some are unfamiliar I contemplate leadership expectations/plan how to use concepts, decision making strategies I talk informally, "water cooler talk" about leadership I compare leadership knowledge with scholarly work, test for validity, evidence, robustness
Applications Development	I have confidence in my ability to lead I self-analyze, self-develop, self-evaluate how/when to lead, grow, change I motivate self to pursue more leadership as others rely on me I admit when wrong, share apologies I lead from experience I survey stakeholders assessing my leadership to set goals I receive compliments, expressions of confidence by others I discuss leadership with my family
Values Development	I construct personal meaning of values by expression and quoting others I consider depth of personal relevance, benefits/values, costs/risks of leadership I self-direct leadership development activities I develop original ideas and share those with stakeholders I establish new boundaries in professional, personal relationships by taking risks to advance leadership I inspire others to lead, I identify leadership disposition strengths/weaknesses I self-reflect on depositions and aligning actions of leadership problems

The area of Leadership Development was assessed to determine survey participants' overall leadership, how participants view their current leadership, and to assess their skill development in knowledge, application, and values.

Table 2 shows the survey items that corresponded with the construct of Leadership Challenges.

The area of Leadership Challenges was assessed to determine survey participants' individual, employee based, and direct report level of leadership.

Table 3 shows the survey items that corresponded with the construct Impact of Lean Leadership.

In the area of Lean Leadership Impact, survey respondents were assessed on their level of Lean experiences and competency levels prior to and after completing the Lean Leadership Certificate program. Additionally, participants were assessed on utilization of Lean tools, asked to provide program recommendations, and identify program strengths and opportunities.

Table 2. Correspondence of survey items to leadership challenges

Material Being Assessed	Survey Items
Individual Assessment of Leadership	I like setting up goals and targets It bothers me when my decision-making abilities are questioned I respond fairly to issues on the team I am open to suggestions from employees and coworkers I take responsibility when a team member fails against expectations A leader should not hold grudges or biases against anyone on the team I feel employees should take orders without asking questions I am satisfied with my time and priority management skills I am comfortable with my level of stress
Employee Based Assessment of Leadership	I receive regular constructive feedback from management I understand how my performance is measured Management cares about my leadership development Management cares about me as a person Management clearly communicates expectations Management explains the reasons behind decision making Management handles disagreements professionally Management creates a trusting and open environment Management dedicates time to coach/mentor me as a leader I have regular schedule meetings with management to discuss goal alignment/targets I received a formal orientation to leadership
Direct Reports Assessment of Leadership	Employees retention is high in my organization Leader retention is high in my organization Employee engagement is high in my organization Employees should be supervised closely, or they are not likely to do their work In complex situations, leaders should let subordinates work problems out on their own Providing guidance without pressure is the key to being a great leader Leadership requires staying out of the way of subordinates as they do their work Most workers want frequent and supportive communication from their leaders Effective leaders give orders and clarify procedures

Table 3. Correspondence of survey items to impact of lean leadership

Material Being Assessed	Survey Items
Lean Context – experience, certifications, and exposure to Lean	• I had experience with Lean prior to program start • I earned a green belt or higher prior to program start • My current/previous employer has a formal Lean program
Lean Competency Level pre/post Assessment	What level of change leadership do you identify with • What level of critical thinking do you identify with • What level of communication leadership do you identify with
Utilization of Lean Tools	1. Mapping – process/current/future state 2. 5 Why/Root Cause Analysis 3. Ishikawa 4. A3 Project Template 5. 5S – Sort, Set in Order, Shine Standardize, Sustain 6. Mentor/Mentee Relationship 7. Value-based Mapping 8. Mindfullness Activities 9. Facilitator of Meetings 10. Identifying Waste 11. Visual management/displays/controls 12. Plan/Do/Check/Act
Program Recommendations	How likely are you to recommend this program How would you rate this program
Program successes and opportunities	• What was your greatest success in the program • What opportunities do suggest for this program

Participation in the Study

The survey was made available to participants from June 2, 2020, through June 23, 2020. A disclaimer was provided to indicate the purpose of the research project as an outcome evaluation on the effectiveness of the Lean Leadership certificate program. Participants were informed that the survey was optional and responses were confidential, with no identifying information such as name, email address, or IP address collected. Participants were also informed that the survey questions would be about their Lean Leadership experiences before and after program completion and would take approximately 10 minutes to complete.

Basic demographics were collected to determine the average age and gender of students enrolled in the Lean Leadership certificate program. This study identified the race and ethnicity of each respondent. A total of seven previously enrolled or graduated students and two currently enrolled students were asked to participate in the survey. Of those seven, six students completed the survey (86% return rate).

Results of the Study

The average age of survey participants were 43 years old with a minimum age of 34 and a maximum age of 47 years old. When looking at gender, females made up the largest percent of the sample and accounted for 67% while males accounted for 33%. In determining race and ethnicity, none of the respondents identified with Asian American, Hispanic American, Native American or other. However, 50% of the respondents reported being black or African American and 50% reported being White.

This survey wanted to assess the socioeconomic status of students enrolled in the Lean Leadership certificate program. Table 4 shows the distribution of the participants marital status, educational level, income, and the employment status.

Table 4. Distribution of student status of participants

Marital Status		**Education**	
Married	33%	Some Post Bachelors less than Masters	40%
Divorced	17%	Master's Degree	20%
Never Married	50%	Some Masters less than Doctorate	40%
Employment Status		**Income Status**	
Full-Time	66%	$30,000 - 60, 000	17%
Part-time	17%	$60,000 - $90,000	50%
Self-Employed	17%	Above $150,000	33%

Looking at marital status, half the sample of participants (50%) had never been married. No respondents reported a status of widowed, separated or other. Students enrolled in the Lean Leadership certificate program who participated in this survey were mostly employed full-time (66%) while attending an American midwestern university with the rest of the participants having worked part time hours or were self-employed. No respondents selected unemployed, a homemaker, student only, retired, or other. The educational background level of the participants was fairly evenly split between post BA (less than a Masters) and master's level (less than a doctorate). This finding is fairly consistent with the intended educational level of the program.

The educational background level of participants was 33% with some post Bachelor's, but less than a Master's, and 33% with some Master's, but less than a Doctorate. A total of 17% of student respondents had a Bachelor's or Master's degrees prior to enrollment. No student respondents had a Doctorate, Juris Doctorate or Medical Degree prior to program start.

The income status of participants ranged from $30,000 to above $150,000. Half of respondents reported incomes between $60,000 and $90,000 while another third of participants reported an income above $150,000. No participants reported incomes between $120,000 - $150,000, $90,000 - $120,000 or less than $30,000.

Findings/Results

The survey data were first used to analyze the survey respondent's leadership development. Table 5 shows the participants' overall percentage in years as a leader in an organization.

Table 5. Percentage of years as a leader

Overall Years as a Leader in an Organization	%
0 - 2 years	16.67%
3 - 5 years	16.67%
5 - 7 years	16.67%
Over 7 years	50.00%

The overall percentage of years as a leader in an organization ranged from 0 – 2 years to greater than 7 years. What is interesting to note is that half of the respondents reported having greater than 7 years of leadership and the remaining half of participants reported under 7 years of overall leadership in an organization.

The survey data were also used to analyze leaders' current level of oversight. Table 6 shows the percentage of survey participants' level of leadership.

Table 6. Percentage level of leadership

Current Level of Leadership	%
Supervisor - no direct reports	33.33%
Not in a leadership position	16.67%
Manager/Director - with direct reports	16.67%
Executive - oversees managers/directors	33.33%

One third of survey participants reported supervisory level of leadership and another third of survey participants reported having executive level leadership. The remaining survey participants reported as non-leaders or manager or director.

The average number of direct reports ranged from zero to 46 people. Two respondents reported no direct reports at 33% at a supervisory level while one participant reported having 12 direct reports at 17%, but not in a leadership position. Another respondent reported having 12 direct reports at 17% in a manager or director role, while two people identified with 46 direct reports at 33% in an executive level leadership position. In general, the number of direct reports increased as the level of leadership increased.

This study also aimed to determine the average number of departments overseen by survey participant leaders which ranged from zero to 15 departments. Two respondents reported having zero oversight of departments and two other respondents reported having at least one departmental oversight. One participant reported having 8 departments while another participant reported overseeing 15 departments. In general, departmental oversight increased as the level of leadership increased.

Additionally, survey respondents reported spending on average 20 – 100% of their time with front-line staff on the work floor overseeing work as it was being delivered to the customer. One respondent reported spending 100% of their time directly overseeing work, while two people reported overseeing work 80% of the time. Other survey respondents reported time spent at 50%, 25% and 20%. Participants were asked how they would rate themselves as a leader and how others would rate their leadership. Table 5 shows in the firm column, the mean and the standard deviation in the second column of how participants rated themselves as a leader compared to how others would rate their leadership abilities. In interpreting the mean and standard deviation, the range of responses was from 1 (lowest) to 10 (highest).

Table 7. Leadership rating comparison

Field	Mean	Std Deviation
How would you rate yourself as a leader	7.67	0.94
How would others rate you as a leader	8.00	0.58

It is interesting to note that participants estimated that other people would rate their leadership higher than they did of themselves. In addition, with a standard deviation of about half a point, compared to a standard deviation of almost a full point on the self-assessment question, the responses were more similar for how others would rate their leadership than how they would rate themselves. Survey participant respondents rated their leadership at an average of 7.76 with a standard deviation of 0.94. Participants rated their leadership abilities higher at an average of 8 and a standard deviation of .058 when asked how others would rate their ability to lead.

The survey data assessed participants Leadership development to determine the mean and standard deviation of their most desired and frequently used learning behaviors in knowledge, application, and values. Table 6 shows survey participants mean knowledge development. This table shows the mean in the first column and the standard deviation in the second column. In interpreting the mean, the range of responses was from strongly disagree (1) to strongly degree (4).

Survey participants had the highest agreement with knowledge items concerning talking with others about leadership, seeking support or criticism and examining leadership mistakes and how to improve. Their lowest responses concerned examining leadership with other sources and taking notes about their own leadership. They were most similar in their responses to participants' knowledge to discuss potential results of new concepts with family, peers, allies, leaders and their understanding of leadership constructs

Table 8. Mean knowledge development

Field	Mean	Std Deviation
Talk with others about leadership and seek support, criticism	3.50	0.50
Examine leadership mistakes and how to improve	3.50	0.76
Seek to understand what others know about leadership	3.33	0.75
Lead in new ways based on theory, research, or documentation	3.33	0.75
Use new language, strategies, ideas to solve, mediate, manage, change or connect theory to general settings	3.17	0.69
Learn from a mentor	3.17	0.69
Discuss potential results of new concepts with family, peers, allies, leaders	3.17	0.37
Understand leadership constructs to influence others, share information, where some are unfamiliar	3.17	0.37
Contemplate leadership expectations/plan how to use concepts, strategies, decision-making strategies	3.17	0.69
Talk informally, "water cooler talk" about leadership	3.00	1.15
Read, research, talk, observe, if leadership theories have personal relevance and relate to my prior knowledge	2.83	0.69
Compare leadership knowledge with scholarly work, test for validity, evidence, robustness	2.67	0.94
Takes notes, journal, self-talk about leadership	2.33	1.37

to influence others and share information where some are unfamiliar. The lowest three items were also those with the most heterogeneous responses.

Table 9 shows survey participants mean application development. This table shows the mean in the first column and the standard deviation in the second column. In interpreting the mean, the range of responses was again from strongly disagree (1) to strongly degree (4).

Table 9. Mean application development

Field	Mean	Std Deviation
Lead from experience	4.00	0.00
Self-analyze, self-develop, self-evaluate how/when to lead, grow/change	3.50	0.50
Receive compliments, expressions of confidence by others	3.50	0.50
Admit when wrong, share apologies	3.50	0.50
Have confidence in my ability to lead	3.50	0.50
Motivate self to pursue more leadership as others rely on me	3.17	0.69
Discuss leadership with my family	3.17	0.69
Survey stakeholders assessing my leadership to set goals	3.00	0.82

Survey participants had the highest agreement in application development in leading from experience. In fact, all the participants marked themselves as "4", giving no standard deviation value for this response. Surveying stakeholders to assess participant leadership to set goals had the lowest mean with the widest range of responses.

Table 10 shows survey participants mean values development. This table shows the mean in the first column and the standard deviation in the second column. In interpreting the mean, the range of responses was again from strongly disagree (1) to strongly degree (4).

Table 10. Mean values development

Field	Mean	Std Deviation
Inspire others to lead	3.67	0.47
Develop original ideas and share those with stakeholders	3.50	0.76
Talk with others noting benefits of sharing about leadership problems	3.33	0.94
Identify leadership disposition strengths/weaknesses	3.33	0.75
Self-direct leadership development activities	3.33	0.75
Self-reflect on depositions and aligning actions of leadership	3.17	0.69
Construct personal meaning of values by expression and quoting others	3.00	0.82
Establish new boundaries in professional, personal relationships by taking risks to advance leadership	3.00	1.15
Consider depth of personal relevance, benefits/values, costs/risks of leadership	2.83	0.90

Survey participants had the highest agreement in values development in inspiring others to lead. Considering depth of personal relevance, benefits/values, costs/risks of leadership had the lowest mean, however, establishing new boundaries in professional, personal relationships by taking risks to advance leadership had the widest range of responses.

Next, Leadership challenges were assessed to determine the mean for participant challenges as it relates to their own leadership, the person they report to, and their direct reports. Table 11 shows survey participants mean leadership challenges as it relates to their own leadership. This table shows the mean in the first column and the standard deviation in the second column. In interpreting the mean, the range of responses was again from strongly disagree (1) to strongly Agree (4).

Table 11. Mean of personal leadership challenges

Field	Mean	Std Deviation
I am open to suggestions from employees and coworkers	3.67	0.47
A leader should not hold grudges or biases against anyone on the team	3.67	0.47
I take responsibility when a team member fails against expectations	3.33	0.94
I like setting up goals and targets	3.33	0.75
It bothers me when my decision-making abilities are questioned	3.17	1.07
I respond fairly to issues on the team	3.17	0.37
I am satisfied with my time and priority management skills	2.67	0.94
I am comfortable with my level of stress	2.50	1.26
I feel employees should take orders without asking questions	1.33	0.47

Survey participants had highest agreement in personal leadership challenges in not holding grudges or biases against anyone on the team and being open to suggestions from employees and coworkers. Survey participants strongly disagreed that employees should take orders without asking questions showing the lowest mean, however, comfort in stress level had the widest range of responses.

Table 12 shows survey participants mean leadership challenges as it relates to the person they report to. This table shows the mean in the first column and the standard deviation in the second column. In interpreting the mean, the range of responses was again from strongly disagree (1) to strongly degree (4).

Table 12. Mean of employer leadership challenges

Field	Mean	Std Deviation
I understand how my performance is measured	3.00	0.82
I receive regular constructive feedback from management	2.50	0.50
Management cares about me as a person	2.50	0.76
Management cares about my leadership development	2.50	0.96
Management explains the reasons behind decision making	2.17	1.07
Management handles disagreements professionally	2.17	0.69
Management creates a trusting and open environment	2.17	1.07
Management dedicates time to coach/mentor me as a leader	2.17	1.07
I have regularly scheduled meetings with management to review my goals/targets	2.17	0.90
Management clearly communicates expectations	1.83	0.69
I received a formal orientation to leadership	1.33	0.47

Survey participants had the highest agreement in employer leadership challenges in understanding how their performance is measured. Their lowest response concerned leadership challenges with receiving a formal orientation to leadership. They were most similar in their responses to employer leadership chal-

Table 13. Mean of direct report leadership challenges

Field	Mean	Std Deviation
Most workers want frequent and supportive communication from their leaders	3.33	0.47
Employee engagement and morale is high in my organization	3.00	0.82
Effective leaders give orders and clarify procedures	3.00	1.00
In complex situations, leaders should let subordinates work problems out on their own	2.83	0.69
Employee retention is high in my organization	2.67	1.11
Providing guidance without pressure is the key to being a great leader	2.67	0.75
Leader retention is high in my organization	2.50	1.26
Leadership requires staying out of the way of subordinates as they do their work	2.33	0.47
Employees should be supervised closely, or they are not likely to do their work	1.33	0.47

lenges in management explaining the reasons behind decision making, creating a trusting environment, and dedicating time to coach/mentor employees as a leader.

Table 13 shows survey participants mean leadership challenges as it relates to direct reports. This table shows the mean in the first column and the standard deviation in the second column. In interpreting the mean, the range of responses was again from strongly disagree (1) to strongly degree (4).

Survey participants had the highest agreement in direct leadership challenges in their understanding that most workers want frequent and supportive communication from their leaders. Survey participants strongly disagreed that employees should be supervised closely, or they are not likely to do their work. The widest range of responses in direct report leadership challenges were in high leader organizational retention.

Lastly, the impact of Lean Leadership was analyzed to assess the survey participants' percentage of Lean exposure prior to enrolling in the Lean Leadership Certificate program. 67% of participants reported having some informal experience with Lean prior to enrolling into the Lean Leadership Certification program while one third reported no prior experience with Lean. Of the 67% of participants who reported they have some informal experience with Lean prior to enrollment, only a third reported earning a Green Belt or higher in Lean prior to enrolling in the Lean Leadership Certification program. When asked if participants' current or former employers had a formal Lean program within their organization, only one third could affirm.

Participants were asked to rate their mastery level in Change Leadership, Communication, and Critical Thinking competencies prior to program enrollment. Self-Mastery refers to being aware of the need for growth and change; believing in growing and learning; and understanding the incentive, goal, plan, skills and resources of all projects prior to initiation. Confidant-Mastery refers to using needs assessments and reasoning to support the Lean initiatives, listen to peer's opinions and use sound reasoning in decision making, realize what is ending, what occurs in transition and what the possibilities are/new beginnings are. Team-Mastery refers to collaborating with the team to employ all ideas and engage all members to facilitate more efficient change, being unconsciously competent with the process of change, and being able to transfer the change matrix or model to any situation. Figure 1 shows a comparison with survey participants' level of mastery in the three competencies prior to enrolling in the Lean Leadership program and after program completion.

Figure 1. Competency mastery level comparison

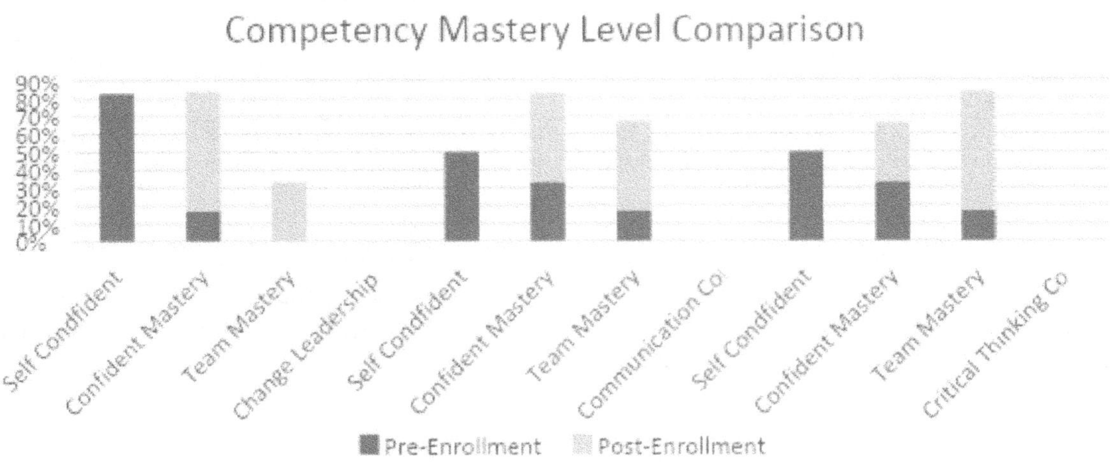

In interpreting the pre-enrollment data, survey participants rated their change leadership competency level the highest at 83% for self-mastery and 17% for confidant-mastery. No participants rated their change leadership competency at the team-mastery level. In both the communication and critical thinking competencies, survey participants rated the highest at 50% for self-mastery, 33% for confidant-mastery, and 17% for team mastery. In general, pre-enrollment competency level in change leadership, communication, and critical thinking were the highest in the self-confidant mastery levels.

In interpreting the post-enrollment data, survey participants rated their change leadership competency level the highest at 67% for confidant-mastery and 33% for team mastery. No participants rated their change leadership competency at the self-mastery level. In the communication competency, participants equally rated their competency level at 50% for both the self- and team-mastery levels. No participants rated their communication competency at the self-mastery level. In the critical thinking competency, no participants rated their competency level at the self-mastery level. However, participants rated their confidant-mastery level at 33% and team-mastery level at 67%. In general, post-enrollment competency level in change leadership, communication, and critical thinking were the highest in the confidant- and team-mastery level.

Survey participants were asked to consider the impact of multiple lean tools and rate the likelihood of utilizing these tools in their current or future organizations. Table 12 shows the mean in the utilization of Lean tools presented in the Lean Leadership certificate program.

Table 14. Mean of lean tools utilization

Field	Mean	Std Deviation
Process/Current/Future state Mapping	3.83	0.37
Visual management/displays/controls	3.67	0.47
Plan/Do/Check/Act	3.67	0.47
5 Why/Root Cause Analysis	3.50	0.76
Identifying Waste	3.33	0.75
Mentor Relationship	3.17	0.37
A3 Project Template	3.17	1.07
Facilitator of Meetings	3.17	1.07
5S - Sort, Set in Order, Shine, Standardize, Sustain	3.00	1.15
Mentee Relationship	3.00	0.58
Mindfulness Activities	2.67	0.94
Ishikawa	2.50	0.76
Value-based Mapping	2.33	1.37

Survey participants had highest agreement in utilization of Lean tools in the use of process, current, and future state mapping. They were most similar in their responses to the use of visual management displays and controls, PDCA, A3, and facilitating meetings. The lowest scoring item was value-based mapping with the widest range of responses.

The survey wanted to assess the participants greatest successes and opportunities in the Lean leadership certificate program. In identifying successes, respondents reported that the program supported structured thinking about leadership that was, and is applicable in daily relationships and work. Another participant indicated a better understanding with complex problem solving. Other respondents expressed success with understanding Lean tools and being able to apply them in varying situations. Additionally, respondents appreciated that the program was an acceptable cognate for the Ed.D program at an American midwestern university. In identifying opportunities of the Lean leadership program, respondents expressed class days were extensive and had a major impact on work-life balance. Other respondents commented that the concepts were abstract which made applying them difficult.

Finally, respondents were asked to rate the Lean leadership program on a scale of 1 (lowest) to 10 (highest) on how likely they were to recommend the Lean leadership program and the overall rating of the program. Respondents rated a mean of 6.17 on their likelihood to recommend the program and a mean of 5.83 on their overall rating of the program. In general, survey participants rated their likelihood to recommend and overall rating of the program above average.

Findings of the Study

The results of the Lean Leadership program implementation survey indicated that the evaluation outcome on program effectiveness was perceived differently by various individuals. Enrollment in the Lean Leadership program identified middle-aged White and African American women as the target population of students. It is interesting to note that the majority of Lean students were employed full-time, had never been married, were Bachelor's-prepared, with an average salary of $75,000 annually.

In evaluating the program effectiveness in Leadership development, results indicated that the Lean Leadership students had seven or more overall years as a leader in an organization at an executive level of leadership. Leadership development findings also indicated that Lean students were confident in their individual leadership abilities as well as their ability to lead others. In relation to knowledge development, Lean Leadership enrolled students are likely to talk with others about leadership and seek support and or criticism as well as examine leadership mistakes and improvement measures. As it relates to application development, the students mostly lead from their experiences as a leader and value their ability to inspire others to lead.

Leadership challenges were also used to evaluate the outcome of program effectiveness. Finding showed that the students were open to suggestions from employees and coworkers and were unbiased as it related to personal leadership.

With respect to employer related leadership challenges, the Lean Leadership students were supportive in receiving constructive feedback and understood how leadership performance was measured. In relation to leadership challenges with direct reports, the students affirmed the need to have frequent and supportive communication from leaders.

The impact of the Lean Leadership Graduate Programs themselves was the last aspect in evaluating program effectiveness. Results indicated that most of the students are prepared with some informal Lean experience prior to program enrollment. Findings include mastery level of competency for the students' pre-enrollment is mostly self-confident in change leadership, and equally distributed between self-confident and confident mastery related to communication and critical thinking competencies. It is interesting to note that mastery level of competency findings post-enrollment is mostly confident mastery in change leadership, and equally distributed between confident and team mastery related to communication and critical thinking competencies. Lastly, findings for the impact of Lean tools that the students would utilize include current and future state mapping in addition to visual management, displays, and controls.

Implications of the Findings

The findings from this outcome evaluation on the Lean Leadership Graduate Programs indicated a need to market to a larger group of emerging leaders. This would include engaging newer leaders with different backgrounds and ethnicities who have zero to minimal direct reports. As Glick and Kulbok (2001) stated, "Three essential areas should be addressed when monitoring a program: (1) the degree to which a program reaches the target population; (2) the consistency of service delivery with the original design criteria; and (3) resource expenditure in the implementation of the program" (p. 39). Students enrolled in the Lean Leadership programs lacked male representation and were dominated by female students. Additionally, students with different socioeconomic backgrounds and ethnicities were not enrolled in the program, resulting in a lack of Asian, Hispanic, and Native American program perspectives.

Leadership development findings suggest the need to further advance knowledge development in the Lean Leadership students in areas of taking notes, keeping journals, self-talking about leadership, and comparing leadership knowledge with scholarly work. Irwin and Supplee (2012) stated, "To reach the goals of implementation science, the field needs strong methods to identify questions, test theory, analyze data, and synthesize information to develop knowledge" (p 340). In terms of application development, findings included a lack of leader ability to survey stakeholders assessing leadership ability to set goals. Longenecker and Yonker (2013) reported, "Feedback is critically important during periods of rapid change to help individuals properly align their actions while making adjustments to improve their leadership performance" (p. 160). Findings in values development included the need for leaders to consider depth of personal relevance, benefits/values, and costs/risks of leadership. Hernez-Broome and Hughes (2004) argued that trends in leadership development in the past 20 years include the proliferation of leadership development methods and the importance of a leader's emotional resonance with and impact on others. In general, continuous assessment of Lean students' knowledge, application, and values development throughout the Lean Leadership Graduate Programs could provide deeper impact in strengthening leadership development.

Other implications from this case study suggested a wide range of personal leadership challenges in managing stress levels, direct reporting challenges with leader retention, and employer challenges with meeting expectations that are not clearly communicated. Longenecker and Yonker (2013) suggested that "There is substantial research that makes it clear that feedback is critically important during periods of rapid change to help individuals properly align their actions while making adjustments to improve their leadership performance" (p. 160). These authors also suggested developing proficiency in time and priority management around the actions and activities that will help them get better results as a key factor for managing leadership challenges. Lastly, Longenecker and Yonker (2013) argued that effective leadership requires leaders to "Stay calm and demonstrate the ability to handle and absorb the stresses and strains that are part of leading organizational transformation" (p. 164). In general, the Lean Leadership Graduate Programs should consider program goals that include strategies that would have a significant impact on the student's emotional well-being.

Lastly, the outcome evaluation on the Lean Leadership Graduate Programs included findings on how students are informed by Lean. To gain buy-in from the students, the program could promote more the already offered Green Belt certification as an outcome to support Lean practice. While a large majority of students have been informally introduced to Lean prior to enrollment, a formal belt certification at the program's end could strengthen leader development. Reflecting on the role of competencies in the program, findings included a continuous need to reinforce these competencies as a core dimension of the program. In fact, Hernez-Broome and Hughes (2004) cited Barrett & Benson (2002) who argued that leadership competencies will become more competitive in changing environments in areas of "Global competition, information technology, the need for rapid and flexible organizations, teams and different employee needs." (p. 5)

Finally, implications related to Lean tools utilization and overall program opportunities emerged. Program consideration for value-based mapping was found to be widely underutilized. Additionally, considerations should be given to the timing of the program to meet the needs of potential students who work full-time jobs. Irwin and Supplee (2012) suggested, "The goals of implementation research include understanding the factors that impede or promote effective implementation, testing new approaches, and determining causal relationships between implementation characteristics and impact" (p. 340). In general, using these key findings and reevaluating program outcomes to inform future Lean students could impact overall program ratings and recommendations.

Recommendations for Next Steps

The feedback generated from the case study provided valuable feedback regarding program outcomes. Glick and Kulbok (2001) described "three essential areas when monitoring a program: (1) the degree to which the program reaches the target population; (2) the consistency of service delivery with the original design criteria; and (3) resource expenditure in the implementation of the program" (p. 39). The first suggestion is to continuously promote the benefits of Lean to various industries and leaders within those industries. Outcomes of this study identified that only a small percentage of the Master's-prepared leaders were enrolled in the Lean Leadership Graduate Certificate program. However, an approach to increase enrollment is suggested by showing how competency mastery levels significantly increase from pre-enrollment to post-enrollment in the areas of change leadership, communication, and critical thinking competencies.

Another suggestion to improve outcomes include examining program feasibility for potential students interested in enrolling in the program. Promoting program success stories from post-graduate students on how Lean improves the overall development of leader's knowledge, application, and values development provides a deeper context for potential students. Additionally, re-evaluating the need to adjust the original program design and measure the outcome results against this survey provides a basis for continuous program improvement.

A third suggestion based on this case study is to tap into new growing markets where Lean is utilized. Glick and Kulbok (2001), stated that "Involvement of the stakeholder community organizations and grassroots members of the community in the planning, evaluation, and revision of a program fosters ongoing community acceptance, client participation, and organizational support for the program" (p. 38). In general, continual implementation success will rely on the program's ability to interpret impact, build on organizational relationships, and understand outcomes presented in this research.

In addition, there are several recommendations that can be made for lean education in general. For example, the use of both content and competency mastery as the basis for the program's curriculum should be given consideration on a larger scale in lean education. The American Society for Mechanical

Engineering, for example, in their Vision 2030 called for four goals, including the need to "re-examine the engineering practice perspective and broadly define the engineering knowledge, skills and competencies should have" (Flumerfelt, S. Kahlen, F.J., Alves, A.C. & Siriban-Manalang, A., 2015, Preface). And further, they called for "the development of professional skills and competencies in engineering graduates," and "move to change degree program accreditation standards" (Flumerfelt, S. Kahlen, F.J., Alves, A.C. & Siriban-Manalang, A., 2015, Preface). In other words, the program in this case study provides a solid framework in terms of how to design, develop and deploy improved engineering education and lean education.

CONCLUSION

Based on the results of this case study, the Lean Leadership Graduate Programs will need to continuously promote the benefits of Lean within various industries and leaders within those industries using storytelling and featuring student outcomes. Other findings included re-evaluating the need to adjust the original program design and measuring outcomes as a basis for continuous program improvement. Findings also indicated that continual implementation success will rely on the program's ability to interpret impact, build on organizational relationships, and understand the findings presented in this research. Further, in terms of a program and curriculum model, there are possible important considerations as to enriched program design for the future of Lean education.

REFERENCES

Alves, A. C., Flumerfelt, S., & Kahlen, F. J. (2017). *Lean education: An overview of current issues.* Springer. doi:10.1007/978-3-319-45830-4

Amaro, A. P., Alves, A. C., & Sousa, R. M. (2019). Lean Thinking: A transversal and global management philosophy to achieve sustainability benefits. In A. C. Alves, F.-J. Kahlen, S. Flumerfelt, & A. B. Siriban-Manalang (Eds.), *Lean Engineering for Global Development* (pp. 1–31). Springer International Publishing. doi:10.1007/978-3-030-13515-7_1

Association for Manufacturing Excellence. (2007). *Lean administration: Case studies in leadership and improvement.* Productivity Press.

Bolman, L. G., & Deal, T. E. (2013). *Reframing organizations: Artistry, choice and leadership* (5th ed.). Jossey-Bass.

Dahl, J. (2020). *Leading lean: Ensuring success and developing a framework for leadership.* O'Reilly Media.

Deming, W. E. (1993). *The new economics: For industry, government, education.* Massachusetts Institute of Technology.

Dombrowski, U., & Mielke, T. (2013). Lean leadership–fundamental principles and their application. *Procedia CIRP, 7*, 569–574. doi:10.1016/j.procir.2013.06.034

Emiliani, B. (2008). *Practical lean leadership: A strategic leadership guide for executives*. The Center for Lean Business Management, LLC.

Flumerfelt, S. (2017). *Lean Leadership at Oakland University* [online video]. Retrieved from: https://drive.google.com/drive/u/3/search?q=owner:haughey%40oakland.edu

Flumerfelt, S., Alves, A. C., Leao, C. P., & Wade, D. L. (2016). What do organizational leaders need from lean graduate programming. *European Journal of Training and Development*, *40*(5), 302–320. doi:10.1108/EJTD-01-2015-0005

Flumerfelt, S., Kahlen, F. J., Alves, A. C., & Siriban-Manalang, R. (2015). *Lean Engineering Education: Driving Content and Competency Mastery*. The American Society of Mechanical Engineers. doi:10.1115/1.860502

Flumerfelt, S., Ross, W., & Wade, D. (2017). Lean learning at Oakland University. In A. C. Alves, S. Flumerfelt, & F. J. Kahlen (Eds.), *Lean Education: An Overview of Current Issues*. Springer. doi:10.1007/978-3-319-45830-4_6

Flumerfelt, S., Schwartz, K. G., Mavris, E., & Briceno, S. (2019). *Complex systems engineering: Theory and practice*. AIAA Press. doi:10.2514/4.105654

Flumerfelt, S. & Soma, P. (n.d.). *Transforming the way we do business*. Destin, FL: Charactership Lean Publishing.

Glick, D. F., & Kulbok, P. A. (2001). Program revision: A dynamic outcome of evaluation. *Quality Management in Health Care*, *10*(1), 37–44. doi:10.1097/00019514-200110010-00005 PMID:11702469

Goodridge, D., Westhorp, G., Rotter, T., Dobson, R., & Bath, B. (2015). Lean and leadership practices: Development of an initial realist program theory. *BMC Health Services Research*, *15*(1), 362. doi:10.118612913-015-1030-x PMID:26345184

Hernez-Broome, G., & Hughes, R. J. (2004). Leadership development: Past, present, and future. *Human Resource Planning, 27*(1).

Higgs, M., & Rowland, D. (2005). All changes great and small: Exploring approaches to change and its leadership. *Journal of Change Management*, *5*(2), 121–151. doi:10.1080/14697010500082902

Irwin, M., & Supplee, L. H. (2012). Directions in implementation research methods for behavioral and social science. *The Journal of Behavioral Health Services & Research*, *39*(4), 339–342. doi:10.100711414-012-9293-z PMID:22907687

Liker, J. K., & Convis, G. L. (2012). *The Toyota way to lean leadership: Achieving and sustaining excellence through leadership development*. McGraw-Hill.

Longenecker, C. O., Papp, G., & Stansfield, T. (2009). Quarterbacking real and rapid organizational improvement. *Leader to Leader*, *2009*(51), 17–23. doi:10.1002/ltl.313

Longenecker, C. O., & Yonker, R. D. (2013). Leadership deficiencies in rapidly changing organizations: Multisource feedback as a needs assessment tool–Part I. *Industrial and Commercial Training*, *45*(3), 159–165. doi:10.1108/00197851311320577

Merriam, S. B., & Tisdell, E. J. (2015). *Qualitative research: A guide to design and implementation.* John Wiley & Sons.

Ptacek, R., Coats, R., & Ptacek, T. (2012). *Today's lean leader.* MCS Media, Inc.

Riordan, C. M. (2013, June 4). Work-life balance isn't the point. *Harvard Business Review.* Retrieved from: https://hbr.org/2013/06/work-life-balance-isnt-the-poi

Sawhney, R., & Chason, S. (2005). Human behavior based exploratory model for successful implementation of lean enterprise in industry. *Performance Improvement Quarterly, 18*(2), 76–96. doi:10.1111/j.1937-8327.2005.tb00334.x

SHRM (Society for Human Resource Management). (2008, March 1). *Leadership competencies.* Retrieved from: https://www.shrm.org/ResourcesAndTools/hr-topics/behavioral-competencies/leadership-and-navigation/Pages/leadershipcompetencies.aspx

Sperl, T. (2017). *Lean Leadership at Oakland University (online video).* Oakland University. Retrieved from https://drive.google.com/drive/u/3/search?q=owner:haughey%40oakland.edu

Chapter 11
The Challenges of Industrial Engineer Management Skills in Industry 4.0

Carina M. O. Pimentel
University of Aveiro, Portugal

Anabela C. Alves
https://orcid.org/0000-0002-2926-4187
University of Minho, Portugal

João C. O. Matias
University of Aveiro, Portugal

Susana Garrido Azevedo
University of Coimbra, Portugal

ABSTRACT

Industrial engineering and management (IEM) is considered a softer type of engineering. IEM professionals have been slow in implementing many changes that have occurred in production, ranging from mass production to mass customization paradigms embedded in Industry 4.0. This chapter introduces and discusses the role of IEM professionals in dealing with all the changes required for the implementation of these paradigms. This chapter discusses the training of these professionals that demands more applied research, and, at the same time, it seeks to instigate their curiosity and creativity to generate new solutions based on fundamental research. A semi-systematic literature review was used. The results indicate that an IEM professional needs a strong leadership style and ethical sense to lead multidisciplinary teams and should also be a systems, lean, and sustainability thinker, who has the technological, digital, and transversal skills to face the current and future challenges of the successive industrial revolutions.

DOI: 10.4018/978-1-7998-8816-1.ch011

INTRODUCTION

Industrial Engineering and Management (IEM) has been evolving from Mechanical Engineering Elsayed (1999), growing and making space for itself in the complex world of engineers. Nevertheless, the creation of goods has existed since humanity started making objects to cover its skin or killing animals to survive. With the advent of machines, mechanical production replaced handicraft production with one-of-a-kind style and customized design. The role of the human being was as important as before, although probably not to perform the job but to organize it. The second and third industrial revolutions were dominated by machines, energy, computers, and information and communication technologies (Kaya, 2019) that enabled the high production of one-fit all products (mass production). Consequently, high consumption patterns developed, endangering the planet's capacity to restore the resources. Currently, the fourth industrial revolution is dominated by IoT technologies (Chou, 2018), where mass customization (Tien, 2020) and niche production (Rolfes et al., 2019) are the production paradigms. Indeed, the shift from mass production to customized production opens new entry opportunities for latecomer firms aimed at niche production, which are less dependent on scale economies (Corrocher et al., 2020). Increasingly, the mass market is turning into a mass of niches (Anderson, 2008; R. Armstrong, 2008) despite few examples presented by Elberse (2008) in the same entertainment (video and music) industry.

Nevertheless, producers face the challenges to ensure sustainable production and more efficient production systems that "do more with less", as Lean production of the Toyota Production System teaches us. At the same time, production needs to be flexible and agile to provide a new value proposition that customers expect with different way of organizing companies' resources to solve their problems and eliminate waste. Being the organization skills the core of an IEM professional this issue will be discussed in this chapter. Also, knowing that the main challenges for them in an industry 4.0 context is their education to provide them with the suitable skills, this issue will be also addressed.

In the literature some authors (e.g. Mourtzis and Doukas (2014)) discuss the challenges faced by an IEM professional in an industry 4.0 environment, starting with the evolution from handicraft production to Industry 4.0. Thus, this chapter sought hereby to identify the current and future skills required of an IEM professional in the context of Industry 4.0. Although there is some published literature on the skills required of professionals in general, and, engineers in particular, evidence of the skills required of IEM is scarce (Sackey & Bester, 2016). Santandreu-Mascarell et al. (2011) analyzed the skills proposed as ideals for an IEM degree and identified them in the current study plans implemented at Spanish universities. Also, Pais-Montes et al. (2019) analyzed the employability traits of engineers (computer engineers, naval engineers and industrial engineers) who had recently graduated and concluded there is a significant gap between the skills learnt in higher education and those needed at the workplace. None of those studies focuses on the new industrial era. Finally, Piwowar-Sulej (2021) addresses this topic from the perspective of IEM professionals in Poland.

Another contribution of our research is related to the role that Higher Education Institutions may play in the development of the skills mentioned previously. In that regard, Gowripeddi (2021) studied the role of advanced and collaborative learning practices in the education of future construction engineers and researchers in the fields of data-driven technologies. Moreover, Bonnaud and Bsiesy (2020) address the topic from the perspective of microelectronic engineering.

Furthermore, Hadgraft and Kolmos (2020) discuss the topic of engineering education from the point of view of the emerging learning environments. In the area of IEM, some recent contributions can be found, such as Tan et al. (2020) and Gupta et al. (2019) who focus on the learning factory concept, and

Gupta, Mukhawana and Mashinini (2019) that address project-based learning for IEM education to support Industry 4.0 skill acquisition.

Thus, the main objectives of this chapter are threefold: 1) to summarize progress from handicraft production to the fourth industrial revolution and the role of IEM professionals herein; 2) to identify the challenges for the IEM professional workforce with the advent of Industry 4.0; 3) to establish adequate skills to be acquired by IEM students in an Industry 4.0 environment and the role of education in providing them. To achieve these objectives, a narrative literature review and the authors own knowledge and experience in researching and teaching in IEM was used.

This chapter is structured into five sections. After this introduction, section two provides a background on the topics under research. The third section is dedicated to the results, while the fourth section discusses these results. Conclusions are presented in section five.

METHODOLOGY

This chapter focuses on competences to be acquired by IEM professionals due to the emergence of Industry 4.0, customized production, and niche production. Furthermore, it highlights the role of education in developing these skills. To help to frame those themes it also includes a summary of the approaches of the IEM field along the several Industrial Revolutions. A narrative review was used as main instrument of research. According to Snyder (2019), this methodology is indicated to map a field of research, synthesize the state of knowledge, and create an agenda for further research or the ability to provide an historical overview or timeline of a specific topic. This methodology is also adequate for topics that have been conceptualized differently and studied by different groups of researchers (Snyder, 2019; Wong et al., 2013). Given the scope of the topic under analysis, a systematic review of the literature was not considered adequate. As stated by Snyder (2019), a strict systematic review approach may not be suitable or even possible when the review is aimed to summarize or assess a large field of research or several research areas.

Unlike systematic reviews that benefit from guidelines such as the PRISMA (Preferred Reporting Items for Systematic Reviews and Meta-Analyses) statement, there are no guidelines for narrative reviews (Ferrari, 2015; Gregory & Denniss, 2018).

Qualitative content analysis was the main technique used. However, since the authors add some reflection based on their experience in teaching, research, and in projects with companies in the IEM field, a simple review was not carried out.

The selection of content analysis provided a systematic and objective means to make valid inferences from written data to describe our phenomena (Downe-Wambolt, 1992).

Regarding the search strategy, a literature search in scientific databases, including Scopus and Web of Science, was carried out. Furthermore, a Google search to identify relevant research projects or reports was also carried out. References in the selected articles and reports were also scanned to identify other relevant articles. The following keywords were searched, either individually and in various combinations: 'Industrial Engineering and Management', 'Industry 4.0', 'factory of the future', 'competencies', 'skills', 'IEM profession/professional'.

Regarding the delimitations and inclusion/exclusion criteria, only English-language original research studies were considered in the sample and neither restrictions were assumed in terms of their design or research methodology, nor their year of publication. Articles where the focus of the study was not directly linked with the topics under research in this study were also excluded. The screening of articles was carried out manually in two stages: first by reading the title and abstract; and then reading the full text of potentially relevant studies. The final sample was equal to 107 papers.

RESULTS

This section describes the progress from handicraft production to the fourth industrial revolution, pointing out the role and challenges for Industrial Engineering Managers. Next, the role of education in providing the adequate skills to IEM managers is focused.

Technological and Organizational Changes since the First Industrial Revolution

There is a relationship between technological change and organizational change since the first industrial revolution, highlighting the importance of labor, capital, and information processes. There appears to be a link between the three previous industrial revolutions, as also seems to be the case in this new Industry 4.0 (Tunzelmann, 2003).

So, what has changed since the late 18th century to the present day? Several factors, but mainly the growing importance of technology, increasing worldwide competition, and the growing focus on customers. The first revolution brought mass production, putting the emphasis on quantity. Since then, and through the increased importance of the quality control, nowadays the aim is to produce and deliver the right quantity to the customer, with the expected quality, at the right time, and at the lowest possible cost.

The companies feel that this is the only way to survive, even if they do not grow. That is, on the one hand, companies can only, at least, survive if they have the capacity to produce marketable goods in global markets. On the other hand, in this knowledge economy, it is not enough to know how to produce products; skills both upstream and downstream the value chain are necessary, i.e. from the design of products and/or services to the end of their life cycle. More than ever, the client/customer has become a strategic stakeholder of the company, being present from the beginning of the design concept of the product and/or service, in a logic of co-production. In other words, clients/customers cannot be ignored, and they play a crucial role in determining the demand for Product-Service Systems, alongside the remaining stakeholders, such as society with its laws and rules governing the environment and society. In this context, what is the role of the industrial engineer and manager?

The Institute of Industrial and Systems Engineers (IISE) has defined industrial engineering as a discipline "concerned with the design, improvement and installation of integrated systems of people, material, information, equipment and energy. It draws upon specialized knowledge and skills in the mathematical, physical and social sciences, together with the principles and methods of engineering analysis and design to specify, predict and evaluate the results to be obtained from such systems." (Institute of Industrial and Systems Engineers, 2020a).

In this sense, the main role of the Industrial Engineering and Management professional is to increase productivity by eliminating non-value added operations and improving the effective utilization of resources (Kumar, 2020). For that, it is necessary to have knowledge and skills to master immaterial competitiveness factors, and not just technology.

The history of Industrial Engineering is associated to the many Industrial Revolutions and it has gone through several phases from the middle of the eighteen century to the current Industry 4.0. Frederick

W. Taylor is considered to be the father of the Industrial Engineering discipline (Elsayed, 1999). At the same time, it is possible to talk about the history of Industrial Engineering and Management considering its association with the history of Management Approaches. Table 1 presents the evolution of Industry, Industrial Engineering and Management Approaches.

There are some slightly different approaches related to the four main columns (separated by double continuous line) of Table 1, mainly related to manufacturing paradigms, industrial strategic methodologies, principles, and tools, not in its existence, but at the moment of its exact appearance.

The first main column introduces the two main logics: the first centered in production and in the market after. Before introducing the other ones, it is important do say that all these paradigms, approaches, and even methodologies, principles, techniques, tools, etc. are not sequential and unique in time. They can coexist and the transitions are not disruptive and radical. In this sense, and for this reason and to give a clear example, Table 1 presents a transition period. On the other hand, the main lines in the table are not straight and uniformly stopped. Still, for instance and as one can see in Table 1, the Industry 2.0 (main column 2) and corresponding 2nd industrial revolution (main column 3) share some of the characteristics of the Classical Approach and Engineer Technologist; and Modern approach and IEM as Applied Science.

Main column 2 presents the four main paradigms until now: from industry 0.0 to 4.0. These paradigms are corresponding to so-called four industrial revolutions (main column 3). The main fourth one presents the main theory on management approaches: No Management approach; Pre-Classical approach; Classical approach (Engineer Technologist); Neo-Classical approach (Marketer). Each main column is divided into two other columns. The first shows the name of the paradigm / approach and the second its main characteristics or relevant marks.

At the end of main columns 2 and 3, two main aims of the near future are set forth: the maturing of the Smart Manufacturing and personalized production, two examples of dynamic and continuous progress and not radical and abrupt ones.

Whatever the type of organization, business or other type of non-profit organization, there is a vital premise: efficiency in all processes and systems, in order to achieve the respective objectives using the least number and amount of resources. And, in this context, where systems are seen as a set of interlocutors and rules, people assume a central role. On the other hand, and taking into account the increasing complexity and greater specialization of tasks, an effective communication with the other functions in a context of concurrent or simultaneous engineering is required (Miranda et al., 2019). In such context, Industrial Engineering and Management emerges as an applied science, and not just as a simplistic intersection between engineering and management, tackling all the above-mentioned aspects and providing the technical and management support to the integrated system, due to its inherent characteristic of understanding the whole system.

Engineering, once only associated with ingenuity and technique, has extended its scope to other areas through IEM, such as logistics, planning, decision support methods, quality, environment, design, sustainability, information systems, systems, processes, among others. Thus, it has not only increased its importance in the industry, for example, in multidisciplinary teams, but it has also expanded to services, even in a modern Product Service System logic.

Table 1. Evolution of industry, industrial engineering and management approaches theory

	Developments in Industry		Developments in Industrial Engineering		Theory of Management Approaches	
		• No established Industry	Pre-Industrial Revolution (Handicraft Age)	• Handicrafts and • Craftsman	No Management	
	Industry 1.0 (mid-18th century)	• Introduction of mechanical manufacturing systems using water and steam power • Use of data for planning purposes	1st Industrial Revolution (Machine Age)	• Emergence of plants • Production of goods • Interchangeable parts	Pre-classical Approach	• Rules of thumb
	Industry 2.0 (Started in the late nineteenth century)	• Mass production • Intensive use of electrical power	2nd Industrial Revolution (Machine Age)	• Use of scientific management to analyze work in production processes. • Increase in productive efficiency. • Scientific selection and training of workers	Classical Approach Engineer Technologist	• Scientific management Employee only complements the machine (Taylor theory) • Division of work and methods and time study, and the employee follows the boss (Fayol theory) • The employee follows organizational rules (Webber theory) • Techno-Centric System (TCS)
Logic of Production Transition	Industry 3.0 (Began in the middle of the twentieth century)	• Automation and microelectronic technology in manufacturing • Information and Communication technologies • Computer-aided design (CAD) and Computer-aided manufacturing (CAM) • Computer-aided process planning (CAPP) and Computer-integrated manufacturing (CIM) • Flexible Assembly Systems (FAS) • Flexible Manufacturing Systems (FMS) • Balanced Automation Systems (BAS)	3rd Industrial Revolution (Information Age)	• Production defense related goods • Use of Operations Research techniques • Statistical Process Control (SPC) • Deming Cycle (PDCA) • Toyota Production System (TPS) • Lean Manufacturing (JIT Just-In-Time, Kanban,...) • Project techniques as CPM and PERT • Material- Requirement Planning//Manufacturing Resource Planning/Enterprise Resource Planning-Customer Focus • Total Quality Management (TQM) • Six Sigma • - Agile Manufacturing (AM)	Behaviorist Approach Social Sciences	• Opposition to Classical Theories • Teamwork is at the center • - Socio-Technical System (STS)

Continued on following page

Table 1. Continued

						• The organization is an integrated system • Cha-ordic System Thinking (CST)
Market Logic	Industry 4.0 (Began in the begin of the twenty-first century)	• Cyber-physical systems (CPS) • Sensors and Internet of things (IoT) • Virtual Agents/Visual Technology • Big Data and Machine Learning • Smart products/parts/materials • Cloud computing • Additive manufacturing	4th Industrial Revolution (Big Data Age)	• Customer Focus • Integrating and extending manufacturing processes at both intra- and inter-organisational levels • Digitalization of manufacturing • Operational Excellence • Reconfigurable manufacturing systems	Neo-classical Pragmatic Approach Marketer	• Adapting the classic approach to the time • Flexible rules • Efficient production
		• Artificial Intelligence and optimized hardware • Decision Management • Collaborative robots SMART Manufacturing		• Information technology, telecommunications and manufacturing are merging • Mass Customization • Production is increasingly autonomous • Product Service Systems Personalized Production	Modern Approach Industrial Engineering and Management as Applied Science	• First idea about customers • Technology + social and behaviorist sciences + market knowledge • System Management • Decision Theory • Quantitate approach • Contingency and Situational approach.

Source: Based on Salvendy, (2001); Zandin (2001); Alves et al. (2012); Ravi (2015); Amaral (2017); Xu et al. (2018); Carvalho et al. (2019) and Zonnenshain and Kenett (2020)

On the other hand, companies are organizing themselves to take account of their customers. Customers and other stakeholders are their catalysts, in so far as the goal is to ensure their full satisfaction. Now, innovation is fundamental and systematic. But even if the winner-order is the product and/or service provided, resulting in a product and/or marketing-innovation as priority, organizational and process innovation are also important for organizational efficiency, which can be vital for effectiveness (stakeholder satisfaction). In short, there is an industrial revolution before and another after.

Before there was a zero industry where Handicrafts and Craftsman predominated and there was no programmed management. After the first industrial revolution, when the steam engine was invented everything changed. Mechanical manufacturing systems were introduced, and mass production started. This was followed by the second industrial revolution with electricity and the intensification of mass production through the intensive use of electrical energy. At the same time, scientific management was used to analyze work in production processes for increasing productive efficiency. The third industrial revolution introduced electronics, automation and robotics, and subsequently microelectronic technology, into manufacturing, which was computer-assisted. The companies became an integrated system. Progress in Information Technology and Communications lead to a paradigm shift in manufacturing automation systems. Based on the Internet of Things (IoT) and Smart Technologies, manufacturing industries are witnessing the fourth industrial revolution (called Industry 4.0) (Kagermann et al., 2013; Nakayama et al., 2020). Mainly, this new era is characterized by digitalization of manufacturing industries.

However, integrating and extending manufacturing processes at both intra- and inter-organizational levels is paramount, as is customer focus. In this way, goods cannot be separated from services. Thus, the Product Service System appears as fundamental in this context. Service Systems feature an intense collaborative relationship with the customer. In fact, the customer also provides significant input to the service providing process as well, which anticipates the proper adaptation to a class of services - which normally appears in the design process as customer feedback. To export the concept of service to other sectors, such as manufacturing (Dutra & Silva, 2016), is key. This means we are currently experiencing the fourth industrial revolution, Smart Manufacturing. Some futurists even discuss what the fifth industrial revolution will be. There are a few visions for Industry 5.0. One emerging theme is human-robot co-working. In recent years, significant progress in robotics and artificial intelligence (AI) research (Demir et al., 2019) was achieved. However, we are only talking about the future, which carries high risks.

Challenges for the IEM Professional Workforce in the New Industrial Age

A European Union initiative aimed at developing Curriculum Guidelines for Industry 4.0 needs predicted a global decline in total manufacturing and production roles (EU, 2020). According to this report, this decline is driven by labor-substituting technologies such as additive manufacturing, and an increase in sustainable, resource-efficient product use, slower growth in demand in ageing societies and threats to global supply chains due to geopolitical volatility.

Also, manufacturing is expected to become a highly advanced sector, where there will be high demand for qualified engineers in the labor market. Due to the complex interaction between humans and machines in Industry 4.0, the labor market is expected to change in terms of content and new types of jobs are also being created leading to new skill requirements (Galati & Bigliardi, 2019; Hartmann & Bovenschulte, 2013). In a recent EU publication (European Union, 2019), a set of eight key competencies for lifelong learning was established, including mathematical and scientific skills, technology and engineering, which highlights the importance of an engineering training for tackling the future needs of society.

Moreover, the continuous adoption of advanced manufacturing technologies in industrial processes and products leads to a factory of the future that is projected to be highly innovative, networked, and flexible (Gehrke et al., 2015; Mckinsey, 2019). Companies will need a skilled workforce to develop and run advanced manufacturing tools and systems, to analyze the data received from machines, consumers and global resources, and to predict system behavior (Gehrke et al., 2015; Vrchota et al., 2019).

Thus, workers must assume cross-functional coordination, with the ability to grasp all elements of the entire production system and dynamics of supply chains and operation environments (Pekkanen et al., 2020). Human operators are expected to experience increased complexity in their daily tasks that will require high flexibility on their side and the ability to demonstrate adaptive capabilities in a very dynamic working environment (Longo et al., 2017). Consequently, there will be more demand for better qualified employees on the shop floor trained in cross-functional areas and with the ability to manage new processes and information systems (Gehrke et al., 2015).

So far, there is little experience in corporate practice with respect to focused and successful Industry 4.0 implementation (Veile et al., 2019). Nonetheless, some studies in the literature seek to highlight the necessary skills for workers in the new industrial age. Despite that, there is currently a limited understanding about the skills needed to work in Industry 4.0, particularly from the industrial stand point (Pinzone et al., 2017; Universities of the Future, 2019).

Next, we discuss some skills that workers, particularly, engineers, must have in the context of Industry 4.0, as identified in some studies found in the literature. Competencies are defined as underlying characteristics of a person, such as skills, abilities, knowledge, attitudes and motivations that result in effective or higher performance in job-related tasks (Armstrong & Taylor, 2014).

Gehrke et al. (2015) recommended a set of qualifications and skills for the worker in a factory of the future. The qualifications and skills are mentioned in order of importance as "must be included", "should be included" and "could be included", and are clustered into two categories: technical and personal qualifications and skills (Q&S). Concerning the technical Q&S, workers must possess knowledge and skills related to IT, data and information processing and analytics, statistical knowledge, organizational and processual understanding, and the ability to interact with modern interfaces. On the personal side, these authors include self and time management, adaptability and ability to change, teamwork skills, and social and communication skills.

Moreover, in a EU project aimed at identifying the educational needs arising from Industry 4.0 in Europe (Universities of the Future, 2019), a set of Industry 4.0 related competencies were drawn from a literature review focusing on work life skills and skills for professional work. In their research, the authors organized the required set of competencies in discipline-specific competencies and transferable skills. The discipline-specific competencies were divided into engineering, business and design competencies. In the field of engineering, the following set of competencies were included: data science and advanced analytics; novel human-machine interfaces; digital-to-physical transfer technologies; advanced simulation and virtual plant modelling; data communication and networks and system automation; artificial intelligence; robotics; programming skills; closed-loop integrated product and process quality control/ management systems; and real-time inventory and logistics optimization systems. Also, the following set of transferable competencies was identified: problem-solving skills, soft skills, systems thinking, business thinking and technological literacy.

In a report from Eurofound (2019), in what concerns the required skills for advanced manufacturing, it is stated that there will be growing demand for high-skilled workers, including people with a more traditional engineering profile, like process engineers, quality control, chemical, electronic, mechanical or mechatronic engineers. According to the same study, due to the centrality of data and information on new technologies, there will also appear newer skill sets, like those of designers, industrial data scientists, 'big data' statisticians/mathematicians and data security analysts – to deal with the increasing data-intensiveness of production processes. Additionally, the most searched profile will likely include some combination of engineering and ICT skills.

According to Veile et al. (2019) in Industry 4.0 workers must have a fundamental understanding of ICT systems, automation technology and data analysis. They must also be aware of issues concerning data abuse and ICT security, have interdisciplinary knowledge and an understanding of interconnected systems. Also, further decision-making competencies are required due to the increase in planning and controlling tasks.

Pinzone et al. (2017) submitted a set of 35 technical skills for Industry 4.0 related to five organizational areas affected by Industry 4.0, – i.e. operations management (10), supply chain management (6), product-service innovation management (4), data science management (7) and information technology-operational technology integration management (8).

Finally, Hecklau et al. (2016) developed a competence model with the required core worker competencies in Industry 4.0, organized in four categories: technical competencies, methodological competencies, social competencies and personal competencies (Table 2).

Table 2. Employee competencies in Industry 4.0

• Technical competencies	State-of-the-art knowledge, technical skills, process understanding, media skills, coding skills and understanding IT security
• Methodological competencies	Creativity, entrepreneurial thinking, problem solving, conflict solving, decision-making, analytical skills, research skills and efficiency orientation
• Social competencies	Intercultural skills, language skills, communication skills, networking skills, ability to work in a team, ability to be compromising and cooperative, ability to transfer knowledge and leadership skills
• Personal competencies	Flexibility, ambiguity tolerance, motivation to learn, ability to work under pressure, sustainable mindset, and compliance

Source: (Hecklau et al., 2016)

The literature on Industry 4.0, which is directly related to IEM, is building up (Sackey & Bester, 2016). In their research based on literature review, Sackey and Bester (2016) highlighted the following implications of the Industry 4.0 for IEM professions: soft skills will become more important than ever; IEM professionals will need to develop skills in deploying and optimizing augmented-reality systems; knowledge on techniques for optimizing human-robot collaboration will become important for tackling problems, such as physically-demanding and ergonomically-inconvenient jobs; and IEM engineers will need IT skills in advanced analytics for simulation that leverages real-time data to produce virtual models of plants for insight and real-time decision-making. Furthermore, the same authors argued that of all the engineering disciplines, the IEM would be the most adversely affected by an Industry 4.0 environment, unless steps are taken to redesign it taking into consideration the competencies required by Industry 4.0.

IEM professionals must also have a holistic view. According to Smith (2001), industrial engineers have been focusing on improving labor productivity on the shop-floor, which is a narrow outlook on productivity. The scope should be broadened and better understood, as IEM engineers can play an essential role in using their competencies to make improvements and help others achieve the same. According to the same author, in the past, the mission of industrial engineers was to increase the output of all the available resources, maximizing machine utilization and suggesting layout and method improvements to produce more. Now, with a shift from mass production concepts to lean production, the IEM should focus on client value – added activities, and flow efficiency, thus eliminating waste by producing exactly what the client needs and requires, when it needs and with the best quality at a competitive price.

This must be accomplished by attending to the same priorities of in-house clients, ensuring the best and safest work conditions and encouraging them to give their best. IEM professionals could continue using many tools learned at the universities because they will remain the same, but the context in which they are applied has changed (Smith, 2001). The ultimate goal is to have one-of-a-kind product for each customer without incurring in extra economic costs for them or the planet. This is the path of mass customization that denotes the organization's ability to provide customized products and services that fulfill each customer's individual needs without substantial trade-offs in cost, delivery and quality performance in relation to standardized offerings (Pine, 1993; Squire et al., 2006). Involving the customer in the design phase is a step further in the direction of Personalization, which aims at Value differentiation (Mourtzis & Doukas, 2014).

Chedid et al. (2018) conducted research about the essential generic skills sought in current IEM graduates aimed at gathering the perspectives of teachers, students, and industrial professionals. The skills were categorized into four groups: scientific and technical, engineering practice, business practice and transversal (Figure 1).

Figure 1. Generic and specific skills for current IEM graduates
Source: Chedid et al., 2018

Generic skills	**Specific skills belonging to the scientific and technical category**
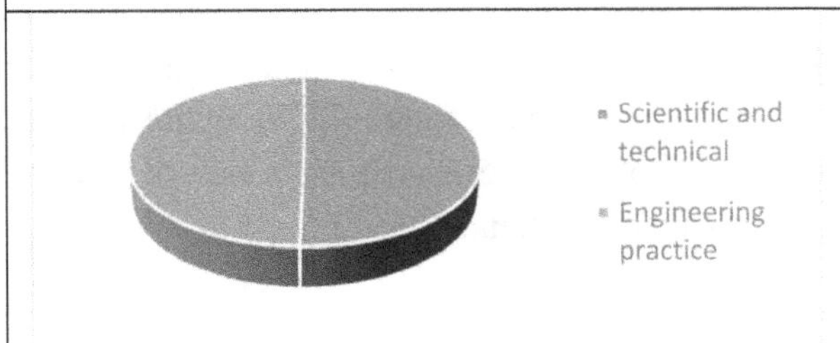	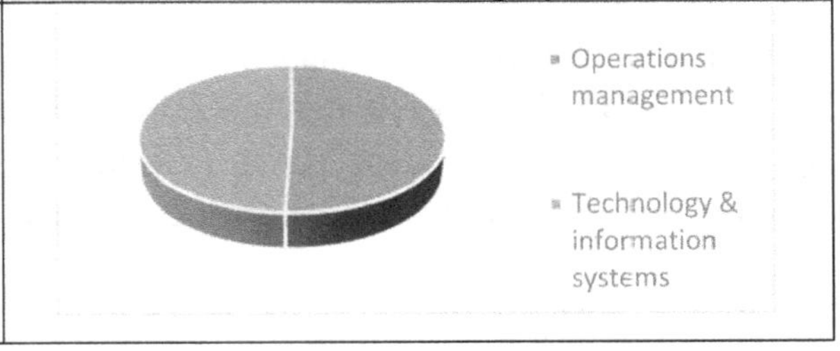

Figure 1 shows clearly that respondents value both the scientific and technical skills and engineering practice. Among the specific skills belonging to the scientific and technical category, operations management is valued slightly more than the skills in technology & information systems.

Using an alternative approach, Lima et al. (2017) developed a longitudinal study, from 2007 to 2013, based on newspaper advertisements to identify trends in professional practice and transversal competency needs in the field of IEM. The authors considered the following eleven professional practice areas, and 26 transversal competencies present in Table 3.

Table 3. Trends in professional practice and transversal skills need in IEM

Professional practice areas	Transversal skills
1. Production Management 2. Automation 3. Quality 4. Economics Engineering 5. Marketing 6. Industrial Optimization 7. Ergonomics and Human Factors 8. Supply Chain Management 9. Maintenance 10. Project Management 11. Product Design.	1. Foreign Languages 2. Information and Communications Technology 3. Teamwork 4. Communication 5. Leadership 6. Planning/Organization 7. Initiative 8. Dynamism 9. Interpersonal relationship 10. Autonomy/Liability 11. Goal Orientation 12. Analytical & Research Skills 13. Negotiation 14. Problem-solving 15. Motivation 16. Ability to adapt 17. Entrepreneurship 18. Dealing with pressure and stress 19. Critical thinking 20. Creativity 21. Openness to learning 22. Accuracy/Rigor/Exactness 23. Strategic Vision 24. Self-reliance 25. Maturity 26. Ethics

Source: (Lima et al., 2017)

From this analysis, it is concluded that Production Management, Supply Chain Management, Project Management and Quality account for 72.3% of all job advertisements in the field of IEM. Quite surprisingly Automation and Industrial Optimization account for 0.9% of all advertisements, which shows that employers refer explicitly to these practice areas in IEM job advertisements.

Skills for IEM Students: The Role of Education

The production of goods and services requires group organization and collaborative behavior, as well as the need for stakeholders to acquire technical knowledge in the learning and teaching process. Therefore, it is natural that education influences and is influenced by this process. Schools were established as privileged places for members of these social groups to learn basic reading and writing skills and to develop vocational skills later in their learning process. Nevertheless, in production, as in schools, the Cartesian thought was dominant and group organization was not present. According to Flumerfelt (2008), what many industrial and teaching environments have taught is division, specialization and departmentalization, as proposed by Taylor and Ford.

The Toyota Production System (TPS) represented a "turning point" in the factory production models. Following a distinctive path from the traditional Ford model, it influenced employee behavior into new and different attitudes and routines, involving competencies such as systemic and ethical thinking. According to Suzaki (1993) this system promotes management that prioritizes "making people before making products". A range of outlooks contributed to the success and knowledge of the TPS (Imai, 1986; Monden, 1998; Ohno, 1988; Shingo, 1989; Sugimori et al., 1977).

The Toyota Education Model (TEM) emerged from the need to adapt the learning process to these new demands, featuring a learning system that exists in the factory in all activities and at all levels. It challenges workers' daily routines and empowers them to make decisions and take responsibility for continuously improving processes and operations driven by customer needs through problem-solving. Such expectations allow for motivating, meaningful and fulfilling work.

By teaching people how to learn to solve problems scientifically, the Toyota group uses a learner-leader-teacher model, encouraging people to act as consultants. Becoming a Toyota worker means becoming a lifelong learner. The following are successful approaches taught in the TEM (Spear, 2004): 1) there is no substitute for direct observation; 2) proposed changes should always be structured as experiences; 3) workers and managers should experiment whenever possible; and 4) managers should advise and question (facilitate learning), not correct.

In order to engage people and promote expected behaviors, such as observation, experimentation, problem-solving and mentoring, the TEM has several tools, including continuous improvement, the Plan-Do-Check-Act (PDCA) cycle, the 5Whys, among others (Takeuchi et al., 2008). The most basic tool is the continuous improvement process, which is defined as a focused and sustained incremental organization innovation process (Bessant & Francis, 1999; Imai, 1986). PDCA is a visual management tool used to organize and implement proposals to solve the problems raised by teams. The PDCA cycle is a continuous loop of thinking and doing invented by Walter Andrew Shewhart (1891-1967) and disseminated in Japan by W. Edwards Deming (1900-1993). For this reason it is often known as the Deming cycle.

These concepts have greatly influenced in-school learning, especially higher education, whose accreditation systems, such as the American ABET system (ABET, 2011), highlighted. Additionally, Park et al. (2013) reported how continuous improvement has manifested itself in education. They identify and describe the commitment of some organizations to continuous improvement. Also, the TEM and the concept of continuous improvement were the basis of the proposal of Flumerfelt et al. (2015), Lean Engineering Education (LEE), which applies the principles of Lean Thinking to the education of engineers. According to the authors, the LEE model is based on the double-stranded mental model of the DNA strand, representing the mastery of content and competencies that promote learning outcomes in three key competencies: 1) ethics (Flumerfelt et al., 2012, 2013), 2) system-thinking (Flumerfelt et al., 2014), and 3) sustainability (Alves et al., 2014; Alves, Kahlen, et al., 2019; Moreira et al., 2010). Given the multi-disciplinarity of Lean Production (Amaro et al., 2019), extending the Lean principles to education means getting students to learn the most appropriate skills in the time available for today's challenges (Alves et al., 2017a). As customers of an apprenticeship system, it is up to them to choose what they need to learn.

To transform IEM students into lean IEM professionals with the skills described above, Higher Education teachers (or even in the lower levels) must implement active teaching/learning methodologies, in which the student takes the central role and the teacher is a facilitator of effective learning (Alves et al., 2017b). Additionally, by providing an experiential learning context, students could also develop such competencies (Fernandes, Cunha, Torres et al., 2020). Project-Based Learning (PBL) is one such example and the results are visible (Alves et al., 2017; Alves et al., 2016; Alves, Moreira, et al., 2019). These projects in an industrial setting (Benis et al., 2020; Lima et al., 2017; Mesquita et al., 2012) foster the learning-by-doing model advocated by Ford and TEM, where by adopting an action research methodology, students learn, interact with people at different levels and contribute to the company's growth (Alves et al., 2014; Alves et al., 2011; Kahlen et al., 2011).

Kilpatrick (1918) was the PBL promotor. He was inspired by Dewey's (1918) concept of democracy and education. PBL is much more than just a project, it allows students to build and learn for themselves, without being limited to the boundaries of the discipline. A real-world project driven by a question, problem, or a challenge in a teamwork environment, where students collaborate and learn with teachers, but also with their peers and anyone contributing to the development of the project. PBL could be case-based learning, community-based learning, game-based learning, passion-based learning, service-based learning, team-based learning that fit into the general category of PBL: challenge-based learning, problem-based learning, place-based learning, activity-based learning, and design-based learning (Pecore, 2015).

PBL was considered a best practice that engineering teachers must adopt in European universities in the broad study of IEM curriculum design (Mesquita et al., 2019). IEM curriculum design is being discussed by authors in different parts of the world (Benis et al., 2020; Chedid et al., 2018; Chen et al., 2005; Hunt et al., 2004; Lima et al., 2017; Lima et al., 2012, 2019; Santos et al., 2017; Zunk, 2018). These authors discussed IEM considering the areas of knowledge and skill development. Although skills are very important, they resemble the discipline-based design favoring the compartmentalization of curriculum design. This was discussed at length in the approaches presented in the previous section.

Nevertheless, engineering students need a holistic view of the system that enables the connections between bodies of knowledge (Flumerfelt, Alves, et al., 2014). In 2007, Grasso and Martinelli (2007) defined a new kind of engineer: "In this evolving world, a new kind of engineer is needed, one who can think broadly across disciplines and consider the human dimensions that are at the heart of every design challenge. In the new order, narrow engineering thinking will not be enough.".

Black and Phillips (2013) goes beyond that and defined a new Industrial Engineer profile which it called the Lean Engineer: "The Lean Engineer (LE) has an Industrial Engineering foundation, enhanced with lean tools, six sigma capabilities and a lean-to-green outlook to take the plant to zero waste going to the landfill." To create this engineer profile, it is necessary to equip students with lean competencies, as the ones discussed above. Also, it is important to recognize Lean holistic approach in the organizations and importance of this for the success of Industry 4.0. Some authors are working to create a center of excellence to help students learn lean competencies suitable for Industry 4.0 (Dombrowski et al., 2019). Learning factories, with or without companies as partners, are being set-up as a useful active learning methodology to learn Lean and Industry 4.0 (Abele et al., 2017, 2019; Bauer et al., 2018; Enke et al., 2018; Gento et al., 2020; Goerke et al., 2015; Pascual et al., 2020; Tan et al., 2020).

Beyond Industry 4.0 or Industry 5.0 there is another sounding concept that originated recently from Japan, i.e. Society 5.0 (Ferreira & Serpa, 2018; Keidanren, 2018; Komiyama & Yamada, 2018; Pereira et al., 2020). According to a publication in the World Economic Forum Meeting "People will be expected to exercise rich imaginations to identify a variety of needs and challenges scattered across society and the scenarios to solve them, as well as creativity to realize such solutions by using digital technologies and data. Society 5.0 will be an Imagination Society, where digital transformation combines with the creativity of different people to bring about "problem-solving" and "value-creation" that lead us to sustainable development. It is a concept that can contribute to the achievement of the Sustainable Development Goals (SDGs) adopted by the United Nations." (Nakanishi, 2019). This concept is a strategy to deal with the impact of an ageing population (Pereira et al., 2020) that could immensely benefit from Industry 4.0 technologies and IEM students, e.g. through service-based learning projects (Braun, 2013) and professionals with the lean competencies referred above.

Nevertheless, many academics continue to ignore the need to teach Lean thinking (Alves et al., 2020). Even the studies mentioned previously about the knowledge areas do not address Lean as having a proper Body of Knowledge, although it is incorporated into Operations Management or distributed by different knowledge areas, i.e. Quality Management and Training, Supply Chain Management and Design and Manufacturing Engineering (Institute of Industrial and Systems Engineers, 2020b). Lean thinking could be the "glue" that holds all these knowledge areas together, being established as a vision that puts people, and not the system or processes or departments or technology, in the center. Every activity performed by students, and later professionals, should create value for the client, while always eliminating waste and doing more with less human effort, less development time, less resources, less stocks, less of everything. Operational excellence (Shingo Institute, 2020) and Agility Manufacturing (Kidd, 1994), among other concepts, are very important and complement this need.

DISCUSSION

Mass customization, flexibility, agile, green, mass of niches, and other ones, brought and will continue to bring new needs and requirements to the IEM profession. Anderson (2008) has called it as the "long tail" related to the U.S. entertainment industries, despite some focus on "hits, but lesser less economically attractive, due to the new digital economy paradigm (Armstrong, 2008). In this sense competencies, such as mind flexibility, customer orientation, negotiation, relationship orientation, team work, creativity and innovative thinking, information seeking, efficiency orientation, analytical thinking, and pattern recognition (Forza & Salvador, 2006; Trentin et al., 2019) will continue to be important and need to be developed further in the IEM curricula to prepare IEM professionals for the mass customization environment. An environment characterized by, for example, product modularity, postponement, product configuration systems, robust process design, cross-functional integration and coordination (Trentin et al., 2019), demanding human resource management practices and a holistic view of the system.

The integration of Lean Manufacturing, Agile Manufacturing, Mass Customization, mass niches and Industry 4.0, among other concepts, will be at the heart of enterprises. Thus, IEM curricula must be able to equip IEM students for such an environment. To achieve that, active learning methodologies, such as the ones previously mentioned, will be mandatory.

Considering the Industrial Engineering Body of Knowledge, although some of the Lean Manufacturing, Agile Manufacturing, Mass Customization, Personalization, and Industry 4.0 concepts are already part of the IEM curricula, there is room for improvement and new contents. To this end, HEIs must become more agile.

In addition to the core knowledge, the modern business world increasingly requires of engineering graduates a wide range of more practical and generic skills (Pekkanen et al., 2020). Modern IEM engineers will face the challenge of having to improve or equipping themselves with the previously identified set of technical competencies due to the emergence of Industry 4.0, but also with transversal/generic skills, since IEM professionals work in the interface between engineering and management. Besides, the classic production job profile, such as that of the industrial engineer, must increasingly dive into the world of information and communication technology (Abele et al., 2015). In a more recent publication (Abele et al., 2019), these authors presented a domain-specific competency model for Lean 4.0 at the Process Learning Factory CiP. In this model, they advocated the role of Industry 4.0 technologies for supporting the vision of Toyota´s True North (0 defects, 100% value-added, one-piece flow, and respect for the employees). To achieve this, the IEM needs to be a Lean thinker (Alves et al. 2012).

The IEM professional will need to be a sustainability thinker and consider the full product-service life cycle in the decision-making process. The multidisciplinary profile of the IEM professional allows for the inclusion of all dimensions of sustainability (People / Planet / Profit - Environment / Economy / Society). These are included at the level of the systems, through the integration of the human being into the work system to provide safety and high moral, at the management level, through their focus on integrated systems management, and at the quality level, by guaranteeing a customer-centric approach that delivers adequate performance. Although sustainability concepts are already included in the IEM curricula they must be addressed more explicitly and with greater depth.

In this sense the Industrial Engineer and Manager plays and will continue to play a relevant and central role in intra- and inter-communication of an organization and, therefore, professionals are increasingly needed in this area. For that, given that the economy is moving towards greater digitalization, the IEM professional must progress to a new level of technological and digital skills, although not neglecting the natural evolution of people, ethics and system management, among others.

CONCLUSION

This research presents a contemporary outlook on the challenges that IEM professionals must face in the Industry 4.0 concept or the concept after this. Before becoming professionals, IEM will have to develop suitable skills in formal environments such as Higher Education Institutions or in informal settings, e.g. as members of a Union or voluntary works, among other environments where they must learn the adequate competencies. At the same time, they should have physical contact with companies and real environments earlier in their academic career. On the other hand, societal challenges will have to be incorporated from the start, not only from the point of view of solving concrete corporate problems, but always framed in SDGs.

As mentioned, in addition to the core knowledge, the modern business world increasingly requires of engineering graduates an extensive set of more practical and generic skills. Not only an identified set of actual and future technical competencies, but also transversal/overarching skills. In this sense, Industrial Engineering and Management emerges as an applied science, and not just as a simplistic connection between engineering and management. On the other hand, it has not only increased its importance in the industry, but it has also expanded to services, even in a modern Product Service System logic.

External communication is important, but internal communication is, more than ever, paramount. Considering the increasing complexity and greater specialization of tasks, effective communication with the other functions is required in a context of concurrent or simultaneous engineering. In this context such cross-functional knowledge is found in no other professional other than IEM.

Moreover, the contexts in which companies operate will be more unpredictable. Contingencies, as the COVID-19 pandemic, brings news challenges for society in general, but not only. As it was previously observed by the current authors, this carries tremendous problems for global enterprises and the impacts on the production systems and supply chains are a great example of this. This requires greater flexibility from enterprises, and consequently IEM professionals. In this sense, future works should not only research these impacts, but also look into how Higher Education Institutions can prepare IEM professionals for the future.

ACKNOWLEDGMENT

This work was supported by the FCT – Fundação para a Ciência e Tecnologia under the following Projects: UIDB/00319/2020, UIDB/04058/2020 and Project UIDB/05037/2020.

Funding: This research received no specific grant from any funding agency in the public, commercial, or not-for-profit sectors.

REFERENCES

Abele, E., Chryssolouris, G., Sihn, W., Metternich, J., ElMaraghy, H., Seliger, G., Sivard, G., ElMaraghy, W., Hummel, V., Tisch, M., & Seifermann, S. (2017). Learning factories for future oriented research and education in manufacturing. *CIRP Annals - Manufacturing Technology, 66*(2), 803–826. doi:10.1016/j.cirp.2017.05.005

Abele, E., Metternich, J., & Tisch, M. (2019). *Learning Factories*. Springer International Publishing. doi:10.1007/978-3-319-92261-4

Abele, E., Metternich, J., Tisch, M., Chryssolouris, G., Sihn, W., ElMaraghy, H., Hummel, V., & Ranz, F. (2015). Learning Factories for Research, Education, and Training. *Procedia CIRP, 32*, 1–6. doi:10.1016/j.procir.2015.02.187

ABET. (2011). *Criteria for accrediting engineering programs: Effective for Reviews During the 2012-2013 Accreditation Cycle*. ABET.

Alves, A. C., Flumerfelt, S., Moreira, F., & Leão, C. P. (2017). Effective Tools to Learn Lean Thinking and Gather Together Academic and Practice Communities. *Education and Globalization, 5*. doi:10.1115/IMECE2017-71339

Alves, A. C., Moreira, F., Carvalho, M. A., Oliveira, S., Malheiro, M. T., Brito, I., Leão, C. P., & Teixeira, S. (2019). Integrating Science, Technology, Engineering and Mathematics contents through PBL in an Industrial Engineering and Management first year program. *Production, 29*, 1–19. doi:10.1590/0103-6513.20180111

Alves, A. C., Sousa, R. M., Dinis-Carvalho, J., Lima, R. M., Moreira, F., Leão, C. P., Maia, L. C., Mesquita, D., & Fernandes, S. (2014). Final year Lean projects: advantages for companies, students and academia. *Project Approaches in Engineering Education*, 1–10. http://hdl.handle.net/1822/30172

Alves, A. C., Flumerfelt, S., & Kahlen, F.-J. (2017). Lean Education: An Overview of Current Issues. In Lean Education: An Overview of Current Issues. Springer International Publishing. doi:10.1007/978-3-319-45830-4

Alves, A.C., Kahlen, F.-J., Flumerfelt, S., & Siriban-Manalang, A.-B. (2014). Fostering Sustainable Development Thinking Through Lean Engineering Education. *Education and Globalization, 5*. doi:10.1115/IMECE2014-38192

Alves, A. C., Dinis-Carvalho, J., & Sousa, R. M. (2012). Lean production as promoter of thinkers to achieve companies' agility. *The Learning Organization, 19*(3), 219–237. doi:10.1108/09696471211219930

Alves, A. C., Dinis-Carvalho, J., Sousa, R. M., Moreira, F., & Lima, R. M. (2011). Benefits of Lean Management: results from some case studies in Portugal. *CLME'2011, 1977.*

Alves, A. C., Leão, C. P., Uebe-Mansur, A. F., & Kury, M. I. R. A. (2021). The knowledge and importance of Lean Education based on academics' perspectives: An exploratory study. *Production Planning and Control, 32*(6), 497–510. doi:10.1080/09537287.2020.1742371

Alves, A. C., Sousa, R. M., Fernandes, S., Cardoso, E., Carvalho, M. A., Figueiredo, J., & Pereira, R. M. S. (2016). Teacher's experiences in PBL: Implications for practice. *European Journal of Engineering Education, 41*(2), 123–141. doi:10.1080/03043797.2015.1023782

Alves, A. C., Kahlen, F.-J., Flumerfelt, S., & Siriban-Manalang, A. B. (2019). Lean Engineering for Global Development. Springer International Publishing. doi:10.1007/978-3-030-13515-7

Amaral, L. M. (2017). The Industrial Engineering Management. *Ingenium, 158*(March/April), 30–33.

Amaro, A. P., Alves, A. C., & Sousa, R. M. (2019). Lean Thinking: A Transversal and Global Management Philosophy to Achieve Sustainability Benefits. In A. C. Alves, F.-J. Kahlen, S. Flumerfelt, & A. B. Siriban-Manalang (Eds.), *Lean Engineering for Global Development* (pp. 1–31). Springer International Publishing. doi:10.1007/978-3-030-13515-7_1

Anderson, C. (2008). *The long tail: Why the Future of Business Is Selling Less of More*. Hyperion e.

Armstrong, M., & Taylor, S. (2014). *Armstrong's Handbook of Human Resource Management Practice* (13th ed.). Kogan Page.

Armstrong, R. (2008). The Long Tail: Why the Future of Business Is Selling Less of More. *Canadian Journal of Communication, 33*(1). Advance online publication. doi:10.22230/cjc.2008v33n1a1946

Bauer, H., Brandl, F., Lock, C., & Reinhart, G. (2018). Integration of Industrie 4.0 in Lean Manufacturing Learning Factories. *Procedia Manufacturing, 23*, 147–152. doi:10.1016/j.promfg.2018.04.008

Benis, A., Nelke, S. A., & Winokur, M. (2020). *Upgrading Industrial Engineering and Management curriculum to Industry 4.0*. IEEE.

Bessant, J., & Francis, D. (1999). Developing strategic continuous improvement capability. *International Journal of Operations & Production Management, 19*(11), 1106–1119. doi:10.1108/01443579910291032

Black, J. T., & Phillips, D. T. (2013). *Lean Engineering The Future Has Arrived.* Virtualbookworm.com Publishing.

Bonnaud, O., & Bsiesy, A. (2020). Adaptation of the Higher Education in Engineering to the Advanced Manufacturing Technologies. *Advances in Technology Innovation, 5*(2), 65–75. doi:10.46604/aiti.2020.4144

Braun, J. M. (2013). Teaching Lean Six Sigma with Service-Learning. *2013 ASQ Advancing the STEM Agenda Conference.*

Carvalho, A. M., Sampaio, P., Rebentisch, E., Carvalho, J. Á., & Saraiva, P. (2019). Operational excellence, organisational culture and agility: The missing link? *Total Quality Management & Business Excellence, 30*(13–14), 1495–1514. doi:10.1080/14783363.2017.1374833

Chedid, M., Teixeira, L., & Rosa, M. J. (2018). The Industrial Engineers' Essential Skills: Students, teachers and industry professionals' perspectives. *Proceedings of the INTED 2018 Conference*, 1815–1825. 10.21125/inted.2018.0322

Chen, C., Jiang, B. C., & Hsu, K. (2005). An empirical study of industrial engineering and management curriculum reform in fostering students' creativity. *European Journal of Engineering Education*, *30*(2), 191–202. doi:10.1080/03043790500087423

Chou, S. (2018). The Fourth Industrial Revolution: Digital fusion with internet of things. *Journal of International Affairs*, *72*(1), 107–120. https://www.jstor.org/stable/26588346

Corrocher, N., Mavilia, R., & Giorgio, M. (2020). The Sino-German alliance for the fourth industrial revolution: Dynamics and policy implications. *Journal of Economic Policy Reform*, *23*(4), 426–446. do i:10.1080/17487870.2018.1547639

Demir, K. A., Döven, G., & Sezen, B. (2019). Industry 5.0 and Human-Robot Co-working. *Procedia Computer Science*, *158*, 688–695. doi:10.1016/j.procs.2019.09.104

Dombrowski, U., Wullbrandt, J., & Fochler, S. (2019). Center of Excellence for Lean Enterprise 4.0. *Procedia Manufacturing*, *31*, 66–71. Advance online publication. doi:10.1016/j.promfg.2019.03.011

Downe-Wambolt, B. (1992). Content analysis: Method, applications and issues. *Health Care for Women International*, *13*(3), 313–321. doi:10.1080/07399339209516006 PMID:1399871

Dutra, D. S., & Silva, J. R. (2016). Product-Service Architecture (PSA): Toward a Service Engineering perspective in Industry 4.0. *IFAC-PapersOnLine*, *49*(31), 91–96. doi:10.1016/j.ifacol.2016.12.167

Elberse, A. (2008, July). The Long Tail Debate: A Response to Chris Anderson. *Harvard Business Review*.

Elsayed, E. A. (1999). Industrial Engineering Education: A Prospective. *European Journal of Engineering Education*, *24*(4), 415–421. doi:10.1080/03043799908923576

Enke, J., Glass, R., Kreß, A., Hambach, J., Tisch, M., & Metternich, J. (2018). Industrie 4.0 - Competencies for a modern production system: A curriculum for Learning Factories. *Procedia Manufacturing*, *23*, 267–272. Advance online publication. doi:10.1016/j.promfg.2018.04.028

EU. (2020). *Skills for Industry Curriculum Guidelines 4.0: Future-proof education and training for manufacturing in Europe*. doi:10.2826/69418

Eurofound. (2019). *The future of manufacturing in Europe*. Author.

European Union. (2019). *Key competences for lifelong learning*. doi:10.2766/569540

Fernandes, S., Cunha, R., Torres, D. P. C., Fernandes, S., Cunha, R., Torres, D., & Pimentel, C. (2020). JELA: An Alternative Approach to Industrial Engineering and Management student's Lean Management Education. In *Proceedings of 6th European Lean Educator Conference (ELEC2019)* (pp. 79–87). Springer International Publishing.

Ferrari, R. (2015). Writing narrative style literature reviews. *Medical Writing*, *24*(4), 230–235. doi:10.1179/2047480615Z.000000000329

Ferreira, C. M., & Serpa, S. (2018). Society 5.0 and Social Development: Contributions to a Discussion. *Management and Organizational Studies*, *5*(4), 26. doi:10.5430/mos.v5n4p26

Flumerfelt, B. S. (2008). *Lean Thinking for schools*. Academic Press.

Flumerfelt, S., Alves, A. C., & Kahlen, F.-J. (2013). What Lean Teaches Us About Ethics in Engineering. *Education and Globalization, 5*. doi:10.1115/IMECE2013-62393

Flumerfelt, S., Alves, A. C., Kahlen, F.-J., & Siriban Manalang, A. B. (2012). Why Require Ethics in Engineering? *Volume 5: Education and Globalization. General Topics*, *5*, 495–502. doi:10.1115/IMECE2012-89392

Flumerfelt, S., Alves, A. C., & Khalen, F.-J. (2014). Lean Engineering Education: The DNA of Content and Competency. *Proceedings of the 2014 IIE Engineering Lean and Six Sigma Conference, 2014 Lean Educators Conference*.

Flumerfelt, S., Kahlen, F.-J., Alves, A. C., & Siriban-Manalang, A. B. (2015). *Lean Engineering Education: Driving Content and Competency Mastery*. ASME Press. doi:10.1115/1.860502

Flumerfelt, S., Kahlen, F.-J. F.-J., Alves, A., Calvo-Amodio, J., & Hoyle, C. (2014). Systems Competency for Engineering Practice. *Systems, Design, and Complexity, 11*. doi:10.1115/IMECE2014-40142

Forza, C., & Salvador, F. (2006). HRM policies for mass customization: understanding individual competence requirements and training needs for the mass customizing industrial company. In Mass Customization: Challenges and Solutions (pp. 251–269). Springer.

Galati, F., & Bigliardi, B. (2019). Industry 4.0: Emerging themes and future research avenues using a text mining approach. *Computers in Industry*, *109*, 100–113. doi:10.1016/j.compind.2019.04.018

Gehrke, L., Kühn, A. T., Rule, D., Moore, P., Bellmann, C., Siemes, S., Dawood, D., Singh, L., Kulik, J., & Standley, M. (2015). *A Discussion of Qualifications and Skills in the Factory of the Future: A German and American Perspective*. Academic Press.

Gento, A. M., Pimentel, C., & Pascual, J. A. (2020). Lean school: An example of industry-university collaboration. *Production Planning and Control*, 1–16. doi:10.1080/09537287.2020.1742373

Goerke, M., Schmidt, M., Busch, J., & Nyhuis, P. (2015). Holistic Approach of Lean Thinking in Learning Factories. *Procedia CIRP*, *32*, 138–143. doi:10.1016/j.procir.2015.02.221

Gowripeddi, V. V., Bijjahalli, M. C., Janardhan, N., & Bhimavaram, K. R. (2021). Role of Education 4.0 Technologies in Driving Industry 4.0. *17th International Conference on Remote Engineering and Virtual Instrumentation, REV 2020: Advances in Intelligent Systems and Computing*, 576–587. 10.1007/978-3-030-52575-0_48

Grasso, D., & Martinelli, D. (2007). Holistic engineering. *The Chronicle of Higher Education. The Chronicle Review*, *53*(28), B8–B9.

Gregory, A. T., & Denniss, A. R. (2018). An Introduction to Writing Narrative and Systematic Reviews—Tasks, Tips and Traps for Aspiring Authors. *Heart Lung and Circulation*, *27*(7), 893–898. doi:10.1016/j.hlc.2018.03.027 PMID:29857977

Gupta, K., Mashinini, M., Mukhawana, D., & Louw, A. (2019). Development of industry 4.0 virtual lab for manufacturing engineering education. *Proceedings of the International Conference on Industrial Engineering and Operations Management*, 1857–1867.

Gupta, K., Mukhawana, D., & Mashinini, M. (2019). A project based learning tool for industry 4.0 manufacturing engineering education. *Proceedings of the International Conference on Industrial Engineering and Operations Management*, 978–984.

Hadgraft, R. G., & Kolmos, A. (2020). Emerging learning environments in engineering education. *Australasian Journal of Engineering Education, 25*(1), 3–16. doi:10.1080/22054952.2020.1713522

Hartmann, E. A., & Bovenschulte, M. (2013). Skills Needs Analysis for Industry 4.0 Based on Roadmaps for Smart Systems. *Using Technology Foresights for Identifying Future Skills Needs. Global Workshop Proceedings*, 24–36.

Hecklau, F., Galeitzke, M., Flachs, S., & Kohl, H. (2016). Holistic Approach for Human Resource Management in Industry 4.0. *Procedia CIRP, 54*, 1–6. doi:10.1016/j.procir.2016.05.102

Hunt, I., O'Sullivan, D., Rolstadas, A., Horan, M., & Precup, L. (2004). Survey of manufacturing curricula from around the world. *Production Planning and Control, 15*(1), 71–79. doi:10.1080/09537280410001662583

Imai, M. (1986). *Kaizen: The Key to Japan's Competitive Success*. McGraw-Hill.

Institute of Industrial and Systems Engineers. (2020a). *Origins of IISE*. https://www.iise.org/details.aspx?id=295

Institute of Industrial and Systems Engineers. (2020b). *The industrial engineering body of knowledge*. IISE Body of Knowledge. https://www.iise.org/

Kagermann, H., Wahlster, W., & Helbig, J. (2013). *Recommendations for Implementing the Strategic Initiative INDUSTRIE 4.0*. Academic Press.

Kahlen, F.-J., Flumerfelt, S., Sinban-Manalang, A. B., & Alves, A. (2011). Benefits of lean teaching. *ASME 2011 International Mechanical Engineering Congress and Exposition, IMECE 2011, 5*, 351–358.

Kaya, R. (2019). The Conceptual Framework of the Industry 4.0 Revolution. In F. Ayhan (Ed.), *Several Dimensions of Innovation, Technology and Industry 4.0* (pp. 13–271). Peter Lang. doi:10.3726/b16202

Keidanren. (2018). *Society 5.0: co-creating the future*. www.keidanren.or.jp/en/policy/2018/095_outline.pdf

Kidd, P. T. (1994). *Agile Manufacturing forging new frontiers*. Addison Wesley Publishers.

Kilpatrick, W. H. (1918). The project method. *Teachers College Record, 19*(4), 319–335.

Komiyama, H., & Yamada, K. (2018). *New Vision 2050*. Springer Japan. doi:10.1007/978-4-431-56623-6

Kumar, R. (2020). *Industrial Engineering* (2nd ed.). Jyothis.

Lima, R. M., & Mesquita, D. C. R., & M., R. (2017). Defining the Industrial and Engineering Management Professional Profile: A longitudinal study based on job advertisements. *Production*, •••, 27.

Lima, R. M., Dinis-Carvalho, J., Sousa, R. M., Alves, A. C., Moreira, F., Fernandes, S., & Mesquita, D. (2017). Ten Years of Project-Based Learning (PBL) in Industrial Engineering and Management at the University of Minho. In PBL in Engineering Education (pp. 33–51). SensePublishers. doi:10.1007/978-94-6300-905-8_3

Lima, R. M., Mesquita, D., Amorim, M., Jonker, G., & Flores, M. A. (2012). An analysis of knowledge areas in industrial engineering and management curriculum. *International Journal of Industrial Engineering and Management*, *3*(2), 75–82.

Lima, R. M., Mesquita, D., Sousa, R. M., Monteiro, M. T. T., & Cunha, J. (2019). Curriculum Analysis Process: analysing fourteen Industrial Engineering programs. *PAEE/ALE2019*, 92–101.

Longo, F., Nicoletti, L., & Padovano, A. (2017). Smart operators in industry 4.0: A human-centered approach to enhance operators' capabilities and competencies within the new smart factory context. *Computers & Industrial Engineering*, *113*, 144–159. doi:10.1016/j.cie.2017.09.016

McKinsey. (2019). Industry 4.0: Capturing value at scale in discrete manufacturing. *Advanced Industries*. doi:10.1007/s12599-014-0334-4

Mesquita, D., Salimova, T., Soldatova, E., Atoev, S., & Lima, R. M. (2019). What can be recommended to engineering teachers from the analysis of 16 European teaching and learning best practices? *Proceedings of the SEFI 47th Annual Conference*, 770–779.

Mesquita, D., Lima, R. M., & Flores, M. A. (2012). *Developing professional competencies through projects in interaction with companies : A study in Industrial Engineering and Management Master Degree Project Based Learning (PBL) and Professional Practice*. Academic Press.

Miranda, J., Pérez-Rodríguez, R., Borja, V., Wright, P. K., & Molina, A. (2019). Sensing, smart and sustainable product development (S 3 product) reference framework. *International Journal of Production Research*, *57*(14), 4391–4412. doi:10.1080/00207543.2017.1401237

Monden, Y. (1998). Toyota Production System: An Integrated Approach to Just-In-Time (3rd ed.). Engineering and Management Press.

Moreira, F., Alves, A. C., & Sousa, R. M. (2010). Towards Eco-efficient Lean Production Systems. In *IFIP Advances in Information and Communication Technology* (Vol. 322, pp. 100–108). Balanced Automation Systems for Future Manufacturing Networks. doi:10.1007/978-3-642-14341-0_12

Mourtzis, D., & Doukas, M. (2014). Design and Planning of Manufacturing Networks for Mass Customisation and Personalisation: Challenges and Outlook. *Procedia CIRP*, *19*, 1–13. doi:10.1016/j.procir.2014.05.004

Nakayama, R. S., de Mesquita Spínola, M., & Silva, J. R. (2020). Towards I4.0: A comprehensive analysis of evolution from I3.0. *Computers & Industrial Engineering*, *144*, 106453. doi:10.1016/j.cie.2020.106453

Ohno, T. (1988). *Toyota Production System: Beyond large-scale production*. Productivity Press.

Pais-Montes, C., Freire-Seoane, M. J., & López-Bermúdez, B. (2019). Employability traits for engineers: A competencies-based approach. *Industry and Higher Education*, *33*(5), 308–326. doi:10.1177/0950422219854616

Park, S., Hironaka, S., Carver, P., & Nordstrum, L. (2013). *Continuous Improvement In Education*. papers3://publication/uuid/2260843A-153F-4D8F-9206-FC474ED2972C

Pascual, J. A., Pimentel, C., Mateo, M., Hoyuelos, I., Matias, J., & Gento, A. M. (2020). A Learning Factory for Remanufacturing: A New Configuration at Valladolid Lean School. In *Proceedings of 6th European Lean Educator Conference (ELEC2019)*. Springer International Publishing. 10.1007/978-3-030-41429-0_1

Pecore, J. L. (2015). From Kilpatrick's Project Method to Project-Based Learning. In M. Y. Eryaman & B. C. Bruce (Eds.), International Handbook of Progressive Education (pp. 155–171). Academic Press.

Pekkanen, P., Niemi, P., Puolakka, T., Pirttilä, T., & Huiskonen, J. (2020). Building integration skills in supply chain and operations management study programs. *International Journal of Production Economics, 225*, 107593. doi:10.1016/j.ijpe.2019.107593

Pereira, A. G., Lima, T. M., & Charrua-Santos, F. (2020). Society 5.0 as a Result of the Technological Evolution: Historical Approach. In T. Ahram, R. Taiar, S. Colson, & A. Choplin (Eds.), *Advances in Intelligent Systems and Computing* (pp. 700–705). Springer., doi:10.1007/978-3-030-25629-6_109

Pine, B. J. (1993). *Mass Customization: The New Frontier in Business Competition*. Harvard Business School Press.

Pinzone, M., Fantini, P., Perini, S., Garavaglia, S., Taisch, M., & Miragliotta, G. (2017). Jobs and Skills in Industry 4.0: An Exploratory Research. *IFIP International Conference on Advances in Production Management Systems*, 282-288. 10.1007/978-3-319-66923-6_33

Piwowar-Sulej, K. (2021). Human resources development as an element of sustainable HRM – with the focus on production engineers. *Journal of Cleaner Production, 278*, 124008. doi:10.1016/j.jclepro.2020.124008 PMID:32901179

Ravi, V. (2015). *Industrial Engineering and Management*. PHI Learning Private Limited.

Rolfes, K., Perry, L., & Shunk, D. (2019). *Industry 4.0 and people-centric leadership*. AME. https://www.ame.org/target/articles/2019/industry-40-and-people-centric-leadership

Shingo. (1989). *A study of Toyota Production System from an Industrial Engineering*. Academic Press.

Sackey, S. M., & Bester, A. (2016). Industrial engineering curriculum in Industry 4.0 in a South African context. *South African Journal of Industrial Engineering, 27*(4), 101–114. doi:10.7166/27-4-1579

Salvendy, G. (2001). Handbook of Industrial Engineering: Technology and Operations Management (3rd ed.). John Wiley & Sons.

Santandreu-Mascarell, C., Canós-Darós, L., & Pons-Morera, C. (2011). Competencies and skills for future Industrial Engineers defined in Spanish degrees. *Journal of Industrial Engineering and Management, 4*(1). Advance online publication. doi:10.3926/jiem.2011.v4n1.p13-30

dos Santos, P. F., Simon, A. T., Guimarães, G. E., Amorim, M., & Vieira, M., Junior. (2017). Analyzing the competences of production engineering graduates: An industry perspective. *Production, 27*. Advance online publication. doi:10.1590/0103-6513.005317

Shingo Institute. (2020). *The Shingo Model for Operational Excellence*. https://shingo.org/model

Smith, K. E. (2001). The concept and importance of productivity. In *Maynard's Industrial Engineering Handbook* (p. 2567). McGraw Hill Standard Handbooks.

Snyder, H. (2019). Literature review as a research methodology: An overview and guidelines. *Journal of Business Research, 104*, 333–339. doi:10.1016/j.jbusres.2019.07.039

Spear, S. J. (2004). Learning to lead at Toyota. *Harvard Business Review, 82*(5), 78–86, 151. https://www.ncbi.nlm.nih.gov/pubmed/15146738 PMID:15146738

Squire, B., Brown, S., Readman, J., & Bessant, J. (2006). The impact of mass customisation on manufacturing trade-offs. *Production and Operations Management, 15*(1), 10–21. doi:10.1111/j.1937-5956.2006.tb00032.x

Sugimori, Y., Kusunoki, K., Cho, F., & Uchikawa, S. (1977). Toyota production system and Kanban system Materialization of just-in- time and respect-for-human system. *International Journal of Production Research, 15*(6), 553–564. doi:10.1080/00207547708943149

Suzaki, K. (1993). *The new shop floor management: Empowering people for continuous improvement*. Academic Press.

Takeuchi, H., Osono, E., & Shimizu, N. (2008). The Contradictions That Drive Toyota's Success. *Harvard Business Review*.

Tan, H.-S., Ivander, Oktarina, R., Reynaldo, V., & Sharina, C. (2020). Conceptual development of learning factory for industrial engineering education in Indonesia context as an enabler of students' competencies in industry 4.0 era. *IOP Conference Series. Earth and Environmental Science, 426*, 012123. doi:10.1088/1755-1315/426/1/012123

Tien, J. M. (2020). Toward the Fourth Industrial Revolution on Real-Time Customization. *Journal of Systems Science and Systems Engineering, 29*(2), 127–142. doi:10.100711518-019-5433-9

Trentin, A., Somià, T., Sandrin, E., & Forza, C. (2019). Operations managers' individual competencies for mass customization. *International Journal of Operations & Production Management, 39*(9/10), 1025–1052. doi:10.1108/IJOPM-10-2018-0592

von Tunzelmann, N. (2003). Historical Coevolution of Governance and Technology in the Industrial Revolutions. *Structural Change and Economic Dynamics, 14*(4), 365–384. doi:10.1016/S0954-349X(03)00029-8

Universities of the Future. (2019). *Industry 4.0 Implications For Higher Education Institutions: State of Maturity and Competency Needs*. Author.

Veile, J. W., Kiel, D., Müller, J. M., & Voigt, K.-I. (2019). Lessons learned from Industry 4.0 implementation in the German manufacturing industry. *Journal of Manufacturing Technology Management*. doi:10.1108/JMTM-08-2018-0270

Vrchota, J., Maříková, M., Řehoř, P., Rolínek, L., & Toušek, R. (2019). Human Resources Readiness for Industry 4.0. *Journal of Open Innovation, 6*(1), 3. doi:10.3390/joitmc6010003

Wong, G., Greenhalgh, T., Westhorp, G., Buckingham, J., & Pawson, R. (2013). RAMESES publication standards: Meta-narrative reviews. *BMC Medicine, 11*(1), 20. doi:10.1186/1741-7015-11-20 PMID:23360661

Da Xu, L., Xu, E. L., & Li, L. (2018). Industry 4.0: State of the art and future trends. *International Journal of Production Research, 56*(8), 2941–2962. doi:10.1080/00207543.2018.1444806

Zandin. (2001). *Maynard´s Industrial Engineering Handbook* (5th ed.). McGraw Hill Standard Handbooks.

Zonnenshain, A., & Kenett, R. S. (2020). Quality 4.0—The challenging future of quality engineering. *Quality Engineering, 32*(4), 1–13. doi:10.1080/08982112.2019.1706744

Zunk, B. (2018). Positioning "Techno-Economics" as an Interdisciplinary Reference Frame for Research and Teaching at the Interface of Applied Natural Sciences and Applied Social Sciences: An Approach Based on Austrian IEM Study Programmes. *International Journal of Industrial Engineering and Management, 9*(1), 17–23.

Chapter 12
Team Teaching in PBL:
A Literature Review in Engineering Education

Natascha van Hattum-Janssen
https://orcid.org/0000-0002-2373-191X
Saxion University of Applied Sciences, The Netherlands

Anabela C. Alves
Department of Production and Systems, School of Engineering, University of Minho, Portugal

Sandra R. G. Fernandes
Department of Psychology and Education, Portucalense University, Portugal

ABSTRACT

Project-based learning (PBL) is a challenging learning methodology, also for teachers, questioning common assumptions of teachers, like control over the classroom and reliance on expert knowledge. Most challenging is teamwork. Team teaching has been explored in many disciplinary areas, both in traditional as well as in PBL curricula. Teachers may feel uncomfortable with sharing knowledge and being assessed by students and peers. This chapter explores characteristics of team teaching in a PBL context through two consecutive literature reviews. The first seeks to characterise team teaching and its meaning to teachers, zooming in from team teaching in general to team teaching in a PBL context in engineering education. The second connects this characterization to the experiences of a specific PBL teaching team in an engineering context, resulting in insights in experiences at practitioners' level. The authors argue that successful team teaching is crucial for the success of PBL in engineering education and is important as an example for students to engage in collaboration.

INTRODUCTION

The challenges that engineering education is facing according to Hadgraft and Kolmos (2020) - sustainability, the fourth industrial revolution and employability - have led to four different types of changes in

DOI: 10.4018/978-1-7998-8816-1.ch012

engineering curricula: 1) student-centred learning, 2) contextual and practice based learning, 3) digital learning and 4) professional competencies. Student-centred learning is described as "ways of thinking about teaching and learning that emphasize student responsibility and activity in learning rather than content or what the teachers are doing" (Cannon & Newble, 2000, pp. 16-17) and is based on constructivist learning theory (Hannafin et al., 1997). This means that learning environments are created that "enable and support individual construction by engaging in design and invention tasks. The design task is to create an environment where knowledge building tool (affordances) and the means to create and manipulate artefacts of understanding are provided, not one in which concepts are explicitly taught" (Hannafin et al., 1997). According to Baeten et al. (2010) student-centred learning activities are characterized as 1) an activity and independence of the teacher, 2) a coaching role of the teacher, and 3), knowledge which is regarded as a tool instead of an aim.

Problem and Project-Based Learning (PBL) involve students actively in their own learning and are commonly regarded as a way to make education more student centred (Bagheri et al, 2013), and, especially in the engineering context, to contribute to professional development of students (Johnson et al., 2015; Lattuca et al., 2017). In project-based learning, students work in teams and carry out a project to solve a large-scale complex open-ended problem, through a long period of time (Powell & Weenk, 2003). They are supported by their teachers and the knowledge and skills from supporting courses. Supporting student teams throughout a multi-disciplinary project semester implies strong teacher collaboration. Planning the project theme, defining requirements, establishing supporting knowledge and organizing tutoring demands from teachers to work closely together and discuss about educational issues across boundaries to help teams in this endeavor. It requires teamwork for tutors, teachers, administrators and integration over the traditional subject boundaries, what could be considered a disadvantage of PBL (Powell, 2000).

Teaching in higher education has traditionally been a rather individual activity, autonomous, often monodisciplinary, and sometimes isolated from other parts of the curriculum (Flinders, 1988; Vangrieken & Kindt, 2019). A stronger focus on the role of the student in the learning process and the importance of active learning has changed the role of the teacher and has shifted the focus in higher education from teacher to student (Gaebel & Zhang, 2018). The growing interest in student-centered approaches to learning, including multidisciplinary project approaches, questions the role of the teacher and the position of teachers within the teaching staff of a degree program. In engineering education, the opportunity for more intensive teacher collaboration also increases (Guerra et al., 2017).

Teacher collaboration is described in different ways varying from, for example, team teaching, co-teaching and collaborative teaching (Vesikivi et al., 2019). It can be defined as two or more teachers planning, instructing, and evaluating the learning of a single group of students; co-teaching refers to two or more teachers instructing a multidisciplinary student team in the same classroom and collaborative teaching emphasizing teacher collaboration and co-operative teaching is used. However, despite the number of teachers involved and their role in the teaching and learning process, these authors argue that "the definition of team teaching should be based on the pedagogical approach and grounded in learning theory" (Vesikivi et al., 2019).

No single definition of team teaching is agreed upon, but characteristics of team teaching are identified, like by Minett-Smith and Davis (2020) who refer to the involvement of two or more teachers, the degree of interaction between the teachers, the resources they share and the interdependence between the teachers. Interdisciplinarity is highlighted by Walsh and Davis (2017), Salonen and Savander-Ranne (2015) and Li (2020), the latter explaining team teaching as a specific form of interdisciplinary teacher collaboration.

Benefits of team teaching are acknowledged by Jones and Harris (2012), who found that team teaching can be positive for both students and teachers, especially when having uniform purposes and expectations and really working as a team. They also acknowledge the reflection on teaching and assessment that team teaching can encourage, especially when mixing teaching assistants and more experienced teachers. The authors have identified several requirements for team teaching to be successful, like continuity in the team, a great deal of mutual trust and the absence of hierarchical relationships within the team. Vesikivi et al. (2019) refer to team teaching as the effort of, at least, two teachers and identify a number of challenges to overcome in their study at a Finnish university of applied sciences. Teachers may fear a loss of autonomy when shifting to team teaching and face a lack of time when planning and preparing team teaching efforts. The authors point out that this requires leadership that answers the teacher concern. Team teaching is not always welcomed by every single team member, so institutional support can help to overcome the challenges. Money and Coughlan (2016) show that benefits from a teacher point of view do not necessarily coincide with the benefits as identified by students. The joint effort that can result in a decreased workload for teachers may also result in duplicated and conflicting content for students.

Although research on team teaching is carried out in a range of contexts, e. g. in initial teacher education, it is not clear yet how research in higher education in general is connected to the practice of team teaching in PBL in engineering education. Having more insight in what it means for teachers to effectively work in a team that coordinates a PBL project helps teachers to overcome the difficulties of team teaching, set the example and become the inspiration for their students, who also need to work together in a team effectively. Therefore, this chapter aims to explore to what extent characteristics of research on team teaching in higher education are related to the characteristics of team teaching in a PBL context in engineering education and as such, identify what possible research gaps exist in the latter context. The following research questions were defined:

- What are the characteristics of team teaching in Higher Education in general?
- What are the characteristics of team teaching in PBL?
- What are the characteristics of team teaching in Engineering Education focused in PBL?
- What lessons can Engineering Education learn from other fields in Higher Education?

This chapter is structured in five sections. In the following section, the research methodology is explained. The subsequent section focuses on the results and discussion. The penultimate section is dedicated to the lessons learnt for engineering education from other fields whereas the last section contains the final remarks with regard to the research questions, as well as suggestions for future research.

RESEARCH METHODOLOGY

The characterization of team teaching, its meaning to teachers and the connection between team teaching and PBL, especially in the context of engineering education is carried out through a literature review in three steps (Miles & Huberman, 1994), seeking to provide data to answer each of the research questions defined. The kind of literature review could be referred to as a qualitative systematic review (Grant and Booth, 2009) as the review is not a meta-analysis, but seeks to find broader themes or constructs in the literature that was analyzed and that had in many cases a more qualitative nature. The Scopus database was used to collect data, as all relevant journals in engineering education are indexed in Scopus.

To develop our analysis, three independent searches were carried out in Scopus database, guided by the first three research questions of our study.

The first search aimed to analyse the **characteristics of team teaching in higher education in general.** For this, a search was developed using the terms "higher education" and "team teaching". The search for the term "higher education" was held within all fields, while the search for the term "team teaching" was held within Article title, Abstract and Keywords. This decision was due to the fact that we intended to include all existing studies developed in higher education, but we wanted to refine our search on the specific topic of team teaching, so limiting the search to only identifying this term in the title, abstract or keyword would exclude studies that did not focus directly on our research topic.

The same procedure was held for the search related to the second research question, aimed to analyse **characteristics of team teaching in Project-based Learning (PBL),** where the terms used for the search were "higher education", "team teaching" and "project-based learning".

The third research question aimed to analyse the characteristics of **team teaching in Engineering Education focused in PBL** and the search used the terms "engineering education" (within all fields), "team teaching" and "project-based learning" or "PBL" (within Article title, Abstract and Keywords).

Table 1 shows a summary of the three searches carried out during the literature review, including the number of papers that were analyzed in each independent search.

Table 1. Searches carried out to answer the research questions

Research Question	Keywords for search	Total search results	Number of publications analyzed
#1	Higher education + team teaching	117	65
#2	Higher education + team teaching + project-based learning	16	15
#3	Engineering education + team teaching + project-based learning + PBL	13	13

Data Analysis

For each step, an overall analysis of the bibliometric results retrieved from the Scopus database search was summarized and presented. Based on this, the authors identified three different perspectives that emerged from the analysis of the publications: the student perspective, the teacher perspective and the multiple perspective, the latter including teacher, student, and/or organizational perspective. Apart from the three perspectives, a thematic content analysis was carried out (Bardin, 2011) identifying themes that are found in at least two publications. Eight common themes were identified into which most of the publications could be included. The eight categories are the following:

1. **Perception of team teaching** is aimed at identifying how students, teachers or both perceive team teaching after they have experienced it.
2. **Experiences with team teaching** describe how team teaching was implemented in a specific course in higher education. It is usually a rather step-by-step description of the process not based on a specific research question.

3. Publications focused on **strengths and weaknesses**, or benefits and pitfalls aim to provide systematic insight in advantages and disadvantages of different types of team teaching.
4. A number of publications are aimed at **blended and online learning**, especially the very recent ones, but also the role of video support is included in this theme.
5. Both from a student as well as a teacher perspective, **team teaching** is compared with **individual** teaching, in experimental settings as well as through literature reviews.
6. **Interdisciplinarity** is discussed especially in active learning contexts like problem and project-based learning as a key feature of team teaching.
7. **Reflections** on the motives for the implementation of team teaching, the benefit of team teaching and the research needed also form a category.
8. **Professional development**, also from an organizational point of view is the final category of themes.

These eight categories were used to guide the analysis and discussion of results.

RESULTS AND DISCUSSION

This section presents the results and discussion of the three literature reviews performed to answer to the questions raised in this research. The first subtopic discusses the characteristics of team teaching in higher education in general, followed by the analysis of team teaching in the context of PBL and, finally, team teaching in PBL approaches in the context of Engineering Education.

Team Teaching in Higher Education

First, the literature is explored to make a general characterization of team teaching by using publications found through SCOPUS search. When analyzing 172 publications on team teaching in SCOPUS, of which 117 (66.7%) are articles, 33 (18.6%) are conference papers and 22 are book chapters (12.4%) using ALL ("higher education") AND TITLE-ABS-KEY ("team teaching")) PUBYEAR > 2009 AND (LIMIT-TO (LANGUAGE, "English")), the number of publications seems to increase, starting with 7 papers in 2010 and resulting in 23 papers in 2020. More than one third of the papers comes from North America. Figure 1 presents the publication year and Figure 2 represents country/territory of the publications. Looking at the subject areas, the social sciences represent 53% of the publications, followed by computer science (9.3%) and engineering (9.0%).

Most of the publications, 114 (65%), are articles, 33 (20%) are conference papers and 26 (15%) are book chapters. Of the articles, conference papers and book chapters, 65 are focused on team teaching in higher education as the central concept, whereas the others have a focus on approaches to teaching, forms of students learning or educational innovation in a broader sense and do not focus specifically on team teaching. The largest number of publications, 29, focus on the student perspective, whereas 25 focus on the teacher perspective. Only a small number is aimed at teachers and students, students and the institution, teachers and the institution or students, teachers and the institution. Around 40% of the publication was based on a qualitative approach, whereas 23% used a quantitative approach and 12% described a mixed method research paradigm. The publications also included a small number of literature reviews. The remaining publications were describing a specific team-teaching experience, guided by a literature review, but not including original research connected to the experience.

Table 2 shows the categorization of the publications in defined categories. Apart from the wide variety of research methods used, varying from self-evaluation, to experiments, to multiple case studies and reflective journals, the results were categorized according to the three different perspectives (student, teacher or multiple) and the eight categories identified in the methods section.

Figure 1. Publications team teaching in higher education by year
Source: Scopus

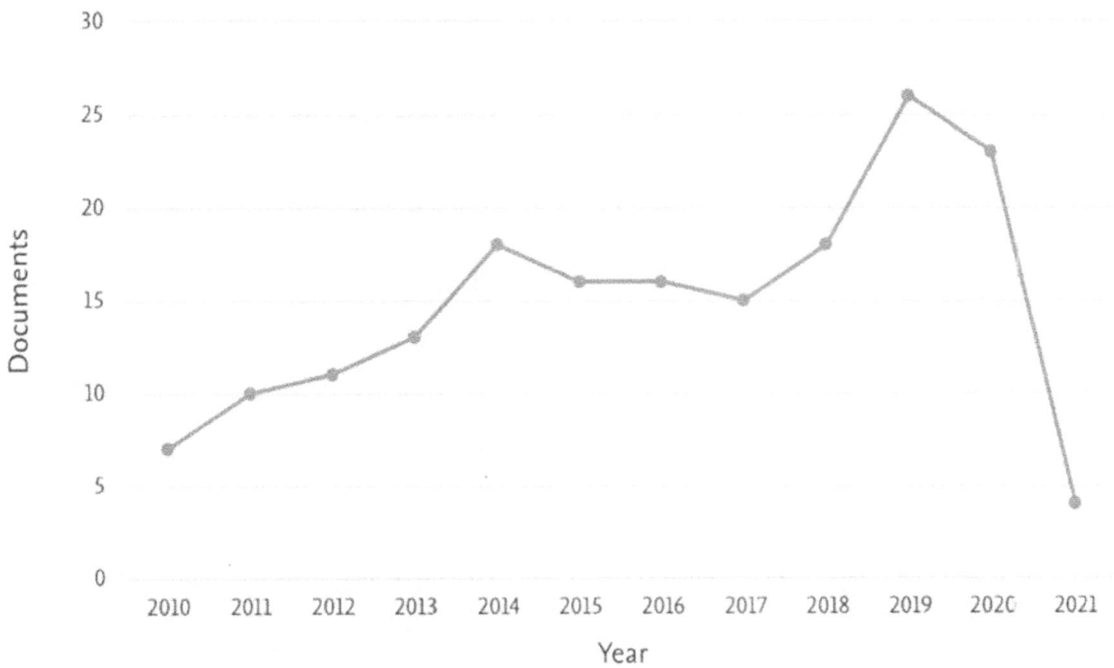

The student and teacher perspectives show overlaps in the themes that are described. Comparisons between a team-taught course and a sole-taught course are made from both perspectives, specific experiences are described from a student as well as a teacher perspective and blended/online learning and long-term strengths and weaknesses are identified from a student, a teacher and a mixed perspective. Publications on reflections on team teaching and the reasons to consider team teaching are written mainly from a teacher perspective, whereas publications on professional development are elaborated from the teacher or the organizational perspective, that is included in the multiple perspectives view. Most publications are focused on either the teacher or the student perspective and a multiple perspective of an organizational perspective is rather scarce.

Figure 2. Publications team teaching in higher education by country/territory
Source: Scopus

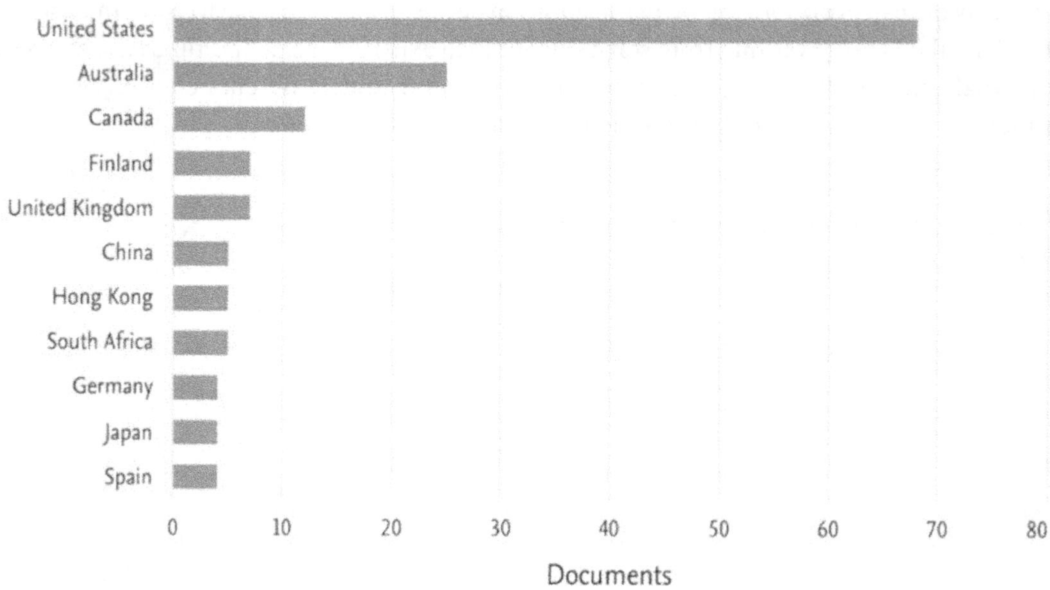

Table 2. Categorization of publications into themes and perspectives

	Student	Teacher	Multiple
Perceptions of team teaching	(Gladman, 2015) (Killingsworth & Xue, 2015) (Minichiello et al., 2011) (Money & Coughlan, 2016) (Saeed et al., 2018) (Schmulian & Coetzee, 2019)	(Minett-Smith & Davis, 2020) (Ott & Meek, 2019) (Vesikivi et al., 2019)	(Kostko, 2019) (Rickard & Walsh, 2019)
Experiences with team teaching	(Baeten & Simons, 2014) (Flint Jr & Dorr, 2010) (Maletina et al., 2015) (Tsybulsky, 2019) (Wöllner & Ginsborg, 2011)	(Conn, 2010) (Fauvel et al., 2010) (Smith & Winn, 2017)	(Jones & Mezo, 2014)
Strengths and weaknesses	(Huerta & McMillan, 2012)	(Hellier & Davidson, 2018) (Monteblanco, 2020)	(Seymour & Seymour, 2013)
Blended, online and other ICT supported learning	(Crawford & Jenkins, 2017) (Egbert & Camp, 2016) (McKenzie et al., 2020) (Tan et al., 2020; Vesikivi et al., 2016)	(Rajamma & Sciandra, 2020)	
Team taught vs sole taught learning	(Aliakbari & Nejad, 2013) (Khoshnodifar et al., 2020) (McDonald et al., 2021)	(Devlin-Scherer & Sardone, 2013) (Hughes et al., 2020)	(Jones & Harris, 2012)
Interdisciplinarity	(Walsh & Davis, 2017)	(Kleyn & Valle, 2014) (Li, 2020) (Pope-Ruark et al., 2019) (Salonen & Savander-Ranne, 2015)	---------
Reflections on team teaching	---------	(Salifu, 2021) (Selkrig & Keamy, 2015; Vesikivi et al., 2016)	(Lasagabaster, 2018)
Professional development	---------	(Clarke & Winslade, 2019) (Gast et al., 2017)	(Kulski & Kerr, 2012)

Team Teaching in PBL

This section presents the analysis based on the results focused on the characteristics of team teaching in PBL at Higher Education. The search in Scopus database produced 16 results. Of these results, only one publication was not available to authors (Lutsenko & Lucenko, 2018).

A few considerations about the overall analysis of the bibliometric results retrieved from the Scopus database search will be presented.

According to Figure 3, the year 2019 (five publications) and 2018 (four publications) are the years which include more publications under this topic. The countries that lead the publications in this research, see Figure 4, topic are the United States (four publications) and Finland (three publications).

Figure 3. Publications team teaching in PBL by year
Source: Scopus

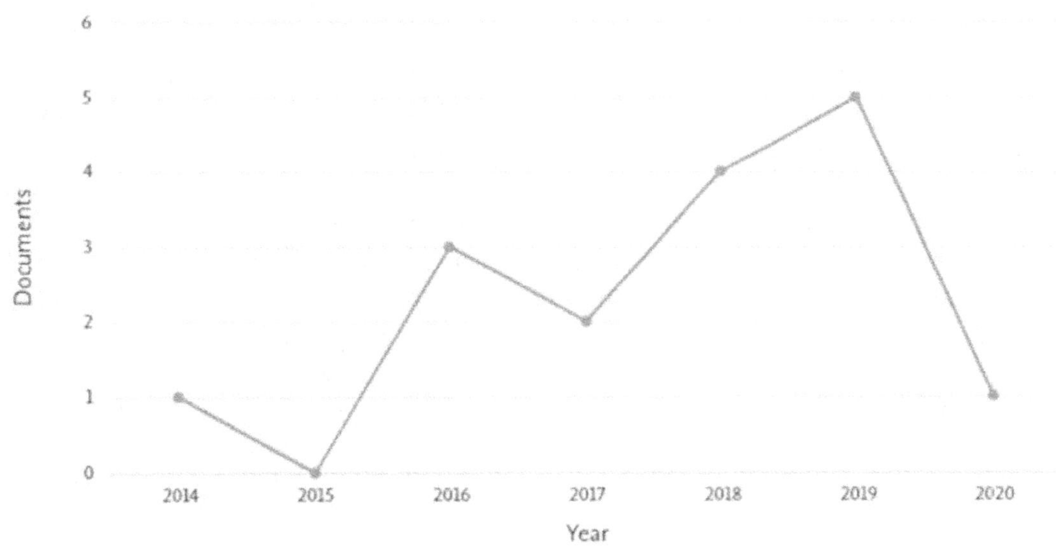

Regarding the type of documents published, eight are articles, six are conference proceedings and only two are book chapters.

The main three subject areas of the publications are Social Sciences (53.8%) and Engineering (19.2%). Other subject areas include Computer Sciences (7.7%), Environmental Science (7.7%)

Mathematics (3.8%), Business, Management and Accounting (3.8%), Earth and Planetary Sciences (3.8%), see Figure 5.

The identification of these papers, by subject area, is provided in Table 3, as well as the categories explored in the publications in this field.

Experiences with team teaching and professional development are the main categories explored in the 14 publications classified under the subject area of Social Sciences. Some of these studies were, at the same time, classified also under the subject area of Engineering (Angelva et al., 2017; Lutsenko & Lucenko, 2018; Vesikivi et al., 2016; Vesikivi et al., 2019). This may be due to the interdisciplinary nature of the sources where the documents were published. These include, in the case of article publications, the following sources: Canadian Journal of Education; European Journal of Engineering Education; IOP Conference Series Earth and Environmental Science; Issues in Educational Research; Journal of Education for Teaching; Journal of Higher Education Theory and Practice; Journal of Management Education; Primus; Teaching and Teacher Education; World Sustainability Series; Advances in Intelligent Systems and Computing. In the context Teacher Education, the studies reported by Tsybulsky (2019) and Tsybulsky & Muchnik-Rozanov (2019) provide evidence on the importance of team teaching and project-based learning in courses such as pedagogical practicum module where student teachers experience team teaching conducted in elementary schools. Findings from this study on the team teaching of science courses contribute to encourage educators to implement team teaching as part of student teachers' practicum involving a project based, student centred methodology (Tsybulsky & Muchnik-Rozanov, 2019).

In regard to the studies published in the subject area of Engineering, the studies focus mainly on interdisciplinarity, experiences and perceptions of team teaching. The study developed by Angelva et al., (2017), carried out in an ICT Department of a University located in Finland, concluded that team-teaching and project-based learning (PBL) give more opportunities to develop industrial cooperation and make it easier for students to get a professional job after graduation. Docherty & Gaubinger (2018), in turn, discuss a holistic system to integrate English language classes into English-medium instruction courses in European Higher Education and investigate how, using team-teaching, inverted classrooms and active classroom techniques, can allow students and teachers to be prepared for the new challenges of teaching and learning. The strengths, weaknesses and challenges of team teaching in interdisciplinary teams are also one of the topics explored in publications found from the engineering field. Vesikivi et al. (2019) for instance, refer that team teaching represents a substantial change for teachers, raising several concerns about time management, getting enough compensation for the work and possible loss of teacher autonomy. Amongst the benefits, the authors identify the opportunity to provide students with the skills they need and also a way to enhance the teacher's own professional development (Vesikivi et al., 2019). In the next section, a deeper analysis into the Engineering Education field, with experiences of team teaching and project-based learning, will be discussed.

Figure 4. Publications team teaching in PBL by country/territory
Source: Scopus

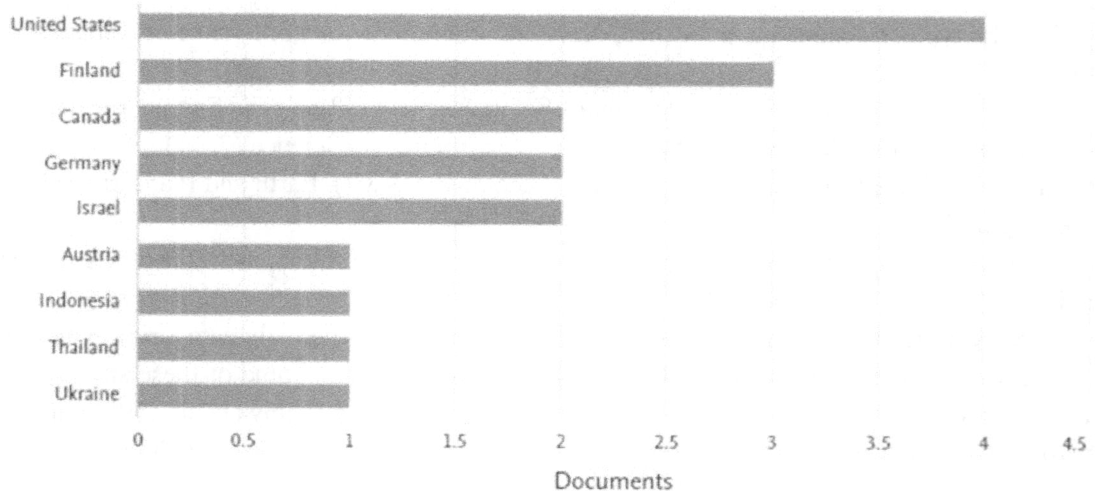

Figure 5. Publications team teaching in PBL by subject area
Source: Scopus

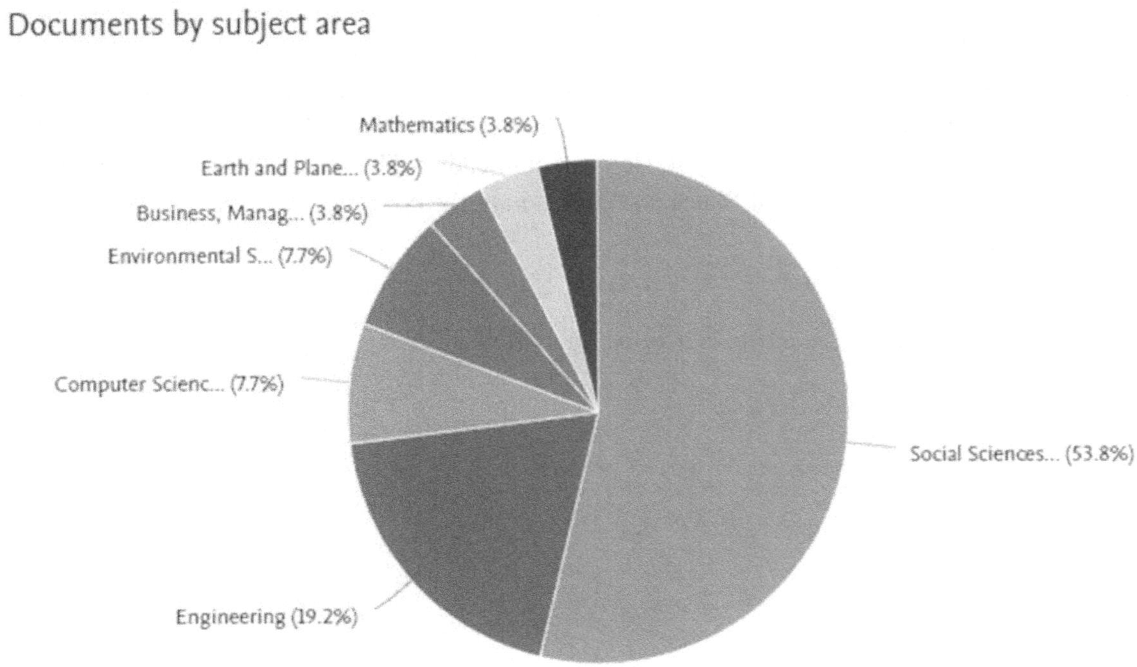

Table 3. Publications by subject area

Subject areas	Publications	Categories explored
Social Sciences (14)	(Angelva et al., 2017) (Demoret, 2020) (Fischer & Golden, 2018) (Hendry et al., 2017) (Kodkanon & Pinit, 2018) (Kröber & Münster, 2014) (Kröber & Münster, 2016) (Ludwig et al., 2018) (Lutsenko & Lucenko, 2018 (Pfeifer & Rosbach, 2016) (Tsybulsky, 2019) (Tsybulsky & Muchnik-Rozanov, 2019) (Vesikivi et al., 2016) (Vesikivi et al., 2019)	• Experiences with team teaching; • Professional development;
Engineering (5)	(Angelva et al., 2017) (Docherty & Gaubinger, 2018) (Lutsenko & Lucenko, 2018) (Vesikivi et al., 2016, 2019)	• Interdisciplinarity; • Experiences with team teaching; • Perceptions of team teaching; • Strengths and weaknesses;
Computer Science (2)	(Docherty & Gaubinger, 2018) (Kröber & Münster, 2014)	• Experiences with team teaching; • Interdisciplinarity
Environmental Science (2)	(Pfeifer & Rosbach, 2016) (Taharu et al., 2019)	• Interdisciplinarity; • Strengths and weaknesses;
Mathematics (1)	(Ludwig et al., 2018)	• Experiences with team teaching
Business, Management & Accounting (1)	(Hendry et al., 2017)	• Experiences with team teaching
Earth and Planetary Sciences (1)	(Taharu et al., 2019)	• Strengths and weaknesses

Team Teaching in Engineering Education Focused on PBL

Finally, the results of the literature review focused on the characteristics of team teaching in engineering education focused in PBL are presented in this section through an overall analysis of the bibliometric results retrieved from the Scopus database search. The search in Scopus database produced 13 results. The main descriptive statistics are reflected in the graphs of Figure 6 and Figure 7 that present, respectively, the documents by year and country/territory of publication. The year 2016 and 2019 are the years with more publications. The United States and Finland are the countries that lead the publications in this research topic.

Figure 6. Publications on engineering education focused on PBL
Source: Scopus

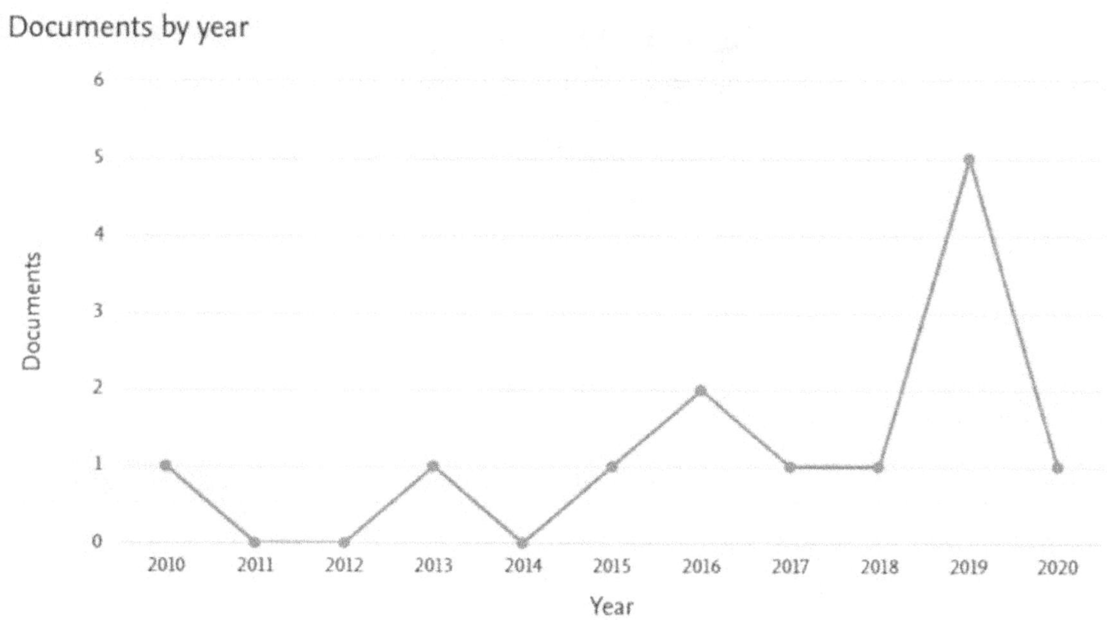

From these results, two papers were excluded from the analysis as one paper was not in English and the full text of three papers was not accessible to the authors, leaving a total of 11 papers to be analyzed in detail. After this analysis, just six were considered as discussing the third question raised in this paper: What are the characteristics of team teaching in Engineering Education focused in PBL? The identification of these papers is provided in Table 4, following the categories defined in the methodology section.

Figure 7. Documents by country/territory of publication
Source: Scopus

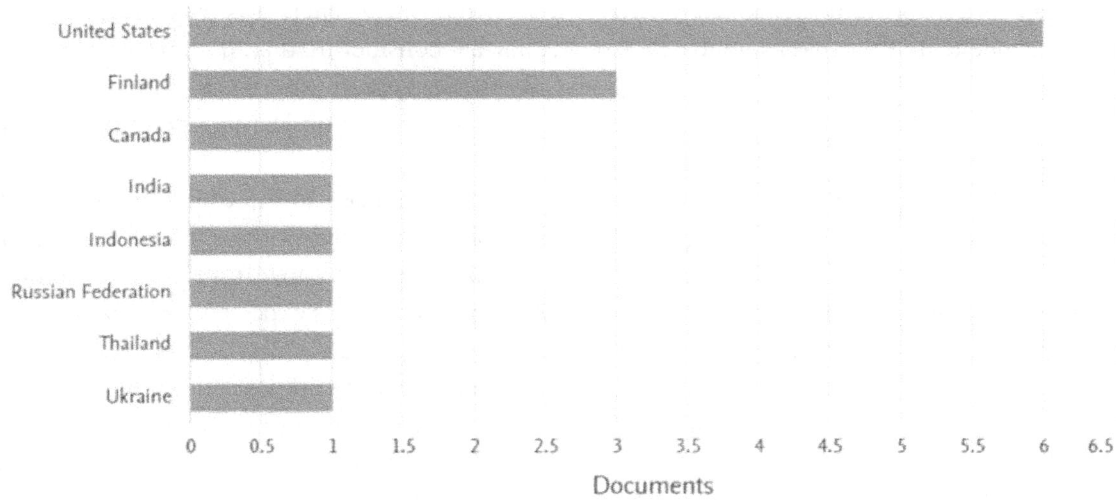

Table 4. Categorization of publications from Engineering Education focused on PBL

	Student	Teacher	Multiple
Perceptions of team teaching	---------	(Vesikivi et al., 2019) (Kodkanon et al., 2018) (Vesikivi et al., 2016)	---------
Experiences with team teaching	---------	(Vesikivi et al., 2019) (Kodkanon et al., 2018) (Angelva et al., 2017)	(Baligar et al., 2018)
Strengths and weaknesses	---------	(Vesikivi et al., 2019) (Kodkanon et al., 2018) (Angelva et al., 2017) (Vesikivi et al., 2016)	---------
Blended, online and other ICT supported learning	---------	---------	---------
Team taught vs sole taught learning	---------	---------	---------
Interdisciplinarity	---------	(Vesikivi et al., 2019) (Kodkanon et al., 2018) (Angelva et al., 2017) (Vesikivi et al., 2016) (Garcia et al., 2015)	(Baligar et al., 2018)
Reflections on team teaching	---------	(Vesikivi et al., 2019)	---------
Professional development	---------	(Vesikivi et al., 2019)	(Baligar et al., 2018)

From the blanks in the table, it is possible to conclude that team teaching is not a theme discussed *per se* in the context of engineering but more as a part of the PBL implementation. From the publications identified just the ones from Vesikivi et al. (2016) and Vesiviki et al. (2019) and Kodkanon et al. (2018) are focused on discussing the theme.

When this is discussed, it seems that just the perspective of the teachers was presented, as the column of the students in the table is empty. This seems normal considering that, if engineering teachers do not discuss this, students will discuss it even less. By their nature, these students think about going to companies, normally they do not like to write anything. Reflecting on learning practices is almost impossible, unless this is promoted, and even so, it is difficult.

Another interesting aspect are the empty cells in the table, related to the categories such as the blended, online and ICT supported learning and the discussion team taught vs sole taught learning. The first probably was not found because when the review was made just one paper from 2020 was found so, it may be too early to expect more publications. The other is, more or less expected, as if team teaching is not a theme *per se*, discussing both approaches is totally out of mind.

It is most common to find publications discussing the experiences with team teaching, as well the strengths and weaknesses of team teaching in EE implementing PBL, as reported by Alves, van Hattum-Janssen and Fernandes (2021). As referred, the category of strengths and weaknesses corresponds also to the benefits and challenges authors face when implementing PBL, which was discussed many times by many authors (Alves et al., 2021). This should be a reason for teachers in engineering to be interested in implementing it, although it continues to be a challenge that needs more investment (UNESCO, 2021). Of course, for those implementing PBL, it is always a challenge, in spite of how many years teachers are doing it. This happens because PBL is a project and by its own nature, it is a temporary unique event involving many stakeholders and requiring the managing of time, costs, resources and risks. For example, in Rugarcia et al. (2000), Felder et al. (2000), Powell & Weenk (2003), Alves et al. (2009), Alves, Sousa, Fernandes, et al. (2016), Alves, Sousa, Moreira, et al. (2016) and Alves et al. (2019) difficulties in implementing PBL and the high workload are discussed as major challenges of PBL approaches.

LESSONS LEARNT BY ENGINEERING EDUCATION FROM OTHER FIELDS

After addressing the main results presented in the literature review with regard to team teaching in Higher Education, it is possible to draw some conclusions and lessons learnt to improve the work of teachers, researchers and professionals in the field of Engineering Education. These conclusions can be summarised in three main ideas.

First, this study shows that the literature analysis on team teaching in Higher Education in general, more specifically in PBL, and then narrowed down to Engineering Education, is mainly focused on the teachers' perspective in PBL and on the teachers' perspective in PBL in Engineering Education. The multiple perspective and a specific focus on the organizational perspective is yet lacking. The literature from the student and teacher perspective is convincing as far as the benefits of student learning and teacher and student perception are concerned but does not yet focus clearly on the implications of team teaching in a PBL context for the organization.

Secondly, the scarce existing literature on professional development and organizational aspects, provide useful insights in what team teaching means for the organization (Gast et al., 2017; Kulski & Kerr, 2012; Lasagabaster, 2018) and the support that is needed from the organization. This may be a clear motive for more research into the reflections on team teaching as well, a category that is also less represented. The publications in this category are dedicated to the motives for implementing team teaching and focus on the connection between the teaching and learning philosophy of the institution and practice of team teaching. Connecting research on team teaching in PBL in Engineering Education to the broader perspective of the institution's teaching and learning philosophy may lead to research that goes beyond specific experiences of teaching teams in specific courses, beyond a focus on a single theme. This can enrich the current debate on the impact of team teaching in this specific context and support teaching staff as well as institutions of higher education to maximize the benefits of team teaching for students, teachers, and the institution.

Thirdly, the studies report the importance of interdisciplinary teaching teams and the role of educational researchers to provide support, mentoring and feedback to the work developed by engineering teachers. Using PBL and team teaching also helps to shape teachers' professional identity (Tsybulsky, 2019; Tsybulsky & Muchnik-Rozanov, 2019) which is developed and enhanced through social interactions, meaningful experiences and reflections / self-evaluation on the interpretation of those experiences to oneself. In the field of teacher education, team teaching has demonstrated several advantages, including emotional and professional support, increased professional learning and personal development (Birrell & Bullough, 2005; King, 2006). It is an opportunity for teachers to critically review their own practices as they learn to teach and improve their pedagogical skills. This is one of the best practices that can be learnt from other fields, as engineering teachers initial training background do not include educational or didactic studies. The quality of academic teaching is decisively determined by the qualifications and competencies of the teaching staff, which also include university pedagogical capabilities (Kersten, 2018).

FINAL REMARKS

This chapter contributed to identify existing research gaps in the context of engineering education and team teaching in PBL when compared to the characteristics of research on team teaching in higher education in general. Team teaching in project-based learning experiences in engineering education is a specific form of team teaching in higher education. It shows several additional features that go beyond teaching and planning for teaching and learning with two or more teachers. The benefits and challenges of this specific form of team teaching are, therefore, also more elaborate. As can be seen in the third part of the literature review, teachers in PBL need to facilitate learning activities from the start till the end of a project and need to be aware of the learning processes of their students. This requires more communication and agreement within teaching teams than the more traditional forms of instruction. The interdisciplinary nature of projects and the different role that teachers in PBL teams have when supporting students also asks from teachers to be not only informed about the expertise their colleagues have, but also the specific planning of content delivery and the contribution of disciplinary areas to the final solutions proposed by the student teams needs to be in the center of their joint attention.

Although the scope of the current study is limited due to the use of Scopus as the single database for the literature review and the narrow definition of team teaching in the keywords, this limited exploration already shows that the role of the teacher as a team member in PBL is crucial to make the learning experience successful for students. Further exploration of the literature on specific roles of teachers in team building, team organization and management and team communication is needed to get a better insight on how to improve teaching teams for working with student teams. An interesting idea for future work could be to carry out a structured survey on the students' perception regarding the collaborative work of their teachers. The authors also suggest more research into the relationship between institutional policies on team teaching and the enhancement of active learning methods like PBL.

ACKNOWLEDGMENT

This work has been supported by FCT – Fundação para a Ciência e Tecnologia within the R&D Units Project Scope: UIDB/00319/2020.

REFERENCES

Aliakbari, M., & Nejad, A. M. (2013). On the effectiveness of team teaching in promoting learners' grammatical proficiency. *Canadian Journal of Education*, *36*(3), 5–22. https://www.scopus.com/inward/record.uri?eid=2-s2.0-84938081687&partnerID=40&md5=0650d3478590ff7bab1661befb620342

Alves, A., Sousa, R., Moreira, F., Carvalho, M. A., Cardoso, E., Pimenta, P., Malheiro, T., Brito, I., Fernandes, S., & Mesquita, D. (2016). Managing PBL difficulties in an industrial engineering and management program. *Journal of Industrial Engineering and Management*, *9*(3), 586–611. doi:10.3926/jiem.1816

Alves, A. C., Moreira, F., Leão, C. P., Pereira, A. C., Pereira-Lima, S. M. M. A., Malheiro, M. T., Lopes, S. O., & Oliveira, S. (2019, June 10-12). *Industrial engineering and management PBL implementation: An effortless experience?* 11th International Symposium on Project Approaches in Engineering and 16th Active Learning in Engineering Education Workshop, PAEE/ALE, Hammamet, Tunisia.

Alves, A. C., Moreira, F., Sousa, R., & Lima, R. M. (2009, May 15-17). Teachers' workload in a project-led engineering education approach. *International Symposium on Innovation and Assessment of Engineering Curricula*, Valladolid.

Alves, A. C., Sousa, R. M., Fernandes, S., Cardoso, E., Carvalho, M. A., Figueiredo, J., & Pereira, R. M. S. (2016). Teacher's experiences in PBL: Implications for practice. *European Journal of Engineering Education*, *41*(2), 123–141. doi:10.1080/03043797.2015.1023782

Alves, A. C., van Hattum-Janssen, N., & Fernandes, S. (2021, July 7-9). *Teacher collaboration in PBL: setting the example for the students.* PAEE/ALE'2021, International Conference on Active Learning in Engineering Education,13rd International Symposium on Project Approaches in Engineering Education (PAEE) and 18th Active Learning in Engineering Education Workshop (ALE), Braga, Portugal.

Angelva, J., Tepsa, T., & Mielikäinen, M. (2017). Team teaching experiences in engineering education a project-based learning approach. *Proceedings of the 45th SEFI Annual Conference 2017 - Education Excellence for Sustainability, SEFI 2017.*

Baeten, M., Kyndt, E., Struyven, K., & Dochy, F. (2010). Using student-centred learning environments to stimulate deep approaches to learning: Factors encouraging or discouraging their effectiveness. *Educational Research Review*, *5*(3), 243–260. doi:10.1016/j.edurev.2010.06.001

Baeten, M., & Simons, M. (2014). Student teachers' team teaching: Models, effects, and conditions for implementation. *Teaching and Teacher Education*, *41*, 92–110. doi:10.1016/j.tate.2014.03.010

Bardin, L. (2011). *Análise de conteúdo*. Edições 70.

Birrell, J. R., & Bullough, R. V. Jr. (2005). Teaching with a Peer: A Follow-Up Study of the 1st Year of Teaching. *Action in Teacher Education*, *27*(1), 72–81. doi:10.1080/01626620.2005.10463375

Cannon, R., & Newble, D. (2000). *A handbook for teachers in universities and colleges. A guide to improving teaching methods* (4th ed.). Kogan Page.

Clarke, D., & Winslade, M. (2019). A school-university teacher education partnership: Reconceptualising reciprocity of learning. *Journal of Teaching and Learning for Graduate Employability*, *10*(1), 138–156. doi:10.21153/jtlge2019vol10no1art797

Conn, C. E. (2010). Learning the hard way (but still learning!): Using team teaching as a vehicle for pedagogical change. *Business Communication Quarterly*, *73*(1), 87–91. doi:10.1177/1080569909358098

Crawford, R., & Jenkins, L. (2017). Blended learning and team teaching: Adapting pedagogy in response to the changing digital tertiary environment. *Australasian Journal of Educational Technology*, *33*(2), 51–72. doi:10.14742/ajet.2924

Devlin-Scherer, R., & Sardone, N. B. (2013). Collaboration as a Form of Professional Development: Improving Learning for Faculty and Students. *College Teaching*, *61*(1), 30–37. doi:10.1080/87567555.2012.714815

Docherty, M., & Gaubinger, K. (2018). Topic Oriented Mixed-Method System (TOMMS): A Holistic System to Integrate English Language Classes into English-Medium Instruction Courses in European Higher Education. In M. E. Auer, D. Guralnick, & I. Simonics (Eds.), *Teaching and Learning in a Digital World. Proceedings of the 20th International Conference on Interactive Collaborative Learning* (Vol. 1, pp. 840–851). Springer.

Egbert, S. C., & Camp, S. (2016). Practical strategies for team teaching in the distance education (IVC) context. In Interdisciplinary Approaches to Distance Teaching: Connecting Classrooms in Theory and Practice (pp. 141-153). doi:10.4324/9781315694412

Fauvel, A. M., Miller, L. K., Lane, P., & Farris, J. (2010). Reflections on an interdisciplinary, community-based, team-taught adventure. *The Journal of Continuing Higher Education*, *58*(1), 40–46. doi:10.1080/07377360903528089

Felder, R. M., Stice, J. E., & Rugarcia, A. (2000). The future of Engineering Education VI. Making Reform Happen. *Chemical Engineering Education*, *34*(3), 208–215.

Flinders, D. J. (1988). Teacher isolation and the new reform. *Journal of Curriculum and Supervision*, *4*(1), 17–29.

Flint, R. W. Jr, & Dorr, N. (2010). Social neuroscience at the college of saint rose: The art of team teaching in emerging areas of psychological science. *Journal of Undergraduate Neuroscience Education*, *8*(2), 122–127. https://www.scopus.com/inward/record.uri?eid=2-s2.0-77956286723&partnerID=40&md5=62cb459ca0248b76923d46df794d0d4e PMID:23493798

Gaebel, B. M., & Zhang, T. (2018). *Learning and teaching in the European Higher Education Area.* E. U. Association. https://eua.eu/downloads/publications/trends-2018-learning-and-teaching-in-the-european-higher-education-area.pdf

Garcia, J. M., Homkes, R., Carnes, M. T., & Taylor, K. D. (2015, June 14-17). *Lessons Learned from Team-Teaching a PBL Robotics Course with Multi- Disciplinary Instructors and Students. 122nd ASEE Annual Conference & Exposition*, Seattle, WA.

Gast, I., Schildkamp, K., & van der Veen, J. T. (2017). Team-Based Professional Development Interventions in Higher Education: A Systematic Review. *Review of Educational Research*, *87*(4), 736–767. doi:10.3102/0034654317704306 PMID:28989192

Gladman, A. (2015). Team teaching is not just for teachers! Student perspectives on the collaborative classroom. *TESOL Journal*, *6*(1), 130–148. doi:10.1002/tesj.144

Guerra, A., Ulseth, R., & Kolmos, A. (2017). *PBL in Engineering Education.* Sense Publishers., doi:10.1007/978-94-6300-905-8

Hadgraft, R. G., & Kolmos, A. (2020). Emerging learning environments in engineering education. *Australasian Journal of Engineering Education*, *25*(1), 3–16. doi:10.1080/22054952.2020.1713522

Hannafin, M. J., Hannafin, K. M., Land, S. M., & Oliver, K. (1997). Grounded practice and the Design of Constructivist Learning Environments. *Educational Technology Research and Development*, *45*(3), 101–117. doi:10.1007/BF02299733

Hellier, S., & Davidson, L. (2018). Team teaching in nursing education. *Journal of Continuing Education in Nursing*, *49*(4), 186–192. doi:10.3928/00220124-20180320-09 PMID:29596706

Huerta, D., & McMillan, V. (2012). Reflections on collaborative teaching of science information literacy and science writing: Plans, processes and pratfalls. In *Libraries within their Institutions* (pp. 19–28). Creative Collaborations. doi:10.1300/J121v17n01_03

Hughes, D. L., Whitburn, L. Y., Colasante, M., McDonald, A. C., McGowan, H., Zacharias, A., & Green, R. A. (2020). Developing anatomy demonstrators of the future: The role of team-teaching. *Journal of University Teaching and Learning Practice, 17*(5), 1-16. https://www.scopus.com/inward/record.uri?eid=2-s2.0-85098052551&partnerID=40&md5=824aed685d13d46534990e0c8a74849a

Johnson, B., Ulseth, R., Smith, C., & Fox, D. (2015, October 21-24). The impacts of project based learning on self-directed learning and professional skill attainment: A comparison of project based learning to traditional engineering education. 2015 IEEE Frontiers in Education Conference (FIE).

Jones, F., & Harris, S. (2012). Benefits and Drawbacks of Using Multiple Instructors to Teach Single Courses. *College Teaching*, *60*(4), 132–139. doi:10.1080/87567555.2012.654832

Jones, J. W., & Mezo, M. (2014). Capstone = team teaching + team learning + industry. *ASEE Annual Conference and Exposition, Conference Proceedings.*

Kersten, S. (2018). Approaches of Engineering Pedagogy to Improve the Quality of Teaching in Engineering Education. In J. Drummer, G. Hakimov, M. Joldoshov, T. Köhler, & S. Udartseva (Eds.), *Vocational Teacher Education in Central Asia. Technical and Vocational Education and Training: Issues, Concerns and Prospects* (Vol. 28, pp. 129–139). Springer.

Khoshnodifar, Z., Abbasi, E., Farhadian, H., Sadighi, H., & Pouratashi, M. (2020). Comparative comparison of lecture and team member teaching design methods in agricultural higher education system of Iran. *Journal of Agricultural Science and Technology*, *22*(4), 891–904. https://www.scopus.com/inward/record.uri?eid=2-s2.0-85090734994&partnerID=40&md5=af92d3714e28f235ccf6309cdf93b16d

Killingsworth, B. L., & Xue, Y. (2015). Investigating factors influencing students' learning in a team teaching setting. *International Journal of Cognitive Research in Science, Engineering and Education, 3*(2), 9-16. https://www.scopus.com/inward/record.uri?eid=2-s2.0-85010411070&partnerID=40&md5=f81164f563a66e1ae216bfa8fc88cbd6

King, S. (2006). Promoting Paired Placements in Initial Teacher Education. *International Research in Geographical and Environmental Education, 15*(4), 370–386. doi:10.2167/irg201.0

Kleyn, T., & Valle, J. (2014). Modeling collaborative teaching in teacher education: Preparing pre-service teachers to teach all students. In Advances in Research on Teaching (Vol. 21, pp. 137-164). Academic Press.

Kodkanon, K., Pinit, P., & Murphy, E. (2018). High-school teachers' experiences of interdisciplinary team teaching. *Issues in Educational Research, 28*(4), 967–989. https://www.scopus.com/inward/record.uri?eid=2-s2.0-85056302680&partnerID=40&md5=c9af039af6d318345af03df05938c2cc

Kostko, A. (2019). The impact of team teaching on student attitudes and classroom performance in introductory philosophy courses. *Teaching Philosophy*, *42*(4), 329–354. doi:10.5840/teachphil20191017111

Kröber, C., & Münster, S. (2016). Educational App Creation for the Cathedral in Freiberg. In J. Spector, D. Ifenthaler, D. Sampson, & P. Isaias (Eds.), *Competencies in Teaching, Learning and Educational Leadership in the Digital Age* (pp. 303–318). Springer. doi:10.1007/978-3-319-30295-9_19

Kulski, M. M., & Kerr, R. (2012). Managing teaching teams effectively. *International Journal of Learning, 18*(9), 149–160. doi:10.18848/1447-9494/CGP/v18i09/47730

Lasagabaster, D. (2018). Fostering team teaching: Mapping out a research agenda for English-medium instruction at university level. *Language Teaching*, *51*(3), 400–416. doi:10.1017/S0261444818000113

Lattuca, L. R., Knight, D. B., Ro, H. K., & Novoselich, B. J. (2017). Supporting the Development of Engineers' Interdisciplinary Competence. *Journal of Engineering Education*, *106*(1), 71–97. doi:10.1002/jee.20155

Li, Y. (2020). Language–content partnership in higher education: Development and opportunities. *Higher Education Research & Development*, *39*(3), 500–514. doi:10.1080/07294360.2019.1685947

Lutsenko, G. V., & Lucenko, G. V. (2018). Project-based learning in automation engineering curriculum. Proceedings of the 46th SEFI Annual Conference 2018: Creativity, Innovation and Entrepreneurship for Engineering Education Excellence.

Maletina, L., Karmanova, Y., & Kashpur, V. (2015). Implication of interdisciplinary team teaching of an ESP course. *Mediterranean Journal of Social Sciences*, *6*(5), 556–561. doi:10.5901/mjss.2015.v6n5p556

McDonald, A. C., Green, R. A., Zacharias, A., Whitburn, L. Y., Hughes, D. L., Colasante, M., & McGowan, H. (2021). Anatomy Students That are "Team-Taught" May Achieve Better Results Than Those That are "Sole-Taught". *Anatomical Sciences Education*, *14*(1), 43–51. doi:10.1002/ase.1954 PMID:32145155

McKenzie, S., Hains-Wesson, R., Bangay, S., & Bowtell, G. (2020). A team-teaching approach for blended learning: An experiment. *Studies in Higher Education*, 1–15. Advance online publication. doi:10.1080/03075079.2020.1817887

Miles, M. B., & Huberman, A. M. (1994). *An Expanded Sourcebook Qualitative Data Analysis*. SAGE Publications.

Minett-Smith, C., & Davis, C. L. (2020). Widening the discourse on team-teaching in higher education. *Teaching in Higher Education*, *25*(5), 579–594. doi:10.1080/13562517.2019.1577814

Minichiello, A., Blake, T. D., Goodridge, W. H., & Sam, D. D. (2011). Team teaching that goes the distance: Team instruction for a broadcast introductory engineering course. *ASEE Annual Conference and Exposition, Conference Proceedings*.

Money, A., & Coughlan, J. (2016). Team-taught versus individually taught undergraduate education: A qualitative study of student experiences and preferences. *Higher Education*, *72*(6), 797–811. doi:10.1007/s10734-015-9976-5

Monteblanco, A. D. (2020). Power Dynamics, Common Pitfalls, and Successful Strategies Associated with Co-Teaching. *College Teaching*, *69*(2), 63–68. doi:10.1080/87567555.2020.1810610

Ott, C., & Meek, N. (2019). Pair Teaching in Action. *ACM International Conference Proceeding Series*.

Pope-Ruark, R., Motley, P., & Moner, W. (2019). Creative innovation takes a (team-teaching) family. *Teaching & Learning Inquiry*, *7*(1), 120–135. doi:10.20343/teachlearninqu.7.1.8

Powell, P. (2000). From classical to project-led education. In A. S. Pouzada (Ed.), *Project Based Learning. Project-led Education and Group Learning* (pp. 11–42). Universidade do Minho.

Powell, P., & Weenk, W. (2003). *Project-led Engineering Education*. Lemma.

Rajamma, R. K., & Sciandra, M. R. (2020). Planning and Implementing a Graduate Online Team-Taught Marketing Course. *Journal of Marketing Education*, *42*(2), 108–122. doi:10.1177/0273475318786336

Rickard, A., & Walsh, T. (2019). Policy, practice and process in team teaching: A pilot project with co-operating teachers and student teachers on school placement. *Irish Educational Studies*, *38*(3), 309–326. doi:10.1080/03323315.2019.1625798

Rugarcia, A., Felder, R. M., Woods, D. R., & Stice, J. E. (2000). The future of Engineering Education I. A vision for a new century. *Chemical Engineering Education*, *34*(1), 16–25.

Saeed, A. A., Mutashar, A. M., & Aldakheel, A. (2018). The impact of applying collaborative team teaching method on students' outcomes. *Indian Journal of Forensic Medicine and Toxicology*, *12*(4), 139–143. doi:10.5958/0973-9130.2018.00212.8

Salifu, I. (2021). Exploring Coteaching as a Trend in Higher Education. *College Teaching*, *69*(3), 150–160. doi:10.1080/87567555.2020.1838426

Salonen, A. O., & Savander-Ranne, C. (2015). Teachers' Shared Expertise at a Multidisciplinary University of Applied Sciences. *SAGE Open*, *5*(3). Advance online publication. doi:10.1177/2158244015596206

Schmulian, A., & Coetzee, S. A. (2019). To Team or Not to Team: An Exploration of Undergraduate Students' Perspectives of Two Teachers Simultaneously in Class. *Innovative Higher Education*, *44*(4), 317–328. doi:10.100710755-019-9466-2

Selkrig, M., & Keamy, K. (2015). Promoting a willingness to wonder: Moving from congenial to collegial conversations that encourage deep and critical reflection for teacher educators. *Teachers and Teaching*, *21*(4), 421–436. doi:10.1080/13540602.2014.969104

Seymour, M. W., & Seymour, D. (2013). Are two professors better than one? Student and faculty perceptions of co-teaching. *International Journal of Learning*, *20*(1), 39–52. https://www.scopus.com/inward/record.uri?eid=2-s2.0-85016634914&partnerID=40&md5=b695b3a8588e5f115bf26a82947cc565

Smith, K. K., & Winn, V. G. (2017). Co-teaching in the college classroom. *Teaching Education*, *28*(4), 435–448. doi:10.1080/10476210.2017.1325863

Tan, H. R., Chng, W. H., Chonardo, C., Ng, M. T. T., & Fung, F. M. (2020). How Chemists Achieve Active Learning Online during the COVID-19 Pandemic: Using the Community of Inquiry (CoI) Framework to Support Remote Teaching. *Journal of Chemical Education*, *97*(9), 2512–2518. doi:10.1021/acs.jchemed.0c00541

Tsybulsky, D. (2019). The team teaching experiences of pre-service science teachers implementing PBL in elementary school. *Journal of Education for Teaching*, *45*(3), 244–261. doi:10.1080/02607476.2019.1599505

Tsybulsky, D., & Muchnik-Rozanov, Y. (2019). The development of student-teachers' professional identity while team-teaching science classes using a project-based learning approach: A multi-level analysis. *Teaching and Teacher Education*, *79*, 48–59. doi:10.1016/j.tate.2018.12.006

UNESCO. (2021). *Engineering for Sustainable Development*. United Nations.

Vangrieken, K., & Kindt, E. (2019). The teacher as an island? A mixed method study on the relationship between autonomy. and collaboration. *European Journal of Psychology of Education*, *35*(1), 177–204. doi:10.100710212-019-00420-0

Vesikivi, P., Holvikivi, J., Lakkala, M., & Bauters, M. (2016). Teacher collaboration in IT project courses: Resistance and success. *44th Annual Conference of the European Society for Engineering Education - Engineering Education on Top of the World: Industry-University Cooperation, SEFI 2016*.

Vesikivi, P., Lakkala, M., Holvikivi, J., & Muukkonen, H. (2019). Team teaching implementation in engineering education: Teacher perceptions and experiences. *European Journal of Engineering Education*, *44*(4), 519–534. doi:10.1080/03043797.2018.1446910

Walsh, E. O., & Davis, E. C. (2017). The geology and sociology of consumption: Team-teaching sustainability in an interdisciplinary first-year seminar. *Journal of Geoscience Education*, *65*(2), 126–135. doi:10.5408/16-172.1

Wöllner, C., & Ginsborg, J. (2011). Team teaching in the conservatoire: The views of music performance staff and students. *British Journal of Music Education*, *28*(3), 301–323. doi:10.1017/S0265051711000222

Chapter 13
i9MASKS Project:
A Learning-by-Doing Summer Experience With Engineering Students

Violeta Meneses Carvalho
University of Minho, Portugal

Cristina Rodrigues
University of Minho, Portugal

Rui A. Lima
University of Minho, Portugal

Graça Minas
University of Minho, Portugal

Senhorinha F. C. F. Teixeira
https://orcid.org/0000-0002-7464-3944
University of Minho, Portugal

ABSTRACT

Engineering education is a challenging topic that has been deeply explored in order to provide better educational experiences to engineering students, and the learning by doing approach has been appraised. Amidst a global pandemic, an engineering summer program denominated i9Masks emerged and aimed to create transparent facial masks for preventing the virus spreading. This project had the participation of 21 students from different engineering areas, as well as professors and monitors whose guidance and commitment were of great importance for its success. Aiming to understand the importance of this engineering hands-on project for students' training, two inquiries were applied, being one for students and the other for professors and monitors/researchers. Students described this initiative as an amazing and innovative experience that they would like to repeat and considered useful for their careers. Regarding the impact perceived by the teaching staff, the results proved that they enjoyed participating in the i9MASKS project and sharing knowledge with students in a practical way.

DOI: 10.4018/978-1-7998-8816-1.ch013

INTRODUCTION

Engineering and science are complex and demanding topics to teach since the theoretical concepts have to be taught and well assimilated by students, but the practice is the key in engineering education (Carlson and Sullivan 1999; Saravanan et al. 2021). For this reason, engineering teachers must have the ability to go beyond theory teaching and try whenever possible to propose practical activities so that students can visualize and apply what they learn in the classroom (Hackathorn, Solomon, Erin D. Blankmeyer, and Tennial, Rachel E. Garczynski 2011). In this sense, different methodologies have been implemented in order to improve and better prepare engineers for professional challenges that will eventually emerge in the future, such as active learning pedagogical frameworks (Freeman et al. 2014; López-Fernández et al. 2019). Students are engaged in the learning process and develop teamwork skills and the ability to operate in technology-enabled environments. These practices have been applied and investigated in different engineering areas, such as electrical engineering (Saravanan et al. 2021), aerospace engineering (López-Fernández et al. 2019), materials engineering (Renuka 2021), and software engineering (Ma et al. 2014), among others, and for engineering education in general (Arulanand, Babu, and Rajesh 2020; Freeman et al. 2014; Ricaurte and Viloria 2020).

Overall, it has been shown that active learning has a positive impact on students motivation and learning (López-Fernández et al. 2019), supporting the use of active learning in regular classrooms, and therefore, the continued use of traditional lecturing have been raised questions (Freeman et al. 2014). Unfortunately, with the pandemic crisis of COVID-19, educational institutions were forced to close and recreate/improvise their teaching methods promptly with the rapid transition to online learning which extremely affected students' learning in higher education (Bestiantono, Agustina, and Cheng 2020; Chaturvedi, Vishwakarma, and Singh 2021; Khan et al. 2020). Despite the endeavor and work of professors to teach and capture the attention and interest of students, some investigations have shown that the quarantine caused disturbances on students such as stress, frustration, and in many cases, depression. Due to the impossibility of carrying out in-person practical activities, the learning process was hindered more and affected the assimilation of the concepts taught. Thus, to resume in-person activities in higher education institutions, the Foundation for Science and Technology – FCT created an initiative denominated "Summer with Science". The special support provided by FCT allowed researchers and professors to develop innovative projects related to the COVID-19 pandemic. This initiative was very enriching for the students but also for the teaching team, researchers, and monitors, as they had a different summer with their hands-on science and developing tools to help fight the pandemic. i9Masks was one of the projects developed at the University of the authors of this paper. As the name indicates, this project intended to develop innovative masks for preventing the virus' spreading. Given that traditional masks hide our facial expressions, emotions, and hinder communication, the i9Masks project aimed to develop transparent facial masks by using a silicone elastomer, frequently used in several biomedical applications (Carvalho et al. 2020; Carvalho, Gonçalves, et al. 2021; Carvalho, Rodrigues, et al. 2021; Catarino et al. 2019; Faustino et al. 2016; Rodrigues et al. 2020; Sadek et al. 2021), while ensuring the protection effectiveness against the virus.

i9Masks project joined students of electrical, mechanical, physical, biomedical, and industrial engineering from two universities and Polytechnic institute, and they worked as a team with researchers and professors, acquiring multidisciplinary and ground-breaking knowledge in various scientific areas in a "learn by doing" way. They had the opportunity to learn about rapid prototyping, computer-aided design, 3D printing, microfabrication and microfluidics techniques, numerical simulations, microelectronics and optics, and nanotechnology and nanoparticle synthesis, all emerging areas in engineering. Given that, the participating students belonged to the first, second and third cycles of various scientific areas, a non-accredited course was created at the School of Engineering of the authors of this paper composed of four main modules: Seminars in Engineering Science, Introduction to Research in Engineering Sciences, Fabrication and Tests of the Silicone Elastomer Masks, and Numerical Simulations. These modules allowed students to acquire essential capabilities for the project implementation and research initiation. The final works presented by students were undoubtedly exciting and proved that combining different researchers, students, and professors of different scientific areas is important to attain good results faster. As a result, some publications emerged from the students' work (Teixeira et al. 2021) and also a book about the i9Masks course (Lima et al. 2021). Experiences similar to the i9Masks summer project have also shown that when students are involved in initiation scientific activities, they gain critical analysis, and intellectual experience in science insights (de Araújo et al. 2016; Fava-de-Moraes and Fava 2000; Neves et al. 2006; Saliba et al. 2019). Furthermore, some investigations have shown that these experiences are also useful for students to acquire knowledge and to develop networking prospects and team spirit (Gunel 2006; Lemberger, Webb, and Moore 2011; Moeed 2013; Scerri, Presbury, and Goh 2020).

Since it was one of the kind experience, the main purpose of this paper is to present the results of two surveys concerning the course's impact among students namely through the evaluation of the attitudes towards research, subjective norms for doing research, and perceived barriers to conducting research, and also the course's impact among the teaching staff by assessing their involvement in the course and its meaningfulness. Based on the literature, the team identified scales developed and validated, such as the Attitudes Toward Research (ATR) scale (Papanastasiou 2005), and adapted several others scales to accomplish the survey objectives. In order to accurately assess the results, the student research questionnaire was implemented among participants and non-participants and collected 47 responses. The professors and monitors questionnaire collected 17 responses. The outcomes are expected to contribute to a better understanding of the impact of a summer project-based learning.

The paper is organized into four sections: "i9Masks project" explains the summer course preparation, and how the project worked. "Survey # 1 – Students perceptions" presents the students' survey, focusing on methodology and selected results and discussion. "Survey # 2 – Professors and monitors perceptions" presents the teaching staff survey, centering on methodology, results, and discussion. The final section, "Conclusions", presents an overview of the results gathered in the present work and the meaningfulness of the present project for engineering education.

i9Masks PROJECT

As previously presented, i9MASKS was one of the engineering projects approved and aimed to create transparent facial masks, a personal protective equipment indispensable to control both public and personal health, for preventing the COVID-19 spreading rate. Appealed by the theme and the desire to have a different summer, several students from different engineering areas applied for this project, but only 21 were approved (6 Mechanical Engineering students, 5 Electronic Engineering students, 3 Biomedical Engineering students, 2 Materials Engineering students, 2 Industrial Engineering students and, 3 students from other engineering courses/ areas). The selection of students resulted from the analysis of the curriculum, namely the academic merit and the student's scientific experience in the area of the course. After the selection process, students had the opportunity to indicate, in order of preference, the works they would like to develop in the project. Based on their preferences, groups of four students maximum were organized in order to promote team spirit. Then, they carried out the assignments integrated into the project chosen at the beginning of the course. All of them were supervised by professors/researchers and monitors whose support was indispensable. The monitors were volunteers, PhD students, and researchers, who followed the project from the beginning giving additional support during the laboratory work. In addition to the project implementation, a non-accredited course was created and implemented at the School of Engineering of the authors of this paper. Apart from the four modules taught during the course, different researchers, teachers, and monitors have given several seminars: engineering science, introduction to research in engineering sciences, fabrication and tests of the silicone elastomer masks, and numerical simulations. These seminars were fundamental for students to acquire basic knowledge for the project development, and to hear different perspectives of the topic by other invited speakers. The three research centers where the course was associated were also presented to the students by their directors, motivating and inspiring students to the research topic. Then, a project-based learning methodology was addressed in which real-world problems and meaningful activities are used to students deepen their knowledge, and acquire effective professional competencies, namely work with a team spirit and deal with unexpected problems (Alves et al. 2018). What better way to capture students' interest and promote thinking deep and out of the box than with one of the worst pandemics the world has ever fought? In the i9Masks project, all the topics covered were related to the development and manufacture of facial masks that became mandatory in the context of the COVID-19 pandemic, and this fascinated the students and made the learning process easier. Each group had a hands-on experience while performing their tasks in the research laboratories, which in turn, supplemented their engineering courses. In the end, interesting works were presented including numerical and experimental research, both valuable for the manufacturing of the transparent masks. More information about the projects can be found at http://i9masks-1.mozello.com/. Moreover, this course created valuable prospects for students to start research activities giving them the opportunity of working in different laboratories.

Motivated by the satisfying outputs that the i9Masks project brought, the present work aims to understand the impact of this project-based learning on engineering students and their opinion about the activity. The importance of this project from the professors', researchers', and monitors' point of view was also evaluated. For this purpose, two different surveys were implemented to assess each perspective, from both students and teaching staff. The next sections explain and present the results obtained at each survey.

SURVEY # 1 – STUDENTS PERCEPTIONS

This section intends to present and explore the students' perceptions and perceived importance of this engineering hands-on project for students' training. Based on literature, several scales were identified and adapted. Results are presented and discussed considering two groups of students: 18 participants and 29 non-participants.

Methodology and Sample

The student questionnaire was developed in two parts. The first part considered the characterization of the respondent, such as age, gender, and course. The second part addressed the attitude towards research, whose preliminary results were analysed in another study (Carvalho, Teixeira, et al. 2021), the subjective norms and barriers towards research, and the perception and behaviour questions regarding the course ("What did you like most/least about the i9Masks course?"). The questionnaire was developed in English and previous to the questionnaire administration, a pre-test was driven with a pilot group of four engineering students who did not participate in the course. After minor changes, the final questionnaire was made available for implementation.

All participants who had finished the i9Masks summer course received one invitation email to participate in the survey with a link through Google Forms. Furthermore, in order to accurately assess the results, the questionnaire was sent by the participants to other colleagues from the School of Engineering who did not participate in the course (based on a snowball sampling: "answer and then send to other colleagues to respond"). The non-participants were considered as a control group, that is, an unexposed reference group used as a baseline for comparison. A total of 47 responses were obtained and considered valid, 18 from participants and 29 from non-participants. The profile of respondents is summarized in Table 1.

Table 1. Survey #1: Profile of respondents (n=47)

	Subgroup	Frequency	%
Type	Participant	18	38,3
	Non-participant	29	61,7
Gender	Female	29	61,7
	Male	18	38,3
Qualification Level	Undergraduate	16	34,0
	Bachelor´s degree	23	48,9
	Master's degree	8	17

Selected Results and Discussion

1. Attitude toward research

To assess the attitudes toward research, the authors used the version of the Attitudes Toward Research (ATR) scale (Papanastasiou 2005) subdivided into five subscales: Research usefulness for career, Research Anxiety, Positive attitudes toward research, Relevance to life, and Research difficulty. Before analysis, items that sounded discordant with the majority of the statements of the subscale were reverse coded. Cronbach's alpha coefficient was used to test the reliability of the subscales. All subscales presented a very good reliability, except the relevance to life with an alpha coefficient of 0.316 with the four initial items. After deleting two items, the alpha coefficient increased (0.683) to an acceptable level. The Cronbach's alpha coefficient and the number of the final items are presented in Table 2.

Table 2. Attitudes toward research (ATR) scale (Cronbach Alpha)

Subscales	Cronbach Alpha	# final items
Research usefulness for career	0.907	9
Research Anxiety	0.896	7
Positive attitudes toward research	0.948	8
Relevance to life	0.922	2
Research difficulty	0.683	3

After reliability analysis, the authors calculated a mean indicator of each attitude subscale with the retained items. Given the nature of the subscale, the closer the indicator to 5, the greater the respondent perceived research usefulness for career, research anxiety, positive attitudes toward research, relevance to life, and research difficulty. Means for the two subsamples (the i9MASKS participants and the unexposed reference group, that is, the non-participants) are presented in Table 3, and the results for the Mann-Whitney non-parametric tests.

Table 3. ATR subscales results (Mann-Whitney)

Attitudes Toward Research (ATR) scale (Papanastasiou 2005)	Participants Mean	Non-participants Mean	Mann-Whitney U	p-value
Research usefulness for career	4.33	3.97	191.5	0.127
Research Anxiety	2.83	3.35	161.0	0.028
Positive attitudes toward research	4.05	3.41	152.5	0.017
Relevance to life	3.83	3.10	168.5	0.039
Research difficulty	2.41	2.64	219.5	0.458

The Mann-Whitney results show that there are significant differences between participants and non-participants in:

a. Research anxiety (U=161.0, p=0.028), with non-participants presenting higher anxiety compared to the group that participated in the I9MASKS;
b. Positive attitudes toward research (U=152.5, p=0.017), with I9MASKS participants showing higher positive attitudes towards research
c. Relevance to life (U=168.5, p=0.039), with the group of I9MASKS participants revealing higher perceived research relevance to life

It is also interesting to notice that the Mann-Whitney test found no statistically significant differences between the two groups in research usefulness for career (both groups reveal high perceived research usefulness for career) and in research difficulty (both groups with mean values lower than 3, which suggests no major problems in terms of mathematics or research).

2. Subjective norms for doing research

Following the work of (Icek 2005), the authors considered it important to include a measure of the perceived social pressure from others for an individual to behave in a certain manner. Thus, the questionnaire measure students perceived subjective pressure for doing research considering the effect of the opinion of salient referents:

a. "Most people who are important to me support me doing research at the university";
b. "Most people who are important to me think it would be a good idea to do research at the university";
c. "Most people who are important to me want me to do research at the university";
d. "It is expected of me to do research at the university";
e. "I feel under social and peer pressure to do research at the university".

The statements were presented with a dichotomous scale ("disagree" or "agree") and the measure of subjective norms was obtained by the sum of "agree" answers. Thus, the higher the value, the greater the perceived pressure to carry out research. Despite the fact that the non-participant group presents a higher mean than the participants, the Mann-Whitney test does not found any statistically significant differences between the two groups in subjective norms for doing research (Table 4).

Table 4. Subjective norms for doing research results (Mann-Whitney)

	Participants Mean	Non-participants Mean	Mann-Whitney U	p-value
Subjective norms for doing research	2.61	2.71	245.5	0.881

3. Perceived barriers to Conducting Research

Based on the assumption that a behavior can be affected by the perceived barriers to that behaviour, each subsample of respondents was asked a question: "Reasons for not doing research more often?". The questionnaire contains seven possible answers to this question, being the respondent able to choose all the answers that applied:

a. "Lack of interest";
b. "Lack of time";
c. "Inadequate knowledge about research methodology";
d. "Lack of guidance by faculties";
e. "Lack of research curriculum";
f. "Inadequate facilities for research"
g. "Inadequate financial support"

Looking at the results in Table 5, it can be observed that participants are the ones who indicate the greatest number of reasons for not doing research more often. They point out with great expressiveness the “Inadequate knowledge about research methodology” (66.7%), the “Lack of guidance by faculties” (61.1%), and the “Inadequate financial support” (61.1%).

In turn, the answers of non-participants suggest “Inadequate financial support” (55.2%) as the main reason, followed by a set of reasons with the expressiveness of less than 50%: “Lack of research curriculum” (44.8%), “Inadequate knowledge about research methodology” (41.4%) and “Lack of guidance by faculties” (41.4%).

Table 5. Perceived barriers results by respondent type

	Participants		Non-participants	
Perceived barriers	Count	%	Count	%
Lack of interest	6	33.3%	10	34,5%
Lack of time	2	11.1%	4	13,8%
Inadequate knowledge about research methodology	12	66.7%	12	41,4%
Lack of guidance by faculties	11	61.1%	12	41,4%
Lack of research curriculum	7	38.9%	13	44,8%
Inadequate facilities for research	7	38.9%	9	31,0%
Inadequate financial support	11	61.1%	16	55,2%

A 2x2 contingency table analysis was conducted to determine whether there was an association between the subsamples and the perceived barriers. Results suggest a significant dependency between the subsamples and:

a. the perceived barrier of “Inadequate knowledge about research methodology”
b. (the “Lack of guidance by faculties” (Fisher’s Exact Test, p=0.013)
   d. the “Inadequate facilities for research” (Fisher’s Exact Test, p=0.049)

Based on the answers given, a measure of perceived barriers was obtained by the sum of reasons selected (Table 6). Thus, the higher the value, the greater the perceived barriers to conducting research more often. Although participants perceive on average more barriers, the Mann-Whitney test found no statistically significant differences between the two groups.

Table 6. Perceived barriers to conducting research results (Mann-Whitney)

	Participants Mean	Non-participants Mean	Mann-Whitney U	p-value
Perceived barriers to conducting research	3.11	2.62	213.0	0.278

Results suggest that:

a. respondents students may not feel methodologically prepared to carry out research activities, which consequently, hinders their motivation;
b. respondents students may perceive faculties do not provide the necessary support to conduct research. This is worrying and must be taken into consideration by universities. They may have this idea due to the lack of interest from professors themselves and also maybe because they are unaware of the existence of some ongoing activities;
c. respondents students assume that they do not have a good enough curriculum to pursue research activities. This alert that students need to enrich their curriculum during the course and not only during the dissertation. This can be achieved by encouraging them to carry out research activities earlier, and here the role of professors and tutors is very important.

4. Suggestions for improving student participation in research

Each subsample was invited to reflect about suggestions for improving student participation. Based on (Unnikrishnan et al. 2014), the questionnaire considered five possible answers to this question, being the respondent able to choose all the answers that applied:

a. "Increase research funding";
b. "Incentives for students and teachers";
c. "Research related workshops for faculty";
d. "Research related workshops for students";
e. "Additional hours for teachers supervising research".

The top answers are similar for the 2 subsamples: "Incentives for students and teachers", "Research related workshops for students" and "Increase research funding". No student ticked the suggestion "Additional hours for teachers supervising research".

Table 7. Suggestions results by respondent type

	Participants		Non-participants	
Suggestions for improving research participation	Count	%	Count	%
Increase research funding	13	72.2%	18	62.1%
Incentives for students and teachers	17	94.4%	23	79.3%
Research related workshops for faculty	12	66.7%	12	41.4%
Research related workshops for students	15	83.3%	21	72.4%
Additional hours for teachers supervising research	0	0.00%	0	0.00%

The 2x2 contingency table analysis only identified a significant dependency between the subsamples and the suggestion "Research related workshops for faculty"
This shows that:

- Respondents students need not only incentives but also more information about what research is, and about the projects taking place in their training areas;
- Respondents students already have the perception of lack of research funding for conducting research projects. The option with the lowest frequency was the "research-related workshops for faculty". This emphasizes that students think that the university does not need these workshops so much, but they do.

5. **How did participants hear about the summer project?**

To learn about the effectiveness of the course's dissemination channels and identify how students learned about the course, the questionnaire included the question: "How did you hear about this course?" with five possible answers:

- "Email notification";
- "Professor/Researcher";
- "Web search";
- "Social media";
- "Other".

Considering only the participants' subsample, the answers suggest that professors were the main responsible for conveying the message (Table 8). This is very important because they can get too many students in a close and more effective way. However, only the students of the professors involved in the project were aware, that is, many potential candidates may not have been informed of the course in this way. The email was also a good way to disclose the project, but it is not enough. Although professors and email play an important role, it is necessary to improve the disclosure method in future events, as promising students may not have been aware of the project due to the lack of adequate dissemination in the university faculties and media.

Table 8. Dissemination channels

	Participants	
Dissemination channels	Count	%
Email notification	5	27,8%
Professor/Researcher	14	77,8%
Web search	0	0,0%
Social media	0	0,0%
Other	3	16,7%

6. Opinion of participating students about the course

For a complete evaluation of the course, the authors consider it important to include an open question: “What did you like most about the course?”. The answers are represented on a word chart (Figure 1) constructed with the introduction of students’ complete answers into a free word cloud generator software (https://www.jasondavies.com/wordcloud/).

Figure 1. Word chart with positive aspects pointed out by students

The most mentioned words were “research”, followed by “opportunity”, “innovation”, and “knowledge”. Other words that stand out are “interesting,”, "experience", and “team”. In general, students described this initiative as an innovative research experience. They valued working in such different areas as a team, considered it an interesting and useful learning project. Besides, they liked to see what the investigation consists of and seeing the theoretical concepts learned put into practice, considered it an excellent opportunity to learn new skills and concepts.

Results confirm a positive impact of this summer course, in particular with an original project and theme, which are not covered in some engineering courses.

The questionnaire also included an open question "What can be improved in the course?". In addition to the positive aspects pointed out, some students suggested improvement ideas to the present project such as: they would like to have had more supervision during the experimental activities, more time to carry out the projects, and considered that more funds are needed for assuring the performance of the experimental activities. Moreover, students would like to have more seminars during the course and more practical activities. Nevertheless, some students also alleged that the project had low visibility by the university. The last statement corroborates the previous results in which the project was not properly disclosed. Since professors cannot reach all potential candidates, our recommendation for future initiatives is that new dissemination tools must be thought of.

SURVEY # 2 – PROFESSORS AND MONITORS PERCEPTIONS

This section intends to present and explore the impact perceived by teachers and monitors who participated in the summer project. As in the previous section, scales were identified and adapted from literature. Results are presented and discussed considering two groups of teaching staff: 13 professors and 4 monitors or researchers.

Methodology and Sample

The professors and monitors questionnaire was organized in two parts: the characterization of the respondent-type (professor, monitor or researcher) and the involvement towards the i9MASKS summer course. The questionnaire was developed in English and previous to the questionnaire administration, a pre-test was driven with a pilot group of two professors and one researcher who participate in the course. Considered completed, the questionnaire was implemented in June 2021: all professors, monitors or researchers involved in the summer project, in its seminar or laboratory modules, received an invitation email to participate in the survey, with a link through Google Forms. A total of 17 responses were obtained, all considered valid. The profile of respondents is presented in Table 9.

Table 9. Survey #2: Profile of respondents (n=17)

Type	Frequency	%
Professor	13	76.5
Monitor or researcher	4	23.5

Selected Results and Discussion

1. Involvement in the course

To evaluate how much the professors and monitors enjoyed participating in the course, the authors decided to adapt and extend the scale of "Involvement in the study" from (Mehta, Hoegg, and Chakravarti 2011). Therefore, the questionnaire requires an answer for the following questions:

a. "Did you enjoy participating in this summer course?";
b. "How much did you enjoy participating as a course monitor for this summer course?";
c. "How motivated were you in participating as a course monitor for this summer course?";
d. "How much effort did you spend in participating as a course monitor for this summer course?";
e. "How interesting do you think this summer course was?";
f. "As a researcher, do you think your performance was enhanced with this experience?".

All questions used a 7 point scale, ranging from "not at all" (-3) to "Very much" (+3). Cronbach's alpha coefficient was used to test the reliability of the scale. With an initial alpha coefficient of 0.669 with the six initial items, it was considered necessary to delete question six, and consequently, the alpha coefficient increased to a good level (0.745). After reliability analysis, the authors calculated a mean indicator of the involvement with the five retained items and the results are present in Table 10. As expected, the closer to +3, the greater the respondent's involvement in the course. Both subsamples present high levels of involvement in the course, and the Mann-Whitney test found no statistically significant differences between the two groups.

Table 10. Involvement in the course results (Mann-Whitney)

	Professors Mean	Monitors or researchers Mean	Mann-Whitney U	p-value
Involvement in the course	2.33	2.29	23.5	0.774

Results confirm that both subgroups, professors and monitors or researchers, enjoyed participating in the i9MASKS project. This is even more important if one considers the fact that the course took place during the professors and researcher's summer vacations.

2. Meaningfulness of the course

The authors also considered it relevant to include a measure of meaningfulness, that is, the extent to which a person perceives a stimulus to be relevant and important. Thus, based on the Meaningfulness scale of (Mano and Oliver 1993), the authors developed a semantic scale with nine statements to measure the meaningfulness of the course. The question presented in the survey was "For you, this experience was/is...":

a. "unimportant"/ "important"
b. "of no concern"/ "of concern"
c. "irrelevant"/ "relevant"
d. "of no consequence"/ "seriousness"
e. "worthless"/ "valuable"
f. "minor"/ "major"
g. "insignificant"/ "significant"
h. "not meaningful"/ "meaningful"
i. "not replicable"/ "replicable"

The answers used a 7 point scale, ranging from 1 (negative meaning) to 7 (positive meaning). The reliability analysis was conducted and obtained an excellent alpha coefficient of 0.960. Then, a mean indicator of the meaningfulness of the course was calculated and it is presented in Table 11. As expected, the closer to 7, the greater the meaningfulness attributed to the course. Although both identify high levels of meaningfulness of the course, monitors or researchers have the highest average, very close to the maximum of 7. However, the Mann-Whitney test found no statistically significant differences between the two groups.

Table 11. Meaningfulness of the course results (Mann-Whitney)

	Professors Mean	Monitors or researchers Mean	Mann-Whitney U	p-value
Meaningfulness of the course	5.77	6.44	17.5	0.332

This result is very interesting because it appears that there is no distinction of relevance or importance of the course between professors and monitors or researchers. Both groups consider the course "important", "of concern", "significant" and ... "replicable"! This is a very promising result and suggests that this experience may be the first of its kind at the University but may not be unique!

3. Opinion of professors, researchers, and monitors about the course

Similarly to the first questionnaire, for a complete evaluation of the course, the authors included three open questions: "What did you like most about the summer course?", "What did you like least about the summer course?", and "What can be improved in the course?"

The responses to the first question are represented on a word chart depicted in Figure 2. This word chart was constructed as previously mentioned, resorting to the same software and by introducing the complete responses of professors, monitors and researchers.

The words that stand out are "students", "knowledge", "involvement", followed by "science", "experience", "practical", and "communication".

The results stress the importance of this summer experience not only for students but also for all teaching staff. As students, the opportunity to share knowledge was valued by professors, monitors, and researchers, mainly because the course was carried out in the context of the COVID-19 pandemic, and the opportunity to contribute to society through small discoveries during this difficult situation. In addition, the teaching staff appraised the multidisciplinarity of the course, and the possibility to communicate and interact with students outside the formal context of classes, that is, in a friendly environment, and show that science is everywhere in a practical way. The integration of monitors, professors, and researchers together in the preparation and implementation of the summer project was an aspect appreciated. One can see that the results were very positive, emphasizing that this experience can be implemented in the future, but in different areas of research, mainly involving current issues in need of urgent scientific progress.

Regarding the second question, "What did you like least about the summer course?" the short duration and the period of the course stood out. The respondents considered that because of the short duration, the application of the knowledge was limited, and dislike the fact that the course was carried out during the summer vacations (the course started in mid of July and extended to October). In the future, these two points may be reconsidered, and the earlier start of the course combined with a break period of the course is a good option to allow the vacations of both professors, monitors/ researchers, and students without compromising the ongoing research. Furthermore, an interesting topic that was also stressed in the responses was the bureaucracy and management of the course. This suggests that similar future activities should be less demanding in terms of administration issues. Besides, some respondents considered that the course was planned at the last minute. This may have happened because it was a completely novel activity implemented in higher education institutions and consequently implied a lot of stress, and extraordinary work to prepare all bureaucracies.

Figure 2. Word chart with positive aspects pointed out by professors, monitors, and researchers

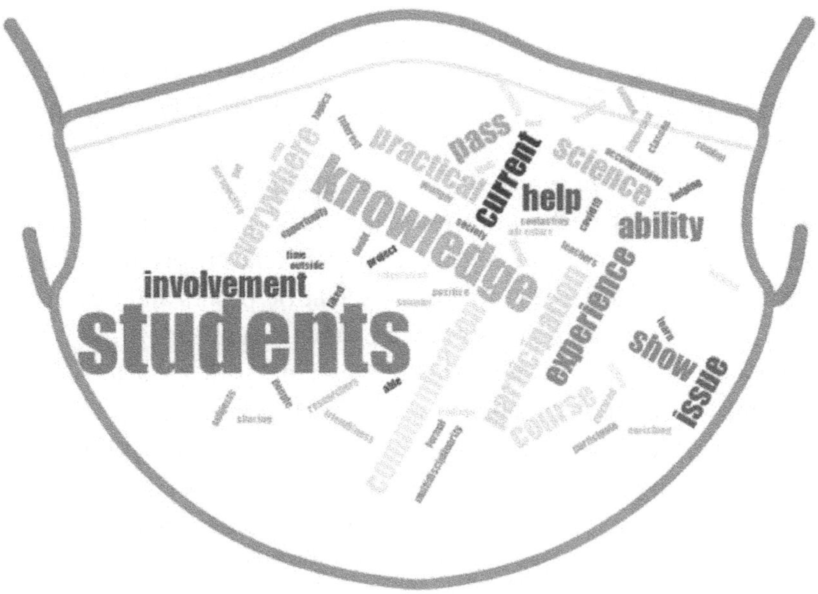

The final question allowed the teaching staff to share their improvement proposals, which agreed with the previous statements. Amongst the received answers, some respondents suggested that the course duration should be studied and that more contact hours are needed for the completion of the activities.

The insufficient financial support was also pointed out, and it is worth mentioning that this worry about the financial support was also present in the student's questionnaire, emphasizing that this, indeed, is an issue that should be taken into account in future activities. Other respondents referred that more equipment is needed, and this may be related to the lack of funding, which may have compromised some research activities.

CONCLUSION

In the present work, a project-based learning was exposed, and the results were undoubtedly encouraging and rewarding for the whole team, as they saw that the effort was worthwhile. The students' observations were very important to understand what actually went well and what needs to be improved in future activities. Effectively, students valued the opportunity to carry out practical activities in multidisciplinary areas and they found this initiative quite enriching and important to their lives.

Results from the second survey showed that with high levels of involvement and meaningfulness of professors and monitors/researchers, the course had the best conditions to be successful.

Despite the positive aspects pointed out by both groups, for future activities several aspects need to be improved, such as the period of the course (during academic holidays) and its short duration. Moreover, it was found that the University needs to improve the tools for disseminating projects and encourage students to enhance their curriculum as well as increase their research interest. The latter can be attained by preparing active-learning activities related to current problems and visits to the various laboratories where PhD students and researchers carry out their works and showing some works being developed in research projects and their impact on the real world.

In everything, there are positive and negative points, but undoubtedly, in this case, the positive points outweigh the negative ones. These findings showed that more initiatives like the i9Masks should be increasingly implemented in higher education institutions, as its impact was quite positive for both teachers and students. As expected, the engagement with the learning process helped students to develop transversal skills and the ability to face real-world problems from another point of view. Furthermore, the experience of the i9Masks provided an unique relation theory-practice, professor-student with paired discussions, the opportunity to work with colleagues of different grades and scientific areas, and the contact with research activities such as the writing of scientific papers and treatment of the data were of utmost importance for the learning process of the upcoming engineers.

In the future, it would be interesting to follow the professional path and academic performance of the students who participated in the project and verify whether this initiative, in addition to having given them a deep knowledge in several scientific areas and experience, also influenced their choices to enter in the professional life and their performance. Moreover, it should be noted that, in future experiences it is intended to use qualitative methodologies such as interviews and focus-groups to deepen the conclusions.

To conclude, in some way, the i9Masks summer project contributed to the preparation of the next generation of engineers by giving them the great opportunity of working in several laboratories, meeting researchers, developing new skills, and team spirit.

REFERENCES

Arulanand, N., Ramesh Babu, A., & Rajesh, P. K. (2020). Enriched Learning Experience Using Augmented Reality Framework in Engineering Education. *Procedia Computer Science*, *172*, 937–942. doi:10.1016/j.procs.2020.05.135

Bestiantono, D. S., Putri, Z. R. A., & Cheng, T.-H. (2020). Online Learning amid the COVID-19 Pandemic: Students' Perspectives Muhammad. *Journal of Pedagogical Sociology and Psychology*, *2*(1), 45–51. doi:10.46627ilet.v1i3.46

Carlson, L. E., & Sullivan, J. F. (1999). Hands-on Engineering: Learning by Doing in the Integrated Teaching and Learning Program. *International Journal of Engineering Education*, *15*(1), 20–31.

Carvalho, V., Gonçalves, I. M., Lage, T., Rodrigues, R. O., Minas, G., Teixeira, S. F. C. F., Moita, A. S., Hori, T., Kaji, H., & Lima, R. A. (2021). 3D Printing Techniques and Their Applications to Organ-on-a-Chip Platforms : A Systematic Review. *Sensors (Basel)*, *21*(9), 3304. doi:10.339021093304 PMID:34068811

Carvalho, V., Maia, I., Souza, A., Ribeiro, J., Costa, P., Puga, H., Teixeira, S. F. C. F., & Lima, R. A. (2020). In Vitro Stenotic Arteries to Perform Blood Analogues Flow Visualizations and Measurements: A Review. *The Open Biomedical Engineering Journal*, *14*(1), 87–102. doi:10.2174/1874120702014010087

Carvalho, V., Rodrigues, R. O., Lima, R. A., & Teixeira, S. (2021). Computational Simulations in Advanced Microfluidic Devices: A Review. *Micromachines*, *12*(10), 1149. doi:10.3390/mi12101149 PMID:34683199

Carvalho, V., Teixeira, S., Lima, R. A., Minas, G., & Rodrigues, C. (2021). Learning Science during Summer Vacations and Its Effects on Attitude and Anxiety towards Research. *International Conference on Active Learning in Engineering Education*.

Catarino, S. O., Rodrigues, R. O., Pinho, D., Miranda, J. M., Minas, G., & Lima, R. (2019). Blood Cells Separation and Sorting Techniques of Passive Microfluidic Devices : From Fabrication to Applications. *Micromachines*, *10*(9), 593. doi:10.3390/mi10090593 PMID:31510012

Chaturvedi, K., Vishwakarma, D. K., & Singh, N. (2020, July). 2021. "COVID-19 and Its Impact on Education, Social Life and Mental Health of Students : A Survey. *Children and Youth Services Review*, *121*, 105866. doi:10.1016/j.childyouth.2020.105866 PMID:33390636

de Araújo, J. A. R., Judson da Cruz, G., Vieira da Silva, W., Deretti, S., Dalazen, L. L., & Pereira da Veiga, C. (2016). Quality Evaluation in Post-Graduate Diploma Courses from the Students' Perspective: An Exploratory Study in Brazil. *International Journal of Management Education*, *14*(3), 454–465. doi:10.1016/j.ijme.2016.10.003

Faustino, Catarino, Lima, & Minas. (2016). Biomedical Microfluidic Devices by Using Low-Cost Fabrication Techniques: A Review. *Journal of Biomechanics, 49*(11), 2280–92. . doi:10.1016/j.jbiomech.2015.11.031

Fava-de-Moraes, F., & Fava, M. (2000). A Iniciação Científica: Muitas Vantagens e Poucos Riscos. *São Paulo em Perspectiva*, *14*(1), 73–77. doi:10.1590/S0102-88392000000100008

Freeman, S., Eddy, S. L., McDonough, M., Smith, M. K., Okoroafor, N., Jordt, H., & Wenderoth, M. P. (2014). Active Learning Increases Student Performance in Science, Engineering, and Mathematics. *Proceedings of the National Academy of Sciences of the United States of America*, *111*(23), 8410–8415. doi:10.1073/pnas.1319030111 PMID:24821756

Gunel, M. (2006). *Investigating the Impact of Teachers' Implementation Practices on Academic Achievement in Science during a Long-Term Professional Development Program on the Science Writing Heuristic*. Iowa State University. doi:10.31274/rtd-180813-12062

Hackathorn, J., Solomon, K. L., Blankmeyer, E. D., Tennial, A. M., & Garczynski, R. E. (2011). Learning by Doing: An Empirical Study of Active Teaching Techniques. *The Journal of Effective Teaching*, *11*(2), 40–54.

Icek, A. (2005). *Attitudes, Personality and Behaviour* (2nd ed.). Milton-Keynes: Open University Press.

Khan, A. H., Sultana, M. S., Hossain, S., Hasan, M. T., Ahmed, H. U., & Sikder, M. T. (2020). The Impact of COVID-19 Pandemic on Mental Health & Wellbeing among Home-Quarantined Bangladeshi Students: A Cross-Sectional Pilot Study. *Journal of Affective Disorders*, *277*(May), 121–128. doi:10.1016/j.jad.2020.07.135 PMID:32818775

Lemberger, M., Webb, L., & Moore, M. (2011). Student Success Skills : An Evidence-Based Cognitive and Social Change Theory for Student Achievement Linda Webb, Florida State University, and Molly m. Moore, University of Northern Colorado. *Journal Od Education*, *192*(2), 88–99. doi:10.1177/0022057412192002-311

Lima, Teixeira, Minas, Rodrigues, & Carvalho. (2021). *I9MASKS A Micro Step towards the Development of PDMS Masks for the Protection of COVID-19*. Universidade do Minho – MEtRICs.

López-Fernández, D., Ezquerro, J. M., Rodríguez, J., Porter, J., & Lapuerta, V. (2019). Motivational Impact of Active Learning Methods in Aerospace Engineering Students. *Acta Astronautica*, *165*, 344–354. doi:10.1016/j.actaastro.2019.09.026

Ma, K., Teng, H., Du, L., & Zhang, K. (2014). Project-Driven Learning-by-Doing Method for Teaching Software Engineering Using Virtualization Technology. *International Journal of Emerging Technologies in Learning*, *9*(9), 26–31. doi:10.3991/ijet.v9i9.4006

Mano, H., & Oliver, R. L. (1993). Assessing the Dimensionality and Structure of the Consumption Experience: Evaluation, Feeling, and Satisfaction. *The Journal of Consumer Research*, *20*(3), 451–466. doi:10.1086/209361

Mehta, R., Hoegg, J., & Chakravarti, A. (2011). Knowing Too Much: Expertise-Induced False Recall Effects in Product Comparison. *The Journal of Consumer Research*, *38*(3), 535–554. doi:10.1086/659380

Moeed, A. (2013). Science Investigation That Best Supports Student Learning: Teachers Understanding of Science Investigation. *International Journal of Environmental and Science Education*, *8*(4), 537–559. doi:10.12973/ijese.2013.218a

Neves, Cunha, Arezes, Leao, Teixeira, Lobarinhas, & Teixeira. (2006). *IN2TEC: A Multidisciplinary Research Project Involving Researchers, Students and Industry*. Academic Press.

Papanastasiou, E. C. (2005). Factors Structure of The Attitudes Toward Research Scale. *Statistics Education Research Journal*, *4*(1), 16–26. doi:10.1037/t64085-000

Renuka, G. (2021). Innovative Teaching and Learning Strategies for Materials Engineering Education. *Materials Today: Proceedings*. Advance online publication. doi:10.1016/j.matpr.2021.04.350

Ricaurte, M., & Viloria, A. (2020). Project-Based Learning as a Strategy for Multi-Level Training Applied to Undergraduate Engineering Students. *Education for Chemical Engineers*, *33*, 102–111. doi:10.1016/j.ece.2020.09.001

Rodrigues, R. O., Sousa, P. C., Gaspar, J., Bañobre-López, M., Rui, L., & Minas, G. (2020). Organ-on-a-Chip: A Preclinical Microfluidic Platform for the Progress of Nanomedicine. *Small*, *16*(51), 2003517. doi:10.1002mll.202003517 PMID:33236819

Sadek, S. H., Rubio, M., Rui, L., & Vega, E. J. (2021). Blood Particulate Analogue Fluids: A Review. *Materials (Basel)*, *14*(9), 2451. Advance online publication. doi:10.3390/ma14092451 PMID:34065125

Saliba, T. A., Borges de Mattos Custodio, L., Adas Saliba, N., & Adas Saliba Moimaz, S. (2019). The Impact of Scientific Initiation on the Scientific Production of Graduate Students. *Revista Da ABENO*, *19*(1), 127–133. doi:10.30979/rev.abeno.v19i1.873

Saravanan, S., Mahesh, V., Kumar, D. G., & Kshatri, S. S. (2021, March). Improving Student's Learning with Efficient Learning Techniques: A Case Study of First Year Basic Electrical Engineering Course. *Materials Today: Proceedings*. Advance online publication. doi:10.1016/j.matpr.2021.02.784

Scerri, M., Presbury, R., & Goh, E. (2020). An Application of the Mentoring Framework to Investigate the Effectiveness of Mentoring Programs between Industry Mentors and Student Mentees in Hospitality. *Journal of Hospitality and Tourism Management*, *45*(August), 143–151. doi:10.1016/j.jhtm.2020.08.010

Teixeira, I., Correia, T., Carvalho, V., Rodrigues, N., Lima, R. A., & Teixeira, S. (2021). Numerical Simulation: Study on the Path of Expelled and Exhaled Water Vapor Particles. *5th International Conference on Numerical and Symbolic Computation Developments and Applications*.

Unnikrishnan, B. (2014). Medical Students' Research - Facilitators and Barriers. *Journal of Clinical and Diagnostic Research*, *8*(12), XC01–XC04. doi:10.7860/JCDR/2014/10223.5291 PMID:25654016

Chapter 14
COVID-19 Pandemic Effects on Brazilian Postgraduate Research:
An Exploratory Study at Two Federal Institute Programmes

Andre Fernando Uébe-Mansur
https://orcid.org/0000-0003-1909-7300
Federal Fluminense Institute, Brazil

Giselle Rôças
Federal Institute of Rio de Janeiro, Brazil

Eduardo dos Santos de Oliveira Braga
Federal Institute of Rio de Janeiro, Brazil

Neila Ferreira da Silva Jesus
Federal Fluminense Institute, Brazil

Lohaine Miguez Martins
Federal Fluminense Institute, Brazil

ABSTRACT

The education area is being deeply affected by COVID-19, and Brazilian students are trying to adapt. This chapter aims to research how postgraduate students are dealing with the challenges of the pandemic. From the following research question, "How did COVID-19 impact different dimensions of students´ lives enrolled at master and doctorate programmes?" the chapter describes the challenges that students from Master and Doctorate programmes of two federal institutes are dealing with and the future perspectives in the context of the pandemic. The research methodology is based on an exploratory approach, grounded on a survey for data regarding the impacts of COVID-19 in three dimensions: private life, professional life, and academic life, aiming to understand if and how their research and educational products development were affected. The results show that, despite stress and efforts, the students could adapt their research for the pandemic situation.

DOI: 10.4018/978-1-7998-8816-1.ch014

INTRODUCTION

Contemporary society is suffering from disruptive changes in its various areas since the coronavirus outbreak in China (Hua et al., 2020), and the SARS-COV-2 pandemic (also called COVID-19 pandemic) was officially declared on 11th March 2020 by WHO (2020). To avoid the virus' dissemination, business, schools and other institutions from the entire world became a rampage lockdown. Nowadays, many people are compulsorily in their homes, avoiding agglomerations and often even without human contact (Kanungo & Sharma, 2021). For this reason, world education has been profoundly affected. An anonymous survey involving 3,000 Brazilians over 18 years old reported that among the participants, 46.4% of them showed depression symptoms, 39.7% anxiety, and 42.2% stress (Serafim et al., 2021). According to Usmani et al. (2021), nearly 90% of the world's student population is experiencing disruptive changes in their learning. For example, Hussain et al. (2021) reported that the COVID-19 pandemic promoted a world run to offer courses using online technology at High Education Institutions. This need provided an increase from 25% to 85% (meaning a 60% growth) of distance teaching classes. In Brazil, institutions at their different educational levels were in lockdown. From the complete classroom suspension right at the beginning of the pandemic to the distance emergency classes, there were many challenges to provide student academic and pedagogical assistance.

The Federal Network of Vocational Education (FNVE) is a huge public network of vocational education from the Brazilian government that promotes education from high school to the doctorate level, including 643 campuses around many Brazilian states. The Rio de Janeiro state FNVE offers 5 doctorate and 27 master programmes in different institutes developing research in many areas, and all of them have been disrupted by the COVID-19 pandemic. Within this scenario, the following question is set: "How did the COVID-19 outburst impact on different dimensions of the students' lives enrolled in master and doctorate programmes?". Prompted by this question, the present chapter aims to map, share and analyse how the COVID-19 pandemic has impacted the research in specific doctorate and master programmes of the two federal institutes of the Rio de Janeiro state, called Fluminense Federal Institute (IFF) and Federal Institute of Rio de Janeiro (IFRJ).

Federal institutes are institutions based on scientific and technological education. They offer courses in vocational secondary education, technologists, engineering, and teacher training. The current study arises from an initial reflection on how the pandemic affects two postgraduate programmes aimed at training trainers in the exact sciences, evidencing an intimate relationship between training courses with a scientific and technological bias and the research carried out.

The chapter is structured in seven sections. Following this first section, which presents the context and objectives of the study, the second section outlines the Federal Institutes (FI) general aspects, detailing characteristics of the two Rio de Janeiro state's institutes focused on this research. The third section presents the survey from master's and doctorate programmes' students from the two institutes. Two case studies from an IFF master student and an IFRJ doctorate student according to the described methodological stages are also presented. The fourth section discusses future research directions, followed by the last section, in which overall conclusions are presented.

POSTGRADUATE PROGRAMMES AT THE FEDERAL INSTITUTES

The Federal Institutes (FI) have been founded in 2008, by a federal Law that created the Federal Network of Vocational Education (FNVE), providing studies from technical high school to doctorate level in different learning modalities ("Lei 11892/2008", 2008), for example on-site, online and young and adult education (Aguiar; Pacheco, 2017; Braga *et al.*, 2019).

The contemporary FI began its roots in 1909 when the Brazilian president Nilo Peçanha founded 19 vocational schools focused on poor and less-educated people. In its early days, these vocational schools provided only elementary school. In 1942 President Getúlio Vargas enacted a law expanding vocational elementary education to a new 24 vocational high schools. Around 1959, these high schools became Federal Vocational High Schools (FVHS), gaining didactic and management autonomy. In 1978, the first high schools were renamed Federal Centres of Vocational Education (FCCE) and began to offer higher education in the next stage of the Brazilian educational system. In 1999, the change of all FVHS into FCCE was intensified by the President Fernando Henrique Cardoso intending the FVHS growing and interiorisation.

After 20 years of growth, in December of 2019 the 38 FVHS's institutes had nine doctorate programmes (one in the vocational modality) and 181 master programmes (including 139 master programmes in the vocational modality). In the Rio de Janeiro state, there are two Federal Institutes: The Fluminense Federal Institute (IFF) and the Federal Institute of Rio de Janeiro (IFRJ).

Brazil has two modalities for master (MSc) and doctorate (PhD) programmes: academic and vocational programmes (Pereira; Rôças, 2017; Rôças *et al.*, 2020). The academic programme's orientation falls within the general research of universities and research institutes. The vocational programme's orientation is applied research, focusing on investigation, understanding and proposing solutions to problems arising on the daily basis experience of different professions and branches of industry and service sectors. Due to these characteristics, the vocational programmes demand the development of a technical product, named Education Products, for programmes related to the Educational Area. (Rôças *et al.*, 2020). Table 1 is based on data from Plataforma Nilo Pecanha (2021) and presents an overview related to IFF and IFRJ institutions:

Table 1. Campuses overview

	IFF	IFRJ
Foundation	2008	2008
Modality	Public	Public
Education Levels	From High School to Postgraduation	From High School to Postgraduation
Campuses	12	15
2020s enrolled students	39,420	36,096
2020s enrolled master and doctorate students	280	762
Vocational High School Programmes	99	70
Engineering Graduate Programmes	8	1
Master & Doctorate Postgraduates	6	12

Each institutions have several postgraduate programmes. The research is focused on two of them (MPET & PROPEC).

Vocational Master in Teaching and Its Technologies (MPET)

The IFF programme is named Vocational Master in Teaching and Its Technologies (MPET) and was authorised in 2016, and began its academic activities in 2017. MPET is located in the Campos dos Goytacazes (north region of Rio de Janeiro state). Campos dos Goytacazes is the largest municipality of the state with approximately 511,000 inhabitants. The MPET target audience is mainly teachers and other education professionals from Campos dos Goytacazes municipality and Rio de Janeiro state schools. In 2021, 28 students from the 2017 and 2018 cohorts concluded their research, and 37 students from the 2019 and 2021 cohorts are on their way. The programme has two research lines: Production and experimentation of technologies, as well as Teacher training and technological resources, which promotes a significant scientific and social contribution, as exemplified by Uébe-Mansur et al. (2019) and Uébe-Mansur et al. (2016). Table 2 presents the MPET research areas and their correlated research projects:

Table 2. MPET's research area and its research projects

Research Area	Research Project
Production and Experimentation of Technologies It aims to research, test, adapt and validate technologies with the aim to improve students' learning processes aligned with pedagogical approaches.	**Learning through Portable Devices** Researching how portable technologies can positively impact in education (m-learning), promoting ubiquity and more effectiveness in the learning process
	Digital Immersion in the Educational Context Promoting new learning environments through simulators and augmented reality
Teaching Training and Technological Resources It aims to research how technological resources can scaffold teacher training and how to promote a dialogical relation between teachers and educational technologies.	**Social Media as a Teaching Strategy** Researching social media adoption by the teachers as a way for collaboratively creating and sharing pedagogical knowledge to the students.
	Technological Resources for Distance Education Promoting research related to the Virtual Learning Environments for on-line teaching.

Vocational Programme in Teaching of Science (PROPEC)

PROPEC's goal is vocational postgraduate master and doctorate in Science Education prepared to flexibly deal with various subjects within Science. The educational product is the main objective of the professional modality, but in PROPEC it must be complemented by a dissertation or thesis that describes the educational products' development, as a means of guaranteeing quality and rigour.

So far (May 2021), PROPEC has had 140 master dissertations approved, 29 students enrolled in the professional master programme and 27 in the professional doctorate programme. When authorised by the CAPES (Brazilian postgraduate quality-regulatory bureau) in 2018, PROPEC was the first doctorate professional programme in the Science Education field of all Brazilian´s federal institutes and one of the four in the entire Brazil.

The PROPEC's audience is professionals holding a graduate/postgraduate diploma (recognised by the Brazilian MEC/CAPES) who work in formal education (elementary, secondary or higher education) or non-formal education. The PROPEC research projects must refer to one of the three different research areas, composed of six research projects (table 3).

Table 3. PROPEC's research area and its research projects

Research Area	Research Project
Science Teacher Education Discusses the initial and continuing education of science teachers, based on curriculum and historical analysis as well as the processes of dissemination and popularisation of science involved in this process of teacher education Master/Doctorate programme	**Scientific Literacy Science Teachers Continuing Education** It investigates the presence of science education developed in formal and non-formal spaces, especially Science Centres and Museums, dedicating itself to the processes of elaboration and literacy of scientific knowledge, in addition to evaluating the process of popularising science during initial and continuing education of science teachers.
	Science Teacher Education It investigates the science teaching actions developed in formal teaching spaces, analysing aspects related to the initial and continuing education of science teachers.
Teaching and Learning Processes and Production of Science Education Material It aims to promote studies in science education in all its aspects, addressing the pedagogical practices developed in the classroom and the production of didactic material, among other elements. Master/Doctorate programme	**Health and Environmental Education for Teaching and Learning Processes** Evaluates science teachers' interventions in the curricular guidelines, associating aspects of the environment and health, in addition to reflecting on the historical and sociological elements of the construction of scientific knowledge, with respect to the themes aforementioned.
	Natural Sciences Education for Teaching and Learning Processes Discusses the physics and chemistry interventions, complementing and deepening knowledge acquired during graduation, with a view to assessing the use of computer, multimedia, and video resources, in addition to the development of teaching materials and experiments.
Reflections on Contemporary Science Education It reflects the current challenges of Science Education, discussing both the theorization and application in Basic Education and Higher Education, as well as the contents that circumscribe it: the teaching work; new frontiers and development of research and theories; the use of new procedures and technologies; the new pedagogical practices; public policies; the elements about inclusion and exclusion; inter and transdisciplinarity; sociological, philosophical and epistemological challenges. Doctorate programme exclusively	**Science Education Innovative Aspects** It investigates the most recent and potentially innovative aspects of Science Education. It looks for cutting-edge content carried out in the area of Sciences with the intention of reflecting them in Science Education and Learning, as well as in relation to public policies.
	Science History, Philosophy and Sociology It investigates the relations between Human Sciences and Philosophy, and the Natural Sciences area, mainly in the educational aspect, but also in relation to the interdisciplinary content and its epistemological implications. It considers both the historical process of this relationship and the challenges of contemporary times.

METHODOLOGICAL DEVELOPMENT

The COVID-19 pandemic is affecting Brazilian learning and teaching in different ways. For this reason, it is necessary to establish boundaries on the research hereby presented, since it is not possible to cover all of the aspects in this chapter. An exploratory and pilot study was set, using phenomenon investigation as a methodological approach, in order to serve as a trial unit on a small scale for methodology experimentation (Swedberg, 2020). This approach intends to establish the research design and for that purpose, a small sample that allows the investigators to define the research problem and formulate hypotheses more accurately is recommended. Considering these preliminary research aspects the postgraduate students from MPET master and PROPEC doctorate programmes were chosen as a study group.

The research brings the following question: "How did the COVID-19 outburst impact on different dimensions of the students' lives enrolled in these master and doctorate programmes?". Based on the hypothesis that the impacts caused by COVID-19 are still preliminary, only initial pandemic aspects and effects in the students' lives are studied. In such a way, Swedberg's (2020) brings the perspective that a phenomenon can be better understood when the researchers are connected with the context of their research focus, and thus analysing it from an integrated point of view, the aforementioned research approach seems to be the logical way.

Two students (from IFF and IFRJ) narrated their impressions and situations concerning the impact of COVID-19 on their research for the case study as part of the methodological step. The first student is nicknamed *IFF_student* and is 32 years old. This master's student has a bachelor in Vernacular Letters and Pedagogy, from Federal University of Bahia and has been operating in education since 2009, at the levels of Basic Education and Secondary Education. The student's research started in 2019, and it is associated with the MPET's research line Teaching Training and Technological Resources. The second student is nicknamed *Student-IFRJ* and is 30 years old. This doctorate student has a degree in Mathematics since 2014 and has been working in the education area since 2014, also at the levels of Basic Education and Secondary Education. This student's research also started in 2019, and it is associated with the PROPEC's research line: Reflections on Science Teaching in Contemporary Times.

Despite the fact that IFF and IFRJ are independent institutes, there are several characteristics in common between them since they are from the same network. It will allow a comparative analysis or even the grouping of data. For this reason, students from both institutions (IFF and IFRJ) participated in the same anonymous survey to map the COVID-19 effects in their lives. The survey sought to verify how the master and doctorate students from MPET and PROPEC have driven their research projects on the new disruptive educational scenario. To this end, a three sections Google Form questionnaire was structured: (a) Questions about student's general data - containing three closed questions and four open questions; (b) Questions about the progress of the research and teaching process - containing four closed questions, and (c) Questions about the progress of the educational products' development - containing five closed questions.

Two methods for dataset analysis are being adopted as tools: Numerical analysis and Bricolage to related research works. The interactions between the researchers' expertise combined with the understanding of the results analysed in the light of the chosen theoretical-methodological framework allow the development of a better picture of the participants' message in the research.

The Bricolage method can allow the understanding of the speeches and the constructions of the participants of the study, questioning the essential social agents without reproducing what is imposed by hegemonic discourses (Sharp, 2019, Denzin, 1994).

Among other objectives, Bricolage seeks to analyse research subjects' responses to a given investigation (Sharp, 2019). Bricolage is a tool for qualitative data analysis that reckon with the researcher's non-neutral interpretation based on theoretical references on the collected data. For this, Bricolage is a non-traditional path that allows researchers confidence for choosing the appropriate research methods.

The two followed chapter's sections shown the results according to the Bricolage analysis methodology. In the section below, the present section is related to two MPET and PROPEC students full narratives in a free translation from Portuguese. In the followed section, the authors adopt the Bricolage to connect the numerical data from the survey (graphs), students' narratives and related researches from other authors.

NARRATIVES FROM THE MPET AND PROPEC STUDENTS' EXPERIENCES

The authors decided to bring the whole students' narratives to allow future researchers to explore and compare the students' perceptions. Furthermore, it will aid the reader's better understanding of the Bricolage.

MPET Student's Narrative

"Despite the optimistic outlook and focus on treating challenges as opportunities in my academic life, a pandemic has also brought mourning for lost loved ones and the resilience of seeing family and friends lose their health, increasing the statistics of people affected by the virus. In addition, the stress of 12 months of social isolation and, also the distressing expectation for the vaccine.

The pandemic contributed to my MPET research autonomy and my participation in extension events like Webinars, Lectures, and more.

Despite these good aspects, I need to make adjustments in my underway research project since I could not foresee the adaptations demanded in the context of social isolation. In spite of these aspects, the adjustments were reasonable and relatively simple to implement since it was an online application project for a pedagogical intervention (PI), through a didactic sequence on teacher training firstly expected to occur on-site. In the beginning of the project, I had no clear idea that I could adopt blended approaches. I just expected normal life back and on-site classes, which did not happen.

For the moment, a persistent pandemic and the government guidelines keep preventing on-site classes. Thus, in addition, to simply adapt the main means of communication for the development of PI, it was necessary to reflect on the professional experience acquired in the pandemic. We feel the need to reflect on the entire project, especially on the methodology and the course planning, in line with the current context. To develop an online experience that was previously planned for an on-site setting is not an easy task. I need to rethink the project foundation, mainly the delineation of the educational product that came to be a proposal for teacher training, based on Edgar Morin's Complex Thinking, personalised teaching and the pedagogical use of adaptive virtual learning environments. The course syllabus was re-thought, highlighting the evidence and experiments concerning the teaching practices on Emergency Distance Learning (EDL) for pandemic situations. I adjusted the proposals for activities and teachers reception, aiming to increase the exchange, and the dialogue.

Furthermore, if I were to keep the original project schedule, I would be very insensitive to the overburdened teachers involved in the project from 2021. These teachers were conducting other teacher training classes at their institutions and were overwhelmed by the parents' and institutional manager expectations related to the new pedagogical practices' success compared to 2020. Furthermore, I felt in tune with the overwhelming stress that the whole pandemic situation entailed by realising the teachers' exhaustion due to the questionable hours/classes via web conferencing platforms (contradicting the World Health Organization on face-to-face classes guidelines).

Concerning the project adjustment, a professional aspect contributed to the need for this adjustment: during the applied research in a specific previously chosen school, other schools contacted me asking for teacher training and partnerships focused on preparing its teachers for distance classes because of the pandemic.

Within this context, a professional and academic movement proposing continuous education for teacher training became apparent beyond the education level considered in the research project (Elementary and Middle School). Higher Education institutions began to demand teacher training, because of the greater complexity related to the classes' framework as well as the number of teachers. Examples of subjects related to the teacher training are the pedagogical use of Technologies for Education and/or Active Methodologies in the context of the EDL. To meet these institutional demands, I established a partnership with a program colleague making it possible to adapt the educational product to short training classes for public schools.

All these opportunities helped me to achieve class planning details about teachers' needs and weaknesses, helpful as a reference for my final prototype of educational product and of my professional life".

PROPEC Student's Narrative

"The current COVID-19 pandemic scenario affected my personal, professional, and academic life since it has led institutions to an academic calendar suspension and the professional reorganisation of the entire scholar community. For me, it is difficult to face the reality of seeing so many lives taken by the virus. It became even more painful when I saw family members of my own being affected by the disease, leading to the death of some of them. Similarly, some social aspects are difficult to live with: negationist behaviour from some individuals in society, scientific discredit, the anti-vaccine movement, the spread of fake news defending the idea that the scenario is alarming, and more.

In my professional life, the impacts were also significant. In January of 2019, I was hired as a teacher and began teaching in a high school course in the Youth and Adult Education (YAE), a modality which was already a challenge in itself. So, the pandemic caused an enormous setback in the first months. Recognising my students' characteristics and their struggle for access and permanence at school was another personal challenge. I started my readings and took part in online live sessions to learn better practices from other teachers with YAE, to create my path at YAE's online math classes. I understood the importance of acting with empathy during those difficult times.

From this new personal and professional pandemic life reality, I rebuilt the YAE specifications and the social and life context reflected in my PhD research. By chance, compelled to combine my professor's actions with my researcher's skills, I could detail how my reality changed:

- The data collection for my research, previously thought to be at face-to-face classes, were now established for remote scenarios;
- The theoretical framework previously discussed was Mathematics Education and Digital Technologies for future math teachers, now we chose digital mathematical performance (Scucuglia, 2012), addressing aspects of the Arts and Critical Mathematical Education for YAE;
- The educational product design was formerly a continuing education course for math teachers. By now, the YAE students are producing digital math videos;
- The educational product validation and evaluation, whose previously planned actions were to be carried out by math teachers with theoretical and practical training on the use of technologies in the teaching of mathematics, are now are being conducted by specialists in YAE, by the YAE students, by YAE's math teachers and also math digital videos' judges at online festivals.

The previous list reflects how the SARS-COV-2 pandemic drove my research adjustments. All of them in permanent dialogue with my advisor, causing us to change approaches, review research protagonists, and face the reality of research in an atypical context. All of these factors brought upon us lots of pain, mourning, and struggle. I could also realise an intrinsic struggle of my research audience. My YAE students suffered much more inequalities and incompatibilities, such as the low-quality internet access, the lack of equipment for online education and the lack of technical knowledge (digital literacy). On top of those, lack of home facilities and other difficulties cannot be undervalued. As the teacher of these courses, my professional life also changed. The reality added to my studies, reinforcing my humanistic side. I was disgusted when I saw my YAE students taking online classes while returning from work on crowded buses or trains, watching from home, but preparing dinner simultaneously, taking care of family members, among other daily life demands. All of it disgusted me a lot.

Despite all the difficulties and pains, my research has advanced both in theoretical and practical terms. In a personal aspect, I got prepared to participate in webinars and had more time for my readings to develop the academic writing expected from a PhD student.

Concerning data collection and educational product development, I created videos relating social and scientific knowledge from reality to mathematics and COVID-19 pandemic reality with an exponential function.

The pandemic also affected the Mathematics Education congresses switching them to online events. As a result, I submitted some of my videos to a validation instance. Thus, we (YAE students included) produced short courses based on the theme presenting it in online congresses and in Mathematics Education festivals (GPIMEM UNESP, 2020; Monitoria de Matemática IFRJ-CDUC, 2021). With the replicability of the knowledge from the research in mind, a three-book collection called "Paths of Mathematics at YAE'' is in the press, which will hopefully allow other teachers to reflect and adapt their practice in accordance with the findings that emerged from my research project, providing advances at theoretical and practical terms in the discussions with YAE".

EXPLORING COVID-19 IMPACTS ON RESEARCH PROJECTS OF MPET AND PROPEC

In the next section, the authors adopt the Bricolage to connect the numerical data from the survey (graphs), students' narratives and related researches from other authors.

Although the overall response rates of the survey (39 respondents) were satisfactory, MPET students were better involved (61%) when compared to PROPEC students (33%). Some PROPEC students justified the low feedback by their overloading from plenty of surveys in their emails regarding several types of other research projects. Table 4 below presents the students' feedback numbers and percentages for the survey.

Table 4. Feedback from enrolled students at IFF and IFRJ postgraduate programmes

Aspects	MPET-IFF	PROPEC-IFRJ	Total by category
MSc enrolments	22	29	51
MSc Answers (% - answers/enrolments)	14 (64%)	17 (59%)	31 (61%)
PhD enrolments	N/A	27	27
PhD Answers (% - answers/enrolments)	N/A	9 (33%)	9 (33%)
MSc+PhD enrolments by Institution	22	56	78
MSc+PhD answers by Institution (% - answers/enrolments)	13 (59%)	26 (46%)	39 (50%)

Due to the qualitative and exploratory research characteristics of this research project, the questionnaire addressed to the students was not composed exclusively of closed answers. Some questions like "Positive and negative questions related to the students' graduate studies", allowed open answers, requiring the use of the bricolage method to scaffold the data analysis. After several readings, the authors were impregnated with the words of the research participants, making it possible to identify common patterns among the speeches of students from both institutions. The communicative interaction found in the written data sources (survey and students' narratives), allied to the researchers' knowledge and professional experience, made it possible to find satisfactory results from the qualitative dataset. The interactions between the researchers' expertise combined with the understanding of the results analysed on the methodology allow a better picture of the participants' message to the research (Oliveira, 2021; Teixeira, 2021).

Similar results were obtained by reading exhaustively the cases shared by the Master and PhD students. They had mixed feelings, but they found a way to redesign their projects and reconfigure their research. "Mourning" was a word that appeared in both discourses and actions.

Despite the optimistic outlook and focus on treating challenges as opportunities in my academic life, the pandemic has also brought mourning for lost loved ones and the resilience of seeing family and friends lose their health, increasing the statistics of people affected by the virus. (Student-IFF).

After the changes listed above,it can be noticed that the SARS-COV-2 pandemic made me review my research entirely, always in permanent dialogue with my advisor, causing us to change course, review research protagonists, and face the reality of research in an atypical context, which brought upon us lots of pain, mourning, and struggle. (Student-IFRJ).

The exhaustive reading of the analysed speeches allowed the authors to identify 114 expressions included in a word cloud generator named Jasondavies (2021), ensuring that the research follows the main discourse ideas. Figure 1 below shows it:

Figure 1. Word cloud with expressions highlighted by the students

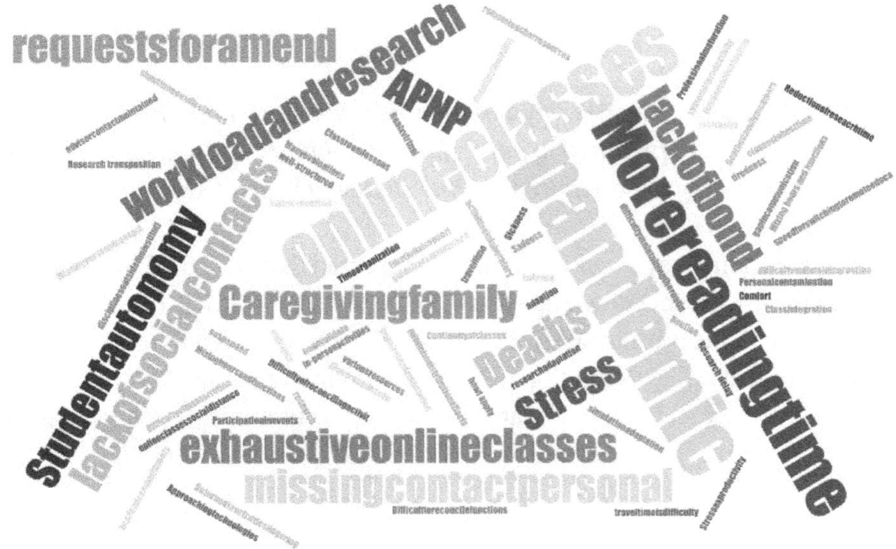

Thirty-nine students from both institutes answered the survey. Twenty-eight (71%) enrolled in 2019. In other words, most of them got their program enrolment before the COVID-19 pandemic. This result shows that contrary to the first general belief, the COVID-19 pandemic did not present an exclusively negative context for them. Research participants were able to identify favourable points, especially about their academic life, such as increased autonomy of them as researchers in training, more investment of time in reading theoretical and methodological references, participation in worldwide congress and colloquia, shorter travel time, the possibility of carrying out simultaneous subjects, in addition to approximation with technologies and other teaching methodologies. The context of positive students' perceptions concerning the COVID-19 pandemic is not a finding exclusive to this research study and can be found in other related studies around the world as reported by Benhima and Ben (2021), Tawafak et al. (2021) and Wang et al. (2020). Some statements from the questionnaire open questions where students are motivated to share their positive and negative perceptions concerning their research experience in the pandemic scenario can better explain the points highlighted:

Online classes - the chance of participating in the classrooms at the best time, since everyone's routine has changed a lot. The possibility of participating in many studies through collaborative platforms. (Student 1)

Our classmates got surprised by the COVID-19 pandemic as we were in the middle of the course. The program was fast in deciding to continue with the activities, and the teachers adapted their classes, giving us full conditions to continue. The attention of the advisor was a very positive point of the program. (Student 6)

I was able to transpose the field research data collection to the virtual mode, which provided me with a reasonable mastery of digital technologies that I did not have before to mediate the teaching or the application of courses and activities in online education. The better integration between classmates, since the distance tasks created new extra classes virtual moments. We were able to carry out doctoral and professional demands besides helping each other with emotional support. My time to remotely meet the demands of the doctorate and remote work got more organised and optimised. I live far from the institute, and the travel time from my residence in the South of the Rio de Janeiro to Nilópolis consumes a considerable amount of time. I was physically exhausted due to traffic, distance, and unforeseen events on the road and mountains. (Student 7)

Despite all the difficulties and struggles, my research has advanced both in theoretical and practical terms. In a personal aspect, I got prepared to participate in webinars and had more time for my readings to develop the academic writing expected from a PhD student. (Student-IFRJ)

In their research, Bourion-Bédès et al. (2021a, p. 4) reported that "[...] 61% of students in the sample experienced anxiety during the lockdown due to the COVID-19 outbreak". The authors also found evidence of student anxiety in the present study. For instance, they noticed dubious feeling and evaluation by the IFF and IFRJ students. They found that IFF and IFRJ migrated quickly to this type of teaching, being important to keep their studies up to date, in addition to minimising the lack and/or absence of social contact.

However, the delay in returning to face-to-face activities, the non-adaptation of the evaluation processes, and the increase in screen time transformed online classes into an exhausting experience. Some comments illustrate it. Furthermore, some IFF and IFRJ students live far from their on-site learning places and may take over three hours in traffic to go and return to classes. Some excerpts can confirm their statements:

With the online classes and virtual work demands, we can act quickly and simultaneously in different situations, but the problem is that the functions and schedules are mixed and not matched. (Student 4)

The continued offer of online classes has minimised our discipline content losses. However, there was a lack of better organization in the use of asynchronous and synchronous time. I believe that asynchronous tasks can stimulate student autonomy, such as reading articles and other materials for analysis). Synchronous classes should be reserved for professors' presentations of discipline concepts and development of students' academic maturity. (Student 8)

The possibility of taking the online classes was good, but the evaluations were not fully adapted to the new study routine. Some professors find it difficult to request a unique evaluation or even shorter one compatible with the time of the course. (Student 11)

As the teacher of these courses, my professional life also changed. The reality added by my studies reinforcing my humanistic side. I was disgusted when I saw my YAE students taking online classes while returning from work on crowded buses or trains, watching from home, but preparing dinner simultaneously, taking care of family members, among other life demands. All of it disgusted me a lot.(Student-IFRJ)

Within the scope of the negative aspects, domestic and emotional overload were the most highlighted points. Caretaking of family members, absence of social contact, stress, death of acquaintances or relatives, sadness, and depression were reported. This was also pointed out in the studies of Faisal *et al.* (2021) and Teotônio *et al.* (2020). The academic field was also impacted by the interruption of practical research activities, or even the lack of face-to-face data collection, the need to adapt and/or stop research.

In what regarded the 39 answers on academic life, thirty-three (85%) claimed to have suffered personal and professional impacts in their life. Twenty-five students (63%) claimed that it was necessary to make partial adjustments to their research, nine (22%) claimed to have made overall adjustments and only six (15%) students claimed that they did not need to adjust their research in any aspect.

A master's and doctorate research usually goes through two phases: project research and final research approval. In this respect, the effects were more evident to the MPET students, since 12/13 of them demanded adjustments to their projects already approved due to the impacts of COVID-19. Regarding PROPEC students, 5/21 had their projects approved, making the necessary adjustments easier for the majority. Another point to highlight is that none of the students took a leave of absence, despite all the difficulties. All of them also indicated that their academic life was impacted in some way, and the authors can infer it by zero responses at the "not impacted" option. The graph in Figure 2 shows in detail which kind of multiple-answer academic impact related to the 34 (75%) students that demanded research adjustments.

Figure 2. COVID-19 academic impacts on students' research

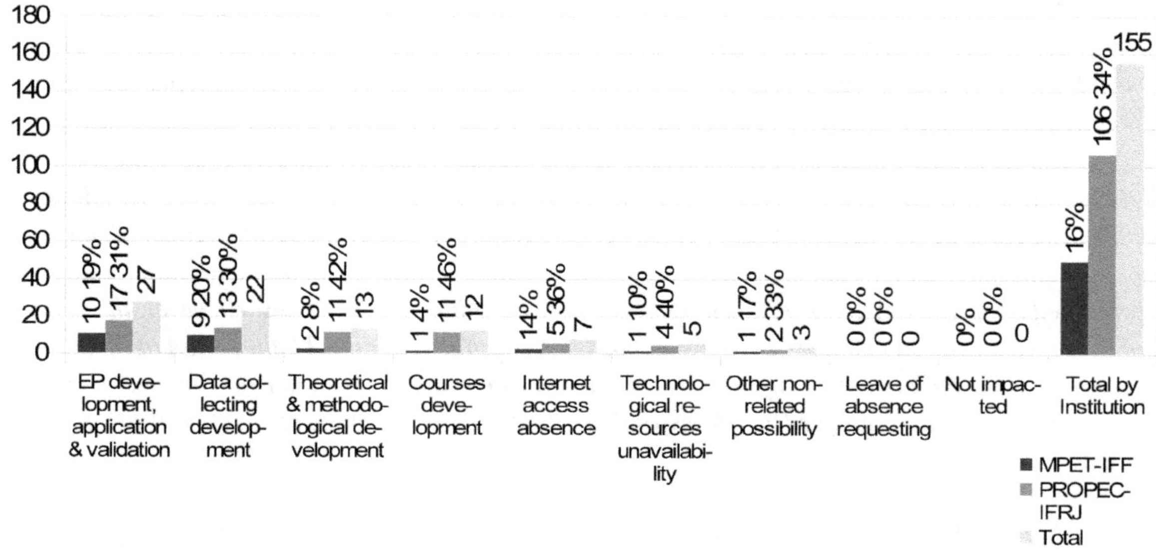

Some students feel anger for continuing their studies and research while many deaths are occurring. Sometimes they criticised a possible excess of demand for productivity by the funding agencies. In the (Bourion-Bédès et al., 2021b) study, the authors report the same kind of students' stress feelings. Some excerpts from their speeches clearly show these anger feelings:

Nevertheless, with the spread of the virus and people's illness close to me, the difficulties to access online classes and the time dedicated to the master's degree increased. Physical and emotional tiredness increased. In addition to the low quality of the internet in the house. (Student 18)

The single positive situation is the absence of the need to travel to City-C, but there are several negative situations. The demands have tripled: home service, children homecare, and online education activities consume all the time available, reducing my research time. (Student 26)

Asymmetrical increase in the demands of all professional and personal levels concerning the imposition of deadlines without respecting a coherent time for carrying out a certain task. In so many cases, there was no understanding that daily time is not infinite. (Student 12)

My project has stages based on face-to-face classroom, by carrying out lab activities. With the classes' suspension by the lockdown, my research implementation is being hampered severely. (Student 36)

Research studies conducted in Brazil like Amorim and Costa (2020) and Fiocruz (2021) as also researchers from other countries like Khan (2021) and Rhoden (2021) seek to understand how the COVID-19 pandemic impacted master and doctoral research and show a similar panorama. In the survey carried out together graduated students form Fiocruz institution, more than 70% of the students had their research projects damaged (slightly or severely) or even did not start the research, due to the pandemic. An investigation conducted by Amorim and Costa (2020) together five master's students from a Brazilian master's program, identified that the pandemic changed the formative process of all of them in relation to time management, production of their research and educational products. Rhcden (2021), on the other hand, narrates the anxieties of being a doctoral student in a scenario of uncertainty, anxiety and confusion. The author also highlights that the delay in accepting the drastic change in the world caused by the pandemic led him to a period of blocking in relation to his research. Khan (2021) reports the difficulties of being resilient in difficult times, away from family members, with a shortage of jobs, all in the midst of his doctoral research.

Concerning the second questionnaire section: "Questions about the progress of the research and teaching process", the first question is related to the adjustment status of the research project. Nine (23%) of the thirty-nine answers declared that they needed to do entire project adjustments, Twenty-four (62%) partial project adjustments, and only six (15%) no need for project adjustments. The next question allows multiple choice answers and it is related to the kind of research project adjustment made by the students. Figure 3 shows the results.

Figure 3. Kind of research project readjustments by institution.

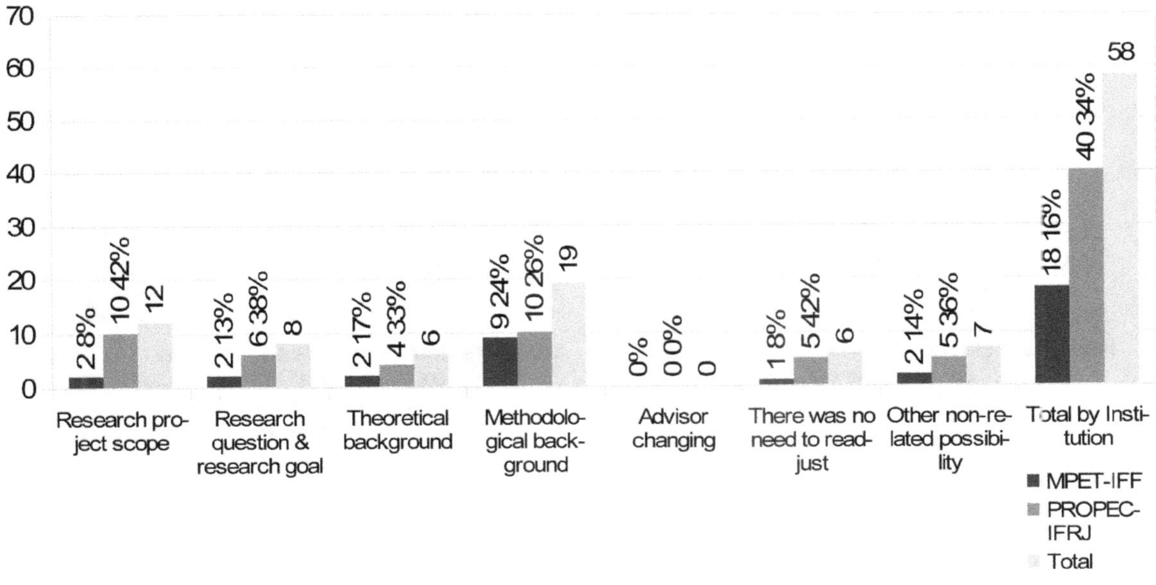

Analysing the MPET's student narrative below, where the student calls our attention to empathy that is necessary with their research's participants:

Furthermore, if I kept the project schedule, I would be very insensitive to the overburdened teacher involved in the project since 2021. These teachers were conducting other teacher training classes at their institutions and were overwhelmed by the parents' and institutional manager expectations related to the new pedagogical practices' success compared to 2020. Furthermore, I got convinced of the necessity of the changes, realising the teachers' exhaustion due to the questionable hours/classes via web conferencing platforms (contradicting the World Health Organization face-to-face classes guidelines (Student-IFF).

Despite the impact of project adjustments, no students declared advisor changing as a direct impact from the pandemic situation, which meant a one less factor for students' stress.

Educational Product (EP) is an obligatory item for Brazilian professional postgraduate courses (Moreira et al., 2016). For this reason, it is necessary to understand how they are affected in their different phases of proposal, development, application, evaluation, and validation. For this reason, the research study sought to obtain information about some characteristics of the educational products developed in the two programs, such as EP's school grade application (Figure 4), EP's learning modality application (Figure 5) and kind of EP developed (Figure 6).

The results in Figure 4 show that the "High School" is the most voted answer, followed by research activities conducted in spaces other than the school for both institutions.

Figure 4. EP's school level application

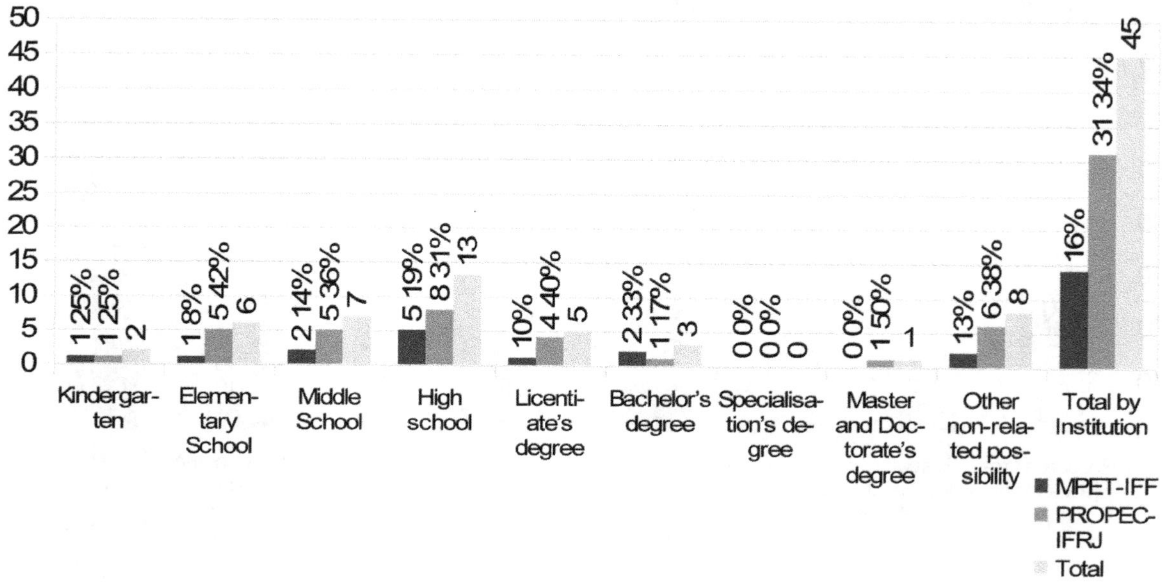

Regarding the teaching modality, Figure 5 shows that classroom teaching is the most common, although the authors found highlights for blended and online classes (also in Figure 5). More specific modalities such as YAE, indigenous and rural were less significant cases and mostly related to the geographic location of each of the federal institutes evaluated in this exploratory study. The authors could not identify EP's application for the Quilombola school education modality. Quilombola school education is a pedagogy focused on Afro-educational concepts of culture, heritage, and lands from black Brazilian slaves' descendants (Monteiro & Reis, 2019).

Figure 5. EP's learning modality application

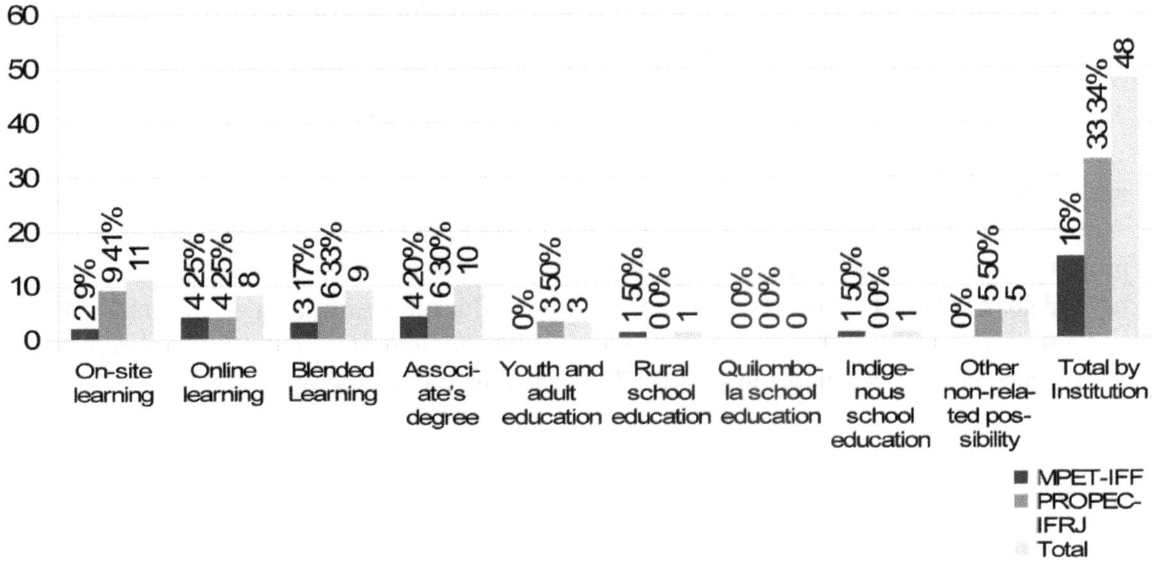

Figure 6. EP's classification

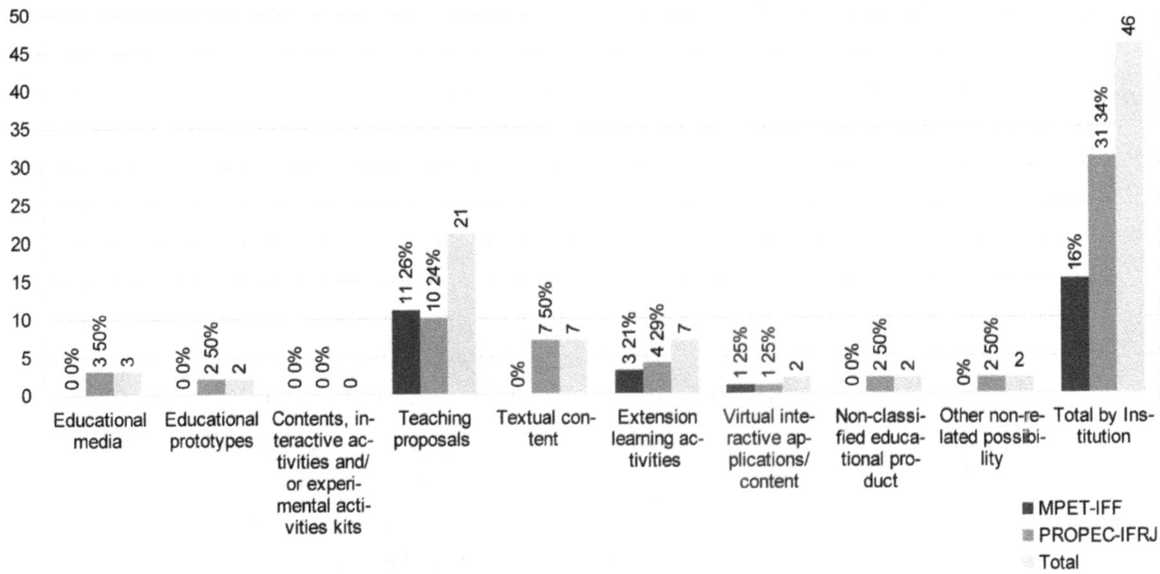

The data shown in Figure 6 refers to a pattern already observed in postgraduate programs in the professional modality (Moreira et al., 2016) that can be: a) Educational media (documentaries, videos, music and others); b) Educational prototypes (maps, board games and others); c) Contents, interactive activities or experimental activities kits; d) Teaching proposals (didactic sequences and others); e) Textual content (didactic and other books); f) Extension learning activities (courses, workshops, exhibitions and others); g) Development of virtual interactive applications/contents (podcast, digital games and others); h) Non-classified educational product; i) Other non-related possibilities. The hegemony of textual knowledge, reinforced by the culture of the textbook, is still imperative. The technical-extension pedagogical activities still appears, though discreetly, as a result of efforts made by the related programs into social insertion actions. Perhaps this trend can be broken in a post-pandemic moment.

The answers regarding the third questionnaire section: "Questions about the progress of the educational products development" is shown in Figure 7 and indicated the same behaviour concerning EP adjustments at both institutions. Thirty-nine students answered the question, and 9% had to make further changes in their projects, the majority (64%) had to adapt their research partially, and a few students (13%) declared that they did not need to adjust the research project. The 13% result is surprising when compared to the results shown in Figure 2, in which no students voted in the "not impacted" option, meaning that some change happened. This is detectable in the MPET's student narrative excerpt:

To meet these institutional demands, I made a partnership with a program colleague making it possible to adapt the educational product in short training classes for public schools.

All these opportunities helped me to achieve class planning details about teachers' needs and weaknesses, which was helpful as a reference for my final prototype of educational product and professional life. (Student-IFF)

Considering that most projects have some level of adjustment, the authors focused on identifying which stage was most affected. Figure 7 presents the EP's typology adjustments considering multiple-answers:

Figure 7. Educational product typology readjustments by institution

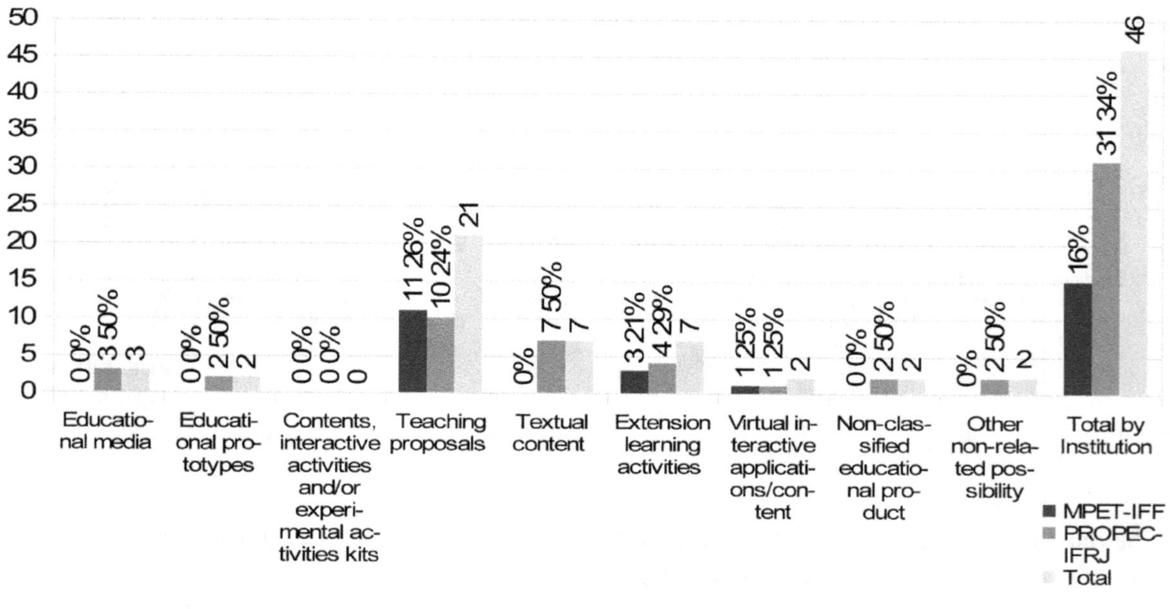

According to the students, the EP application and methodological background were the most affected. Such a reality is illustrated by the excerpt below:

The educational product validation and evaluation stage, consisting of previously planned actions to be carried out by math teachers with theoretical and practical training on the use of technologies in the teaching of mathematics, are now being conducted by specialists in YAE, by the YAE students, by YAE's math teachers and also math digital videos' judges at online festivals. (Student-IFRJ)

These results are consistent with the data presented in Figure 3, where methodological project background followed by project scope and research question & goals were subsequently the project's components suffering more adjustments.

CONSIDERATIONS

Other studies referred on the present research indicate that the disruptive pandemic scenario officially declared by WHO (2020) around the world is breaking down all pedagogical and academic *status quo* that exists for a long time, for example, on-site meetings, on-site classrooms, content-centred on-site assessments, pedagogical approaches based on expositive content, travels for lectures and meetings, and paper-based documentation. In other words, the comfort zone about how teachers teach, students learn, and staff work no longer exists. Despite the first impact, and also how hard the adaptations are in a new scenario, new ways for teaching, learning and establishing relationships are rising.

Regarding the research question presented at the beginning of the text, "*How did the COVID-19 outburst impact on the different dimensions of the students´ lives enrolled in master and doctorate programmes?*", the authors are able to identify that the personal and professional spheres have been more affected in students' lives. Important points of restructuring in academic life were highlighted, however, we can assume the impact was more evident in the professional life domain.

Data from the present research study shows that both IFF's and IFRJ's students had their studies impacted in similar ways. As expected, students enrolled in 2020 were frightened by the pandemic and its impact on their research projects. However, they were at the beginning of their studies, making it easier to change their projects. On the other hand, many students with approved research projects (enrolled in 2019 or before), who were already in the programmes, suffered the impact more abruptly, implicating underway project changes, including preliminary data collected.

Despite the abrupt changes in the 2018 and 2019 MPET cohort and the adjustments described in their research projects and educational products, students adapted their research activities adequate to the new scenario, which, for example, is reflected on the fact that only 2/15 students from the 2018 cohort went beyond 2020 for their master's final defence.

For this reason the questionnaires aimed at the students could be satisfactorily adapted, except for the personal and professional stress caused by the pandemic scenario. Other positive aspects are that once they overcame the first challenges, the teachers are being more confident in adopting digital technologies and active methodologies in their classrooms, students are overcoming the first negative impacts from full-time online classes and assessments and the institutions are finding new ways for staff work interactions and online management processes.

About the proposed methodology, the numerical analysis and bricolage approach seems to handle the aim for a specific and qualitative overview, and also general and quantitative data analysis.

The research study was limited to comparing two programs of two institutions, because of its exploratory nature, but it allowed the collection of data in an unprecedented way, within the network of institutes context. Despite it, the present study brings a lot of factors related to the pandemic Brazilian situation and the difficulties that students have had to overcome in their professional and personal life that are not related in similar studies related to the pandemic.

FUTURE RESEARCH DIRECTIONS

The researchers in the world are concerned with the impacts on students and teachers lives over time elapsed on the pandemic scenario. Especially in the Brazilian pandemic scenario, there is a lack of this related information not because the researchers are not publishing but because of the Brazilian territorial and cultural diversity.

Authors can realise that the COVID-19 impacts seem to be more than a temporary changer in social behaviour, guessing that this pandemic is profoundly changing the social behaviour of the 21st century. In some way, it is possible to say that the 21st century officially started after 11th March 2020, when the World Health Organisation declared the pandemic. For this reason, the pandemic means a social revolution that is boosting educational changes that would only occur five for ten years from now.

Examples of these changes are the ubiquity for students since the virtual meetings are breaking the boundaries of the physical school environment as the boosting of teachers training for the student-centred learning methodologies (more fittable for the blended learning that is being the most usual teaching method in this scenario). The pandemic and its social isolation imposed new educational practices that seem to stay since after the pandemic ends. In some way, it means schools and teachers changes that the students and the society asking for a long time.

Authors expect that the scenario described can be helpful for the readers, teachers and next student generations as a shareable experience related to the academic and pedagogical strategic decisions.

Authors expect that the scenario described can be helpful for the readers, teachers and next student generations as a shareable experience related to the academic and pedagogical strategic decisions.

REFERENCES

Aguiar, L. E. V., & Pacheco, E. M. (2017). Os Institutos Federais de Educação, Ciência e Tecnologia como política pública [The Federal Institutes of Education, Science and Technology as a public policy]. In *As políticas públicas e o papel social dos Institutos Federais de Educação, Ciência e Tecnologia* (Vol. 1). Natal: Editora IFRN.

Amorim, D. C., & Costa, C. J. A. (2020). Impactos da pandemia Covid-19 no processo formativo de professores de Biologia de um mestrado profissional: Desafios em tempos de quarentena [Impacts of the Covid-19 pandemic on the training process of Biology teachers in a professional master's degree: challenges in times of quarantine]. *Devir Educação*, *2*(4), 80–103. doi:10.30905/ded.v4i2.271

Anjos, M. A., Deccache-Maia, E., & Bonfim, A. M. (2013). Os desafios da construção de um Mestrado Profissional: Um panorama dos sete anos do PROPEC [The challenges of building a Professional Master's Degree: an overview of the seven years of PROPEC]. *Polyphonía, 24*(2), 319–337. doi:10.5216/rp.v24i2.37941

Benhima, M., & Ben, S. M. (2021). Moroccan English Department Student Attitudes Towards the Use of Distance Education During COVID-19: Moulay Ismail University as a Case Study. *International Journal of Information and Communication Technology Education, 17*(3), 105–122. Advance online publication. doi:10.4018/IJICTE.20210701.oa7

Bourion-Bédès, S., Tarquinio, C., Batt, M., Tarquinio, P., Lebreuilly, R., Sorsana, C., Legrand, K., Rousseau, H., & Baumann, C. (2021a). Psychological impact of the COVID-19 outbreak on students in a French region severely affected by the disease: Results of the PIMS-CoV 19 study. *Psychiatry Research, 295*, 113559. doi:10.1016/j.psychres.2020.113559 PMID:33189368

Bourion-Bédès, S., Tarquinio, C., Batt, M., Tarquinio, P., Lebreuilly, R., Sorsana, C., & Baumann, C. (2021b). Stress and associated factors among French university students under the COVID-19 lockdown: The results of the PIMS-CoV 19 study. *Journal of Affective Disorders, 283*, 108–114. doi:10.1016/j.jad.2021.01.041 PMID:33540333

Braga, E. S. O., Silva, C. S., & Rôças, G. (2019) É Pau, É Pedra... É o Fim do Caminho ou Começo de um Novo Caminhar? Reflexão sobre os Institutos Federais na Rede Federal de Educação Profissional, Científica e Tecnológica [It's Stick, It's Stone... Is it the End of the Path or the Beginning of a New Walk? Reflection on the Federal Institutes in the Federal Network for Professional, Scientific and Technological Education]. In *Educação, cultura e sociedade*. IFPB. Retrieved from http://editora.ifpb.edu.br/index.php/ifpb/catalog/book/236

Denzin, N. K. (1994). Romancing the Text: The Qualitative Researcher-Writer-as-Bricoleur on JSTOR. *Bulletin of the Council for Research in Music Education, 1*(1), 15–30. https://www.jstor.org/stable/40318652?seq=1

Faisal, R. A., Jobe, M. C., Ahmed, O., & Sharker, T. (2021). Mental Health Status, Anxiety, and Depression Levels of Bangladeshi University Students During the COVID-19 Pandemic. *International Journal of Mental Health and Addiction*, •••, 1–16. doi:10.100711469-020-00458-y

Hodges, C., Moore, S., Lockee, B., Trust, T., & Bond, A. (2020). The difference between emergency remote teaching and online learning. *EDUCAUSE Review, 12*, •••. https://er.educause.edu/articles/2020/3/the-difference-between-emergency-remote-teaching-and-online-learning

Fundação Oswaldo Cruz. (2021). *Análise do impacto da Pandemia causada pelo SARS-Cov-2 sobre PPGBB* [Analysis of the impact of the pandemic caused by SARS-Cov-2 on PPGBB]. Retrieved from https://www.posgraduacao.icc.fiocruz.br/analise-do-impacto-da-pandemia-causada-pelo-sars-cov-2-sobre-o-ppgbb/

GPIMEM UNESP. (2020, May 31). *Mathnews*. https://youtu.be/sz_8EKAuh5Q

Hua, W., Xiaofeng, L., Zhenqiang, B., Jun, R., Ban, W., & Liming, L. (2020). Consideration on the strategies during epidemic stage changing from emergency response to continuous prevention and control. *Zhonghua Liu Xing Bing Xue Za Zhi, 41*(2), 297–300. doi:10.3760/cma.j.issn.0254-6450.2020.02.003 PMID:32088947

Instituto Federal do Rio de Janeiro. (2021, May 15). *Programa de Pós-graduação (Mestrado e Doutorado Profissional) em Ensino de Ciências* [Postgraduate Program (Master's and Professional Doctorate) in Science Teaching]. https://portal.ifrj.edu.br/cursos-pos-graduacao/stricto-sensu/propec

Jasondavies. (2021). Retrieved May 19, 2021 from https://www.jasondavies.com/wordcloud

Kanungo, S., Sharma, D., & Aggarwal, A. (2021). Impact and New Normal. *International Journal of Security and Privacy in Pervasive Computing, 13*(1), 57–68. Advance online publication. doi:10.4018/IJSPPC.2021010103

Khan, T. (2021). How COVID-19 has impacted my PhD research. *Journal of Health and Social Care Improvement, 3*(2), 4–9. Retrieved April 10th, 2021, from https://www.researchgate.net/publication/341727378_How_COVID-19_has_impacted_my_PhD_research

Lei 11892/2008. (2008). Retrieved April 28, 2021, from http://www.planalto.gov.br/ccivil_03/_ato2007-2010/2008/lei/l11892.htm

Monitoria de Matemática IFRJ-BCDUC [IFRJ-BCDUC Mathematics Monitoring]. (2021, March 26). *Behind the smoke screen.* https://www.youtube.com/watch?v=w9eL9WyynR0

Monteiro, E., & Reis, M. C. G. (2019). Afro-brazilian heritage in the context of quilombola school education. *Educação e Realidade, 44*(2). Advance online publication. doi:10.1590/2175-623688369

Moreira, M. C. A., Rôças, G., Anjos, M. B., & Pereira, M. V. (2016) Analysis of educational products in a professional science education master's degree course. *Proceedings of XVII International Organization for Science and Technology Education Symposium.*

Oliveira, A. F. (2021). *Museus, Centros de Ciências e Formação Inicial de professores: contribuições da prática de mediação realizada no Museu Ciência* [Museums, Science Centers and Initial Teacher Training: contributions from the mediation practice carried out at the Science Museum] (Dissertação de Mestrado). Programa de Pós-graduação em Ensino de Ciência, Instituto Federal de Educação, Ciência e Tecnologia do Rio de Janeiro, City-C, Rio de Janeiro, Brasil.

Plataforma Nilo Peçanha. (2021, September 20). Retrieved from http://plataformanilopecanha.mec.gov.br/

Scucuglia, R. (2012). *On the nature of students' digital mathematical performance: When elementary school students produce mathematical multimodal artistic narratives.* Alemanha: Verlag/LAP Lambert Academic Publishing.

Serafim, A. P., Durães, R. S. S., Rocca, C. C. A., Gonçalves, P. D., Saffi, F., Cappellozza, A., … Lotufo-Neto, F. (2021). Exploratory study on the psychological impact of COVID-19 on the general Brazilian population. *PLoS ONE, 16*(2). doi:10.1371/journal.pone.0245868

Sharp, H. (2019). Bricolage research in history education as a scholarly mixed-methods design - CORE Reader. *History Education Research Journal, 16*(1), 50–62.

Swedberg, R. (2020). Exploratory Research. In C. Elman, J. Gerring, & J. Mahoney (Eds.), The Production of Knowledge: Enhancing Progress in Social Science (pp. 17-41). Cambridge University Press. doi:10.1017/9781108762519.002

Tawafak, R. M., College, A. B., Sohail, O., Malik, I., Oman, G., Alfarsi, A. B., & College, O. (2021, July). Impact of Technologies During the COVID-19 Pandemic for Improving Behavioral Intention to Use E-Learning. *International Journal of Information and Communication Technology Education*, *17*(3), 137–150. Advance online publication. doi:10.4018/IJICTE.20210701.oa9

Teixeira, A. L. O. (2021). *"Estamira" e Educação Ambiental Crítica: inspirações na formação de professors* ["Estamira" and Critical Environmental Education: inspirations in teacher education] (Dissertação de Mestrado). Programa de Pós-graduação em Ensino de Ciência, Instituto Federal de Educação, Ciência e Tecnologia do Rio de Janeiro, City-C, Rio de Janeiro, Brasil.

Teotônio, I., Hecht, M., Castro, L. C., Gandolfi, L., Pratesi, R., Nakano, E. Y., Puppin Zandonadi, R., & Pratesi, C. B. (2020). Repercussion of COVID-19 pandemic on brazilians' quality of life: A nationwide cross-sectional study. *International Journal of Environmental Research and Public Health*, *17*(22), 1–11. doi:10.3390/ijerph17228554 PMID:33218087

Uébe-Mansur, A. F., Alves, A. C., & Torres, R. B. (2019). Trello as virtual learning environment and active learning organiser for PBL classes: An analysis under Bloom's Taxonomy. *International Symposium on Project Approaches in Engineering Education*, 9.

Uébe-Mansur, A. F., Leite, F. C., & Bastos, H. P. P. (2016). Lean education for applied science universities: A proposal by federal institutes of applied sciences in Brazil. In Lean Education: An Overview of Current Issues. doi:10.1007/978-3-319-45830-4_2

Usmani, R. S., Saeed, A., & Tayyab, M. (2021). Role of ICT for Community in Education During COVID-19. In N. Zaman, K. Rafique, & V. Ponnusamy (Eds.), *ICT Solutions for Improving Smart Communities in Asia* (pp. 125–150). IGI Global. doi:10.4018/978-1-7998-7114-9.ch006

Wang, Y., Jing, X., Han, W., Jing, Y., & Xu, L. (2020). Positive and negative affect of university and college students during COVID-19 outbreak: A network-based survey. *International Journal of Public Health*, *65*(8), 1437–1443. doi:10.100700038-020-01483-3 PMID:33063142

WHO. (2020, March 29). *WHO/Europe | Coronavirus disease (COVID-19) outbreak - WHO announces COVID-19 outbreak a pandemic*. Retrieved from https://www.euro.who.int/en/health-topics/health-emergencies/coronavirus-covid-19/news/news/2020/3/who-announces-covid-19-outbreak-a-pandemic

KEY TERMS AND DEFINITIONS

APNP: Brazilian acronym for distance pedagogical activities.

Educational Product: Part of the dissertation or thesis Scopus in the Brazilian vocational postgraduate programmes, referring to a Technical-applied product.

Federal Institute (FI): A unit of the Brazilian Federal Network of Vocational Education

Federal Network of Vocational Education (FNVE): Huge Brazilian government public network of vocational education.

IFF: Fluminense Federal Institute located in Campos dos Goytacazes municipality at Rio de Janeiro state.

IFRJ: Federal institute of Rio de Janeiro located in Nilópolis municipality at Rio de Janeiro state.

MPET: Vocational Master program from IFF.

PROPEC: Vocational Master and Doctorate programmes from IFRJ.

YAE: Young and Adult Education.

Chapter 15
Digital Competencies and Transformation in Higher Education:
Upskilling With Extension Actions

Cristine Martins Gomes de Gusmão
https://orcid.org/0000-0001-8831-217X
Federal University of Pernambuco, Brazil

ABSTRACT

Extension action promotes inclusive, equitable, and quality education, a goal of sustainable development that guides educational actions around the world. The development of digital skills is a differential, together with the encouragement of open educational practices. This chapter provides reflections, as well as lessons learned, through experience in a biomedical engineering course. Encouragement, through the development of actions that correlate important skills and abilities, is essential for professional development. The carrying out of teaching, research, and extension activities and actions contributed to promote the interdisciplinarity of the various fields of study, necessary for professional development in the digital age. Thus, the actions developed stimulated the investigation, improvement, and study of topics of interest related to education and health and technology areas related to the role of the biomedical engineer, the main protagonist of this project.

INTRODUCTION

Since March 2020, due to the Covid-19 pandemic, there is a need to address with greater emphasis on Flexible Open Distance Learning (OFDL). The challenge raised by the educational disruption is noticeable. The demand for open, flexible, and remote learning is a reality. Working and teaching online, open and at a distance is considered a milestone in the entire educational system, together with valuable experiences lived by a diversity of professionals.

DOI: 10.4018/978-1-7998-8816-1.ch015

Technological innovations are constantly presented, not only as proof of concepts, but as short, medium, and long-term economic and social benefits for society. Technological mediation in education, as well as innovation in instructional design, is essential in the context of physical distance. The digital transformation encourages the use of Digital Information and Communication Technologies (DICT). DICT combined with educational innovations are used as alternatives in care and educational situations (UNESCO, 2020), in addition to enabling collaboration. The variety of applications can be worked on when there is a need to involve many people, in addition to the difficulty of accessing information.

UNESCO's motto is a timeless inspiration "learning never stops". Lifelong learning must be nurtured and encouraged in vocational training. This action is fully aligned with sustainable development goal 4 (SDG), 2030 agenda (United Nations, 2015).

In 2019, the Ministry of Education of Brazil published a version of the National Curriculum Guidelines (DCN - National Curriculum Guidelines). In this updated version, improvements were introduced with a focus on meeting the demands of more and better engineers (MEC / CNE, 2019). This has been a growing purpose in this pandemic moment, where skills training, curriculum flexibility and pedagogical innovation (Ferguson et al, 2017) with the use of active methodologies are present in open and online education. The training of biomedical engineers covers a wide range of knowledge, such as engineering in general. The pandemic period raised some methodological and pedagogical issues in the view of open educational practices, both for teachers and students. Thus, the development of new skills is an explicit demand both for compliance with curricular guidelines and for meeting the requirements of emergency and distance education.

In this context, the objective is to present an experience report on the work with biomedical engineering students in the development of extension action. The vision will focus on teachers' knowledge needs versus required digital skills. In the presentation and discussion of results, the objectives achieved by the students, as protagonists of the proposed solution, focus on the development of important skills for participation in a dynamic and digital labor market.

This chapter will describe how the extension action, which has been fundamentally based on offline teaching and learning, met the challenge of encouraging young engineering students in the use of technological mediation, for the development of research and extension activities, in continuing education and for professional qualification.

BACKGROUND

This section presents the concepts considered important in the development of the work. In addition to clarifying aspects involving the challenges of teacher education in Brazil and the importance of open educational resources (OER) in the context of this initiative.

Epistemic Fluency

A trend that emerges in the educational environment is the search for comprehensive knowledge through the so-called "Epistemic Fluency". In education, researchers and teachers work on ways to stimulate students' "epistemic cognition" and help them become better trained for knowledge, that is, to develop "epistemic fluency" (Markauskaite; Goodyear, 2018).

According to the Markauskaite & Goodyear (2018), epistemic fluency is the ability to recognize different types of knowledge and to work flexibly with different forms of knowledge. For example, certain effective actions require specialized research knowledge on problems inherent to the topics under investigation, combined with knowledge from areas such as economics, politics, and law (Markauskaite; Goodyear, 2018). In view of this stimulus to knowledge in professional practice, one can observe Epistemic Fluency in (Markauskaite; Goodyear, 2018):

- Knowledge in the workplace - teachers and students learn while carrying out professional and academic activities.
- Work with multidisciplinary teams - the relationship and engagement of multidisciplinary teams are the object of study.

Applied Learning

Applied learning is a concept presented in a document developed by the University of Camosun - Canada (Camosun College, 2018). This approach refers to learning experiences that make the student think, collaborate, communicate. It's an incentive to get involved and contribute to the world around you. These learning experiences can take place in a variety of contexts. Examples can be applied in the classroom, in the workplace, in the community. This approach allows students to apply, share and integrate theoretical knowledge as well as personal, practical, and professional skills. The development of learning activities that simulate everyday situations, a professional practice or that are carried out in a real context are encouraged.

Projects and actions that favor the connection with the job market and the community enrich students' understanding of what is possible and facilitate the transition from academia to professional practice. Extension actions are an example. University extension is identified as a necessary activity for, with social commitment, through the engagement of the academic community. It is a fundamental part of society's transformation process, both in the dissemination of knowledge and in the search for social inclusion. The extension activity must be seen as a development process of an educational, scientific, and intercultural nature, which enables the integration between the university and society (PDI, 2019).

These actions favor, such as internships and social entrepreneurship programs, practices that offer even more opportunities for the development of creativity and innovation in educational environments. In addition to exploring social innovation to solve real world problems in the community or industry. Experiences like these promote professional development and allow students to understand the meaning of their education. The focus is on promoting real situations where the student can be the protagonist of the proposed solution.

Collaboration is a practice recognized by the academic literature as a very efficient means of learning, since the socialization of the content allows the course participants to complement each other's knowledge, and to increase their mastery of the subjects discussed. In digital media, clearly, virtual meetings of discussion, through meeting rooms, have been used for this purpose, constituting the main means for users to engage in a conversation both about content and to establish social ties, which is necessary even in the absence of physical contact (Huang et al, 2020a; Huang et al, 2020b).

Open Educational Resources (OER)

According to UNESCO (OER, 2019):

Open Educational Resources (OER) are teaching, learning and research materials in any medium – digital or otherwise – that reside in the public domain or have been released under an open license that permits no-cost access, use, adaptation, and redistribution by others with no or limited restrictions. OER form part of 'Open Solutions', alongside Free and Open-Source software (FOSS), Open Access (OA), Open Data (OD) and crowdsourcing platforms.

UNESCO is the only international standards-setting framework in this area in the world (OER, 2019). In its recommendation document, UNESCO presents five major areas considered important for the promotion, dissemination, implementation, execution, and follow-up of actions aimed at the use of open educational resources (OER, 2019). With the advent of the pandemic, open education and the use of open educational resources had a considerable prominence and advancement in educational institutions around the world.

When talking about open educational resources, it is important to know about open educational practices (OEP) (Chiappe; Adame, 2018). Licensing of materials and products through the internet is one of them. Creative Commons licenses are used and indicated to support issues related to intellectual property. Creative Commons is a non-profit organization that helps to understand the legal obstacles to sharing knowledge and creativity to meet the challenges in a digital world (CC, 2021).

Active Methodologies

Active methodologies, as the name implies, favor active participation, active learning (Bacich; Moran, 2018). They contribute to creating conditions that allow simulating the participation of students in situations that encourage dialogue, negotiation, and the proposal of solutions. Faced with the digital age, there is a need for new professional skills, such as initiative, pro-activity, versatility, creativity, among others (Soares, 2021).

Pedagogical approaches innovate the design of educational offerings, especially in virtual environments. The use of active methodologies is part of this list of innovations, as it contributes to an active posture of students in the search for and consolidation of knowledge. The activities developed are focused on recognizing differences, resolving conflicts, and building consensus. There are countless active methodologies strategies that contribute to the individual's formation, such as lifelong learning. In this work, only two will be addressed, the flipped classroom and project-based learning.

The expansive role of active learning is important for a better understanding of the educational process, a more comprehensive assessment of the student and, consequently, the personalization of education.

Flipped Classroom

The Flipped Classroom approach is based on concepts such as active learning, conducting experiments, student engagement, and hybrid course design (Bacich; Moran, 2018).

In this pedagogical approach, students receive resources that will introduce them to concepts needed to develop practical activities or discussions bfore the practical class, for example. In the flipped classroom,

students have the flexibility to initiate and complete the learning normally covered in the classroom on their own time, and classroom time is devoted to hands-on activities and interactive, personalized learning, which favors a better understanding of the topics covered. Class time is now used for the application of theory and concepts discussed through the educational resources used and for the use of techniques, including problem-solving and group negotiation.
According to the authors, the flipped classroom is (Karanicolas, S., Snelling C., and Winning, T., 2015):

An engaging series of learning segments, that are closely linked to learning and assessment outcomes, that provide feedback to the learner during each stage. Carefully designed pre-class activities assist students to learn key concepts in a self-paced manner, developing their confidence and motivation to engage in peer-led discussions during class that lead to synthesis and application of these key concepts. Post-class assessment activities are clearly connected to pre-class and face-to-face class learning experiences and address 'capabilities that count,' making the students' learning relevant, real and sustainable.

Project-Based Learning

Working on acquiring knowledge is a process of continuous involvement of the student professional. The learning trails make it possible to design the desired professional profile. Commonly throughout academic life, several projects are developed, such as the course project, the discipline project, the research project. These experiences allow learning about planning, executing, and monitoring activities. In this context, fostering the development of solutions through projects is widely used in the classroom (Soares, 2021).

Project-Based Learning (PBL) allows both the teacher and the student the necessary vision for the construction of knowledge in a limited period with premises and restrictions for its implementation. That's the idea - project-based solutions (Bacich; Moran, 2018).

Thus, Project-Based Learning is perhaps the most popular of the active methodologies. It is an exercise for both individual and collective competences. Therefore, PBL favors the development of fundamental skills beyond the search and analysis of information, stimulates the debate of ideas and promotes the development of communication and negotiation techniques, which can be decisive in a dynamic digital scenario.

Competences

In this context, there is an opportunity to work and encourage the development of both technical and emotional skills. The versatility of the professional in meeting competencies focused on teamwork, zeal, and ethics in the performance of their activities are opportunities for professional training.

According to the United Nations Development Program (UNDP), the United Nations global development network (UNDP, 2008, p.2):

"Competency" is defined as a combination of skills, knowledge, and behaviors that lead to effective performance on the job and are therefore important for the success of the organization in achieving organizational effectiveness, as well as the success of individual staff member. Competencies describe the "how ", i.e., what behaviors, skills and knowledge are necessary to successfully meet the requirements of the post.

Digital skills are extremely necessary, and, in Brazil, we still find a deficiency in the qualification of teachers and educators. The European community has made great progress in this area. The Joint Research Center (JRC Science) structured research in 2016-2017 in three main areas: 21st century skills and abilities; Innovating and modernizing Education and Training; and Open education (EU SCIENCE HUB, 2019).

Faced with the need to address the demands for digital competences, the European community published a study that provides a framework to support educators in the challenge of understanding the necessary competencies in the digital age - a common European Framework for the Digital Competence of Educators (DigCompEdu) (Redecker; Punie, 2017). Through the framework it is possible to frame educators in performance levels. DigCompEdu consists of six areas that focus on different aspects of educators' professional activities: professional engagement; digital resources; teaching and learning; evaluation; training students; and facilitating students' digital competence. The first area is related to the professional skills of educators. Areas 2 to 5 are considered the main areas and work on aspects related to the required pedagogical skills. Finally, the sixth area deals with skills for students.

METHODOLOGICAL APPROACH

One of the consequences of the Covid-19 pandemic is the challenge caused by the educational disruption and the need to work more intensely with the development and use of remote activities and digital resources. In this context, technological mediation favors the performance of educational actions, professional training and provides opportunities for the development of skills in collaboration and teamwork, in creativity to perform tasks, in the interdisciplinary knowledge.

Extension action was proposed with the purpose of allowing the integration of students, teachers, researchers, and professionals from different areas of knowledge. Online digital platforms were used to promote the debate and discussion of topics of interest, transversal, and trends, namely in the areas of activity of Biomedical Engineering. All themes addressed in the planned activities, in the extension action, were aligned with the sustainable development objectives.

Table 1. View of methodological approach

Activity(ies)	Tool(s) and Technique(s)	Intervention provided	Outcomes
Advise on the themes to be dealt with in the semester.	Flipped Classroom Brainstorming	Guidance, Pedagogical	Debates on themes exposed in expository panels and investigated in current literature.
Encourage readings and studies associated with the topic in focus, with support from a supervisor.	Provision of guidelines for investigation and development of skills and abilities necessary for discussion and understanding of the chosen topics.	Intervention, Pedagogical	Discussion session
Encourage the writing of scientific documentation on points of interest - digital information and communication applications and technologies, which are part of the extension action catalog.	Development and learning aimed at writing scientific communication on the topics investigated and discussed.	Supervision	Extended abstracts and booklet
Encourage teamwork and collaboration.	Meetings	Tutorials	Chosen Thematic areas and planning activities
Encourage the practice of discussion and decision on planned activities and adjustments necessary for the smooth running of the project.	Project-Based Learning	Peer review, supervision, group collaboration	Product release

To specifically respond to this objective, a set of activities was planned in line with the promotion of skills in the digital age. Project-based learning was the chosen approach and presented to the team - participating faculty and students. In general, the context of the project brought the importance of curricular components related to teaching, research, and extension practices, so that a set of activities could be carried out, shown in Table 1.

As shown in Table 1, the methodological approach excelled in working the extension action in a large group that involved researchers, teachers, and students. For the development of the action, active learning was used through project-based learning.

EXPERIENCE REPORT

The Higher Education Institutions (HEI) have in extension actions a curricular component necessary for the good fulfillment of the social commitment, foreseeing, for this, the engagement of the academic community. It is seen as a fundamental part of the process of transforming society, both in the dissemination of knowledge and in the search for social inclusion. Due to the complexity of the theme that deals with social inclusion, the extension activity must be seen as a process of educational, scientific, and intercultural development, to allow integration between the university and society (PDI, 2019).

The main goal is to allow and provide students with the possibility of developing research actions in areas of interest, especially Education and Health. The themes were aligned with the sustainable development goals 3 and 4 of the 2030 Agenda (United Nations, 2015) and allowed the development of important skills for professional life.

Planning the Extension Action

The main theme of the action was to favor students in reaching and consolidating important skills for their professional development. The execution plan was structured based on the use of active methodologies, the flipped classroom and project-based learning.

The aim was to allow the student to develop actions and research and extension aimed at a collective interest in a specific area of Biomedical Engineering. For this, the guiding question defined was: ***What knowledge is needed to understand the current needs of public health and how biomedical engineering can help?***

The research focus has been on professional requalification as teachers are helped to prepare for their work in a different context with virtual learning environments that arise specific demands. In these environments, the way to interact is different. Presentation of content, sharing of materials, interactions in chats and forums, evaluative activities differ from the activities practiced in person. In this way, the webinars were designed to help them connect academic knowledge with practice in the workplace and carry out their professional tasks.

For biomedical engineering students, topics such as health education, medical equipment and health technology assessment, digital health, and smart cities, for example, draw attention. The main idea was to promote the search for and advancement in topics of interest aimed at professional practice. Typically, planned assessment activities are designed to support the student in consolidating academic knowledge with practice in the workplace and better accomplishing their professional tasks. In this way they are prepared for a competitive job market and related opportunities.

Teamwork

The team consisted of two professors, two researchers and 16 undergraduate students in Biomedical Engineering. Students were encouraged to form groups with 4 members each. There was no restriction on the student's education level and most of the groups were formed by 1st year engineering students.

The team formed took on specific assignments. Teachers were responsible for coordinating the action, as well as groups, with special attention to team participation in meetings and expository sessions; maintenance of student groups; support in carrying out actions related to the development of projects by the groups; monitoring the attendance and participation of students in activities; monitoring the development and execution of projects.

The researchers were responsible for guiding students on the chosen topics. They also developed the pedagogical interventions necessary for the action to work properly.

All students participating in the action were regularly enrolled in the undergraduate course in Biomedical Engineering. To develop the project, students took on specific and sometimes overlapping roles.

Strategic Actions

Based on the competence frameworks dealt with in the background section, a set of important competences for students was defined and integrated into the actions to carry out the extension action, as shown in Table 2.

Table 2. Competencies and actions

Competence	Action(s)
Ethics and Values	Interaction with faculty advisors; Integrating with professionals and organizational actives.
Institutional Awareness	Presentation and discussion of the Pedagogical Plan of the Biomedical Engineering Course; Presentation of Institutional Rules
Development and Training	Multifaceted work – research and extension activities related to defined thematic areas; Stimulus to improve the oral and written expressions of the developed activities.
Collaborative Work	Students were invited to work in groups and between groups; Teachers and researchers performed tutorials with problem situations for discussion and sharing of ideas.
Communicate and Share Ideas	All actions taken during the action were discussed with the team in fortnightly sessions.
Self-management and Emotional Intelligence	Each group managed their defined assignments and activities and collaborated with the others in managing the action.
Negotiation and Conflict Resolution	Encourage discussion and problem-solving. The practice of complying with project definitions and assumptions.
Continuous Learning	Fostering continuous improvement.
Democratic Decision Making	Democratic policy – all project definitions were voted on by the group.

The competences presented in Table 2 were stimulated through individual and group actions, promoting collaboration among the team. In general, all actions were guided with the objective of favoring:

- Skills related to planning, executing, and monitoring activities.
- Definition and investigation of the chosen themes and in line with the areas of Education, Health and Technology.
- Emphasis on sustainable development goals (SDGs), on SDGs 2, 3, 4, 5, 6, 8, 9, 10, 16 and 17 (United Nations, 2015).
- Definition of the project to be carried out and planning of activities and related tasks.
- Production of scientific documentation on each of the defined thematic areas.
- Carrying out a workshop with the group of students and teachers involved to guide studies, follow up on the planning of actions and discuss the production of technical material related to the development of each project.
- Periodic evaluation of the defined model.

Development Process

Once in online learning, topics related to student and teacher learning guided most of the discussion. Emergency remote teaching brought some questions related to the development of content and activities focused on encouraging student participation. Open Educational Resources (OER), Virtual Environments, Flexible Learning, and Digital Curation were themes addressed with a focus on the importance of licensing and copyright, and the binomial health and education. The process used was composed of the following activities:

1. Understand the needs of the student audience: Due to the digital trend, practices have changed and adjusted. It is important to note how technology is changing classrooms and how to deal with technological mediation. Definition and adoption of innovations and digital inclusion - focusing on the characteristics of groups or individuals specifically in virtual learning environments.
2. Design the objectives according to the identified needs: select, prepare the necessary material - present and guide students. Define and document pedagogical strategies, considering the public - teachers, researchers, and students.
3. Test and Technological Orientation: test the platform to be used to support the activities of the extension action. Definition of guidelines and rules for the correct fulfillment of planned activities.
4. Execution of the extension action: carry out the planned activities and monitor their compliance through the initial schedule developed.
5. Monitoring and evaluation: periodic meetings of the team of professors, researchers, and students to identify and correct any gaps, as well as identify areas for improvement. The team's assessments were carried out as follows:
    a. Alpha - by the pedagogical team - teachers, researchers, and technicians.
    b. Beta - by members of the undergraduate group of students.

During the execution of the action, some results considered important in the student's development throughout the university career were observed. Two major products were developed: holding open sessions aimed at debating emerging areas of Biomedical Engineering, based on debate and discussion of the area of interest; and development and organization of the material investigated and elaborated from the readings, research, and open sessions carried out.

All these activities were carried out with the aim of providing personalized education. The purpose was to improve the quality and efficiency of teaching, learning and the outcome of the project-based learning methodology. Monitoring of activities was carried out periodically. Meetings were held with groups of students and teachers involved. The themes dealt with the orientation of the studies, monitoring of the planning of actions and discussion of the production of the defined technical material.

Active Learning

The extension action was developed following a schedule and in the initial meeting, the subdivision into 4 groups was defined with the group of students. Each group was supervised by a professor and a researcher. Each professor and researcher were responsible for two groups, that is, 8 students. The activities agreed at the kick-off meeting included:

- Define the main areas of interest and thematic axes – each group was responsible for two areas.
- Investigate the topics of interest to promote the necessary initial knowledge – the researcher was responsible for gathering the guidance material to support the investigation. Each group had two orientation meetings.
- Carry out the mapping and selection of professionals of interest for the development of online sessions and the production of research materials.
- Define the schedule of activities necessary to carry out at least 8 sessions during the bimester.
- Produce an extended summary for each theme chosen.
- Develop dissemination material for online sessions and promote dissemination in the academic community across the country.
- Hold at least two monthly meetings with groups of students involved in the project to guide their studies, monitor the activities carried out and discuss the production of technical material related to each session.
- Evaluate and adjust the model according to experimental studies of use, as well as the defined methodology.

Based on the discussions and definitions, a macro schedule was defined by the team to guide the execution of the activities of the groups of students and the performance of teachers and researchers, as can be seen in Table 3.

The project was based on a multidisciplinary strategy, so that the themes worked could be related to curricular components, such as, for example, the subjects of medical equipment or health informatics. In the project presentation, concepts related to the development of a project were initially addressed - scope, deadline and expected results, as well as the relevance of cooperative and collaborative work. The importance of brainstorming meetings and sessions was emphasized, for a better understanding and planning of the work to be developed.

Based on sustainability and open educational practices, students were instructed to indicate sustainable development goals (United Nations, 2015) that could be related to the thematic areas under the responsibility of each group. All products developed were registered under Creative Commons license (CC,

2021). Open session videos, scientific communications were developed as open educational resources. The products produced were registered in the institutional repository of the Institution of Higher Education.

All activities were evaluated according to the scheduled deliveries: scientific communication (text in a specific model and with a limit of 1000 to 2000 words); video script; final video with the domain expert; team performance evaluation; presence at monthly meetings to verify ongoing activities; overall assessment of the team; development of activities by the participants versus difficulties encountered; peer review; suggestions and considerations about the solutions presented by the other groups.

Table 3. Schedule of activities

Activity(ies)	Semester					
	M1	M2	M3	M4	M5	M6
Provide study and research material on the defined themes and studies. Schedule open sessions - 4 in month 3 and 4 in month 4.	X					
First-month orientation activities: in-service training and the importance of seeking knowledge; initial research on areas of interest related to rehabilitation, education in and for health and medical equipment.	X					
Second-month orientation activities: presentation of the Systematic Literature Mapping method to support the investigation of the chosen area of interest for the development of scientific communication; presentation on Open Education and Open Educational Resources.		X				
Definition of document templates: scientific communication (extended summary); cards for the dissemination of online sessions; and structure of the scientific communications booklet.	X					
Online sessions were held weekly for one hour. During monthly meetings, all deliverables were evaluated.			X	X		
Review of developed materials and production of the booklet. All materials are registered under Creative Commons licenses.					X	X

RESULTS AND DISCUSSION

In-service training is a stimulus to continuing education, as it facilitates dialogue and the exchange of experiences between communities in the exercise of their profession (Markauskaite; Goodyear, 2018). The purpose was to allow undergraduate students in Biomedical Engineering to work with real situations. It must be recognized that the pandemic context generated an unthinkable impact, until then, in each country. Public health is the most affected area, as it needs to ensure the most adequate care possible for the infected population. The global economy is already showing some degree of depression and it is necessary to minimize the consequences for the reorganization in the coming years. It will certainly be necessary to carry out social and community assistance actions for populations affected by formal unemployment and growing informality at work, and by pre-existing vulnerabilities, such as inadequate housing conditions in communities, homeless people and even populations of origin.

This rich experience provided members, especially students, protagonists of the extension action, the opportunity to discuss training issues related to technical-scientific development. It was possible to expand the network of contacts, define learning paths, needs, and required skills, and new ways to apply post-pandemic knowledge in a dynamic labor market.

The model presented by DigCompEdu has a degree of proficiency levels for the competences (Section 2.5). Most of our participants were classified as newcomers, level A1. As mentioned, (Redecker; Punie, 2017; page 30)

Newcomers are aware of the potential of digital technologies for enhancing pedagogical and professional practice. However, they have had very little contact with digital technologies and use them mainly for lesson preparation, administration, or organizational communication. Newcomers need guidance and encouragement to expand their repertoire and to apply their existing digital competence in the pedagogical realm.

And a minor group was considered as Explorer, level A2.

Explorers are aware of the potential of digital technologies and are interested in exploring them to enhance pedagogical and professional practice. They have started using digital technologies in some areas of digital competence, without, however, following a comprehensive or consistent approach. Explorers need encouragement, insight, and inspiration, e.g., through the example and guidance of colleagues, embedded in a collaborative exchange of practices.

The use of digital online platforms enables the serialization training processes of professionals from different areas and accelerates the integration and access to technologies. In this light through the main lessons learned, we can highlight:

1. the importance to structure a didactic transition team to support the teachers to produce differentiated didactical materials suitable for online education.
2. the need for teacher's qualifications to participate in the online courses, which should focus not only on how to use virtual learning environments, but also on the differences in the didactic-pedagogical mediation and the evaluation system adopted.
3. differential provided by strategies for periodic monitoring and evaluation of the activities developed by teachers and tutors.

With the completion of the planned activities and fulfillment of the extension action objectives, technological, economic, and social impacts can be presented:

Technological Impact: It allowed access to technology and remote work platforms, contributing to the growth in the use of technology. Stimulus to search and test technological solutions compatible with the demands presented.

Treatment and solution of emerging themes, addressed, including in an international context, thus maintaining a globalized view of technological advances.

Economic Impact: Encouraging the use of open and free technologies. Fostering the practice of open education. Strengthening human capital in the region and communities.

A continuous search for increasing quality and competitiveness in Biomedical Engineering, enabling its use as an instrument to improve the quality of health and education of the population.

Social Impact: It enabled the integration between the academic sector and the market through the transfer of knowledge and innovation resulting from research.

It promoted an improvement in the quality of training and teaching-learning. Students demonstrated proactivity in all planned actions. Encouraged teaching-service-community integration using innovative and digital technologies. The difficulties that teachers and pedagogical teams faced to do this are not exactly "transfer" problems, but the need to understand that the daily practice of actions requires different forms of knowledge to deal with a specific situation.

The team was formed by professionals from different areas, which made academic multidisciplinary possible. This was a challenge faced by the team and experts involved, since the development of the first webinar. The structuring activities of the speakers, the technical-pedagogical team and the time allocated for each topic required coordination and engagement from those involved. Fostering the incorporation of new technologies in daily professional practice and communities assisted by these professionals.

All topics are closely related to SDG 4, with the skills relevant to decent work. According to 2030 Agenda - SDG 4 (United Nations, 2015), the substantial increase in the number of young people and adults with relevant skills, including technical and professional skills, is encouraged. With the coronavirus pandemic, the importance of ensuring diverse learning opportunities, with a focus on the acquisition of specific professional skills, with an emphasis on the development of cognitive and non-cognitive skills, for example problem solving, critical thinking, creativity, teamwork in communication skills and conflict resolution.

During the extension action, 8 open sessions were held with the participation of external professionals - researchers, renowned professionals, and entrepreneurs. All the work was mainly focused on the integration of the areas of health, technology and education that was not necessarily part of the common curriculum of the engineering degree course. Each open session brought discussion and concern about the actions when face-to-face activities were resumed. The themes addressed important points for knowledge sharing and continuous learning in the digital age.

The areas were defined with the main theme "*Technology and Health: what does Biomedical Engineering have in store?*". The themes dealt with in the open sessions were Neural Engineering, Telehealth and Telemedicine, Forensic Engineering, Tissue Engineering, Clinical Engineering, Biomechanics and Rehabilitation, Assistive Technologies, Biomedical Robotics. Practices and solutions from each professional experience were presented, unparalleled experience. Physical detachment brought up important social issues. Creative alternatives in the teacher-student relationship in virtual environments were discussed, as well as trends in the labor market. Technical and emotional skills were highlighted.

Once each group of students defined two specific themes, the artifacts developed were: scientific communication based on conducting research in the chosen area; video script with domain expert; video with the chosen expert; analysis of the team in the roles indicated – facilitators and panelists and expert participation. All students developed study and research activities on the themes defined in the semester, whose materials were made available in a virtual learning environment.

During each session, the students who were part of the responsible group assumed the following roles:

- Chair – with the role of initiating discussion on the topic with up to 2 questions, favor debate and make it possible to resolve doubts on the subject.
- Moderator – this role can be filled by more than two students. Responsibility to prepare the session, check the time issues, record the students who are interested in participating, generate attendance list, control the participation in the virtual room, support with the list of questions from the audience, among others.

- Facilitator – responsible for the documents that will be generated before and after the session. Must review the extended abstract sent by the presenter, receive set of slides, and make it available to the community as an open educational resource, write, together with the other group colleagues, the chapter of the booklet referring to the session under the responsibility of the group.

According to the activities developed, the participation of the Public (Civil Society) is evident - represented by public and private initiative teachers, students - Public Higher Education Institution (HEI), Business Subsystems, to a lesser extent, the State Government. The participation of HEIs and the business/entrepreneurs' sectors was preponderant since the action was carried out for public servants with the participation of professors and researchers from public and private non-profit institutions of higher education.

Civil society had an important prominence, as it was represented by students, diverse professionals. The public consisted mostly of undergraduate students, not just engineers. The media and communication component related to local culture was quite evident in the debates, mainly due to the difficulties faced in ensuring the work was carried out due to the lack of technological infrastructure, as well as access to teacher training.

The integration between professionals working in Biomedical Engineering gave students the opportunity to discuss the importance of professional development and the knowledge needed in real situations. It evidenced concepts needed to understand the current needs of public health and the areas of action of Biomedical Engineering.

FINAL CONSIDERATIONS

The extension action developed promotes inclusive, equitable and quality education, through lifelong learning, a goal of sustainable development - Quality Education, which guides educational actions worldwide. The digital medium, as a work platform for students, allowed the project to expand the teaching-learning possibilities by facilitating the interaction, between a diversity of professionals, variety of experiences, that can be presented. The promotion of the development of digital skills is a differential, together with the encouragement of Open Educational Practices. The reflections, as well as the lessons learned, could be discussed, with a view to the future that is configured, in the labor market, in constant transformation. Stimulus, through the development of actions that correlate important competences and skills, is essential for professional development. The carrying out of activities and actions of teaching, research and extension corroborated to promote the interdisciplinarity of the various fields of study, necessary for professional development in the digital age. Thus, the actions developed stimulated the investigation, improvement, and study of topics of interest related to Education, Health and Technology, areas related to the performance of the Biomedical Engineer, the main protagonist of this project.

These pedagogical activities, promoted by the extension curriculum, have promoted, and still promote the digital inclusion of professionals (teachers and researchers), tutors and students; development of specific tools and applications to support the conduct of open educational activities; production of open educational content and resources; training and professional qualification and research and investigation into related topics.

This exceptional experience in supporting the development and implementation of higher professional education programs favors the continuous improvement of knowledge of the theories and approaches of learning design and encourages teachers to an even greater search for excellence in the courses developed and the constant encouragement to improve their skills.

It is important to highlight that the greatest professional gains reside in the particularity of all this experience having taken place in a developing country like Brazil, in a singular moment, a country that, due to its unique characteristics - has certain aspects of economic development as intense as other countries richer and, at the same time, regions where the human development index reaches very low levels, guarantee a broad vision of the diverse realities of our planet.

ACKNOWLEDGMENT

The author would like to thank the team that contributed to the idealization of this extension action. Biomedical Engineering students who assiduously participated in this professional development initiative, with great enthusiasm and dedication to improve professional practice and performance in an atypical moment such as the coronavirus pandemic, as well as the extension dean's managers and fellow professors for the opportunity to innovate and foster open and continuing education.

REFERENCES

Bacich, L., & Moran, J. (2018). Metodologias ativas para uma educação inovadora: uma abordagem téorico-prática. Porto Alegre: Penso.

Camosun College. (2018). *Applied Learning* https://legacy.camosun.ca/about/applied-learning/index.html

CC – Creative Commons. (2021). https://creativecommons.org/

Chiappe, A., & Adame, S. A. (2018). *Open Educational Practices: a learning way beyond free access knowledge*. https://www.scielo.br/j/ensaio/a/jTpx6MgYFpGbzLgQJ6FY6Hh/?lang=en

EU Science Hub. (2019). *Learning and Skills for Digital Era*. https://ec.europa.eu/jrc/en/research-topic/learning-and-skills

Ferguson, R., Barzilai, S., Ben-zvi, D., Chinn, C. A., Herodotou, C., Hod, Y., Kali, Y., Kukulska-Hulme, A., Kupermintz, H., McAndrew, P., Rienties, B., Sagy, O., Scanlon, E., Sharples, M., Weller, M., & Whitelock, D. (2017). *Innovating Pedagogy 2017: Open University Innovation Report 6*. The Open University.

Huang, R. H., Liu, D. J., Guo, J., Yang, J. F., Zhao, J. H., Wei, X. F., Knyazeva, S., Li, M., Zhuang, R. X., Looi, C. K., & Chang, T. W. (2020). *Guidance on flexible learning during campus closures: Ensuring course quality of higher education in COVID-19 outbreak*. Smart Learning Institute of Beijing Normal University. https://iite.unesco.org/wp-content/uploads/2020/05/Guidance-on-Flexible-Learning-during-Campus-Closures-in-COVID-19-outbreak-SLIBNU-V1.2_0508.pdf

Huang, R. H., Liu, D. J., Tlili, A., Yang, J. F., & Wang, H. H. (2020). *Handbook on facilitating flexible learning during educational disruption: The Chinese experience in maintaining undisrupted learning in COVID-19 outbreak*. Smart Learning Institute of Beijing Normal University. https://iite.unesco.org/wp-content/uploads/2020/03/Handbook-on-Facilitating-Flexible-Learning-in-COVID-19-Outbreak-SLIBNU_V2.0_20200324.pdf

Karanicolas, S., Snelling, C., & Winning, T. (2018). *Flipped Learning in Adelaide*. University of Adelaide. https://www.adelaide.edu.au/flipped-classroom/about/

Markauskaite, L., & Goodyear, P. (2018). *Universities should take stronger leadership on knowledge and how it matters*. University of Sidney. https://theconversation.com/universities-should-take-stronger-leadership-on-knowledge-and-how-it-matters-89849

MEC/CNE. (2019). *Diretrizes Curriculares Nacionais do Curso de Graduação em Engenharia*. http://portal.mec.gov.br/docman/marco-2019-pdf/109871-pces001-19-1/file

OER. (2019). *UNESCO Recommendation on OER*. https://en.unesco.org/themes/building-knowledge-societies/oer/recommendation PDI - Plano de Desenvolvimento Institucional da 2019/2023

Redecker, C., & Punie, Y. (2017). *European Framework for the Digital Competence of Educators: DigCompEdu*. https://ec.europa.eu/jrc/en/publication/eur-scientific-and-technical-research-reports/european-framework-digital-competence-educators-digcompedu

Soares, C. (2021). *Metodologias ativas: uma nova experiência de aprendizagem*. Editora Cortez.

UNDP – United Nations Development Programme. (2008). *Competency Framework*. https://procurement-notices.undp.org/view_file.cfm?doc_id=10206

UNESCO. (2020). *International Commission on the Future of Education, Education in a Post-COVID World: Nine ideas for public action*. https://unesdoc.unesco.org/ark:/48223/pf0000373717/PDF/373717eng.pdf

United Nations. (2015). *2030 Agenda*. https://nacoesunidas.org/pos2015/agenda2030/

Chapter 16
Information and Communication Technologies in the Educational Process:
Mapping the Critical Success Factors

Michelle Merlino Lins Campos Ramos
https://orcid.org/0000-0002-3379-1801
Universidade Federal Fluminense, Brazil

Helder Gomes Costa
https://orcid.org/0000-0001-9945-0367
Universidade Federal Fluminense, Brazil

Glaucia da Costa Azevedo
Universidade Federal Fluminense, Brazil

ABSTRACT

The study aimed to map the critical success factors for the adoption of information and communication technologies (ICTs) in the educational process of educational institutions. Problems related to the adoption of ICT in the educational system stem from the need to adapt to the use of new technologies in the internal processes of institutions and in teaching and learning processes, common to different profiles of educational institutions including of engineering courses with them specificities. To meet the objective, a review of the existing bibliography in the Scopus database was carried out to highlight articles relevant to the topic. Based on the review, 31 articles identified the main factors and effects that influence and impact the process of implementation and continued use of ICTs. The survey generated a broader view of the challenges faced in different dimensions, from SWOT framework, involving different stakeholders. It is suggested in future studies to engineering analyze deeper the complex scenario that involves the theme.

DOI: 10.4018/978-1-7998-8816-1.ch016

INTRODUCTION

The fast and growing evolution of Information and Communication Technologies (ICTs) leads educational institutions to invest resources in research and implementation of innovative technologies to improve the teaching-learning process in order to develop competitive graduates. Universities benefit from ICTs, which offer important economic advantages, accessibility and support the development of specific skills in students (Hernandez-de-Menendez & Morales-Menendez, 2019).

Educational institutions take advantage of the opportunity to use ICTs in the educational area through technologies such as internet, social media, simulations, games, augmented reality and using mobile applications as a means of virtual communication. The use of ICTs for educational purposes engages students and helps to develop creativity and collaboration in addition to oral and written communication skills (Hernandez-de-Menendez & Morales-Menendez, 2019).

According to Martínez-Cerdá et al. (2020), recent studies shows that "current online universities have to take into account many factors related to improve the skills of their students in pursuit of social and economic progress" (p. 8). Their results present the influence on the development of competences for employability, considering the importance of social factors, the time use by students, and the use of ICT for pedagogical practices, with more relevance to Science, Technology, Engineering and Mathematics (STEM) students.

The application of ICT in educational environments, as well as in higher education institutions, is important due to the potential to contribute to the improvement of the teaching and learning process, in addition to stimulating the knowledge construction process. It is necessary to develop educational models focused on information and communication technologies (ICT) as tools to support equal opportunities and social responsibility (Buenaño-Fernandez et al., 2019; Lytras et al., 2018).

The training of individuals for life and the labor market demands preparation for the required level of preparation for the exercise, added experience and constant change in the needs of expertise in technologies imposed by technological evolution. However, the integration of ICTs in the educational process by educational institutions is part of a complex process in which several factors and stakeholders intervene. According to Almerich et al. (2016), the literature presents several factors that influence the integration of these technological resources into the educational process.

The results of the study by De La Iglesia et al. (2018) confirm that the integration of ICTs in the classroom is directly or indirectly influenced by the understanding of factors that interact with each other. Factors such as the availability of means; skills in technical management of ICT, allowing the acquisition of other types of ICT skills, such as teaching skills (main modulating factor); sex and age cause low to moderate effects when interacting with different types of variables (De La Iglesia et al., 2018).

For Langroudi (2015), the main factors that affect the use of ICT in the university are based on the attitude of professors regardless of their age, according to the research carried out with 97 professors. Therefore, to improve the use of ICTs, the attitude of professors and their professional applications of ICT must be modified. On the other hand, the results of several researchs such as that of Radovan & Kristl (2020) support that the key to success of learning and teaching in virtual learning environments (VLE) lies in the creation of effective learning communities. In the study, the authors assume that effective higher education is the result of the functioning of cognitive factors, the professor's intervention and the social dimension that defines the development and functioning of the community.

The interdisciplinarity of new technologies and the impact of cyberinfrastructure demand new paradigms in engineering research and development. Technology is a means to facilitate educational changes derived from global trends that facilitates educators to act and adapt effectively to the needs of the educational community (Hernandez-de-Menendez & Morales-Menendez, 2019).

According to Martínez-Cerdá et al. (2020), a virtual training for employability that goes beyond the economic needs of the labor market and also develops the capacity for social interaction and citizenship of employees will undoubtedly have an effect on the future, in the form of higher rates of employment quality. In their study, they confirmed the importance of the use of new learning digital devices, since continuous mobile learning, ICT for pedagogical practices, and polychronic and multi-tasking ICT uses are relevant latent factors as well, and the effects of ICT use are more important for STEM students.

Due to the complexity of understanding how numerous factors influence the process of adherence to ICTs in the educational process, several authors present studies aimed at understanding the factors that affect different dimensions of the problem applied in different contexts, such as, the factors that influence the implementation of ICTs in e-learning systems in developing countries (Bhuasiri et al., 2012); user satisfaction factors through the design of the website where materials for students and professors access are available (Suárez, 2016); factors that affect the behavioral intention of college students (Ching-Ter et al., 2017).

Many works show important points of implementation of ICT as barriers to adoption or models of ICT application in educational institutions, but in this research were not identified studies that integrate the most complex scenario for its understanding, in order to deal with the sets of influencing and impacting factors to success in different dimensions that surround the problem from different stakeholders (students, professors, educational institutions, regulatory institutions, TI professionals and others).

This study aims to map in the literature the factors and effects that influence and impact the adoption of ICTs in the educational process, through the SWOT perspective, through the Scopus Elsevier platform - According to Rodriguez et al. (2013), the use of a data like Scopus or Web of Sciences (WoS) reduces the possibility of using no blinded-review or even gray literature sources. Thus, making it possible to understand the scenario that permeates the problem so that it is feasible to identify adequate solutions to the context of each profile and reality of different educational institutions and their stakeholders.

This research's goal is to answer following questions: Which are the effects of ICT adoptions in the educational institutions educative process? Which are the external factors that influence ICT adoption identified in the educational institutions educative process? Which are the internal factors that influence ICT adoption identified in the educational institutions educative process? Which are the positive and negative effects associated of ICT adoptions in the higher educational institutions educative process?

BACKGROUND

Information and Communication Technologies (ICTs) in the Educational Process of Engineering Education Institutions

In STEM programs education, some technologies have been used and already present a great impact on the teaching-learning process, such as virtual and remote laboratories, immersive learning environments, machine learning and virtual assistants. However, there are challenges ahead, including the lack of faculty members with the necessary digital skills and the resistance to change (Johnson et al., 2013).

Many factors influence online learning success, but the level of interaction between professors and students was identified as a critical factor and is influenced by social presence, which increases the effectiveness of teaching and building a sense of unity. The authors cite three dimensions of social presence: social context, online communication and interactivity, which are essential elements to build a sense of community for online students (Sulisworo et al., 2020).

Muhammad et al. (2020) presents a perspective on the influence of compatibility of the e-learning system, including the necessity of constant updating of the content in order to keep up with the evolution of the area of study which is highlighted along the accuracy and authenticity of the content focusing on the reliability of the e-learning system to the user, using of a variety of resources and attractive learning material according to the student's needs. Whereas web-based e-learning systems must support the learning process and allow students to use the system efficiently.

Higher education science and engineering curricula are structured to provide students with a mix of theoretical and practical education, ensuring access to general knowledge and training the skills applicable to the field. There are concerns about the shortcomings of current educational practices, as educational resources include the availability of trained professors, laboratory infrastructure for experiments and facilities required to execute them. ICT-enabled Virtual Laboratories, for example, allow students to self-controlled individualistic learning environments, complementing standard laboratory practices (Hurlburt et al., 2010; Shorfuzzaman et al., 2015).

In the knowledge society, key competences include ICTs combined with cooperation, management, organization and other specific subjects and general competences, with the incorporation of ICT being critically significant in the teaching and learning processes. Digital literacy represents an important area that leads to technology proficiency and general skills in specific subjects. Developing skills for lifelong learning and employability involves engaging participation in social media, computer-supported collaborative work and mobile work. (Starčič & Turk, 2010).

The use of blended learning resources is possible and can create changes in learning patterns and practices. Blended learning has been gaining ground, with ICTs being used to complement it rather than replacing traditional forms of learning, offering greater opportunities for understanding and extending the knowledge presented (Cooner, 2010; López-Pérez et al., 2011; Mitchell & Forer, 2010; Osguthorpe et al., 2003).

Combining traditional classroom teaching with ICT-supported activities impacts academic performance using a technology-enhanced modality. The two aspects that define the learning environment in this format are Traditional Classroom Instruction (instructor conducts a face-to-face class with students for a number of hours per week and presents the material through oral presentations, classroom chart lectures and PowerPoint presentations) and the technology-enhanced Course (content delivered efficiently through the web mediated by ICT, including readings, homework, videos, images, links and specific web pages) (Rosero-Zambrano et al., 2018; Rutz et al., 2003).

Another course format used is e-learning which differs from the traditional course in many ways and can represent a complex undertaking, requiring precise planning, monitoring and control to make the conversion efficient and cost-effective. In other words, to provide an effectual solution the advantages must outweigh the disadvantages for both educational institution and students. Some of the benefits of e-learning are: it is generally cheaper to deliver, individualized, faster, provides consistent content, works anywhere and anytime, is updated easily and quickly, which can result in greater retention and greater mastery of the subject, in addition to facilitating the management of large groups of students (Cantoni et al., 2004).

Malathi et al. (2019) highlight on the major challenges to the implementation of e-learning and ICT include work overload on the professors, resistance to change, proper training to professors, limited computing skills, and continuous up gradation and removal of obsolescence in e-infrastructure. The authors include the benefits pointed by Saxena (2017) such as the use of ICT for teacher training and raise of availability of quality education materials.

In their study, López-Pérez et al. (2011) identified the lack of consensus on the effects of the use of technology on academic results, despite the indication of negative impacts of its use among young people. Which is a consequence of the limited number of studies on the educational impacts of the use of technology. However, the studies that identified educational impacts did not contemplate specific types of technology used or were limited to measures or sampling designs adopted. In addition, new technologies were already being created and used at such rapid pace that it was difficult to capture the effects of these technologies in environments with rapidly changing.

Junco and Cotten (2011) identified studies in which the Internet presents positively impact on students, by using social media and student involvement, but some studies indicate that the Internet can negatively affect the progress of some university students. The negative impacts involve their interactions with each other and with the faculty, and their grades, due to excessive time spent on the Internet or even symptoms of psychological dependence to the tool. In contrast, Junco and Cotten (2011) provided a study in which students who attended more connected institutions were more likely to report good educational practices such as student/faculty contact, cooperation between students and active learning compared to less connected institutions.

The new demands of the labor market include up-to-date professional skills, which demand the intensive use of ICTs, requiring many advanced skills from workers. In this sense, lifelong learning becomes the most adequate strategy for the development of new skills to answer the demands of the labor market. Therefore, employability-oriented learning integrated with ICTs is developed through online education by several institutions (Martínez-Cerdá et al., 2020; Spitz-Oener, 2006; X. Wang, 2010).

Engaging and motivating students connected through technology is one of the skills required for instructors of those who will be part of this workforce. Virtual Learning Environments (VLEs) allow professors to update their multimedia pedagogical resources and allow students to filter and direct the focus on their learning processes. Cognitive load management skills are not included in most accreditation bodies for engineering courses but are part of the skills expected of future workers to deal with different technologies, including the ability to discriminate and filter information by importance in their learning process (Rosero-Zambrano et al., 2018).

An online training focused on employability goes further than the standard requirements of the labor market and also develops the workers' capacity for social interaction and social conscience, reflecting in the future higher rates of job quality (Martínez-Cerdá et al., 2020).

Umek et al. (2017) contributes to a large gap in research regarding the contributions of sociodemographic factors of students involved in an e-learning system to their academic performance. Despite indicating the little interference of individual factors, other studies reinforce that age and gender interfere in the online experience, whether due to resistance to use, lack of previous experiences, student maturity, system complexity, issues they consider more relevant in design of the website (Bulić et al., 2017; Callo & Yazon, 2020; Suárez, 2016). However, they indicated that external factors such as family issues, lack of organizational support and workload affect student performance in the e-learning system (Umek et al., 2017).

Sang et al. (2011) observed in the research that if professors adopted favorable attitudes towards ICT in education, they would be more willing to integrate them into teaching, in line with other studies reported by the authors. The need for an appropriate level of school planning to improve the successful integration of ICT in classrooms, combined with the development of a shared vision with professors, is reinforced.

Wasserman and Migdal (2019) found that the ease of use perceived by professors had significant effects on UP and attitudes towards computers. As well as external factors, subjective norm, facilitating conditions and technological complexity were considered significant to the attitudes of professors in training in relation to computer use. It is understood that if professors act positively, they tend to integrate ICTs in teaching and learning processes.

To enable access to materials and collaboration of students and professors, the educational institutions need to adhere to digital platforms and tools. New forms of evaluation and monitoring of student performance are also required, due to the large amount of data that can be collected by institutional learning management systems (Hernandez-de-Menendez & Morales-Menendez, 2019).

Among the technologies related to education, there are those developed particularly for teaching-learning and others adapted to the field. Hernandez-de-Menendez and Morales-Menendez (2019) list the most innovative ones in engineering education, such as remote laboratories, digital simulations, virtual environments and agents. They provide relevant benefits such as increasing student motivation and more efficient implementation of pedagogical strategies. On the other hand, they emphasize the importance of identifying the challenges that the institution may face when adopting those innovative technologies (Hernandez-de-Menendez & Morales-Menendez, 2019).

CRITICAL SUCCESS FACTORS IN ICT ADOPTION IN THE EDUCATIONAL PROCESS

In the context of higher education institutions, the Critical Success Factors (CSFs) constitute a set of parameters that must be ensured and guaranteed by the institution to be successfully implemented in the e-learning approach in its educational process. CSFs can be used to investigate and analyze the reasons why some institutions seem to be more successful than others and can be very useful for organizations working on their strategic plans, as noted in the specific case of e-learning (Benchicou et al., 2010).

Authors such as Benchicou et al. (2010), Hernandez-de-Menendez and Morales-Menendez (2019), and Mahdi (2006) indicate the need of the education system to be student-centered, focusing on the student's perspective of the educational and learning process. However, it requires drastic changes in the current way university courses are designed, taught and evaluated.

Studies revealed that the barriers for the application of e-learning system are heterogeneous and multidimensional in nature. Benchicou et al. (2010) indicated categorization into seven main types of barriers: (1) personal or dispositional; (2) learning style; (3) instructional; (4) situational; (5) organizational; (6) content adequacy; (7) technological barriers.

According to Papaioannou and Charalambous (2011), the success of ICT integration depends on the existence of adequate infrastructure, the support from the Ministry of Education and the existence of quality software aligned with the curricular subjects. Other barriers identified are the lack of special administrative programs on computers designed for beginners and the highly demanding official curriculum, which pressures professors and raise doubts about their attempt to integrate ICT into classroom practice since it is a time-consuming process. Especially professors who lack computer skills because they feel behind in covering the official curriculum.

For Papaioannou and Charalambous (2011), the external factors that facilitate or inhibit the integration of ICT are in-service training in ICT; students' prior knowledge of ICT; technical support and maintenance; the necessary time for directors to prepare for ICT integration; number of computers per class; incentive for innovation by the Ministry of Education and Culture. Whereas the internal factors are: inspiring and competent leadership; in-service training in ICT; trained school's ICT coordinator; incentives for the school's ICT coordinator; involvement of primary education stakeholders in the integration process; collaboration with the district's ICT professor advisor; professors competence and knowledge in ICT; acceptance of innovation by professors.

Based on the research by Bhuasiri et al. (2012), the identified factors referring to student characteristics include computer and Internet self-efficacy, computer and Internet experience, computer anxiety and attitude towards e-learning. The relevant instructor characteristics are timely response, self-efficacy, technology control, focus on interaction, attitude towards e-learning and learners, distributive fairness, procedural fairness, and fairness of interaction. The e-learning environment refers to social influence, students' perceived interactions with others. Motivation is an essential factor in individual attitude and expectations, perceived pleasure and usefulness, clear direction, social pressure and competition.

Directors with experience in the use of computers mentioned the importance of having inspiring and competent leaders during the ICT integration process. According to them, school leaders must believe in the potential of ICT and have a clear vision of how ICT can be integrated into their schools (Papaioannou & Charalambous, 2011).

ICT integration efforts should not be sporadic, as it must be at the top of the school agenda and the planning to achieve success must be continuous. It also emphasizes that the school leader should be the motivating force that would encourage their professors to substitute their traditional way of teaching for the technology to improve the quality of their work (Papaioannou & Charalambous, 2011).

Several universities face the big challenge of taking advantage of emerging technological innovations and advances in e-learning to enhance their teaching programs and improve the quality of teaching, especially in the engineering fields. Higher education institutions are responsible for strategic leadership, commitment, and large-scale investments in infrastructure resources (Benchicou et al., 2010).

Ali (2014) explored the challenges behind implementing ICT facilities in the engineering education system. These include content delivery, the need for hands-on lab experience, inadequate faculty training in the use of pedagogical tools in engineering education, and the lack of clear policies for choosing what type of content delivery mode can benefit the most. The ICT facilities are implemented by the wide availability of free sophisticated simulation software, high-speed Internet access and general ICT tools familiar to the younger generations of students and instructors. To take advantage of these opportunities and address the many challenges facing the full use of ICT tools in education, engineering education policy makers need to plan and make long-term strategic plans.

Ching-Ter et al. (2017) identified factors that positively or negatively influence the experience with e-learning. In the context of e-learning, subjective norm is the student's perception of pressure from his/her peers to use e-learning systems. Perceived pleasure deals with the perception of use of a specific system being enjoyable regardless of system performance. Computer anxiety is about an individual's tendency to be restless, apprehensive, or fearful about using computers in general. Self-efficacy is the belief in the ability to organize and perform actions necessary for specific achievements, not related to the number of skills of the student, but to the belief in what he is able to do under a variety of circumstances (Ching-Ter et al., 2017).

Suárez (2016) identified factors related to students, which influence the way they interact with the website, demonstrating that considering the target audience when designing the website is important for their satisfaction. In addition, Romi (2017) identified that student performance is determined by the e-learning system that consists of a series of interrelated and interdependent factors (instructors, course, students and ICTs).

Umek et al. (2017) contributes to a large gap in research regarding the contributions of sociodemographic factors of students involved in an e-learning system to their academic performance. Despite indicating the little interference of individual factors, other studies reinforce that age and gender interfere in the online experience, whether due to resistance to use, lack of previous experiences, student maturity, system complexity, issues they consider more relevant in design of the website (Bulić et al., 2017; Callo & Yazon, 2020; Suárez, 2016). However, they indicated that external factors such as family issues, lack of organizational support and workload affect student performance in the e-learning system (Umek et al., 2017).

Wasserman and Migdal (2019) found that the ease of use perceived by professors had significant effects on UP and attitudes towards computers. As well as external factors, subjective norm, facilitating conditions and technological complexity were considered significant to the attitudes of professors in training in relation to computer use. It is understood that if professors act positively, they tend to integrate ICTs in teaching and learning processes.

METHODOLOGY

This study is literature review that was applied in three sections: (1) bibliometric review; (2) identification and classification of the results; and (3) SWOT analysis of the results. The development structure of this review followed some steps indicated in the Population, Intervention, Comparison, Outcomes, and Study (PICOS) and Search, Appraisal, Synthesis, and Analysis (SALSA) methods (Mengist et al., 2020), which indicate the structuring of the systematic literature review, in order to represent how the work was organized.

Section 1 - Bibliometric Review

The methodology used in this research literature review was the bibliometric review called Webibliomining (Silva & Costa, 2015), which is a combination of Bibliometrics, Bibliomining (data mining) and Webmetrics with the adoption of methods of preliminary analysis of the bibliographic reference, expanding the investigation of the state of the art on the subject. The model, according to Silva and Costa (2015), considers the six steps presented on Figure 1.

The database used in this research was Scopus Elsevier, as it is considered the largest database of abstracts and citations of the literature with peer review in the sector, such as scientific journals, books, conference proceedings and publications (*Scopus*, n.d.). According to Rodriguez et al. (2013), the use of its reduces the probability of using gray literature' sources. The search in the Scopus was made with the following keywords: digital transformation, new technologies, information and communication technology, educational process, educational institutions, higher education. The research was carried out on 2020 November 23rd and updated on 2021 May 3rd, with results' date filtered until 2020. The steps of the refinement are presented on Figure 2.

Figure 1. Webibliomining method steps
Source: (Silva & Costa, 2015)

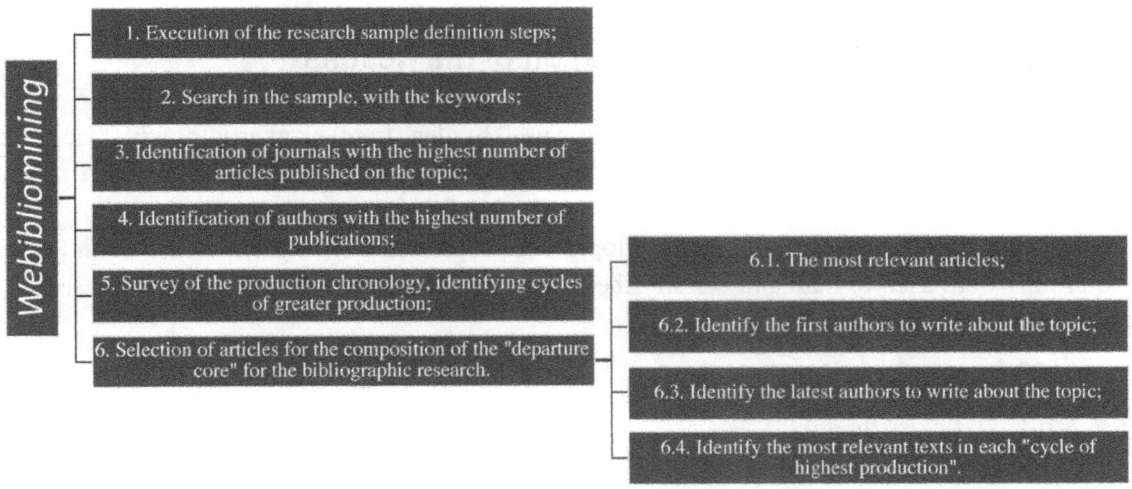

Figure 2. Refinement steps to the research on Scopus database

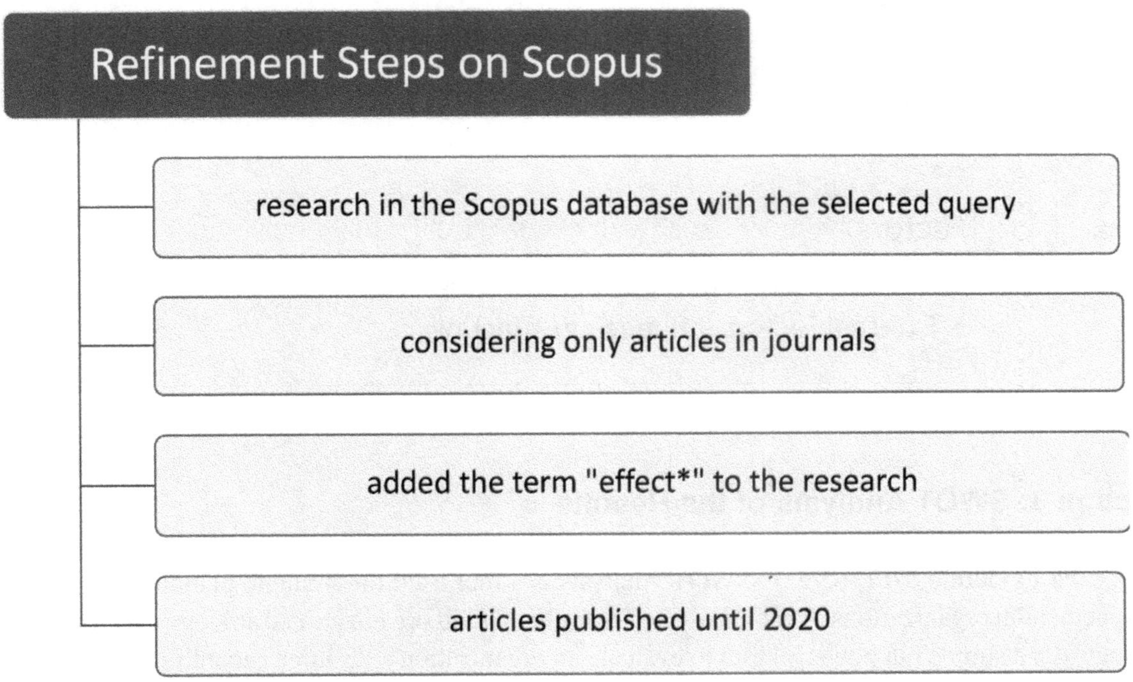

From Bibliometrics, groups of articles were identified, in order to identify the success factors and the effects of adopting ICTs in the educational process. The articles were fully read and analyzed after complete the refinement of the research. Articles that could not be found in full texts and those that did not contribute to parameters of this study were disregarded.

Section 2: Identification and Classification of the Results

After the Scopus refinement, the Mendeley tool was used to identify relevant articles with "factor" and "effect" words in titles, abstracts or keywords of the selected articles. These articles were read to identify which of them was related to the topic.

The articles identified as important to the topic were separated in two groups: factors and effects. These groups were used to map and classify the factors and effects found in the authors' research, as presented in Figure 3.

Figure 3. Article's classification

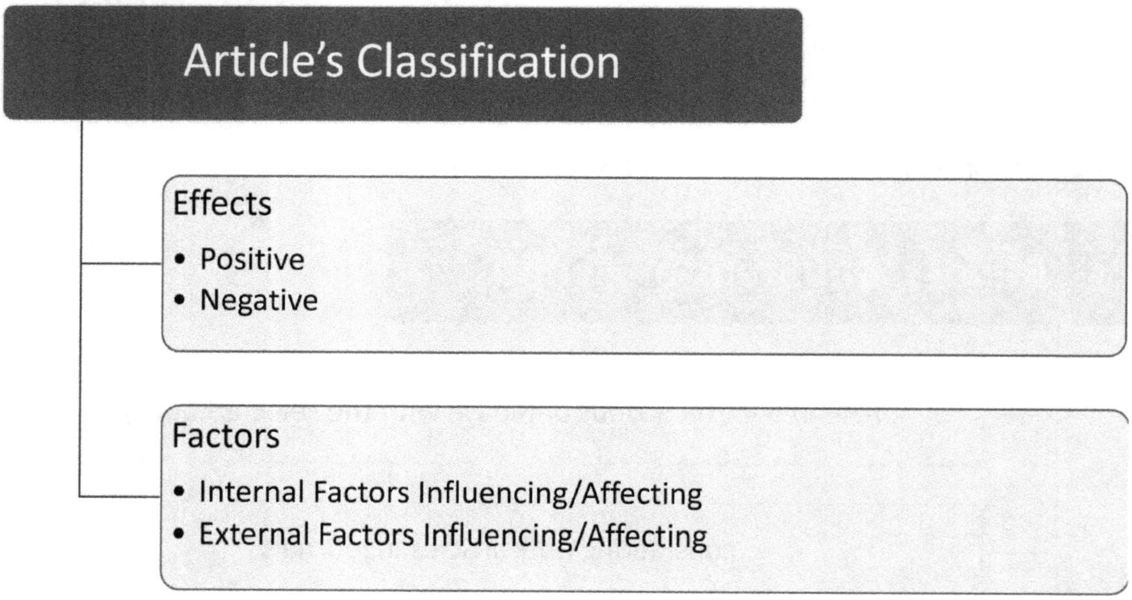

Section 3: SWOT Analysis of the Results

According to Gürel (2017, p.995) "SWOT Analysis is a tool used for strategic planning and strategic management in organizations. It can be used effectively to build organizational strategy and competitive strategy". Is a simple but powerful tool to evaluate an organization's resource capabilities and deficiencies, its market opportunities, and the external threats (Thompson et al., 2007).

From the acronym, SWOT stands for "strengths", "weakness", "opportunities" and "threats" and has two dimensions: Internal and external. Internal dimension includes organizational factors, also strengths and weaknesses, external dimension includes environmental factors, also opportunities and threats (Gürel, 2017).

In this research, the factors mapped were categorized from SWOT framework, allowing to identify in the internal and external factors, which can be the critical success factors in the adoption of the ICT in the educational process in the educational institutions.

The adoptions of the SWOT-based framework is original in reviews on this topic and was inspired in the previous work of (Pereira et al., 2019) and (Pereira et al., 2017) that looks for barriers and opportunities in university patenting process. In addition, Oliveira et al. (2018) and Méxas et al. (2013) were used as source inspirations to organize the present research.

RESULTS AND FINDINGS

In this chapter, the results obtained through the application of Sections 1, 2 and 3 above, are presented.

Section 1: Results of Research Refinement

The first step was the initial search in the Scopus database with the following command: (TITLE-ABS-KEY ("information communication technolog*" OR "ICT" OR "digital tool*") AND TITLE-ABS -KEY ("educational process*" OR "teaching process*" OR "learning system" OR "learning process")). With the described query, 3,411 publications were obtained.

Figure 4. Results of refinement steps on Scopus database
Source: Scopus Data Base

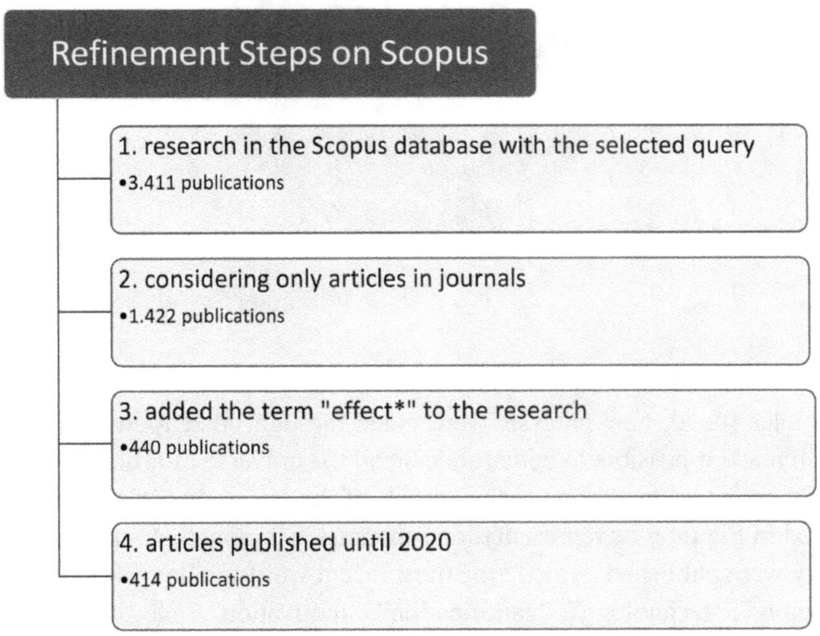

The following step was the first refinement, totaling 41.7% of the total documents, corresponding to 1,422 publications. However, in order make the research more adequate to what is sought through the review of the bibliography, one more refinement was applied to limit the research to articles that addressed the effects of the adoption of ICT in the educational process, as shown in Figure 4. Thus, the command for the new refinement applied was: (TITLE-ABS-KEY ("information communication technolog*" OR "ICT" OR "digital tool*") AND TITLE-ABS-KEY ("educational process*" OR " teaching process*" OR "learning system" OR "learning process") AND TITLE-ABS-KEY ("effect*")) AND (LIMIT-TO (DOCTYPE, "ar")) AND (EXCLUDE (PUBYEAR, 2021)).

With the last refinement made in step 4, limiting to articles published until 2020, 414 publications were found that were the basis for the bibliometric analysis.

As seen in Figure 5, Social Sciences is the Scopus' subject area with more publications with a total of 37.4%. Next, there is the areas of Computer Science, with 19.6% and Engineering, with 11.5%. The other areas, despite being represented by smaller percentages, demonstrate the existing interest in the subject.

Figure 5. Percentage of articles published by Scopus area
Source: Scopus Data Base

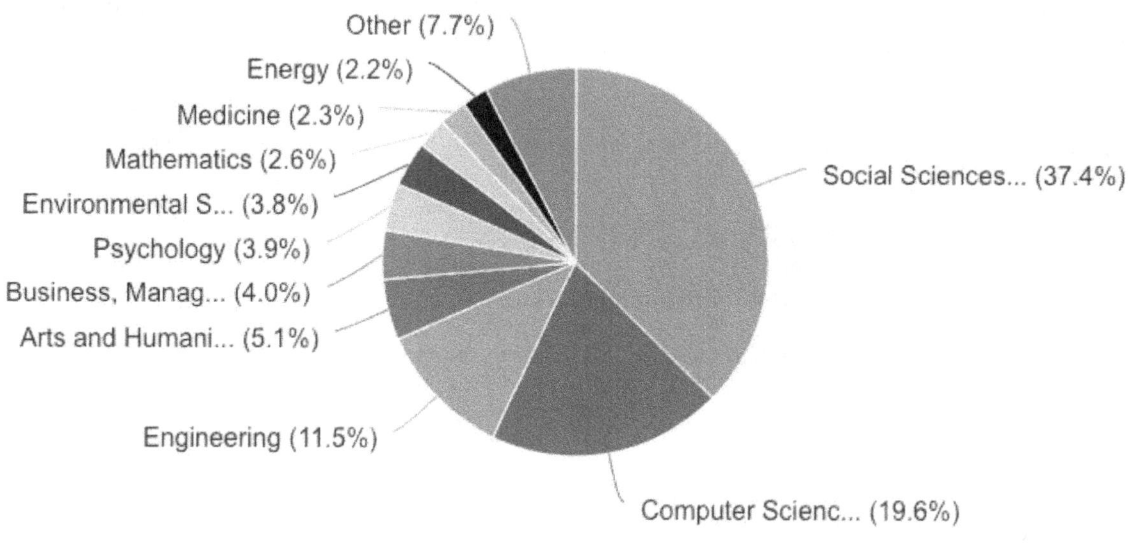

Among the articles found, new analysis were made through other tools, such as VOSviwer and Wordclouds, which made it possible to better understand the universe of articles.

Using the VOSviewer tool to analyze the keywords of the selected articles, which appear at least 5 times, it is observed in Figure 6 its representation separated by colors that represent the clusters of the years in which they were published. Among the most recent words, yellow cluster, are "blended learning", "mobile learning", "technology", "gamification", "motivation", "physical education", "machine learning". In a north of about 4 years, there are light green and dark green clusters, featuring the words "ICT", "teaching and learning", "education", "pedagogy", "educational innovation", "higher education", "teaching, training", "information and communication" and, at about 6 years old, in the pale blue cluster, "education technology", "collaborative learning", "e-learning", "information technology", "communication technologies" and "learning". It is observed that ICTs have been analyzed in research for some years, but in the proposed theme, the keywords are found in a more recent time, covering relations with higher education, technology in education, the education system and learning, e-learning, the pedagogical part, demonstrating the existence of research in the educational process, going against the use of ICT. It is noteworthy that the most prominent keywords are present in the last 5 years of research, going against the recent peaks of publications on the subject.

Figure 6. Author keywords with a minimum of 5 occurrences (VOSviewer tool)
Source: Scopus Data Base

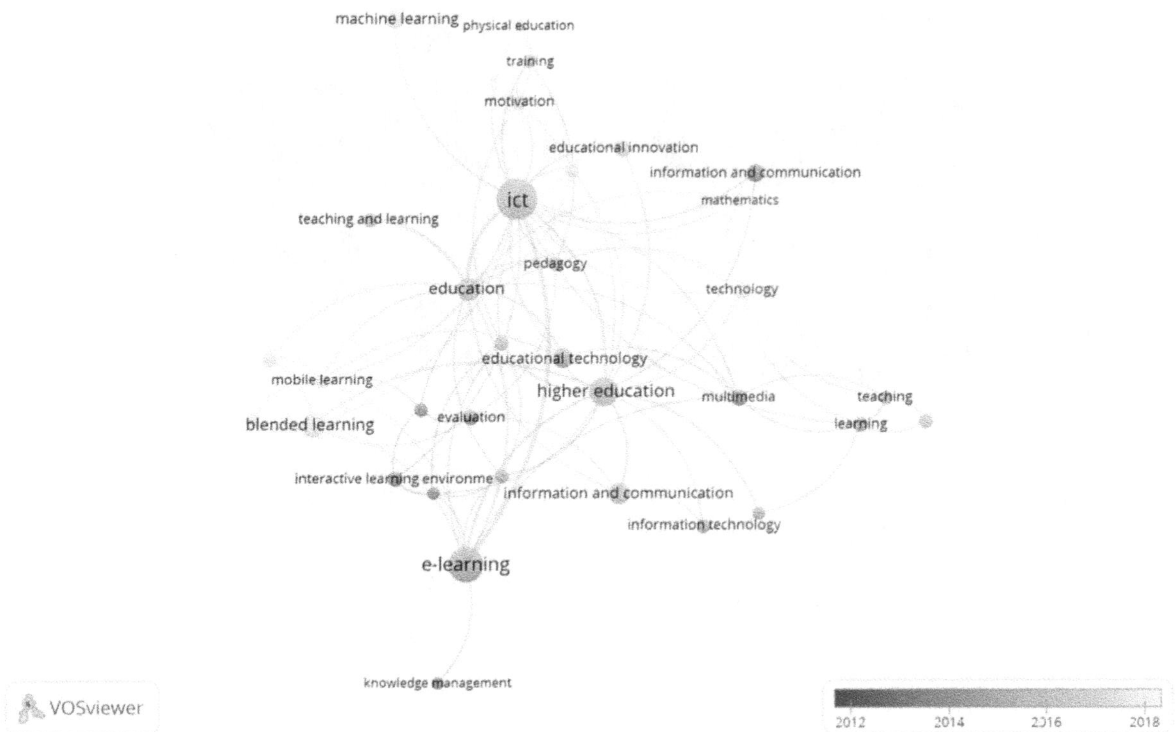

By joining the words present in the titles, abstracts, and keywords of the selected articles, which resulted in 42 articles, word adjustments were performed to standardize or exclude as necessary among prepositions, plurals, pronouns, articles, numerals, among others, to obtain a cleaner text. In figure 7, it is possible to observe the "Word Cloud" built with the adapted text inserted in the online tool Word-clouds referring to the articles in the "Departure Core" of the research. From the cloud, it is possible to observe more clearly the representativeness of the search words defined in those articles, as well as the agreement of the search result with the proposed theme.

To identify the "Departure Core", with the search results in the database, to meet the Webibliomining methodology, the initial selection was (Heymann et al., 2019): the 3 oldest articles by different authors present at the base; the 15 most recent articles by different authors present in the database; the 15 articles with the highest degree of relevance present in the database; and the articles with greater relevance for the years when there were peaks of publication on the subject - correspond to the evolution of mode.

Figure 7. Search core articles word cloud (wordclouds online tool)
Source: Adapted from the Scopus Data Base

Section 2: Identification and Classification of the Results

New refinements on articles found were made through Mendeley tool, which made it possible to separate the research focuses. These focuses were divided between factors (internal and external) and positive and negative effects.

Among the articles located by the "factors" and "effects" terms, 40 articles were selected that presents factors and 20 articles that presents effects. After analyzing each article individually as relevant to the research, 21 articles presented factors that influence and affect the adoption of ICTs in the educational process, and were found 19 articles presenting positive or negative effects related to the ICT adoption in the educational process.

From the literature review, it was possible to perceive the complexity involved in the adoption of ICTs in the educational process, considering the different stakeholders involved. For better understand the influence and impact of each of the factors in the adoption of ICTs in the educational process, through the context of identifying each of the groups of factors (internal and external), the authcrs sought to classify the factors in dimensions, as some authors presents their work, as (Bhuasiri et al., 2012) and (Rezaie et al., 2012).

Bhuasiri et al. (2012) categorizes e-learning success factors into seven dimensions based on various theories, such as social cognitive theory, information systems success model and motivation theory. Among those, the dimensions that portray the external factors are student characteristics; instructor characteristics; e-learning environment; institution and quality of service; system infrastructure and quality; quality of the course and information; and Motivation.

Based on Y. S. Wang (2003) e-learning user satisfaction model, Rezaie et al. (2012) determined four main dimensions: user interface, learning community, content, and personalization. Using pedagogical rules, user interface (friendly e-learning software, compatibility with other software and ease of use), user characteristics (computer self-efficacy and previous computer experience and teaching skills), content (suitability to the needs of the users, usefulness, levels of availability and medium used for presentation), ICT infrastructure (access to the internet and a home personal computer) and interactivity (discussion among students, access to content shared by colleagues, discussion of students with professors and face-to-face classes) as independent variables that influence user satisfaction.

According to Kanwal and Rehman (2014), there are four dimensions of external variables of implementation of an e-learning system. The first dimension is user characteristics, where all users of e-learning systems have different perceptions due to their diverse characteristics and attitudes towards using the same technology. The second is the system context in terms of technology acceptance and use, with system functionality, interactivity and response added as system context for technology implementation. The third dimension is the social context in which is the social influence on the acceptance of e-learning systems, as the subjective norm, which affects the use of the systems through the perceived usefulness. The fourth is the organizational context, where the organization's support encourages the implementation and use of the system, which can influence the acceptance of the technology through the accessibility and visibility of the system.

These dimensions are related to the influencing factors to success of ICT adoption in a sustainable way in the educational institutions. The main influencing factors to success to the adoption of ICTs in the educational process as presented on Tables 2 and 3. The dimensions associated to the external factors, can be seen in table 2, presenting the references where was found the external factors: Student Characteristics, Professor Characteristics, Motivation.

In table 3, the dimensions are associated to the internal factors to educational institutions: e-learning environment, System infrastructure and quality, Site Design, Course and information quality, Institution and service quality, Motivation, Organizational.

These dimensions (tables 2 and 3) helped to the comprehension of the influencing factor origin for the SWOT Matrix application. Besides, showed if the factors could be controlled or not for the educational institution.

Table 2. Dimensions of external factors that influence to ICT adoption in educational process: literature mapping

STUDENT'S CHARACTERISTICS	REFERENCES
Performance expectation	(Callo & Yazon, 2020; Wahid et al., 2019)
Expectation of effort	(Wahid et al., 2019)
Social influence	(Khasawneh, 2010; Wahid et al., 2019)
Facilitating conditions	(Wahid et al., 2019; Wasserman & Migdal, 2019)
Individual learning	(Ahmadpour et al., 2010)
Collective learning	(Wahid et al., 2019)
Attitude towards e-learning	(Bhuasiri et al., 2012; Umek et al., 2017)
Activity level	(Sulisworo et al., 2020)
Student gender	(Bulić et al., 2017; Suárez, 2016)
Student age	(Bulić et al., 2017; Callo & Yazon, 2020)
Personal Factors	(Ahmadpour et al., 2010; Almerich et al., 2016; Kanwal & Rehman, 2014; Wasserman & Migdal, 2019)
Demographic factors	(Umek et al., 2017)
Financial factors	(Ahmadpour et al., 2010)
Support from students	(Bhuasiri et al., 2012)
Subjective Norm (SN)	(Ching-Ter et al., 2017; Kanwal & Rehman, 2014; Wasserman & Migdal, 2019)
Self-efficacy (SE)	(Bhuasiri et al., 2012; Bulić et al., 2017; Callo & Yazon, 2020; Ching-Ter et al., 2017; Kanwal & Rehman, 2014; Umek et al., 2017)
Computer Anxiety (CA)	(Ching-Ter et al., 2017; Kanwal & Rehman, 2014)
Experience	(Callo & Yazon, 2020; Ching-Ter et al., 2017; Kanwal & Rehman, 2014; Umek et al., 2017)
Technical competence and skill	(Ahmadpour et al., 2010; Bulić et al., 2017; Callo & Yazon, 2020; Kanwal & Rehman, 2014; Umek et al., 2017)
PROFESSOR'S CHARACTERISTICS	**REFERENCES**
Personal Factors	(Ahmadpour et al., 2010; Almerich et al., 2016; Kanwal & Rehman, 2014; Wasserman & Migdal, 2019)
Financial Factors	(Ahmadpour et al., 2010)
Self-efficacy	(Almerich et al., 2016; Bhuasiri et al., 2012; Bulić et al., 2017; Callo & Yazon, 2020)
Technology Control	(Bhuasiri et al., 2012; Callo & Yazon, 2020; Langroudi, 2015; Sulisworo et al., 2020)
Chronological Answer	(Bhuasiri et al., 2012)
Focus on interaction	(Bhuasiri et al., 2012; Sulisworo et al., 2020)
Attitude towards the student	(Almerich et al., 2016; Bhuasiri et al., 2012; Callo & Yazon, 2020; Stefanovic et al., 2011; Sulisworo et al., 2020)
Attitude towards ICT	(Almerich et al., 2016; Bulić et al., 2017; Langroudi, 2015; Sang et al., 2011; Stefanovic et al., 2011)
Interaction justice	(Bhuasiri et al., 2012)
Informatics knowledge	(Almerich et al., 2016; Bhuasiri et al., 2012; Bulić et al., 2017; Callo & Yazon, 2020; De La Iglesia et al., 2018; Langroudi, 2015)

Table 2. Continued

STUDENT'S CHARACTERISTICS	REFERENCES
Technical graduation	(Almerich et al., 2016; Bhuasiri et al., 2012; Callo & Yazon, 2020; De La Iglesia et al., 2018; Sulisworo et al., 2020)
Didactic Competence	(Almerich et al., 2016; Bulić et al., 2017; Callo & Yazon, 2020; De La Iglesia et al., 2018; Langroudi, 2015; Sulisworo et al., 2020)
Gender and age	(Almerich et al., 2016; De La Iglesia et al., 2018)
Constructivist Educational Beliefs	(Sang et al., 2011; Wasserman & Migdal, 2019)
Motivation of computer use	(Sang et al., 2011; Wasserman & Migdal, 2019)
Perceptions of school policy related to ICT	(Sang et al., 2011)
Use of ICT support	(Sang et al., 2011)
Use of ICT in the classroom	(Almerich et al., 2016; Bulić et al., 2017; Sang et al., 2011)
MOTIVATION	**REFERENCES**
Perceived pleasure	(Bhuasiri et al., 2012; Ching-Ter et al., 2017; Kanwal & Rehman, 2014)
Attitude and individual expectations	(Bhuasiri et al., 2012; Callo & Yazon, 2020; Umek et al., 2017)
Social pressure and competition	(Bhuasiri et al., 2012; Kanwal & Rehman, 2014)
Perceived usefulness	(Bhuasiri et al., 2012; Khasawneh, 2010; Sanchez-Franco, 2010; Wasserman & Migdal, 2019)
Group interactivity	(Rezaie et al., 2012; Stefanovic et al., 2011; Wasserman & Migdal, 2019)
Perceived Ease of Use (PEOU)	(Bulić et al., 2017; Kanwal & Rehman, 2014; Khasawneh, 2010; Sanchez-Franco, 2010; Wasserman & Migdal, 2019)
Pleasant and stimulating perception	(Sanchez-Franco, 2010)
Interaction between professors and students	(Bulić et al., 2017; Callo & Yazon, 2020; Stefanovic et al., 2011; Sulisworo et al., 2020; Wasserman & Migdal, 2019)
Attractive experiences (Flow)	(Sanchez-Franco, 2010)
Technology	(Ahmadpour et al., 2010; Bhuasiri et al., 2012; Kanwal & Rehman, 2014; Stefanovic et al., 2011)
Culture	(Ahmadpour et al., 2010; Kanwal & Rehman, 2014)
Community	(Kanwal & Rehman, 2014)
Government ICT Policies	(Ahmadpour et al., 2010; Kanwal & Rehman, 2014)
Social environment	(Kanwal & Rehman, 2014)
Social presence (professors and students)	(Bulić et al., 2017; Sulisworo et al., 2020)
E-Learning Industry Guidance	(Muhammad et al., 2020)

Continued on following page

Table 3. Dimensions of internal factors that influence to ICT adoption in educational process: literature mapping

E-LEARNING ENVIRONMENT	REFERENCES
Institution education and support	(Ahmadpour et al., 2010; Bhuasiri et al., 2012; Callo & Yazon, 2020; Kanwal & Rehman, 2014; Umek et al., 2017)
Student assessment	(Ahmadpour et al., 2010; Bhuasiri et al., 2012; Stefanovic et al., 2011)
System communication structure	(Bhuasiri et al., 2012; Bulić et al., 2017)
Accessibility	(Ahmadpour et al., 2010; Bhuasiri et al., 2012; Bulić et al., 2017; Callo & Yazon, 2020; Messing et al., 2008)
Support from instructors	(Bhuasiri et al., 2012; Bulić et al., 2017; Callo & Yazon, 2020)
Student needs	(Ahmadpour et al., 2010; Callo & Yazon, 2020)
User interface	(Rezaie et al., 2012)
User characteristic	(Ahmadpour et al., 2010; Kanwal & Rehman, 2014; Rezaie et al., 2012)
Instructor's role	(Ahmadpour et al., 2010; Bulić et al., 2017)
SYSTEM INFRASTRUCTURE AND QUALITY	**REFERENCES**
Computer labs	(Almerich et al., 2016; Bulić et al., 2017; Callo & Yazon, 2020)
Electronic library	(Callo & Yazon, 2020)
Infrastructure Quality	(Bhuasiri et al., 2012; Rezaie et al., 2012)
System quality	(Bhuasiri et al., 2012; Rezaie et al., 2012)
Internet quality	(Ahmadpour et al., 2010; Bhuasiri et al., 2012; Callo & Yazon, 2020; Rezaie et al., 2012; Stefanovic et al., 2011)
System reliability	(Bhuasiri et al., 2012; Rezaie et al., 2012; Umek et al., 2017)
Ease of use	(Bhuasiri et al., 2012; Rezaie et al., 2012)
Ease of access	(Kanwal & Rehman, 2014; Rezaie et al., 2012)
System accessibility	(Callo & Yazon, 2020; Kanwal & Rehman, 2014; Rezaie et al., 2012)
System functionality	(Bhuasiri et al., 2012; Kanwal & Rehman, 2014; Rezaie et al., 2012)
System interactivity	(Bhuasiri et al., 2012; Bulić et al., 2017; Kanwal & Rehman, 2014; Rezaie et al., 2012; Umek et al., 2017)
System response	(Bhuasiri et al., 2012; Kanwal & Rehman, 2014; Rezaie et al., 2012)
WEBSITE DESIGN	**REFERENCES**
Appearance of the website	(Muhammad et al., 2020; Suárez, 2016)
Website usability	(Muhammad et al., 2020; Suárez, 2016)
Design quality	(Muhammad et al., 2020)
Site content	(Ahmadpour et al., 2010; Muhammad et al., 2020)
Design (attractiveness, compatibility, resources and text)	(Ahmadpour et al., 2010; Muhammad et al., 2020; Stefanovic et al., 2011)
COURSE AND INFORMATION QUALITY	**REFERENCES**
Course quality	(Bhuasiri et al., 2012; Bulić et al., 2017; Stefanovic et al., 2011)
Content relevance	(Bhuasiri et al., 2012)
Content quality	(Ahmadpour et al., 2010; Bulić et al., 2017; Stefanovic et al., 2011; Umek et al., 2017)
Course flexibility	(Bhuasiri et al., 2012; Bulić et al., 2017; Stefanovic et al., 2011)
INSTITUTION AND QUALITY OF SERVICE	**REFERENCES**

Continued on following page

Table 3. Continued

E-LEARNING ENVIRONMENT	REFERENCES
Computer training	(Bhuasiri et al., 2012; Ching-Ter et al., 2017; Messing et al., 2008)
Institution support	(Ahmadpour et al., 2010; Bhuasiri et al., 2012; Callo & Yazon, 2020; Ching-Ter et al., 2017; Messing et al., 2008; Umek et al., 2017)
Program Flexibility	(Bhuasiri et al., 2012; Bulić et al., 2017)
INTERNAL MOTIVATION FACTORS	**REFERENCES**
Setting challenging goals	(Bhuasiri et al., 2012)
Clear direction	(Bhuasiri et al., 2012)
Reward and recognition	(Bhuasiri et al., 2012; Callo & Yazon, 2020)
Punishment/regulation	(Bhuasiri et al., 2012)
ORGANIZATIONAL FACTORS	**REFERENCES**
Pedagogical Regulation	(Ahmadpour et al., 2010; Rezaie et al., 2012)
Teacher Training Programs	(Ahmadpour et al., 2010; Bhuasiri et al., 2012; Callo & Yazon, 2020; Messing et al., 2008; Umek et al., 2017)
Organizational characteristics	(Ahmadpour et al., 2010; Kanwal & Rehman, 2014; Umek et al., 2017)
Resources	(Ahmadpour et al., 2010)
Costs	(Ahmadpour et al., 2010)
Partnerships	(Ahmadpour et al., 2010)
Intellectual property and copyright	(Ahmadpour et al., 2010; Bulić et al., 2017)
Strategies	(Ahmadpour et al., 2010)
Infrastructure	(Ahmadpour et al., 2010; Messing et al., 2008)
Agility to change	(Ahmadpour et al., 2010)

Section 3: SWOT Analysis of the Results

The identified factors were classified by the author in the SWOT matrix logic, that is, strengths, weaknesses, opportunities, and threats, as each factor is perceived by the author (Table 4). The various works analyzed present factors that represent strengths or weaknesses of educational institutions and others that are identified as opportunities or threats to the process of integrating the teaching and learning process with ICTs.

Table 4 presents a compilation of the information collected in each of the selected references, aiming to provide an overview of what is already known in the literature and in which fields.

Some factors appear more frequently, especially when we observe the characteristics of the student and professor, who are more visible and vulnerable in the process and are a latent threat because they cannot be controlled by educational institutions without an action strategy well defined and a consolidated structure in the educational institution involved.

Therefore, external factors are classified as threats or opportunities, as they are not controlled by institutions, but are influenced by the environment and stakeholders, while external factors are classified as weaknesses or strengths, as they are made available by resources, infrastructure, expertise, or interest may be controlled by educational institutions.

Table 4. Previous research information about influencing and impacting factors to success the adoption of ICTs in Educational Process

REFERENCES	STRENGTHS	WEAKNESSES	OPPORTUNITIES	THREATS	AREA
(Ahmadpour et al., 2010)	Institutional instruction and support; Student assessment; Accessibility; Instructor's Role; Site Content; Design (attractiveness, compatibility, resources, and text); Content quality; Pedagogical Regulation; Professor Training Programs; Strategies; Agility to change	Student needs; User characteristic; Internet quality; Organizational characteristics; Resources; Costs; Partnerships; Intellectual property and copyright; Infrastructure	Individual learning; Technical competence and skill; Government ICT Policies	Personal Factors; Financial factors; Personal Factors (professor); Financial factors (professor); Technology; Culture	Agricultural and Biological Sciences
(Almerich et al., 2016)	Computer labs		Self-efficacy (professor); Informatics knowledge; Technical graduation; Didactic Competence; Use of ICT in the classroom	Personal Factors; Personal Factors (professor); Attitude towards the student; Attitude towards ICT; Gender and age (professor);	Computer Science / Social Sciences
(Bhuasiri et al., 2012)	Institutional instruction and support; Student assessment; System communication structure; Accessibility; Infrastructure Quality; System quality; Ease of use; System functionality; System interactivity; Course quality; Content relevance; Course flexibility; Computer training; Setting challenging goals; Clear direction; Reward and recognition; Punishment/ regulation; Professor Training Programs	Support from instructors; Internet quality; System reliability; System response; Program Flexibility	Self-efficacy (SE); Self-efficacy (professor); Technology control; Chronological answer; Focus on interaction; Attitude towards the student; Informatics knowledge; Technical graduation; Perceived pleasure; perceived usefulness	Attitude towards e-learning; Support from students; Interaction justice; Attitude and individual expectations; Social pressure and competition; Technology	Computer Science / Social Sciences
(Bulić et al., 2017)	System communication structure; Accessibility; Instructor's Role; Computing Laboratories; System interactivity; Course quality; Content quality; course flexibility	Support from instructors; Program flexibility; Intellectual property and copyright	Self-efficacy (SE); Technical competence and skill; Self-efficacy (professor); Didactic Competence; Use of ICT in the classroom; Perceived Ease of Use (PEOU); Social presence (instructors and students)	Student's gender; Student age; Attitude towards ICT; Interaction between professors and students	Social Sciences
(Callo & Yazon, 2020)	Institutional instruction and support; Accessibility; Computing Laboratories; System accessibility; Institution support; Reward and recognition; Professor Training Programs	Support from instructors; Student needs; Electronic library; Internet quality	Self-efficacy (SE); Experience; Technical competence and skill; Self-efficacy (professor); Technology control; Attitude towards the student; Informatics knowledge; Technical graduation; Didactic Competence	Performance expectation; Student age; Attitude and individual expectations; Interaction between professors and students	Social Sciences
(Suárez, 2016)	Appearance of the website; Website usability			Student's gender	Business, Management and Accounting / Economics, Econometrics and Finance

Continued on following page

Table 4. Continued

REFERENCES	STRENGTHS	WEAKNESSES	OPPORTUNITIES	THREATS	AREA
(Ching-Ter et al., 2017)	Computer training; Institution support		Self-efficacy (SE); Experience; Perceived pleasure	Subjective Norm (SN); Computer Anxiety (CA)	Computer Science / Social Sciences
(De La Iglesia et al., 2018)			Informatics knowledge; Technical graduation; Didactic Competence	Gender and age (professor)	Computer Science / Social Sciences
(Kanwal & Rehman, 2014)	Institutional instruction and support; Ease of access; System accessibility; System functionality; system interactivity	User characteristic; System response; Organizational characteristics	Self-efficacy (SE); Experience; Technical competence and skill; Perceived pleasure; Perceived Ease of Use (PEOU); Government ICT policies; social environment	Personal Factors; Subjective Norm (SN); Computer Anxiety (CA); Personal Factors (professor); Social pressure and competition; Technology; Culture; Community	Biochemistry, Genetics and Molecular Biology
(Khasawneh, 2010)			Perceived usefulness; Perceived Ease of Use (PEOU)	Social influence	Business, Management and Accounting / Social Sciences
(Langroudi, 2015)			Technology control; Informatics knowledge; Didactic Competence	Attitude towards ICT	Business, Management and Accounting / Decision Sciences / Social Sciences
(Messing et al., 2008)	Accessibility; Computer training; Institution support; Professor Training Programs	Infrastructure			Engineering / Social Sciences
(Muhammad et al., 2020)	Appearance of the website; Website usability; Design quality; Site Content; Design (attractiveness, compatibility, resources, and text)		E-Learning Industry Guidance		Energy / Environmental Science / Social Sciences
(Rezaie et al., 2012)	User interface; Infrastructure Quality; System quality; Ease of use; Ease of access; System accessibility; System functionality; System interactivity; Pedagogical Regulation	User characteristic; Internet quality; System reliability; System response	Group Interactivity		Computer Science / Social Sciences
(Sanchez-Franco, 2010)			Perceived usefulness; Perceived Ease of Use (PEOU); Pleasant and stimulating perception; Attractive experiences (Flow)		Computer Science / Social Sciences
(Sang et al., 2011)			Motivation of computer use; Perceptions of school policy related to ICT; Use of ICT support (professor); Use of ICT in the classroom	Attitude towards ICT; Constructivist Educational Beliefs	Computer Science / Social Sciences

Continued on following page

Table 4. Continued

REFERENCES	STRENGTHS	WEAKNESSES	OPPORTUNITIES	THREATS	AREA
(Stefanovic et al., 2011)	Student assessment; Design (attractiveness, compatibility, resources and text); Course quality; Content quality; Course flexibility	Internet quality	Attitude towards the student; Group Interactivity	Attitude towards ICT; Interaction between professors and students; Technology	Computer Science / Engineering / Environmental Science / Social Science
(Sulisworo et al., 2020)			Technology control; Focus on interaction; Attitude towards the student; Technical graduation; Didactic Competence; Social presence (instructors and students)	Activity level; Interaction between professors and students	Social Sciences
(Umek et al., 2017)	Institutional instruction and support; System interactivity; Content quality; Professor Training Programs	System reliability; Organizational characteristics	Self-efficacy (SE); Experience; Technical competence and skill	Attitude towards e-learning; Demographic factors; Attitude and individual expectations	Business, Management and Accounting / Social Sciences
(Wahid et al., 2019)			Facilitating conditions; collective learning	Performance expectation; Expectation of effort; social influence	Business, Management and Accounting / Social Sciences
(Wasserman & Migdal, 2019)			Facilitating conditions; Motivation of computer use; Perceived usefulness; Group Interactivity; Perceived Ease of Use (PEOU)	Personal Factors; Subjective Norm (SN); Personal Factors (professor); Constructivist educational beliefs; Interaction between professors and students; Technology	Computer Science / Social Sciences

In table 5, can be seen the positive and negative effects identified on literature resulted by the ICT adoption in the educational process of educational institutions.

However, there is an agreement on several surveys that each institution make use of models adapted to their needs, with interpretations of the e-learning concept depending on the need to adjust for its implementation.

Table 5. Previous research information about effects of the adoption of ICTs in educational process

REFERENCES	POSITIVE EFFECTS	NEGATIVE EFFECTS
(Anggeraini et al., 2019)	Development of required skills	
(Condie & Livingston, 2007)	Better student performance; Increased possibilities of working tools for teachers	Resistance to change; Ineffectiveness of the learning process; Impairment of the added value and the learning experience inside and outside the classroom
(Costa et al., 2014)	Better student performance; Development of basic skills and new skills in students related to technology	Ineffectiveness of the learning process; Impairment of the added value and the learning experience inside and outside the classroom; Demotivation of student participation and attendance
(De La Iglesia et al., 2018)	Development of basic skills and new skills in students related to technology; Increased possibilities of working tools for teachers; Development of required skills	
(Deaconu et al., 2018)		Resistance to change
(Dunleavy & Heinecke, 2007)	Better student performance	Unequal student performance
(Hernandez et al., 2011)	Flexibility and development of innovative teaching and learning strategies; Improvement in students' attitudes; Dissemination of knowledge with mobility	Demotivation of student participation and attendance; Dependence on student acceptance and continued use of tools; Lack of integration on the part of students; Worsening student performance; Lack of student interest
(Heinrich et al., 2020)	Better student performance; Boosting the students' learning process; Increased possibilities of working tools for teachers; Development of required skills; More effective teaching in a physical classroom environment; Facilitate student learning; Better access to resources and information; Reduction of student dropout; Improved communication and relationship between teacher and student	Unequal student performance
(Hrast & Savec, 2018)	Better student performance; Development of students' research capacity; Ease of data management	
(Karim & Heckman, 2005)	Better learning environment	Inefficiency in the use of ICT
(Lai & Pratt, 2007)	Better student performance; Boosting the students' learning process; Development of basic skills and new skills in students related to technology; More effective teaching in a physical classroom environment; Development of students' research capacity; Improvement in students' attitudes; Better access to resources and information; Improvement of thinking, questioning, criticism by students	Increased practice of plagiarism by students
(Martínez-Cerdá et al., 2018)	Development of basic skills and new skills in technology-related students	
(Regueras et al., 2009)	Better student performance; Boosting the students' learning process; Development of students' research capacity	
(Sánchez-Polo et al., 2019)	Increased possibilities of working tools for teachers; Development of required skills	Ineffectiveness in supporting required resources, capabilities and skills; Failure in IT technological implementations

Continued on following page

Table 5. Continued

REFERENCES	POSITIVE EFFECTS	NEGATIVE EFFECTS
(Tüzün et al., 2009)	Better student performance; Boosting the students' learning process; Development of basic skills and new skills in students related to technology; Flexibility and development of innovative teaching and learning strategies; More effective teaching in a physical classroom environment; Improvement in students' attitudes; Dissemination of knowledge with mobility	
(T. Wang & Yang, 2016)	Boosting the students' learning process; Flexibility and development of innovative teaching and learning strategies; More effective teaching in a physical classroom environment; Immediate feedback to students of applied tests; Better tracking of student learning status; Reduction of teacher workload; Ease of data management; Facilitate student learning; Wastage reduction	Inefficiency in the use of ICT
(Wood et al., 1999)	Better student performance; Immediate feedback to test students; Better tracking of student learning status	Resistance to change; Ineffectiveness of the learning process; Unequal student performance
(Zidoun et al., 2019)		Resistance to change

The lack of regulation and standardization of curricula and procedures involving ICTs in educational institutions, allow institutions to take paths that correspond to their convenience, aligned with the profile of these institutions and their stakeholders, but without a coordinated and adequate assessment.

The results of the adoption of ICT in educational institutions are directly related to planning to overcome possible problems related to the influence and impact factors identified in the SWOT Matrix as weaknesses and threats, as well as investing in the factors indicated as points of strength and opportunity.

PREVIOUS RESEARCH OF LITERATURE REVIEW

Along the search in the Scopus, a set of 14 articles having the characteristic of review articles were identified. However, among the 14, only 7 articles present information regarding factors or the adoption of ICT in the educational process. Of these, just two involve engineering in the areas related to the study, one from South Africa and the other from Malaysia. Despite it, aiming to increase the covering of the present research, it follows a synthesis of these previous reviews on the subject ICT, including even those that did not address engineering educations.

Hernandez-de-Menendez and Morales-Menendez (2019) review the available ICTs that can be applied in the educational field. In addition, selected universities from the QS World University Rankings by 2018 engineering and technology subjects are reviewed. The objective was to review the state of the art of the technology employed by them and summarize the innovative technological tools, trends, and teaching practices of these organizations for successful engineering education.

Joshi and Vaidya (2013) propose the taxonomy of e-learning. They discuss teaching-learning process models and learning approaches and review various methodologies used in adaptive / personalized e-learning and, in conclusion, present their vision of personalized e-learning.

Oyetade et al. (2020) searched for studies on the adoption of technology in education and carried out a systematic review aiming to classify and evaluate the articles searched in Google Scholar and IEEE.

Of the 132 articles found, 17 were analyzed among the authors' filter criteria. The authors highlighted trends in technology adoption and provided with their study empirical evidence of applications that are being used in the implementation of technology in education. In addition to addressing the topic of COVID-19 in relation to its effect on the academic curriculum around the world.

Qurat-ul-Ain et al. (2019) present a review of teaching methodologies and technological tools (computer-based, game-based, mobile device-based, online learning tools, and multimedia technologies). The study aimed to examine the roles and effectiveness of technological tools in the teaching and learning process, definition of each method, similarities and differences between teaching methodologies and technological tools. The authors identified that the use of technological tools in the teaching and learning process increases the interaction between professors and students.

Rajarapollu and Bhagwatkar (2018) shared the experiences of interacting and communicating effectively with the Millennial Learner. The authors present a discussion of using ICTs and active learning methods to increase student engagement in a course. According to the study, they identified that most of the 35 students in the survey feel comfortable using ICT tools, in addition to improving students' skills. However, they identified challenges such as the speed of the Internet and devices available to students.

Salehi et al. (2014) conducted a survey on the role of e-learning courses and ICTs in the educational environment. The research identified different effects on student learning from the way ICTs are integrated into educational practice. They indicated that the growth of ICT and E-learning courses in education result in opportunities to improve the information process, especially in universities.

Yee et al. (2019) analyzed the ideal model and recommendations for planning future blended learning activities. They analyzed nine articles between 2013 and 2018 from the Google Scholar and Emerald databases that contemplated a framework or models with future suggestions. The authors identified in the research that blended learning goes beyond ICT tools, involving ways to increase its effectiveness in the learning process. They included that blended learning design must involve collaboration between educators and its application to real life.

The authors applied SWOT Analysis to the influencing factors to success identified in the literature in order to offer new perspective to this study, as presented in the next sections.

REGARDING THE PREVIOUS REVIEWS ON ICT

In order to map information that contributed to the identification of influencing and impacting factors to success, each one of the previous reviews were analyzed under a SWOT Matrix framework (strengths, weaknesses, opportunities, and threats) plus mapping positive and negative effects of adopting ICT, which resulted in the elicitation of the positive and negative effects of the adoption of ICTs in the educational process.

Table 6 shows the sources, the aspects of SWOT covered, and the area in which the journal that published the articles is classified in Scopus database, as well as, table 7 shows positive and negative effects, from the databases used in the review.

Table 6. Previous research of literature review identifying factors that influence or affect ICTs adoption in educational process

REFERENCES	STRENGTHS	WEAKNESSES	OPPORTUNITIES	THREATS	DATA BASE	AREA
(Hernandez-de-Menendez & Morales-Menendez, 2019)			Flexible learning; Innovative learning activities; Support to professors; Teaching quality; Student assessment	Specific Digital Skills (professor); Resistance to change; Motivation; Interaction and communication between professors and students; Student satisfaction; Self-efficacy; Technical competence and skill		Social Sciences
(Joshi & Vaidya, 2013)			Personalization in e-learning; Adaptation to the student			Computer Science
(Oyetade et al., 2020)	Content Quality	Communication cost; Design; System navigability	Device portability; Collective capabilities; Control of the device by the student; Mandatory use of e-learning; Learning Style; Mobile devices; Technical and educational support	Economy; Adequacy; Ease of use; Social influence; Gender; Student satisfaction; Learning performance; Skill with technology; Perceived usefulness; Perceived ease of use; Subjective norm; Perceived playfulness (happiness); Internet	Google Scholar and IEEE	Business, Management and Accounting/ Engineering/ Physics and Astronomy
(Qurat-ul-Ain et al., 2019)		Costs	Integration of technology in education; Interactive learning; Communication	Creativity; Learning based on imagination; Critical and cognitive analysis; Educational requirements of students		Mathematics/ Social Sciences
(Rajarapollu & Bhagwatkar, 2018)				Internet; Availability of equipment/devices to students		Computer Science/ Engineering/ Social Sciences
(Salehi et al., 2014)			Develop and present curriculums in education; Feedback; Student follow-up; Accessibility; Support for the learning process; Students' access to administrative processes	Motivation; Academic achievement; platform scalability, Technical support; Standardization and personalization of content; Professor and student interaction; Interaction between students		Arts and Humanities/ Social Sciences
(Yee et al., 2019)		Adequacy of Blended Learning Models; Knowledge and ability to apply the model; Structure of educational institutions	Blended learning model (face-to-face and web-based); Flexibility in the use of technology; Use of mobile devices	Interaction of students; Professional development; Collaboration; Critical thinking; Communication; Creativity	Google Scholar and Emerald	Business, Management and Accounting/ Engineering

Table 7. Previous research of literature review identifying effects of ICTs adoption in educational process

REFERENCES	POSITIVE EFFECTS	NEGATIVE EFFECTS
(Hernandez-de-Menendez & Morales-Menendez, 2019)	Develops Skills; Effectiveness in student learning; Develops critical thinking and analytical resolution; Increases confidence; Develops self-efficacy and procedural knowledge skills; Develops spatial visualization skills; Successful completion of courses; Speeds up student progress; Improves collaboration, communication and problem solving skills	
(Joshi & Vaidya, 2013)	Flexibility in virtual room mode; Satisfaction with virtual learning	
(Oyetade et al., 2020)	Improves the interactivity modules in the students' learning process; Effectiveness of online sources	
(Qurat-ul-Ain et al., 2019)	Increases student motivation; Increases social interactions; Improved learning ability; Increases positive student outcomes; Greater student collaboration; Greater engagement in the learning process	
(Rajarapollu & Bhagwatkar, 2018)	Improved student skills	
(Salehi et al., 2014)	Greater access and equity to information	
(Yee et al., 2019)	Improved student learning	Different interpretations of the concept of "blended learning" by institutions

Cells in blank is these tables means that the information was not available in the source.
Analyzing tables 6 and 7, one can conclude that:

- most of reviews were articles published in journals indexed in the Social Science area by Scopus;
- almost all of them did not approach negative effects of using ICT tools – the only exception is Yee et al. (2019);
- Yee et al. (2019) is also the only one reference that mapped strengths in ICT adoption;
- almost all of the reviews analyzed, mentioned factors that should be classified as opportunities or as threats to the adoption of ICT.

CONTRIBUTION OF THIS PAPER

As shown in this research, the SWOT analysis was not performed in any other review article selected in this research. This research provides the mapping and compilation of the factors and the SWOT matrix classification of those factors to allow a better perspective of possible actions to improve the ICT implementation process in educational institutions, as in engineering courses, with sustainability.

In addition to the influencing factors to success, the research gathered and mapped information regarding the positive and negative effects identified in the literature of the ICT implementation process in the teaching-learning process of educational institutions. The effects mapping allows a better perspective to improve of actions to circumvent the negative effects of the ICT adoption process in the educational process of teaching institutions.

SOLUTIONS AND RECOMMENDATIONS

This work searched for gaps in the literature to contribute with a different perspective, complementing the research that were found. However, there were limitations such as how the database used, the specificity and the refinement of the research.

The authors recommend future research to be carried out with specialists for a better analysis of the factors and effects caused by the adoption of ICTs in different institutional profiles. Also covering the complex scenario that involves the issue of the adoption of ICTs in the educational process to enable the identification of the most fragile and critical points that should have efforts directed as a priority.

The analysis of the dimensions involved separately compromises the understanding of the root causes of the problem, preventing the developed models from being adequate, effective, and probable to be replicated in other profiles of institutions.

FUTURE RESEARCH DIRECTIONS

From the survey carried out in this review, it is feasible and essential to compare information with experts to validate the data with the reality of other educational institutions so that the best solution to the problem is identified.

CONCLUSION

The present study fulfilled its objective, bringing together and mapping the influencing and impact factors to success and the effects of adopting ICTs in the educational process based on a literature review.

The set of articles analyzed comprise a vast sample of one of the main databases, Scopus Elsevier, covering a significant period to understand the evolution and relevance of the theme in the scientific community.

The tools used in the bibliometric review process and in the research refinement ensured a better data analysis within the parameters defined in the research methodology. Through Mendeley and individual reading of the articles, the research was refined to obtain a reliable sample of what has been researched and worked on in the subject.

In this research limits, the small number of articles in the research sample that are classified from Scopus database as engineering area is noteworthy, however, the articles analyzed emphasize the importance of ICTs in STEM courses, which include engineering courses. There is a need to integrate ICT in the teaching-learning process to improve the resources available to students and teachers, as well as to train students for the job market, as technology advances rapidly and is increasingly present in organizations.

Therefore, it is essential that educational institutions give due attention to the need and criticality that involves the inclusion and adequacy of ICTs in the educational process, that is, involving both the institution's internal processes and the process teaching-learning.

However, it is important that the government and the sector's regulatory institutions create policies, regulations, standards that guide and guarantee the minimum structure necessary for the stakeholders involved, including the educational institution.

The pandemic faced worldwide in 2020 and 2021 due to COVID-19, impacted in several sectors, as well as in education and the use of technologies has been extremely necessary while there is no security for the exposure of students and professors. The theme has gained relevance increasing and constant strength in recent years, demonstrating the big impact on education if there is no management focused on the successful integration of ICTs in the educational process.

REFERENCES

Ahmadpour, A., Mirdamadi, M., Hosseini, J. F., & Chizari, M. (2010). Factors influencing the design of electronic learning system in agricultural extension. *American Journal of Agricultural and Biological Sciences*, *5*(2), 122–127. doi:10.3844/ajabssp.2010.122.127

Aleem, M., Qurat-ul-Ain, Q.-A., Shahid, F., Islam, M. A., Iqbal, M. A., & Yousaf, M. M. (2019). A review of technological tools in teaching and learning computer science. *Eurasia Journal of Mathematics, Science and Technology Education*, *15*(11). Advance online publication. doi:10.29333/ejmste/109611

Ali, E. A. (2014). *ICT in engineering educational content delivery: Challenges and opportunities.* doi:10.5339/qproc.2014.wcee2013.4

Almerich, G., Orellana, N., Suárez-Rodríguez, J., & Díaz-García, I. (2016). Teachers' information and communication technology competences: A structural approach. *Computers & Education*, *100*, 110–125. doi:10.1016/j.compedu.2016.05.002

Anggeraini, Y., Faridi, A., Mujiyanto, J., & Bharati, D. A. L. (2019). The teachers' perceptions on digital literacy competences in EFL classroom. *Asian EFL Journal*, *24*(4), 5–12.

Benchicou, S., Aichouni, M., & Nehari, D. (2010). E-learning in engineering education: A theoretical and empirical study of the Algerian higher education institution. *European Journal of Engineering Education*, *35*(3), 325–343. doi:10.1080/03043797.2010.483610

Bhuasiri, W., Xaymoungkhoun, O., Zo, H., Rho, J. J., & Ciganek, A. P. (2012). Critical success factors for e-learning in developing countries: A comparative analysis between ICT experts and faculty. *Computers & Education*, *58*(2), 843–855. doi:10.1016/j.compedu.2011.10.010

Buenaño-Fernandez, D., Luján-Mora, S., & Gil, D. (2019). A hybrid machine learning approach for the prediction of grades in computer engineering students. In *Research and Innovation Forum, Rii Forum 2019* (pp. 125–134). Springer. 10.1007/978-3-030-30809-4_13

Bulić, M., Jelaska, I., & Jelaska, P. M. (2017). The effect of E-learning on the acquisition of learning outcomes in teaching science and biology. *Croatian Journal of Education*, *19*(2), 447–477. doi:10.15516/cje.v19i2.2230

Callo, E. C., & Yazon, A. D. (2020). Exploring the factors influencing the readiness of faculty and students on online teaching and learning as an alternative delivery mode for the new normal. *Universal Journal of Educational Research*, *8*(8), 3509–3518. doi:10.13189/ujer.2020.080826

Cantoni, V., Cellario, M., & Porta, M. (2004). Perspectives and challenges in e-learning: Towards natural interaction paradigms. *Journal of Visual Languages and Computing*, *15*(5), 333–345. doi:10.1016/j.jvlc.2003.10.002

Ching-Ter, C., Hajiyev, J., & Su, C. R. (2017). Examining the students' behavioral intention to use e-learning in Azerbaijan? The General Extended Technology Acceptance Model for E-learning approach. *Computers & Education*, *111*, 128–143. doi:10.1016/j.compedu.2017.04.010

Condie, R., & Livingston, K. (2007). Blending online learning with traditional approaches: Changing practices. *British Journal of Educational Technology*, *38*(2), 337–348. doi:10.1111/j.1467-8535.2006.00630.x

Cooner, T. S. (2010). Creating opportunities for students in large cohorts to reflect in and on practice: Lessons learnt from a formative evaluation of students' experiences of a technology-enhanced blended learning design. *British Journal of Educational Technology*, *41*(2), 271–286. doi:10.1111/j.1467-8535.2009.00933.x

Costa, S., Cuzzocrea, F., & Nuzzaci, A. (2014). Uses of the internet in educative informal contexts. Implication for formal education. *Comunicar*, *22*(43), 163–171. doi:10.3916/C43-2014-16

De La Iglesia, J. C. F., Morante, M. C. F., & López, B. C. (2018). Influence of personal and enviroment factors on classroom ICT integration in Galicia. *Revista de Medios y Educacion*, *53*, 79–91. doi:10.12795/pixelbit.2018.i53.05

Deaconu, A., Dedu, E. M., Igreţ, R. S., & Radu, C. (2018). The use of information and communications technology in vocational education and training-premise of sustainability. *Sustainability (Switzerland)*, *10*(5), 1466. Advance online publication. doi:10.3390u10051466

Dunleavy, M., & Heinecke, W. F. (2007). The Impact of 1:1 Laptop use on middle school math and science standardized test scores. *Computers in the Schools*, *24*(3–4), 7–22. doi:10.1300/J025v24n03_02

Gürel, E. (2017). SWOT analysis: A theoretical review. *Journal of International Social Research*, *87*(1-2), 149–200. doi:10.17719/jisr.2017.1832

Heinrich, C. J., Darling-Aduana, J., & Martin, C. (2020). The potential and prerequisites of effective tablet integration in rural Kenya. *British Journal of Educational Technology*, *51*(2), 498–514. doi:10.1111/bjet.12870

Hernandez, B., Montaner, T., Sese, F. J., & Urquizu, P. (2011). The role of social motivations in e-learning: How do they affect usage and success of ICT interactive tools? *Computers in Human Behavior*, *27*(6), 2224–2232. doi:10.1016/j.chb.2011.07.001

Hernandez-de-Menendez, M., & Morales-Menendez, R. (2019). Technological innovations and practices in engineering education: A review. *International Journal on Interactive Design and Manufacturing*, *13*(2), 713–728. doi:10.100712008-019-00550-1

Hrast, S., & Savec, V. F. (2018). ICT-supported inquiry-based learning. *World Transactions on Engineering and Technology Education*, *16*(4), 398–403. https://www.scopus.com/inward/record.uri?eid=2-s2.0-85062687538&partnerID=40&md5=4649f1374ae29596070d5c0da0be43e7

Hurlburt, G., Voas, J., Miller, K., Laplante, P., & Michael, B. (2010). A nonlinear perspective on higher education. *Computer, 43*(12), 90–92. doi:10.1109/MC.2010.338

Johnson, L., Adams Becker, S., Estrada, V., & Martín, S. (2013). *The Technology Outlook for STEM+ Education 2013-2018: An NMC Horizon Project Sector Analysis* (Vol. 148). The New Media Consortium. https://files.eric.ed.gov/fulltext/ED559376.pdf

Joshi, M., & Vaidya, R. (2013). A review of paradigm shift from conventional to personalized e-learning. *2013 2nd International Conference on Advances in Computing, Communications and Informatics, ICACCI 2013*, 1263–1269. 10.1109/ICACCI.2013.6637359

Junco, R., & Cotten, S. R. (2011). Perceived academic effects of instant messaging use. *Computers & Education, 56*(2), 370–378. doi:10.1016/j.compedu.2010.08.020

Kanwal, F., & Rehman, M. (2014). E-learning adoption model: A case study of Pakistan. *Life Science Journal, 11*(4), 78–86. https://www.scopus.com/inward/record.uri?eid=2-s2.0-84899152280&partnerID=40&md5=08de3e65655d108cf64f5cfdd905d4d3

Karim, N. S. A., & Heckman, R. (2005). Group communication media choice and the use of information and communication technology to support learning: A case study. *Campus-Wide Information Systems, 22*(1), 28–42. doi:10.1108/10650740510574401

Khasawneh, A. M. (2010). Using internet technology to support flexible learning in jordan: A case of Hashemite university. *International Journal of Innovation and Learning, 8*(1), 1–10. doi:10.1504/IJIL.2010.034010

Lai, K.-W., & Pratt, K. (2007). Positive to a degree: The effects of ICT use in New Zealand secondary schools. *Computers in the Schools, 24*(3–4), 95–109. doi:10.1300/J025v24n03_07

Langroudi, Z. K. (2015). Effective factors on information and communication technology usage by university teachers; a case study of kharazmi university. *International Journal of Information Science and Management, 13*(2), 51–59.

López-Pérez, M. V., Pérez-López, M. C., & Rodríguez-Ariza, L. (2011). Blended learning in higher education: Students' perceptions and their relation to outcomes. *Computers & Education, 56*(3), 818–826. doi:10.1016/j.compedu.2010.10.023

Lytras, M. D., Visvizi, A., Daniela, L., Sarirete, A., & De Pablos, P. O. (2018). Social networks research for sustainable smart education. *Sustainability (Switzerland), 10*(9), 2974. Advance online publication. doi:10.3390u10092974

Mahdi, A. E. (2006). Introducing peer-supported learning approach to tutoring in engineering and technology courses. *International Journal of Electrical Engineering Education, 43*(4), 277–287. doi:10.7227/IJEEE.43.4.1

Malathi, M., Paramasivam, C., Sangeetha, S., Gospeline Christiana, S., & Nagalakshmi, K. (2019). Effective implementation of padlet ICT tool for engineering education. *International Journal of Innovative Technology and Exploring Engineering, 8*(8), 1586–1590.

Martínez-Cerdá, J.-F., Torrent-Sellens, J., & González-González, I. (2018). Promoting collaborative skills in online university: Comparing effects of games, mixed reality, social media, and other tools for ICT-supported pedagogical practices. *Behaviour & Information Technology*, *37*(10–11), 1055–1071. doi:10.1080/0144929X.2018.1476919

Martínez-Cerdá, J.-F., Torrent-Sellens, J., & González-González, I. (2020). Socio-technical e-learning innovation and ways of learning in the ICT-space-time continuum to improve the employability skills of adults. *Computers in Human Behavior*, *107*, 105753. Advance online publication. doi:10.1016/j.chb.2018.10.019

Mengist, W., Soromessa, T., & Legese, G. (2020). Method for conducting systematic literature review and meta-analysis for environmental science research. *MethodsX*, *7*, 100777. doi:10.1016/j.mex.2019.100777 PMID:31993339

Messing, J., Altas, I., & Howarth, J. (2008). Implementing Mmicrosoft PiL in Australian schools. *International Journal of Emerging Technologies in Learning*, *3*(2), 63–68. https://www.scopus.com/inward/record.uri?eid=2-s2.0-84862905363&partnerID=40&md5=da7c88fcf192b826630e018d3b55e048

Méxas, M. P., Costa, H. G., & Quelhas, O. L. G. (2013). Avaliação da importância relativa dos critérios para a seleção de Sistemas Integrados de Gestão (ERP) para uso em empresas da construção civil. *Gestão & Produção*, *20*(2), 337–356. doi:10.1590/S0104-530X2013000200007

Mitchell, P., & Forer, P. (2010). *Blended Learning: The Perceptions of First-year Geography Students.* doi:10.1080/03098260902982484

Muhammad, A. H., Siddique, A., Youssef, A. E., Saleem, K., Shahzad, B., Akram, A., & Al-Thnian, A.-B. S. (2020). A hierarchical model to evaluate the quality of web-based E-learning systems. *Sustainability (Switzerland)*, *12*(10), 4071. Advance online publication. doi:10.3390u12104071

Oliveira, A. S., Barros, M. D., Pereira, F. C., Gomes, C. F. S., & Costa, H. G. (2018). Prospective scenarios: A literature review on the Scopus database. *Futures*, *100*, 20–33. doi:10.1016/j.futures.2018.03.005

Osguthorpe, R. T., Osguthorpe, R. T., & Graham, C. R. (2003). Blended Learning Environments: Definitions and Directions. *Quarterly Review of Distance Education*, *4*(3), 227–233.

Oyetade, K. E., Zuva, T., & Harmse, A. (2020). Technology adoption in education: A systematic literature review. *Advances in Science. Technology and Engineering Systems*, *5*(6), 108–112. doi:10.25046/aj050611

Papaioannou, P., & Charalambous, K. (2011). Principals' attitudes towards ICT and their perceptions about the factors that facilitate or inhibit ICT integration in primary schools of Cyprus. *Journal of Information Technology Education*, *10*(1), 349–369. doi:10.28945/1530

Pereira, F. C., Costa, H. G., & Pereira, V. (2017). Patent filings versus articles published: A review of the literature in the context of Multicriteria Decision Aid. *World Patent Information*, *50*, 17–26. doi:10.1016/j.wpi.2017.07.003

Pereira, F. C., Costa, H. G., & Pereira, V. (2019). Stimulating and inhibiting factors of patent filing with Brazilian universities. *International Journal of Entrepreneurship and Innovation Management*, *23*(3), 261–280. doi:10.1504/IJEIM.2019.099844

Radovan, M., & Kristl, N. (2020). Learning and teaching in a virtual learning environment – the importance of community building and collaboration. *Sodobna Pedagogika/Journal of Contemporary Educational Studies, 71*(2), 10–23. https://www.scopus.com/inward/record.uri?eid=2-s2.0-85089198731&partnerID=40&md5=ee91a39cbaf3314f6669d90304fcae70

Rajarapollu, P. R., & Bhagwatkar, S. A. (2018). ICT - An Effective Way for Active and Collaborative Learning Teaching Process in Engineering Education - A Review. *2017 International Conference on Transforming Engineering Education, ICTEE 2017*. 10.1109/ICTEED.2017.8585697

Regueras, L. M., Verdú, E., Muñoz, M. F., Perez, M. A., de Castro, J. P., & Verdú, M. J. (2009). Effects of competitive E-learning tools on higher education students: A case study. *IEEE Transactions on Education, 52*(2), 279–285. doi:10.1109/TE.2008.928198

Rezaie, M., Nosratabadi, H. E., & Fazlollahtabar, H. (2012). Applying KANO model for users' satisfaction assessment in e-learning systems: A case study in iran virtual higher educational systems. *International Journal of Information and Communication Technology Education, 8*(3), 1–12. doi:10.4018/jicte.2012070101

Rodriguez, D. S. S., Costa, H. G., & Carmo, L. F. R. R. S. (2013). Métodos de auxílio multicritério à decisão aplicados a problemas de PCP: Mapeamento da produção em periódicos publicados no Brasil. *Gestão & Produção, 20*(1), 134–146. doi:10.1590/S0104-530X2013000100010

Romi, I. M. (2017). A Model for e-learning Systems Success: Systems, Determinant, and Performance. *International Journal of Emerging Technologies in Learning, 12*(10), 4–20. doi:10.3991/ijet.v12i10.6680

Rosero-Zambrano, C. A., Avila, A., Osorio, L. A., & Aguirre, S. (2018). Impact of Adding Internet Technology on Student Performance and Perception of Autonomy in Fundamentals of Electronics Course. *Journal of Science Education and Technology, 27*(2), 177–187. doi:10.100710956-017-9716-y

Rutz, E., Eckart, R., Wade, J. E., Maltbie, C., Rafter, C., & Elkins, V. (2003). Student performance and acceptance of instructional technology: Comparing technology-enhanced and traditional instruction for a course in statics. *Journal of Engineering Education, 92*(2), 133–140. doi:10.1002/j.2168-9830.2003.tb00751.x

Salehi, H., Shojaee, M., & Sattar, S. (2014). Using e-learning and ICT courses in educational environment: A review. *English Language Teaching, 8*(1), 63–70. doi:10.5539/elt.v8n1p63

Sanchez-Franco, M. J. (2010). WebCT - The quasimoderating effect of perceived affective quality on an extending Technology Acceptance Model. *Computers & Education, 54*(1), 37–46. doi:10.1016/j.compedu.2009.07.005

Sánchez-Polo, M. T., Cegarra-Navarro, J.-G., Cillo, V., & Wensley, A. (2019). Overcoming knowledge barriers to health care through continuous learning. *Journal of Knowledge Management, 23*(3), 508–526. doi:10.1108/JKM-10-2018-0636

Sang, G., Valcke, M., van Braak, J., Tondeur, J., & Zhu, C. (2011). Predicting ICT integration into classroom teaching in Chinese primary schools: Exploring the complex interplay of teacher-related variables. *Journal of Computer Assisted Learning, 27*(2), 160–172. doi:10.1111/j.1365-2729.2010.00383.x

Scopus. (n.d.). Retrieved May 3, 2019, from https://www-scopus.ez24.periodicos.capes.gov.br/term/analyzer.uri?sid=a318f65953790c5f3e34092c19bad69c&origin=resultslist&src=s&s=EID%282-s2.0-78049514934+OR+2-s2.0-79251568005+OR+2-s2.0-80052818916+OR+2-s2.0-84890889040+OR+2-s2.0-84897615891+OR+2-s2.0-84

Shorfuzzaman, M., Alelaiwi, A., Masud, M., Hassan, M. M., & Hossain, M. S. (2015). Usability of a cloud-based collaborative learning framework to improve learners' experience. *Computers in Human Behavior*, *51*, 967–976. doi:10.1016/j.chb.2014.10.002

Silva, G. B., & Costa, H. G. (2015). Mapeamento de um núcleo de partida de referências em Data Mining a partir de periódicos publicados no Brasil. *Gestão & Produção*, *22*(1), 107–118. doi:10.1590/0104-530X792-13

Spitz-Oener, A. (2006). Technical change, job tasks, and rising educational demands: Looking outside the wage structure. *Journal of Labor Economics*, *24*(2), 235–270. doi:10.1086/499972

Starčič, A. I., & Turk, Z. (2010). Powerful learning environments in engineering education. *7th WSEAS International Conference on Engineering Education, EDUCATION'10, International Conference on Education and Educational Technologies*, 409–416. https://www.scopus.com/inward/record.uri?eid=2-s2.0-79958763750&partnerID=40&md5=e8188bff062cfb8cfe6aa6ba889ffafb

Stefanovic, D., Drapsin, M., Nikolic, J., Scepanovic, D., Radjo, I., & Drid, P. (2011). Empirical study of student satisfaction in e-learning system environment. *Technics Technologies Education Management, 6*(4), 1152–1164. https://www.scopus.com/inward/record.uri?eid=2-s2.0-84855521462&partnerID=40&md5=a344298b5a59d03fa3af02b5ea26a6e5

Suárez, L. M. C. (2016). Investigating website appearance and usability effects on student satisfaction with the website: A descriptive analysis in three countries. *International Review on Public and Nonprofit Marketing*, *13*(3), 223–238. doi:10.100712208-015-0147-2

Sulisworo, D., Ummah, R., Nursolikh, M., & Rahardjo, W. (2020). The analysis of the critical thinking skills between blended learning implementation: Google Classroom and Schoology. *Universal Journal of Educational Research, 8*(3B), 33–40. doi:10.13189/ujer.2020.081504

Thompson, A., Strickland III, A. J., & Gamble, J. (2007). *Crafting and Executing Strategy: Text and Reading* (15th ed.). McGraw Hill-Irwin.

Tüzün, H., Yilmaz-Soylu, M., Karakuş, T., Inal, Y., & Kizilkaya, G. (2009). The effects of computer games on primary school students' achievement and motivation in geography learning. *Computers & Education*, *52*(1), 68–77. doi:10.1016/j.compedu.2008.06.008

Umek, L., Keržič, D., Aristovnik, A., & Tomaževič, N. (2017). An assessment of the effectiveness of moodle e-learning system for undergraduate Public Administration education. *International Journal of Innovation and Learning*, *21*(2), 165–177. doi:10.1504/IJIL.2017.081939

Wahid, K. A., Ismail, W. S. A. W., & Numprasertchai, H. (2019). The role of social media in collective learning. *International Journal of Innovation and Learning*, *25*(4), 363–376. doi:10.1504/IJIL.2019.099981

Wang, T., & Yang, K. (2016). Technology-enhanced science teaching and learning: Issues and trends. In Science Education Research and Practice in Asia: Challenges and Opportunities (pp. 461–481). Springer Singapore. doi:10.1007/978-981-10-0847-4_25

Wang, X. (2010). eLearning in Europe: Action plans and implications. *Proceedings - 2010 International Conference on Artificial Intelligence and Education, ICAIE 2010*, 240–245. 10.1109/ICAIE.2010.5641445

Wang, Y. S. (2003). Assessment of learner satisfaction with asynchronous electronic learning systems. *Information & Management*, *41*(1), 75–86. doi:10.1016/S0378-7206(03)00028-4

Wasserman, E., & Migdal, R. (2019). Professional development: Teachers' attitudes in online and traditional training courses. *Online Learning Journal*, *23*(1), 132–143. doi:10.24059/olj.v23i1.1299

Wood, D., Underwood, J., & Avis, P. (1999). Integrated learning systems in the classroom. *Computers & Education*, *33*(2–3), 91–108. doi:10.1016/S0360-1315(99)00027-5

Yee, C. P., Ismail, Z. I. A., & Seman, E. A. A. (2019). A check on planning access for blended learning. *International Journal of Recent Technology and Engineering*, *8*(2), 1338–1345. doi:10.35940/ijrte.B1064.0882S819

Zidoun, Y., Dehbi, R., Talea, M., & El Arroum, F. (2019). Designing a theoretical integration framework for mobile learning. *International Journal of Interactive Mobile Technologies*, *12*(12), 152–170. doi:10.3991/ijim.v13i12.10841

KEY TERMS AND DEFINITIONS

E-Learning: Tool to improve education using computer network technology, usually through electronic means such as internet, intranet, and extranet, providing information to users regardless of time and space limitations, also better use of web-based communication, training, sharing and knowledge transfer.

Information and Communication Technology (ICT): Technology consisting in the collection, organization, storage, and dissemination of information such as sound, image, text or numbers, computers and telecommunications devices.

M-Learning or Mobile Learning: Teaching methodology that provides a new environment for students and professors, using mobile devices as platforms to enable distance learning.

Perception of Ease of Use (PEOU): Computer acceptance factor that affects users' attitude and behavioral intent towards technology adoption and continued use, with respect to their perception of the ease of using technology.

Perception of Utility (PU): Computer acceptance factor that affects users' attitude and behavioral intent towards technology adoption and continued use, with respect to their perception of the technology's usefulness.

Chapter 17
Influence of E-Learning and Project-Based Learning on Engineering Education

Nandhini Vineeth
https://orcid.org/0000-0003-4967-368X
B.M.S. College of Engineering, India

H. S. Guruprasad
B.M.S. College of Engineering, India

Sheetal V. A.
B.M.S. College of Engineering, India

ABSTRACT

Imparting quality technical education and training can be expected to be the vision of most engineering institutions globally to build a healthy society. The major stakeholders who contribute to this are students, teachers, industrialists, researchers, and institutes. The current scenario of rapid technological advancements demands engineering students to be dynamic and novel. Considering the heterogeneous intellectual ability of students, institutions frame time-restricted curriculums. Students who want to outperform have a challenge that they cannot be completely dependent on their academic curriculum. The objective of this chapter is to motivate and bring awareness among engineering students to adapt self-learning to excel in their professions. E-learning and project-based learning are identified as the two significant tools that could help students to self-learn. The influence of these tools on engineering students has been proved in this chapter with a case study, surveys, and feedback from students.

INTRODUCTION

The education system has to move towards igniting curiosity, critical thinking, problem-solving, and learning ability, qualities that can be inculcated 'only by good quality teachers'-**N.R. Narayana Murthy**

DOI: 10.4018/978-1-7998-8816-1.ch017

You can't teach people everything they need to know. The best you can do is position them where they can find what they need to know when they need to know it- **Seymour Papert.**

Engineering is the branch of study which aids in the design and development of efficient products, structures, or systems that could serve mankind by applying scientific and mathematical principles (Lucas, 2014).

Gone are the days when students were trained only using the method of chalk and talk. Digital media is the preferred tool of every student today as it keeps them active and makes learning faster. Any engineering program conducted globally exhibits time-restricted curriculums. The period in between the two phases, the welcome phase, and farewell phase is very critical for an engineering student. As the current technology is rapidly changing and becoming challenging day by day, engineering education requires the students to update their knowledge in their professional life in a continuous manner in the required domain. The students need to gain knowledge, experience, and confidence, to apply their learnings in their careers and become successful engineers professionally.

Self-Learning can be defined as the activity where the students develop the characteristics of learning without depending on another person to acquire knowledge. They become autonomous and active receivers of information from different sources. Teachers play a vital role in motivating the students to develop an affinity towards Self-Learning (SL) with such prevailing challenges (Reyes, 2019).

The institutes and universities that are seen globally finalize the curriculum of every course by considering the heterogeneous intellectual ability of students. Despite students being given autonomy to choose their streams and courses to some extent, SL is a significant tool that influences the successful completion of their courses. SL helps students with a low intellectual ability to cope with the challenges during their course of study and gain better knowledge. SL encourages students with a high intellectual ability by giving them the confidence to learn beyond such a bounded curriculum and exhibit their potential in their domains.

The current era of education experiences that internet/electronic (online) resources act as the friendly, first-hand reference tool for most of the students to clear their queries as against the earlier method of referring to textbooks which are followed by very few students today. The reason being that these resources act as hubs providing the requested information from various web servers distributed globally in a handful of pages.

E-Learning (EL) and Project-Based Learning (PBL) are identified as key techniques for discussions by the authors of this chapter as these two major tools help the students to take off in their sub-areas of interest. These make extensive use of such e-resources that initiate the students to self-learn and apply their learning in real-time implementations through projects. These two tools – EL and PBL are briefly introduced in this section.

The education system at all levels has incorporated a digital transformation in the teaching-learning process which is called E-Learning. EL can be defined as a hybrid of two streams -electronic technology and learning which were independent earlier. Some of the terms that are used to refer to EL are online learning, technology-enhanced learning, and computer-assisted learning (Andrews, 2007).

The different categories identified for learning include online, distance, virtual, MOOC, etc. The characteristics observed in EL are that it is student-centric, learning support can be crowd-driven, updating knowledge with emerging technology, and distributed multi-disciplinary learning. EL helps students in laying a strong foundation by analyzing and designing new processes or products and grow further in their stream of study. EL uses the computerized approach to enable and facilitate the learning process.

It focuses on the various concepts of learning that make use of computers for learning purposes. The myths with e-learning are identified as minimized human interactions, reduction of economic cost, difficulty in defining the infrastructure, decrease absenteeism and dropout rates among university students. (Berrocoso et al., 2020).

The other tool, PBL is a sub-stream of Problem Based Learning. One of the crucial roles of teachers is to make students understand the main objective of education. Engineering students play a major role here as their main objective is to design appropriate solutions to the given problem. When these solutions serve society, their success is exhibited. Problem-based learning (PbBL) can be looked upon as a bottom-up approach that involves various stages like exploring and understanding the existing problems clearly, applying the acquired knowledge to solve such problems, presenting and defending the results obtained, loopback if any changes need to be incorporated. Sustainable development is strongly linked with this as the solutions suggested for the problems need to be long-lasting for many generations (Thakur et al., 2021).

PbBL triggers the students to become active participants in classrooms as well as active learners exhibiting their potential through their work. Challenges faced by such students could be limited resources, economic status among others. Students are expected to establish connections with some factors to succeed:

Naturalism: Curiosity is aroused and the resulting questioning nature
Metaphysics: Critical thinking that helps in solving problems using logic
Rationalism: ensuring strong links between intellect and emotions
Empiricism: Discovery done through deep observations
Existentialism: Independent thinkers born through self-motivation and self-directions

Teachers play the primary role of facilitators here. They facilitate the students by initiating them, giving them the right direction to start, mentoring them when they are stuck with problems, review and evaluate the work done and provide effective feedback which provides the students with a rich learning experience (Thakur et al. 2021).

Project implementation is one of the significant ways to solve the problems defined. The term project can be understood as a significant activity with clearly defined goals achievable through affordable resources. Some factors like duration, cost involved, team size among others influence the success of a project. When the team members possess good coordination, intellectual capacity, self-motivation, participation, communication, etc., the successful completion of the work is foreseen. The strong backbone behind such successful projects is a knowledgeable, efficient, and approachable professor.

Earlier schemes in engineering education were oriented towards one large project just before students become engineers. Now the schemes are modified to include multiple small projects from their first or second year of engineering so that students get exposed to multiple streams available and identify and explore more on their sub-stream of interest.
Projects may fall under different categories

- Real-World Projects which are authentic ones directly linked to health care, agriculture, waste management, one that deals with conservations
- Problem Solving/Decision Making Projects are the ones that may act as a preface for optimizing resources
- Design Projects - are the ones that help the future generation a roadmap

- Major investigation projects - are the ones that encourage an extensive survey on an existing problem, issues faced, and offer an acceptable solution (Seif, 2014).

The outcome of implementing such projects is the learning gained by the students that can be termed as Project-Based Learning (PBL). This gives a pre-hand experience (mock) to engineering students before they land in their place of work.

PBL proves a good partnership with EL as concepts learned theoretically are applied practically. When encouraging results start showing up, students gain more confidence that leads to further success. Information needed and practical implementations are abundantly available in various forms of internet resources like documents, images, animations, videos, and discussion forums of communities.

The learning obtained from EL and PBL can be more effective, quick and can result in providing optimized solutions when the students work under the guidance of an experienced professor. Good teaching is influenced by pedagogical knowledge, institutional and environmental factors, personal attributes, and professional competencies. The performance of teachers can be improved by better training. The methodologies used to assess teachers' performance through feedback from students need to be revised frequently. Interpersonal skills like activating the positive emotions of the students and not encouraging the negative ones are equally important for a teacher. The content of the course plays a significant role in the quality of teaching and learning. Offline courses have a better impact than their online counterparts. Hence the orientation given on online courses was to be looked into. Ratings from students, peers, alumni, employers, administrators, etc., would help in improving the quality of the teaching-learning process (Chaubey et al., 2018).

The objective of this chapter is to motivate and bring awareness among engineering students to adapt Self-Learning to excel in their professions. Any student who understands the importance and is ready to adapt to SL will exhibit gradual progress in performance and can reap success. The two important tools- E-learning and Project-Based Learning have been identified by the authors to help the student community perform better.

The chapter is organized as follows. The second section includes the extensive literature survey done on the two methods – EL and PBL. The third section elaborates on the main focus of this chapter which includes the issues and challenges faced by engineering students, various methodologies followed in the department of the authors to the practice EL and PBL like virtual hackathons, online certification courses, online industrial visits, etc., case studies, the two surveys conducted by the authors and the flow diagram which demonstrates the experience of the authors showing the importance of EL and PBL. The fourth section lists some solutions that are recommended for the challenges. The fifth section includes future research directions and the last section concludes the chapter.

BACKGROUND

This session surveys the existing literature dealing with e-learning and project-based learning.

E-Learning

E-learning has been widely used by many students across the globe. Many challenges have also been observed from the viewpoint of an engineering student. A survey has been conducted among students of various institutions and departments to get clarity on the effectiveness of e-learning. The advantages are many like–during challenging situations like the prevailing pandemic when students are not able to be physically present in classes, online recordings of the faculty are made available to the students all the time so that their doubts can be clarified to a larger extent. The shortcoming here is geographical bias as the survey has been done only in Indian engineering colleges. The conclusion made is that e-learning does not suit courses that deal with lab sessions especially that involve physical equipment or components. Audio-video quality is very challenging on some platforms. Infrastructure development is a quick necessity to make e-learning platforms successful (Thakker et al., 2020).

The experience of teaching faculty who have tried implementing e-learning in a higher education institution is shared. The challenges faced are internet connectivity, shortage in ICT (Information and Communication Technologies) infrastructure, technical assistance/support not being to the required level, and inadequate training for activities in the e-learning platform of the university. The suggestions that have been given as an outcome of the experience after facing challenges are that the broadband Internet access on campus could get better connectivity between the student and the teachers, effective training of teachers keeping them periodically updated, and making the e-learning platform user-friendly. This is a study from one university. An extensive study of methodologies followed by success stories and trials faced by different universities would help in finding better solutions to various challenges faced by institutions (Maphalalal & Adigun, 2020).

At Weber State University, to improve the effectiveness of their academy, the students were given an option to register for an online course. Many assumptions were made before the beginning of the online course which belongs to different categories like online learning - for a specific population or individuals, or as a choice, or for enrollment. These were checked with the experiences obtained when implemented practically. When the survey was conducted more than fifty percent of the students were very satisfied with the online teaching and few students felt that face-to-face teaching gives them better subject knowledge. Feedback from students and faculty was incorporated to improve the online experience (Anderson, 2020).

During this pandemic, the teaching-learning process in many institutes had no choice but to enter online mode. There were a lot of challenges faced by the institutions, students, faculties, and parents i.e., technical problems like internet issues, hardware and software availability, problem-solving skills, and non-technical problems including funds, economic issues, etc. The best solution differs for every individual. A survey was made called a Technological Test i.e. T-Test to check the change in Technological ways incorporated in schools, colleges, and universities. As per the requirement, 500 million new internet connections were made available in India. The outcomes of which have been the following- technology became affordable with reduced rates as it can be affordable by all groups of people, IT sector need to stop outsourcing and need to increase the production without compromising on the quality and made available to every household (Khanna & Prasad, 2020).

The majority of the students have made up their mental and physical stability to get adapted to the online learning process and working out in the right direction to achieve their degree but still, there are few percentages of students who get distracted using laptops and phones for long hours, lack of concentration, lack of physical teamwork during their project, college events, missing friends, etc. are making students mentally and physically weak (Khattar et al., 2020).

In an offline mode, both faculties and students interact physically and the complete learning of the students is observed closely. The COVID situation has made the process of online teaching mandatory. Online teaching is required to be correctly planned and properly deliver the lectures and practical sessions. The required resources and tools should be available at both ends for the required course and extra monitoring of the students' needs to be taken care of, so the complete knowledge on the course is correctly delivered and students understand the concepts completely. The various online teaching modes like Google Meet Link, Microsoft teams, Zoom, etc. are used to deliver lectures. Google Classroom is created to interact with students, share study material, lecture videos, post assignments, conduct quizzes, and schedule tests (Pratama & Surahman, 2020).

Facebook was the other platform used, where some institutes were using Facebook's Social learning as a supporting Course Management Source for the courses they were teaching and the students were quite familiar with the tool since the beginning of the online classes. As the students were used to this platform, they continued using the same (Amir et al, 2020).

In some institutions, MOOC-based teaching models were used for students within a college, across different colleges, and to train teachers online in primary and secondary schools. This model helped to meet all the teaching needs and reduce the gap between face-to-face teaching and online teaching and provided high-quality course resources. Innovations are done using different teaching methods like re-shaping and restructuring the contents in the teaching and evaluation process. The outcomes experienced using such MOOC courses are effective addressing for students across different colleges and satisfied students (Xie.Y., 2020).

Project-Based Learning (PBL)

PBL can be termed as a classroom approach that makes the students dynamically explore the problems and challenges in the real world. This acts as a bridge and helps them relate the theory learned to practical applications. The experiences help them identify their strengths and help them focus on specific areas. The add-on gains here are communication skills, critical thinking, teamwork, creativity, etc. PBL generally is interdisciplinary and student-centered. Real-time projects involve multiple disciplines and hence interdisciplinary projects may give a hands-on practice for real-world application implementation. The wants of society are changing. The four Cs- communication, collaboration, critical thinking, and creativity are the need of the hour. Students can be given small tasks on some survey and analysis from their earlier stages so that they get a sense of responsibility towards society (Brien., 2021)

Innovative teaching-learning practices have led to the transformation in engineering institutes like the University of Minho towards the implementation of active learning for highly motivated, cooperative, and autonomous students. The European Union had initiated the Bologna Process movement to improve their Higher Education system with better quality, employability, and mobility. To incorporate these features into their education system, a group of teachers from the Department of Industrial Engineering and Management (IEM) has played a major role by creating an innovative curriculum. Despite project-based learning not being a part of their curriculum initially, this took a significant role later. The integrated master's degree in IEM (MIEGI – acronym in Portuguese) introduced a novel methodology of integrating the different courses of a semester in one interdisciplinary project during regular semesters, with a team size of around 10. During their seventh semester, students were made to interact with people from industry to work on projects, so that requirements of industry and the scope of their education and implementations became clear to them. This became possible as students started working from their first day of the first semester with projects with their discipline integrated with basic science like the use of organic waste for the production of bio-alcohol, Clean-up and recovery of crude oil from sea spills, Recovery, and transformation of cooking oils waste, Specification of a biodiesel production system, etc. The European Credit Transfer System (ECTS) encouraged the transfer processes which helped the students to get continuation in their education. This PBL approach has helped both students and teachers in continuous knowledge gain, endless effort, and commitment. This has reduced the gap between educational institutes and industry requirements and also aims to reduce the gap between developed and required competencies (Lima et al., 2017).

The methodology used at the Swinburne University of Technology is practice-based education (PBE) which has many commonalities with PBL but with the difference insisting on students becoming professionals taking the responsibility of the society as well working and learning simultaneously. The students who work on projects in industries gain more practical knowledge along with the concepts taught in classes. The experience gained during such learning helps them perform academically well by applying the expanded knowledge as well make them better engineers (Mann et al., 2020).

As heterogeneous PBL implementations have been challenging, the various ways of successfully implementing PBL are suggested. PBL practices are also made at the course level, cross-course level, curriculum level, project level. As a result of these practices, it is suggested that it is better to approach this not on the above levels but instead on individuals i.e., for students or teachers, departments, institutions universities, etc., and culture levels – language, nationality, etc. The combination of strategies used in these levels is also suggested (Chen et al., 2020).

In Southern Federal University, educational standards are framed internally where PBL takes a significant place. The success of PBL is proved as some of them after they graduate, get the confidence to become entrepreneurs, launching start-up companies. The confidence for these initiatives is obtained from working on projects that are very close to real-world requirements. They had targeted products more than processes. The teachers in engineering education have a great role in this as these results are practically possible only with the complete cooperation of students, academicians, industry experts, and customers who give the requirements. The teaching methods need to be an integration of classical methods as well project-based as per industry requirements. Generally, the students of higher semesters get the opportunity to interact with experts from the industry. Students are encouraged by projecting the outstanding works on the university website. The CDIO (Conceive, Design, Implement, Operate) approach helps in tuning the minds of students towards new designs and creations which is a significant implementation towards PBL. A separate department for project support is established to encourage enthusiastic minds that supervises all project-based activity and facilitates the interaction of students with foreign companies and universities (Veselov et al., 2019).

PBL has influenced different streams – like Civil engineering, mechanical engineering, health departments, etc. The assessment methods of some orient on undergraduate students, some on postgraduate students, and some on only teachers like vocational training departments. Such case studies in Universidad de Castilla-La Mancha (UCLM), an Indonesian university, and Adelphi University are discussed here.

Universidad de Castilla-La Mancha (UCLM) has run a survey on their graduation program which included teaching methods with project-based learning with the students who graduated 15 years ago. The results of the survey have shown that PBL has enhanced their civil engineering students to successful professionals with the ability of teamwork, improved communication skills, effective learning, satisfying result-to-effort ratio, and leadership qualities. Future engineers get hands-on experience with challenges in various types of constructions, maintenance, renovations, and planning. The students were made to work in teams on some real-world problems and once complete was asked to present it to the client which could exhibit their oral communication and also had to submit a written report. Such an approach makes learning - student-centered as they become active and express their ideas passionately whereas general learning is termed teacher-centered. The Ministry of Education, Culture, and Sport has surveyed the percentage of graduates holding jobs matching their education level. The results of the survey showed that 94% of professionals who exhibited good performance were from ULCM (Coronado, J.M. et al. 2021).

This case study on PBL involves a Mechanical Engineering Expertise Program from an Indonesian university. PBL is approached through quantitative as well qualitative techniques. The quantitative study describes the different phases like planning, implementation, and evaluation using questionnaires. The qualitative analysis gets deeper considering the results of a quantitative study like various methods of planning that involve lesson plans, worksheets, design suggestions, sketches from the teachers, proposals, intermediate result demonstrations, and reports given by the students. The vocational teachers were involved in this becoming the data sources. The work done only with surveys and simulated demonstrations is termed as Pre-PjBL (Project Based Learning) whereas the ones with simple and complex projects were the other classifications. The teams formed were from different categories -only teachers without students, one being students and teachers, the entire team from the same stream, the team which is a mix of people from different streams (multi-disciplinary), some involving industry or people from other institutes, etc., They were able to observe that multi-disciplinary teams were able to solve complex, diverse projects with societal impacts. The evaluation mainly was on the teachers who were involved in the initial sketches and the assessment was process and product-based (Sudjimat et al. 2021).

The Communication Sciences and Disorders Department of Adelphi University has conducted a comparative study on two research method courses in the clinical field taught online to master-level students during the pandemic. One being lecture-based using the traditional techniques through the presentation, discussions, and final presentation, and the other being a combination of project-based and problem-based learning. The study is based on both clinical and research self-efficacy. The formative assessments involve analytic rubrics from instructors. Three projects were done by the students out of which two involved group projects. PBL group students have expressed that the collaboration with the team has given them confidence and the team that had opted for the traditional course has expressed that the extensive online lectures have given them a good experience. The assessment of students was done using scales mentioned in the General Research Self-Efficacy (GRSE) scale and the Mixed Methods Research Self-Efficacy (MMRSE). The survey was conducted once before the commencement of classes and the other towards the last week of the course. The outcome observed has been that students who took the PBL course have exhibited very high satisfaction and self-efficacy with the course which has been done using project-based learning (Randazzo et al., 2021).

The extensive survey that involves the study of various methodologies and case studies proves that the two methods EL and PBL have an important role for engineering students and are hence used globally. The applications of these two have benefitted stakeholders of the teaching-learning process – both teachers as well students. As an outcome of the survey done, the authors of this chapter have conducted two surveys - one on EL and the other on PBL among the students who are in the pre-final year, the final year of engineering, and passed out graduates. The various ways in which EL and PBL are practiced are discussed in detail and the experience gained is shared in the next session.

E-LEARNING AND PROJECT-BASED LEARNING

Issues, Controversies, Problems

In the current scenario, a large number of engineering students graduate every year from different disciplines. Most of the industries are automated and hence the employment rate is decreasing. As the demand-supply ratio is decreasing, the graduates face challenges with employment. Such an environment demands students to have critical thinking and adapt to dynamic changing requirements to stand out from their peers. Despite the new pedagogical approaches followed by the teaching fraternity, students seeking job opportunities are not able to meet the expectations of employers.
The challenging environments observed in student fraternity are listed here.

- The diverse background of students
  - Students who join an engineering program may be from different countries, villages, different boards where the standard of education and learning vary widely.
- Orientation many times move towards grades
  - Many students aim only for good grades and GPA. They overlook the learning required for applying the techniques to get fruitful outcomes.
- Confidence of students
  - It is observed that some students even with good knowledge, logical and critical thinking lack the confidence to project their findings or learning. Students who join in the middle of the program – due to transfers of institutes, detained, after their diploma courses, etc. make themselves a little isolated from their peers and find it very difficult to regain their confidence levels.
- Self-motivation is observed to be less in some students
  - Only some of the students try to deep dive into the concepts whereas others try to work with the best cases with all good assumptions set in their work. It is observed that the number of students who self-learn and improve their capabilities is very less.
- Communication
  - One of the main reasons for the lack of confidence seen in students is communication. Students who are not comfortable with the language of communication used by most of their peers, hesitate to open up and express their ideas. There is also a possibility that in team projects, the students who are smart with very good communication skills may impress the evaluators more than the ones who were truly involved in core work.
- Learning/working pace of individual students
  - Some students may have good knowledge but may not be quick enough to implement their ideas
- Unequal participation
  - In projects worked in teams, some students show complete enthusiasm in the beginning phase but become lethargic as the work progresses, which may lead to a delay in completion.
- Cost of components
  - Though some students may have good knowledge and will be very enthusiastic to demonstrate their new ideas, due to lack of funds, they may not be able to implement their ideas.

Methodologies

E-Learning: Online Classes Compared with Classroom Teaching Along with its Merits and De-Merits

As known, the subjects in engineering studies include Basic science courses, programming courses, non-programming courses, management courses, etc.

When classes are conducted in the offline mode, there can be a direct interaction between the professors and the students. Professors can make sure that the students have clearly understood the concepts and their doubts are cleared. Students can be given additional assignments on the topics in which more doubts are raised. Here in this scenario, teachers are in regular touch with the students and can easily understand the student's capability levels. The weak students are given additional attention and will be taught the concepts again.

During the laboratory sessions, the faculty will have the individual attention of every student who will be helping the student in analyzing the problem, writing the solution to the problem, conducting and executing the problem, debugging and correcting the errors, and arriving at the solution. This helps the student to understand the concept, analyze, apply, design, and arrive at the solution. Here in this scenario, there is complete live interaction with the students and with the faculty teaching the subject. The students can also learn in groups, think different logic to the problem, a different alternate method to arrive at a solution. This brings confidence in the student concerning the subject. Along with this, the weaker section of the students can be given special attention, and also, they can learn from the other students in the group. This increases their learning capability and confidence in the subject among the weaker section of students.

Since March 2020 the pandemic has brought in a lot of changes in the teaching-learning process; the entire Engineering courses which were taught offline have now been taught in the online mode. Initially, it was very difficult for both the faculties and students to shift from the physical phase to the virtual phase. But slowly as the lockdown period started extending and because of the safety and security of the students and their families, the teaching was made online. The E-learning mode of teaching was adapted.

To give the students a complete feel of the classroom teaching to learn the courses, the practice followed during the conduction of few theory courses, practical sessions, quizzes, online tests, assignments, and evaluation, various methodology used, tools used are discussed here.

The process incorporated to impart the subject knowledge to the students using online mode which has currently become the most widely used tool in all colleges during this pandemic is discussed here. Online e-teaching and learning was the only choice. The calendar events were scheduled and classes were conducted through the google meet link. The students joined the meet link on the respective schedule. Pen Tablet was used so that the student will have the classroom board like teaching and it is visible to the student at the other end and all the lectures were recorded and shared so that the student can view the videos again and again to clarify their doubts. In the case of programming subjects core concepts were taught in-depth, programs and corresponding logic were taught and application-level programs were given to students and monitored at every level. Students were free to ask doubts at any time and were cleared. This made students feel that professors are available to them at all times. The same was followed for other courses.

The limitations on online classes are as follows:

- There was no physical interaction with the students which was the major drawback
- When classroom teaching was done for courses like the Internet of Things, students used to work with the real components like Sensors, Arduino board, etc., but due to the pandemic, all the concepts and designs were taught using the simulation tool i.e., TINKERCAD. Hence, professors feel that students lack the real touch and feel of the components and their execution.
- Evaluation of the tests, quizzes, and other assignments was a challenge. Though different sets of questions were prepared and tests were strictly monitored, the students were found to be involved in discussing the answer on the remote end which was out of the teacher's control.
- Most of the students in rural areas may not have proper internet facilities. Some of the students have lost interest in their studies due to a lack of physical interaction with friends. Lack of group discussion while solving has made students mentally disturbed.

Conduction of Virtual Hackathons

The virtual Hackathon, which is a coding event conducted in the institute of the authors, helps the interested students to improve their programming skills on the respective topic, improve upon and build a new software program. Students work together on collaborative projects, where the competing teams apply their innovative ideas to create several prototypes to the new project or the existing project. This has helped students to a large extent to improve their skills.

Figure 1. Sample screenshot of the hackathon

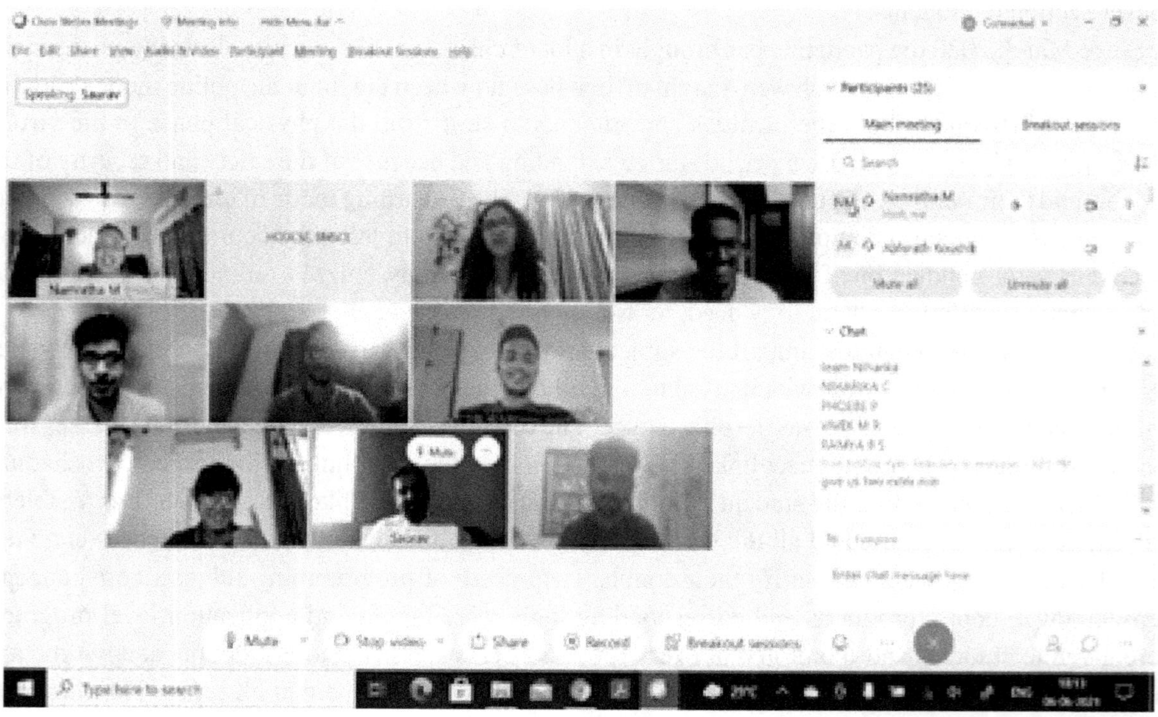

The department of Computer Science and Engineering in association with Riskcovry organized a 24-hours hackathon titled "Ode-To-Code" on 5th - 6th June 2021. Figure 1 shows a sample screenshot of the Hackathon conducted. The evaluators for the Hackathon were from the Technical Team of Riskcovry. Totally around 130 teams from various colleges had participated. The hackathon was conducted online and the evaluations were done online using the Cisco Webex platform. Two problem statements, one on emotion detection and one on sentiment analysis were given and the participants were given the choice to select one among them. The winner of the Hackathon was awarded Rs.10,000 cash prize along with COVID insurance of Rs. 6 Lakh and the runner of the Hackathon was awarded Rs.5,000 Cash Prize along with COVID Insurance of Rs. 6 Lakh. Also, spot award winners were given various Medical Insurance offers from Riskcovry. The events like these encourage students to learn and explore, design solutions and bring innovative ideas. There are some interesting mandated offline physical activities like sports activities, college and inter-college fests where the students need to participate in person.

Figure 2. Virtual industrial tour to Hewlett Packard Enterprise, Bangalore on 18-Dec-2020

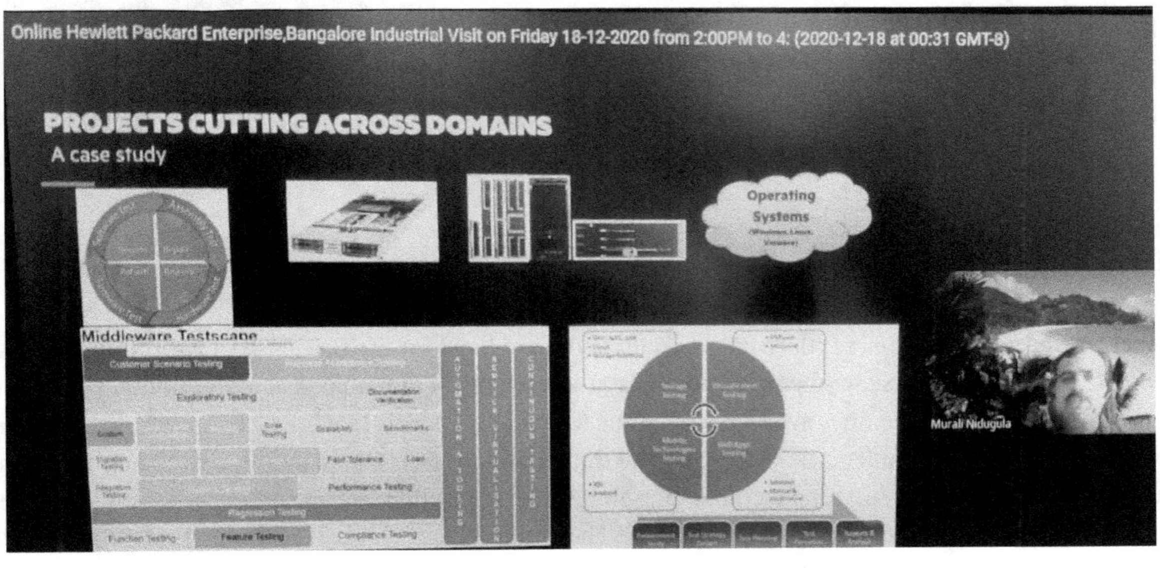

As part of the academic practices, industrial visits will be arranged for the pre-final year students where the students will be taken to a company so that the student gets exposed to the industry environment, current technology, projects, research, work culture, internships, career development, and hiring process, etc. Every year a maximum of three visits will be arranged. As companies are working from home and physical visit to the company was not possible, virtual industrial tours were arranged for the students where experts from the company had given talks on the different approaches through Google meet links. This has benefited the students to a large extent. Generally, the companies restrict the number of students to a maximum of 40, during offline visits. It is observed that more participants were benefitted when the visit was conducted online. The same has been shown in Figure 2 and Figure 3. Online mode has helped in the conduction of visits without giving a stop in the practice even during this pandemic.

Figure 3. Virtual Industrial Tour to Impelsys Technology, Bangalore on 6-Aug-2021

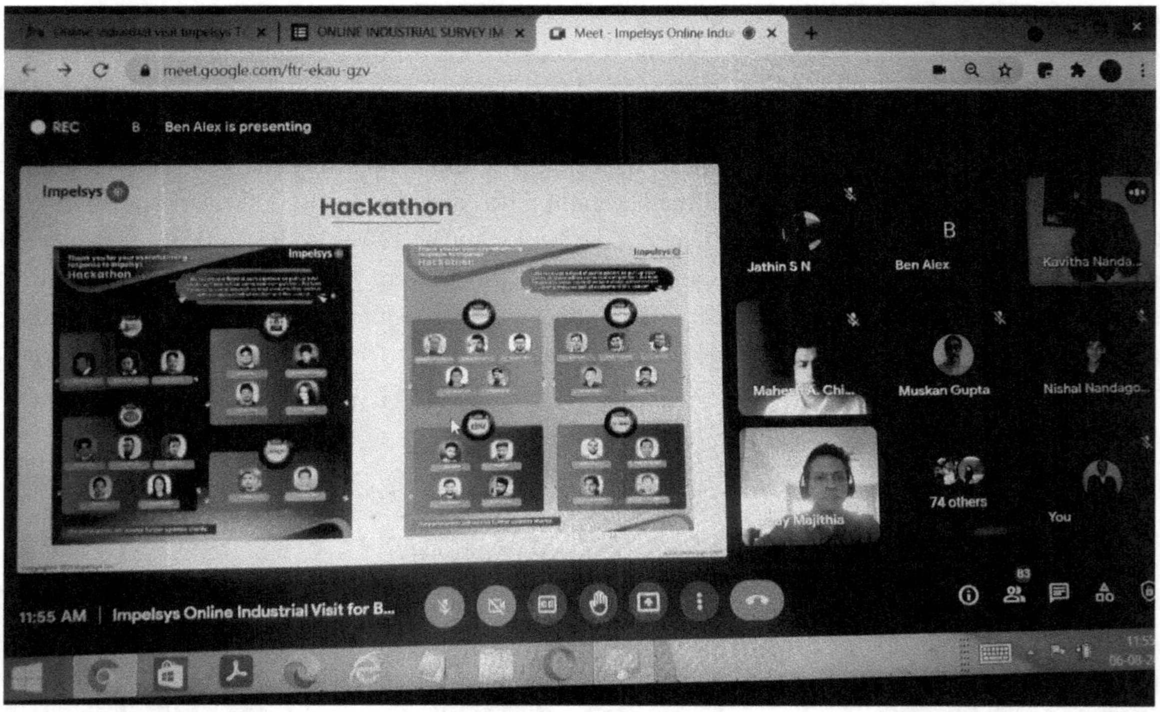

Online Certification Courses

E-learning has become a part of the academic curriculum in Jan-May 2020. Massive Open Online Course (MOOC) one-credit course has been introduced to the students of pre-final year undergraduate students.

MOOC courses are the ones available on online learning platforms like NPTEL, SWAYAM, Course-Era to name a few. The students need to complete the certification course in their interested domain. The students are given access to video lectures, presentations, and study materials of experts from esteemed institutions, and industries teaching the course. The students can choose the level (Basic/ Intermediate/Advanced) in the course of their interest. This helps students to learn and gain knowledge from the course which may not be prescribed in the regular academic curriculum. This helps the student to become independent, understand the concept of self-study and assess their self-learning capability. The student can continue with different levels which can help in further project development in the final semester and can also continue with research if interested. It is also observed that the number of students taking up online courses has increased during this pandemic time.
As an outcome of the course, the student will be able to

1. Understand the basics and domain knowledge through videos, lecture notes, presentations for the course registered during the course period.
2. Answer, solve, develop/implement the solutions with appropriate techniques, resources, and contemporary tools.
3. Allocate time effectively and manage to complete the work allotted and submit the assignment within the given deadline.

4. Demonstrate communication skills with effective oral presentation and the preparation of course reports.

Project-Based Learning

The procedure followed during the phases of projects done by the students in the institution of authors is discussed here. Students work on one credit project every semester starting from their second year (2 semesters are run in an academic year). Generally, the team size followed is two to four. They work on application development projects which are on streams like Web-Based, Database, Mobile, Data Science, and Security-based. The students have the option to decide if they are to work on an extension of a project done in their previous semester or a new one depending on their interests. A professor helps the teams choose their topics, guides them during their work, and evaluates the work based on a rubric.

The criteria for evaluation rubrics generally are on problem formulation, planning, literature survey, high-level design, detailed design, front-end design (if applicable), and back-end design (if applicable), implementation, testing & validation, publication of their work, report, oral communication, group participation, ethics, and peer review.

Week-wise schedule is planned by the professors which include deadlines for team formation, finalization of project title, synopsis submission, intermediate presentations and discussion with their professors, report writing, and final presentation. Such a plan helps the students not to lose their focus, work towards their set milestones and complete their project successfully.

Other than these projects done per semester, students work on their major projects in two phases during the two semesters in their final year.

The project-based courses play a vital role in helping the students become professionals. The outcomes of such courses encourage the student

- Identify a legal Engineering /real-life/ societal problem
- Investigate extensively and apply prior practical knowledge
- Plan and arrive at an innovative design
- Apply latest tools along with project development
- Monitor the resources available and plan for timely completion
- Develop the skill of accepting responsibilities and take initiatives
- Communicate formally and informally with peers and mentors.
- Document the work done professionally

Project-based learning is facilitated to students with the introduction of mini-projects in their curriculum from their second year.

Case Study –Project-Based Learning

One of the motivating factors for this chapter has been the academic results of two different batches of students of a course Web Programming- one of the Engineering undergraduate courses in computer science Engineering program. When the course was conducted in 2013, project implementation was not a part of their curriculum. When the same course was conducted during 2017, the project implementation became a significant part of the scheme.

The course conducted during 2013 had components – theory and lab sessions. When the course was conducted in 2017, a project component was introduced along with the theory and lab components. Students were given the objective of developing a website to explore the practical applications of the concepts learned under the project component of the Web programming course.
The weekly schedule was planned and shared during the commencement of the course which included

- Formation of groups with a size of two to four
- Project topic selection by each group
- Presentation: Student and Project topic introduction by each group
- Design Layout of the Web Pages
- Presentation of Back-end Design by each group
- Presentation of Front-end Design by each group
- Integration of front-end web pages with backend database
- Complete Project Work Demonstration by each group
- Project report preparation

Students were able to apply the theoretical concepts learned, during the development of websites.
The academic results of both these batches are presented in the result section in Figure 5.

Case Study – E-Learning

Most of the engineering students are observed to be active but their learning is restricted based on the academic curriculum. Very few dynamic students, stand out of the crowd by learning beyond their curriculum boundaries. They explore their areas of interest, exhibit their potential, and excel demonstrating self-learning.

The authors of the chapters have observed many such students who are successful in their careers. Feedback has been taken from such students and summarized in the next sub-section "Student Feedbacks on EL and PBL".

Student Feedbacks on EL and PBL

As students are the main stakeholders in the usage of these two tools – EL and PBL, feedbacks have been taken from students who are passed out and also students who are currently pursuing their courses. Four sample feedbacks from students are presented here.

I was working for Wells Fargo after my graduation in BMSCE and am currently doing my masters in Management Information Systems at Texas A&M University, USA. PBL has helped me in my career. Individual projects instead of group ones might have been more effective. In addition to projects, if students have been part of competitions or landed internships that have a higher weightage - **Student1 (Graduated in 2018)**

My current status is that I've completed my bachelors in engineering at BMSCE, India, studied masters in Cass Business School, London, and am currently working for an investment management firm called Ninety-One in London.

Self-learning + Project-based learning - I think this is very crucial and goes hand in hand with practical project-based learning. The masters I have pursued focused on 50% self and project-based learning, 50% taught learning. I think this model is efficient since students do more by learning rather than being taught. The curriculum was such that all subjects were project-based assessments combined with peer review. Due to this, there was no choice for students but to help with projects and therefore learn. I've always condemned the practice where we were made to learn by heart, I think this approach is not right as in the practical work everyone is allowed to use help from open-source/ websites. This approach that my masters used has helped in many ways. Some of the main points I would highlight, would be that it particularly helps understand the dynamics of teamwork & building self-reliance by being accountable for teams performance as well. E-learning - I think this may not be as helpful as one would expect since it does not guarantee students involvement as they may just keep videos open just for the sake of watching. There should be a knowledge check or something similar in my opinion to make sure that the students have paid attention. But either way, students who want to learn may continue learning. - **Student 2 (Graduated in 2018)**

I am working as a Software Engineer in the Infrastructure and Cloud Services team at Cisco. My work is a blend of development and operations-related tasks. I feel that self-learning, e-learning, and Project-based learning are extremely important for engineering students in this day and age, particularly for Computer Science and Engineering students. I have been able to grow and learn more through these modes. However, the importance of these does not undervalue the traditional classroom-based mode of learning. Rather, these should be viewed as enhancements to the traditional mode of learning. - **Student 3 (Graduated in 2021)**

I am pursuing Computer Science and Engineering degree in BMSCE. For an engineering student, e-learning has become the easiest way to acquire knowledge from any place just with a single click. It has helped me to learn things at my own pace, make notes and complete the course whenever I have time without any time constraints. It also saved time and money as I need not travel to any place to listen to classes. - **Student 4** *(A student who is interning and placed in a reputed company during her third year of graduation)*

I am currently pursuing my B.E degree in the field of Computer Science and Engineering in BMSCE. I completed my internship at Twilio during the summer of 2021. Based on my performance I got a job offer to join the company as a full-time employee upon the completion of my graduation in July 2022. Self-learning helps us keep updated and be on track with the ever-changing fast modern world. It keeps us on par with the current requirements which are demanded by the industrial sector. The rise and popularity of the internet has given way to a plethora of platforms and one of the most popular ones in terms of education are e-learning platforms. The spectrum of learning content is very fast. E-learning has made it possible and bridged the gap between students and teachers around the world. E-learning not only offers theoretical knowledge but also provides hands-on knowledge too. - **Student 5** *(Student achiever with the highest pay package for her internship in the third year of graduation and has been proving her academic excellence as on date)*

Demonstration of the Significance of EL and PBL

As an outcome of the experience gained from handling different courses with changing curriculum, case studies observed and feedback received from students, the authors of this chapter demonstrate the importance of these two tools for any engineering curriculum through the flow diagram shown in Figure 4. The flow diagram is a representation of the following:

Figure 4. Flow diagram showing the importance of e-learning and project-based learning in engineering curriculum

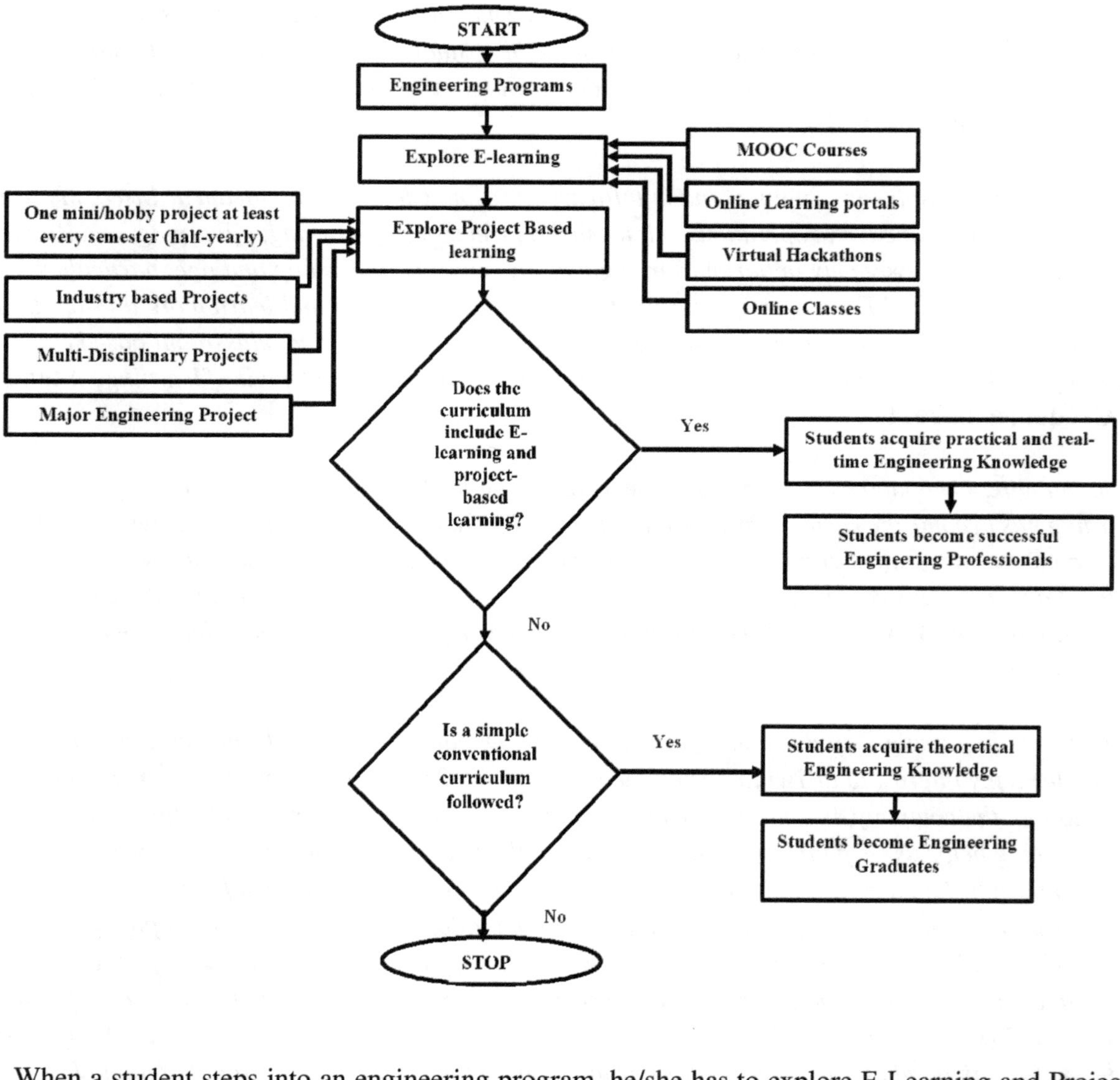

When a student steps into an engineering program, he/she has to explore E-Learning and Project-Based Learning. Some of the prevalent components that promote EL are MOOC Courses, Online Learning portals, Virtual Hackathons, and online classes. PBL is promoted by the components like mini-projects, industry-based projects, multi-disciplinary projects, major projects, etc. An average student would generally be keen on acquiring theoretical knowledge and completing graduation. They aim for a good job opportunity if he/she is interested in starting his/her career immediately, or would aim for his/her postgraduation in a good university. Some may succeed but most of them may fail. Many will be in a position to compromise with allotted designations in their jobs or university seats allocated. EL and PBL would sow the seeds of responsibility in a student towards himself and towards society. When these components are not seen as a part in the curriculum, it becomes optional and may go unattended by many students. Such a scheme would produce simple engineering graduates. When EL and PBL are given a significant part in the core courses in every year of graduation in an engineering curriculum, the students acquire the potential needed to understand the problems of the industry and society and render solutions to them. Such students graduate as successful and confident professionals as against simple engineering graduates. These students need not compromise but can demand their designations in their workplace or succeed in acquiring the post-graduation in their stream of interest from reputed universities.

Surveys

Intending to understand the view of students on the two tools, two surveys based on EL and PBL and were conducted among the students of an engineering under-graduation Computer Science programme. The responses received were analyzed and the results are shown and discussed in the result section in Figure 6, Figure 7, and Figure 8.

SOLUTIONS AND RECOMMENDATIONS

The results of the analysis done with case studies and surveys are presented here.

The Outcome of Case Study on Project-Based Learning

During 2013, in the institute of the authors, the course 'Web Programming' was conducted without the project component and with the changing curriculum, a project component was introduced during 2017. The academic results included grading the students based on the marks scored by the student. The list of grades awarded is- Outstanding, Excellent, Very Good, Good, Above Average, Poor, and fail. The percentages of grades acquired by the two batches of students were compared and are shown in Figure 5. As can be observed, the number of students who have failed is less in the batch with the project component. The number of Outstanding grades observed in this batch is more compared to the batch without the project component. This proves that the students have gained better knowledge as an effect of the project implemented by them.

Figure 5. Result analysis of an undergraduate course – web programming with and without PBL

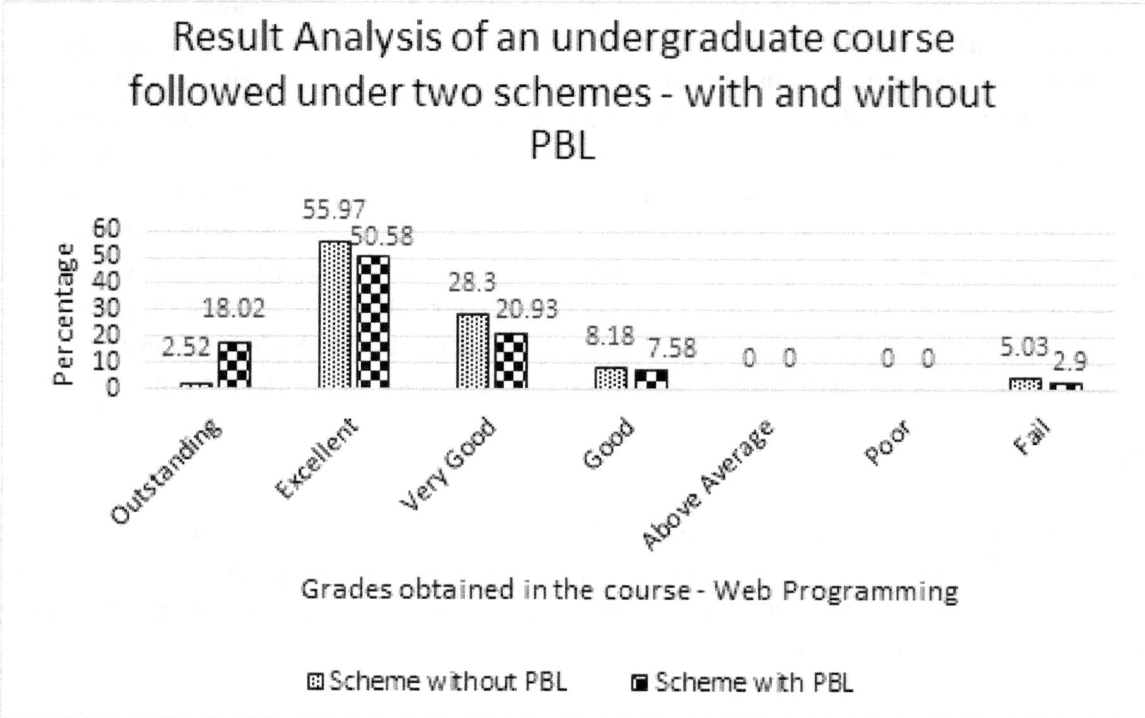

The Outcome of Survey on E-Learning

Figure 6. Influence of MOOC courses

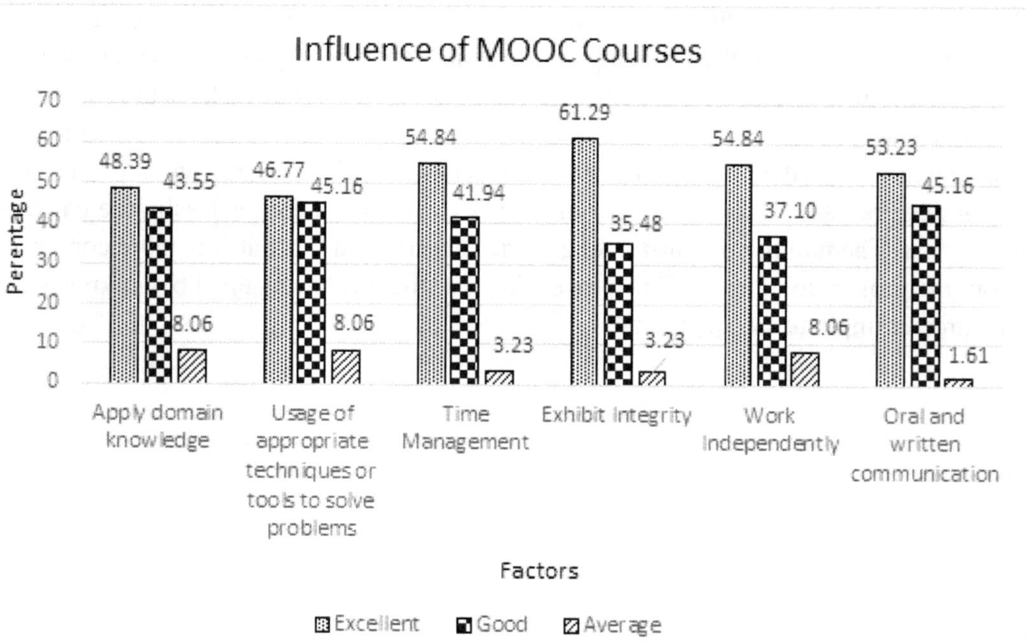

A survey was conducted to get feedback on effective learning through online certification courses. The questionnaire included the following - applying their knowledge, their ability to develop/implement solutions with appropriate techniques, resources, and contemporary tools, time management, exhibit integrity, and ethical behavior, ability to work independently /collaborate in a multidisciplinary environment, oral and written communication. Sixty-two students have participated in the survey.

Figure 6 shows the percentage of grading given by the students for the various factors. It is observed that the grading for Excellent is more compared to the other two. The same is seen observed for all the factors.

The Outcome of Survey on Project-Based Learning

A survey has been conducted to get feedback on the influence of the projects done. The questionnaire included the following factors- understanding the course contents better, apply their learning in their profession, performance in interviews and internships, team participation skills, and communication skills. Fifty-five students have participated in the survey. The results of the analysis done on the feedback of the students on PBL are shown in Figure 7.

Figure 7. Influence of PBL

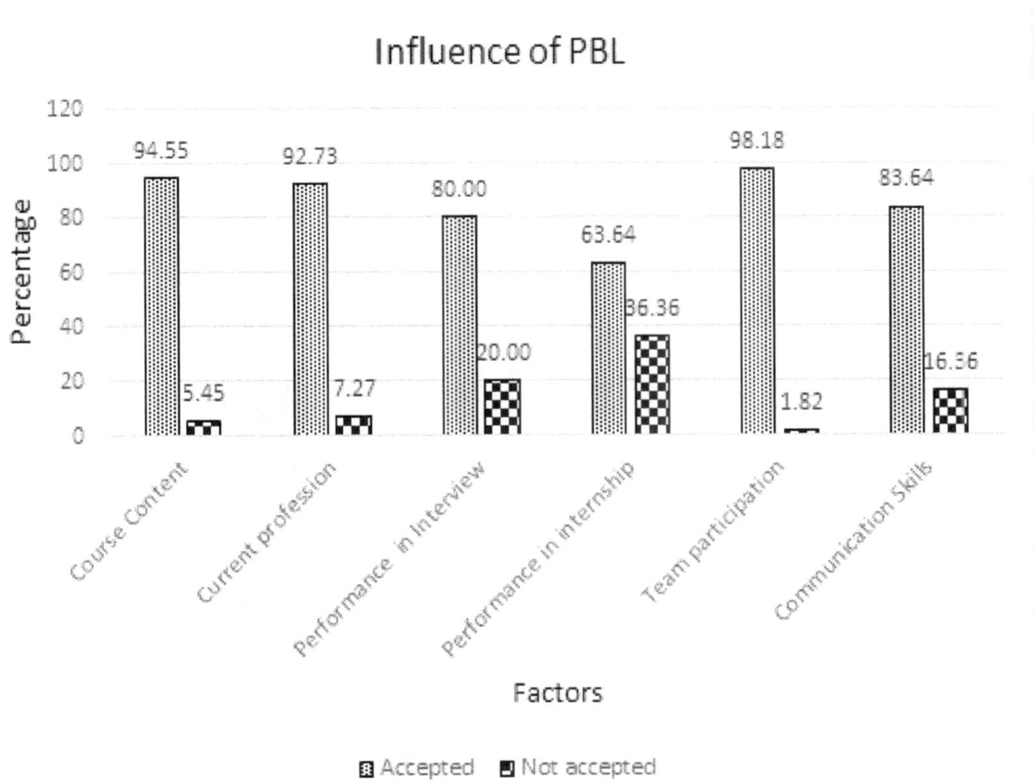

A deeper analysis of the survey results infers the following:
Most of the students agree that

- They can understand the course contents better because of the projects they have worked on. A closed-loop can be seen here as after getting the basics of courses, students start working on projects which again give them a better understanding of the concepts learned earlier.
- PBL has helped them to perform better in the other projects/career. This proves that PBL gives a mock environment for students to learn better when they are students so that they perform better as professionals
- PBL has given the confidence to face the interviews.
- Perform better in internships offered by industries.
- The practice to work in teams has been very helpful when they work in the industry
- Their communication skills have been improved as every student is allowed to demonstrate and explain the module, he/she has designed during the presentation of every project.

As a result of the survey, it can be observed that students consider projects to be a very useful tool to learn and prepare themselves for their careers. It is also observed from Figure 8 that working on projects has been helpful. The numbers saying Helpful outnumber the Very Helpful option.

Figure 8. Rating on PBL

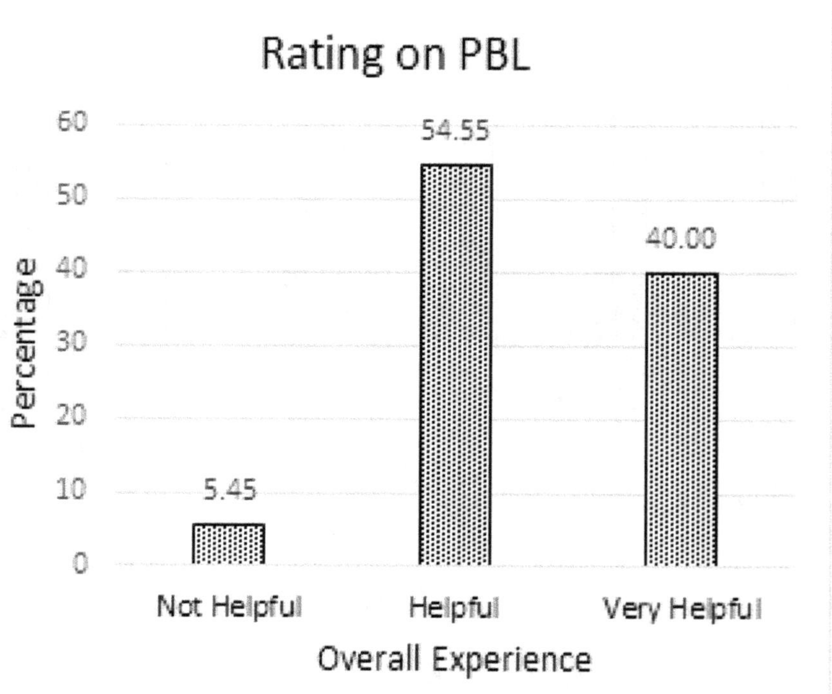

Recommended Solutions

Professors play the role of academic parents and become the first advisers to students. Hence, some recommended solutions are listed here.

- The students should be made to understand the need for critical thinking and the importance of the learning they get from the projects. They must be exposed to the requirements of industry/ expectations from society.
- Students can be encouraged to extend their projects done in their previous semesters. This would give them the sense of working on better applications and improve their perspectives on the projects.
- Encouraging students with awards like Best project awards- department level, institute level, university level, etc.,
- An exclusive panel of counselors (a team of professors) can be formed to mentor and encourage such students to bridge the differences between what they are and what they need to become to for successful completion of their projects
- Professors can interact with such students more even during class hours and give them the confidence that they have the potential to achieve what they target.
- Importance of deadlines and frequent practices required to target milestones need to be given
- Every student needs to be made responsible for a module or a part of the work and insisting on submission of intermediate results individually may boost up their participation.
- An exposure to the details of different funding agencies and the procedures to approach them by the professors can bring in new useful products to the society.

FUTURE RESEARCH DIRECTIONS

1. In future, professors can decide on the project titles after a discussion with people in charge in the industry – alumni or others. Alumni students who are entrepreneurs with startups can offer students with real-time projects during their course of study. This benefits both -as the alumni get the feel of contributing morally to their institute and the students gain a rich real-time experience.
2. A seed needs to be sown in the minds of engineering students to engage themselves in a life-long commitment to servicing society. They need to form groups that stay intact and work together even after the restricted years of their under-graduation / graduation programs and target products through projects that serve society. By then, they would have realized that every individual has a unique potential. A significant successful product can be built when the individuals in the teams contribute to the work with the maximum magnitude of their strengths.
3. Engineering schools can get into Memorandum of Understanding (MoU) with industries which would open doors for enhanced industry-academia interactions, exposure of the expectations of industry from students, industries selecting students, and training them on projects among others. The industries are benefitted as their customized training can be provided during the student's

graduation program and they get industry-ready graduates. Students exhibit confidence because of their hands-on work experience in the industry.

CONCLUSION

Self-Learning (SL) is an important factor to bring in quality, motivation, confidence, and employability in engineering students. A successful career is a guaranteed outcome of a self-learning approach. The significant two streams that support SL are E-learning and project-based learning. These two streams have been chosen for an in-depth study in this chapter as they are highly influential in bringing the potential of engineering students to exhibit novelty and design new products. The importance of these two needs to be given during the initial years of Engineering education. The success of both these techniques depends on the students who understand, accept and apply them to showcase their potential. Various case studies when discussed make an environment prevalent. It becomes challenging for the professors to guide the students in the right direction if they do not update themselves about the happenings and requirements of the industry. To further improve the knowledge, involvement, confidence level of students, the industries' involvement in academia and vice versa is highly required. When students interact with industry personnel during their course of engineering, they will be able to acquire the smart ways of working within a small duration. They would also get a prior-hand experience about the challenges that are to be faced and the attributes required to face them. The role of professors here is vital as they need to cope up with heterogeneous students, see to it that the takeaway from the courses should be maximized as well as think about their future and plan their interactions. They can facilitate meetings with the alumni students who can share their experiences with the juniors who can visualize and train themselves for their successful careers.

ACKNOWLEDGMENT

We would like to express our gratitude to the management, faculty colleagues of B.M.S. College of Engineering, Bangalore for their constant and continuous support in the development of this chapter. The chapter has been written with the rich work experience gained from our institute.

REFERENCES

Amir, S., Kamal, M. S., Shahria, M. T., & Iftekhar, L. (2020). *Facebook's Social Learning Group for Undergraduate Engineering Courses: A Case Study of Emergency Remote Teaching Amid Large Digital Divide* [Paper presentation]. *2020 IEEE International Conference on Teaching, Assessment, and Learning for Engineering (TALE).* 10.1109/TALE48869.2020.9368316

Anderson, N. (2020). *Examining ways COVID-19 Adaption Derailed our Assumptions about Online Learning* [Paper presentation]. IEEE Conference on e-learning, e-management and e-services, Malaysia.

Andrews, R., & Haythornthwaite, C. (2007). Introduction to e-learning research. The SAGE handbook of e-learning research. doi:10.4135/9781848607859.n1

Berrocoso, J. V., Arroyo, M. D. C. G., Videla, C. B., & Cevallos, M. B. M. (2020). Trends in Educational Research about e-Learning: A Systematic Literature Review (2009–2018). *Sustainability*, *12*(5153), 5153. Advance online publication. doi:10.3390u12125153

Brien, M. O. (2021). *Defined Learning-What is Project Based learning.* https://blog.definedlearning.com/blog/what-is-project-based-learning

Chaubey, A., Bhattacharya B., & Mandal, S.K.D. (2018). *Attributes of good teaching in engineering education in Indian subcontinent.* Sadhana, Indian Academy of Sciences. doi:10.1007/s12046-018-0946-3

Chen, J., Kolmos, A., & Du, X. (2020). Forms of implementation and challenges of PBL in engineering education: A review of literature. *European Journal of Engineering Education*. Advance online publication. doi:10.1080/03043797.2020.1718615

Coronado, J.M., Moyano, A., Romero, V., Ruiz, R. & Rodríguez, J. (2021). Student Long-Term Perception of Project-Based Learning in Civil Engineering Education: An 18-Year Ex-Post Assessment. *Sustainability*, *13*(1949). doi:10.3390/su13041949

Khanna, D., & Prasad, A. (2020). *Problems Faced by Students & Teachers During Online Education Due to COVID-19 and How to Resolve Them* [Paper presentation]. *6th International Conference on Education and Technology (ICET)*, Indonesia.

Khattar, A., Jain, P. R., & Quadri, S. M. K. (2020). *Effects of the Disatroue Pandemic COVID 19 on Learning Styles, Activities and mental Health of Young Indian Students- A Machine Learning Approach* [Paper presentation]. International Conference on Intelligent Computing and Control Systems (ICICCS 2020), Secunderabad.

Lima, R. M., Carvalho, J. D., Sousa, R.M., Alves, A.C., Moreira, F., Fernandes, S., & Mesquita, D. (2017). Ten Years of Project-Based Learning (PBL) in Industrial Engineering and Management at the University of Minho in PBL in Engineering Education - International Perspectives on Curriculum Change. Sense Publishers.

LucasJ. (2014). https://www.livescience.com/47499-what-is-engineering.html

Mann, L., Chang, R., Chandrasekaran, S., Coddington, A., Daniel, S., Cook, E., Crossin, E., Cosson, B., Turner, J., Mazzurco, A., Dohaney, J., O'Hanlon, T., Pickering, J., Walker, S., Maclean, F., & Smith, T. D. (2021, January 2). From problem-based learning to practice-based education: A framework for shaping future engineers. *European Journal of Engineering Education*, *46*(1), 27–47. Advance online publication. doi:10.1080/03043797.2019.1708867

Maphalala, M. C., & Adigun, O. T. (2021). Academics' Experience of Implementing E-Learning in a South African Higher Education Institution. *International Journal of Higher Education*, *10*(1), 1–13. doi:10.5430/ijhe.v10n1p1

Pratama, U. N., & Surahman, E. (2020) *Investigating Student Responses of online learning during the Covid Pandemic in Art Education* [Paper presentation]. *6th International Conference on Education and Technology (ICET)*, Indonesia.

Randazzo, M., Priefer, R., & Dakwar, R. K. (2021). Project-Based Learning and Traditional Online Teaching of Research Methods During COVID-19: An Investigation of Research Self-Efficacy and Student Satisfaction. *Frontiers in Education.*, *6*(662850). Advance online publication. doi:10.3389/feduc.2021.66285

Seif, E. (2014). *Seven Types of Projects that Foster Powerful Learning.* https://www.ascd.org/blogs/seven-types-of-projects-that-foster-powerful-learning

Sudjimat, D. A., Nyoto, A., & Romlie, M. (2021). Implementation of Project-Based Learning Model and Workforce Character Development for the 21st Century in Vocational High School. *International Journal of Instruction*, *14*(1), 181–198. doi:10.29333/iji.2021.14111a

Tejada, R. V. (2019). *Self-Learning Techniques to Contribute to the Learning of the English Language of the Students*. https://files.eric.ed.gov/fulltext/ED595374.pdf

Thakker, S.V., Parab, J., & Kaisare, S., (2020). *Systematic research of e-learning platforms for solving challenges faced by Indian engineering students*. doi:10.1108/AAOUJ-09-2020-0078

Thakur, P., Dutt, S., & Chauhan, A. (2021 April) Problem Based Learning in Education– its Need for Sustainable Development. *Journal of Engineering Education Transformations, 34*(4).

Veselov, G., Pljonkin, A., & Fedotova, A. Y. (2019). Project-based learning as an effective method in education. *ICMET 2019: Proceedings of the 2019 International Conference on Modern Educational Technology*, 54-57, 10.1145/3341042.3341046

Xie, Y., Qui, Y., Huang, Y., Li, Y., Liu, A., & Wu, J. (2020), *The Construction and Effect of the MOOC-based One Plus Three Teaching model During the covid 19 pandemic* [Paper presentation]. *2020 Ninth International Conference of Educational Innovation through Technology (EITT)*. 10.1109/EITT50754.2020.00042

KEY TERMS AND DEFINITIONS

E-Learning: A hybrid of two streams -electronic technology and learning which were independent earlier.

Problem-Based Learning (PbBL): A bottom-up approach that involves various stages like exploring and understanding the existing problems clearly, applying the acquired knowledge to solve such problems, presenting, and defending the results obtained, loopback if any changes need to be incorporated.

Project: A significant activity with clearly defined goals achievable through affordable resources. Some factors like duration, cost involved, team size among others influence the success of a project.

Project-Based Learning (PBL): The outcome of implementing such projects is the learning gained by the students.

Self-Learning: The activity where the students develop the characteristics of learning without depending on another person to acquire knowledge.

APPENDIX 1

In connection with the survey conducted by the authors of the chapter on E-Learning (EL), the questionnaire used in the survey is included here in Table 1. The results of the survey are shown in Figure 6 in the chapter.

Table 1. Questionnaire used for survey

1. Are you able to apply the domain knowledge and skills on a registered MOOC course? a. Able to apply domain knowledge and skills completely on a registered course. b. Able to apply domain knowledge and skills in most of the issues during the course. c. Able to apply domain knowledge and skills in specific issues during the course. d. Unable to apply domain knowledge and skills completely on a registered course
2. Are you able to answer, solve, and develop/implement the solutions with appropriate techniques, resources, and contemporary tools during the registered MOOC course? a. Able to answer, solve, and develop/implement all the solutions with appropriate techniques, resources, and contemporary tools during the course. b. Able to answer, solve, and develop/implement most of the solutions with appropriate techniques, resources, and contemporary tools during the course. c. Able to answer, solve, and develop / implement a few specific solutions with appropriate techniques, resources, and contemporary tools during the course. d. Not able to answer, solve, and develop/implement solutions with appropriate techniques, resources, and contemporary tools during the course.
3. Are you able to manage and complete the allocated work within the specified deadline? a. Able to manage and allocate time effectively and complete all the work allotted within the deadline. b. Able to manage and allocate time effectively and complete most of the work allotted within the deadline. c. Able to manage and allocate time effectively and complete only the specific work allotted within the deadline. d. Unable to manage and allocate time effectively and complete only the specific work allotted within the deadline.
4. Are you able to exhibit integrity and ethical behavior while carrying out the course? a. Able to effectively exhibit integrity and ethical behavior while carrying out the course. b. Able to moderately exhibit integrity and ethical behavior while carrying out the course. c. Able to partially exhibit integrity and ethical behavior while carrying out the course. d. Unable to exhibit integrity and ethical behavior while carrying out the course
5. Are you able to work independently /collaborate in a multidisciplinary environment? a. Able to work independently and in a collaboration/multidisciplinary environment. b. Able to work independently with minimal guidance and in a collaboration/multidisciplinary environment. c. Able to work independently with more guidance and in a collaboration/multidisciplinary environment. d. Unable to work independently without guide support and in a collaboration/multidisciplinary environment.
6. Are you able to make an oral presentation and write the report on the registered course? a. Able to demonstrate effective oral and written communication skills. b. Able to demonstrate oral and written communication skills moderately. c. Able to demonstrate oral and written communication skills minimally. d. Unable to demonstrate effective verbal and written communication skills.

APPENDIX 2

In connection with the survey conducted by the authors of the chapter on Project-Based Learning (PBL), the questionnaire used in the survey is included here. The results of the survey are shown in Figure 7 and Figure 8 in the chapter.

1. Name any two courses which involve projects in your curriculum (other than your final year project).
2. Did these projects help you understand the course contents better? Yes/No
3. Are you able to apply the learning from these projects now in other projects/work? Yes / No
4. If yes, brief about the same
5. Did these projects help you in answering questions in your interviews? Yes / No
6. Did these projects help you in getting an internship offer in an industry? Yes / No
7. Did these projects help you in improving your team participation in your current work? Yes / No
8. Did these projects help you in improving your communication skills? Yes / No
9. Rate your overall experience with these projects. Not helpful / Helpful / Very helpful

Compilation of References

Abele, E., Chryssolouris, G., Sihn, W., Metternich, J., ElMaraghy, H., Seliger, G., Sivard, G., ElMaraghy, W., Hummel, V., Tisch, M., & Seifermann, S. (2017). Learning Factories for Future Oriented Research and Education in Manufacturing. *CIRP Annals*, *66*(2), 803–826. doi:10.1016/j.cirp.2017.05.005

Abele, E., Metternich, J., & Tisch, M. (2019). *Learning Factories. Concepts, Guidelines, Best-Practice Examples.* Springer. doi:10.1007/978-3-319-92261-4

Abele, E., Metternich, J., Tisch, M., Chryssolouris, G., Sihn, W., ElMaraghy, H., Hummel, V., & Ranz, F. (2015). Learning Factories for Research, Education, and Training. *Procedia CIRP*, *32*, 1–6. doi:10.1016/j.procir.2015.02.187

ABET. (2011). *Criteria for accrediting engineering programs: Effective for Reviews During the 2012-2013 Accreditation Cycle.* ABET.

ABET. (2019). *Criteria for Accrediting Engineering Programs, 2020 – 2021.* https://www.abet.org/accreditation/accreditation-criteria/criteria-for-accrediting-engineering-programs-2020-2021/

Adams, R. S., & Atman, C. J. (1999). *Cognitive processes in iterative design behavior.* Paper presented at the FIE'99 Frontiers in Education. 29th Annual Frontiers in Education Conference: Designing the Future of Science and Engineering Education 10.1109/FIE.1999.839114

Adderley, K., Ashwin, C., Bradbury, P., Freeman, J., Goodlad, J., Greene, D., Jenkins, J. R., & Uren, O. (1975). *Project methods in Higher Education.* Society for Research into Higher Education.

Aguiar, L. E. V., & Pacheco, E. M. (2017). Os Institutos Federais de Educação, Ciência e Tecnologia como política pública [The Federal Institutes of Education, Science and Technology as a public policy]. In *As políticas públicas e o papel social dos Institutos Federais de Educação, Ciência e Tecnologia* (Vol. 1). Natal: Editora IFRN.

Ahmadpour, A., Mirdamadi, M., Hosseini, J. F., & Chizari, M. (2010). Factors influencing the design of electronic learning system in agricultural extension. *American Journal of Agricultural and Biological Sciences*, *5*(2), 122–127. doi:10.3844/ajabssp.2010.122.127

Ahmed, F., Fernando Capretz, L., Bouktif, S., & Campbell, P. (2012). Soft skills requirements in software development jobs: A cross-cultural empirical study. *Journal of Systems and Information Technology*, *14*(1), 58–81. doi:10.1108/13287261211221137

Aleem, M., Qurat-ul-Ain, Q.-A., Shahid, F., Islam, M. A., Iqbal, M. A., & Yousaf, M. M. (2019). A review of technological tools in teaching and learning computer science. *Eurasia Journal of Mathematics, Science and Technology Education*, *15*(11). Advance online publication. doi:10.29333/ejmste/109611

Ali, E. A. (2014). *ICT in engineering educational content delivery: Challenges and opportunities.* doi:10.5339/qproc.2014.wcee2013.4

Aliakbari, M., & Nejad, A. M. (2013). On the effectiveness of team teaching in promoting learners' grammatical proficiency. *Canadian Journal of Education, 36*(3), 5–22. https://www.scopus.com/inward/record.uri?eid=2-s2.0-84938081687&partnerID=40&md5=0650d3478590ff7bab1661befb620342

Allal-Chérif, O., Bidan, M., & Makhlouf, M. (2016). Using serious games to manage knowledge and competencies: The seven-step development process. *Information Systems Frontiers, 18*(6), 1153–1163. doi:10.100710796-016-9649-7

Almerich, G., Orellana, N., Suárez-Rodríguez, J., & Díaz-García, I. (2016). Teachers' information and communication technology competences: A structural approach. *Computers & Education, 100*, 110–125. doi:10.1016/j.compedu.2016.05.002

Alves, A. C., Dinis-Carvalho, J., Sousa, R. M., Moreira, F., & Lima, R. M. (2011). Benefits of Lean Management: results from some case studies in Portugal. *CLME'2011, 1977.*

Alves, A. C., Flumerfelt, S., Moreira, F., & Leão, C. P. (2017). *Effective Tools to Learn Lean Thinking and Gather Together Academic and Practice Communities.* Paper presented at the ASME 2017 International Mechanical Engineering Congress and Exposition. 10.1115/IMECE2017-71339

Alves, A. C., Kahlen, F.-J., Flumerfelt, S., & Siriban-Manalang, A. B. (2019). Lean Engineering for Global Development. Springer International Publishing. doi:10.1007/978-3-030-13515-7

Alves, A. C., Leão, C. P., Moreira, F., & Teixeira, S. (2018). Project-Based Learning and its Effects on Freshmen Social Skills in an Engineering Program. In Human Capital and Competences in Project Management. InTechOpen. doi:10.5772/intechopen.72054

Alves, A. C., Moreira, F., Leão, C. P., & Fernandes, S. (2020). Ten Years Of Positive Feedback On Project-Based Learning From First-Year Engineering Students' Prespective. *IMECE2020-23212*, 1–10.

Alves, A. C., Moreira, F., Leão, C. P., & Teixeira, S. (2017). Tutoring Experiences in PBL of Industrial Engineering and Management Program: Teachers vs Students. *ASME 2017 International Mechanical Engineering Congress and Exposition Volume 5: Education and Globalization.* doi:10.1115/IMECE2017-71306

Alves, A. C., Moreira, F., Leão, C. P., Pereira, A. C., Pereira-Lima, S. M. M. A., Malheiro, M. T., Lopes, S. O., & Oliveira, S. (2019, June 10-12). *Industrial engineering and management PBL implementation: An effortless experience?* 11th International Symposium on Project Approaches in Engineering and 16th Active Learning in Engineering Education Workshop, PAEE/ALE, Hammamet, Tunisia.

Alves, A. C., Sousa, R. M., Dinis-Carvalho, J., Lima, R. M., Moreira, F., Leão, C. P., Maia, L. C., Mesquita, D., & Fernandes, S. (2014). Final year Lean projects: advantages for companies, students and academia. *Project Approaches in Engineering Education*, 1–10. http://hdl.handle.net/1822/30172

Alves, A. C., van Hattum-Janssen, N., & Fernandes, S. (2021, July 7-9). *Teacher collaboration in PBL: setting the example for the students.* PAEE/ALE'2021, International Conference on Active Learning in Engineering Education,13rd International Symposium on Project Approaches in Engineering Education (PAEE) and 18th Active Learning in Engineering Education Workshop (ALE), Braga, Portugal.

Alves, A.C., Kahlen, F.-J., Flumerfelt, S., & Siriban-Manalang, A.-B. (2014). Fostering Sustainable Development Thinking Through Lean Engineering Education. *Education and Globalization, 5.* doi:10.1115/IMECE2014-38192

Alves, A. C., Dinis-Carvalho, J., & Sousa, R. M. (2012). Lean production as promoter of thinkers to achieve companies' agility. *The Learning Organization, 19*(3), 219–237. doi:10.1108/09696471211219930

Alves, A. C., Flumerfelt, S., & Kahlen, F.-J. (Eds.). (2017). *Lean Education: An Overview of Current Issues.* Springer. doi:10.1007/978-3-319-45830-4

Alves, A. C., Leão, C. P., Uebe-Mansur, A. F., & Kury, M. I. R. A. (2021). The knowledge and importance of Lean Education based on academics' perspectives: An exploratory study. *Production Planning and Control*, *32*(6), 497–510. doi:10.1080/09537287.2020.1742371

Alves, A. C., Moreira, F., Carvalho, M. A., Oliveira, S., Malheiro, M. T., Brito, I., Leão, C. P., & Teixeira, S. (2019). Integrating Science, Technology, Engineering and Mathematics contents through PBL in an Industrial Engineering and Management first year program. *Production*, *29*(e20180111), e20180111. doi:10.1590/0103-6513.20180111

Alves, A. C., Moreira, F., Sousa, R., & Lima, R. M. (2009, May 15-17). Teachers' workload in a project-led engineering education approach. *International Symposium on Innovation and Assessment of Engineering Curricula*, Valladolid.

Alves, A. C., Sousa, R. M., Fernandes, S., Cardoso, E., Carvalho, M. A., Figueiredo, J., & Pereira, R. M. S. (2016). Teacher's experiences in PBL: Implications for practice. *European Journal of Engineering Education*, *41*(2), 123–141. doi:10.1080/03043797.2015.1023782

Alves, A., Moreira, F., & Sousa, R. (2007). *O papel dos tutores na aprendizagem baseada em projectos: três anos de experiência na Escola de Engenharia da Universidade do Minho.* Libr. Actas Do Congr. Int. Galego-Portugués Psicopedag.

Alves, A., Sousa, R., Moreira, F., Carvalho, M. A., Cardoso, E., Pimenta, P., Malheiro, T., Brito, I., Fernandes, S., & Mesquita, D. (2016). Managing PBL difficulties in an industrial engineering and management program. *Journal of Industrial Engineering and Management*, *9*(3), 586–611. doi:10.3926/jiem.1816

Amaral, L. M. (2017). The Industrial Engineering Management. *Ingenium*, *158*(March/April), 30–33.

Amaro, A. P., Alves, A. C., & Sousa, R. M. (2019). Lean Thinking: A transversal and global management philosophy to achieve sustainability benefits. In A. C. Alves, F.-J. Kahlen, S. Flumerfelt, & A. B. Siriban-Manalang (Eds.), *Lean Engineering for Global Development* (pp. 1–31). Springer International Publishing. doi:10.1007/978-3-030-13515-7_1

Amir, S., Kamal, M. S., Shahria, M. T., & Iftekhar, L. (2020). *Facebook's Social Learning Group for Undergraduate Engineering Courses: A Case Study of Emergency Remote Teaching Amid Large Digital Divide* [Paper presentation]. *2020 IEEE International Conference on Teaching, Assessment, and Learning for Engineering (TALE)*. 10.1109/TALE48869.2020.9368316

Ammer, D. S. (1983). Purchasing and Materials Management for Health-Care Institutions (2nd ed.). Lexington Books.

Amorim, D. C., & Costa, C. J. A. (2020). Impactos da pandemia Covid-19 no processo formativo de professores de Biologia de um mestrado profissional: Desafios em tempos de quarentena [Impacts of the Covid-19 pandemic on the training process of Biology teachers in a professional master's degree: challenges in times of quarantine]. *Devir Educação*, *2*(4), 80–103. doi:10.30905/ded.v4i2.271

Anderson, N. (2020). *Examining ways COVID-19 Adaption Derailed our Assumptions about Online Learning* [Paper presentation]. IEEE Conference on e-learning, e-management and e-services, Malaysia.

Anderson, C. (2008). *The long tail: Why the Future of Business Is Selling Less of More.* Hyperion e.

Anderson, M. G., & Katz, P. B. (1998). Strategic sourcing. *International Journal of Logistics Management*, *9*(1), 1–13. doi:10.1108/09574099810805708

Andrews, R., & Haythornthwaite, C. (2007). Introduction to e-learning research. The SAGE handbook of e-learning research. doi:10.4135/9781848607859.n1

Andrews, J., & Higson, H. (2008). Graduate employability,' soft skills' versus 'hard' business knowledge: A European study. *Higher Education in Europe*, *33*(4), 411–422. doi:10.1080/03797720802522627

Angelva, J., Tepsa, T., & Mielikäinen, M. (2017). Team teaching experiences in engineering education a project-based learning approach. *Proceedings of the 45th SEFI Annual Conference 2017 - Education Excellence for Sustainability, SEFI 2017*, 1182–1189.

Anggeraini, Y., Faridi, A., Mujiyanto, J., & Bharati, D. A. L. (2019). The teachers' perceptions on digital literacy competences in EFL classroom. *Asian EFL Journal, 24*(4), 5–12.

Anjos, M. A., Deccache-Maia, E., & Bonfim, A. M. (2013). Os desafios da construção de um Mestrado Profissional: Um panorama dos sete anos do PROPEC [The challenges of building a Professional Master's Degree: an overview of the seven years of PROPEC]. *Polyphonía, 24*(2), 319–337. doi:10.5216/rp.v24i2.37941

Ankrah, S., & Al-Tabbaa, O. (2015). Universities–Industry Collaboration: A systematic Review. *Scandinavian Journal of Management, 31*(3), 387–408. doi:10.1016/j.scaman.2015.02.003

Anthony, E. M., & Kadir, Z. A. (2012). A road not taken: A breakthrough in English for specific purposes via problem-based learning. *Journal of Technical Education and Training, 4*(1).

Aquere, A. L. (2007). Projects Laboratory: building a university through PBL. *Proceedings of the PAEE'2017, 9th International Symposium on Project Approaches in Engineering Education (PAEE): Implementing Problem/Project-Based Learning (PBL) in different contexts of Brazilian Higher Education Institutions*, 72-81.

Araújo, A., Manninen, H., & Peixoto, B. (2011). Estudo Ergonómico de um Posto de Trabalho numa Empresa de Tecelagem. *Livro de Actas Do Encontro Nacional de Engenharia e Gestão Industrial 2011 (ENEGI 2011)*, 219–220.

Araújo, A., & Alves, A. C. (2012). Pull System implementation through FIFO lane to achieve synchronism between lines and assembly cells. *International Conference in Industrial Engineering and Operations Management (ICIOEM2012)*, 351.

Araújo, A., Alves, A. C., & Romero, F. (2021). Barriers to Lean and Pull System implementation: a case study. *9th Manufacturing Engineering International Conference MESIC2021.*

Araújo, A., Manninen, H., Peixoto, B., & Sarmento, A. L. (2010). Análise do Contributo de um Projecto Interdisciplinar na Aquisição de Competências em Processos de Gestão. In R. M. Lima & D. Carvalho (Eds.), *Natascha van Hattum-Janssen* (pp. 195–198). Ibero-american Simposium on Projectt Approaches in Engineering Education. https://repositorium.sdum.uminho.pt/bitstream/1822/19191/1/2010 _book_ed PAEE2010_proceedings.pdf#page=139

Armstrong, M., & Taylor, S. (2014). *Armstrong's Handbook of Human Resource Management Practice* (13th ed.). Kogan Page.

Armstrong, R. (2008). The Long Tail: Why the Future of Business Is Selling Less of More. *Canadian Journal of Communication, 33*(1). Advance online publication. doi:10.22230/cjc.2008v33n1a1946

Arrudo, H., & Silva, É. R. (2021). *Engineering Education Active Learning Maturity Model: a Conceptual Framework.* PAEE/ALE'2021, International Conference on Active Learning in Engineering Education, 13rd International Symposium on Project Approaches in Engineering Education (PAEE) and 18th Active Learning in Engineering Education Workshop (ALE). http://paee.dps.uminho.pt/proceedingsSCOPUS/PAEE_ALE_2021_PROCEEDINGS.pdf

Arulanand, N., Ramesh Babu, A., & Rajesh, P. K. (2020). Enriched Learning Experience Using Augmented Reality Framework in Engineering Education. *Procedia Computer Science, 172*, 937–942. doi:10.1016/j.procs.2020.05.135

Association for Manufacturing Excellence. (2007). *Lean administration: Case studies in leadership and improvement.* Productivity Press.

Autio, T. (2020). *Loppuraportti.* TEHOJA-hanke.

Baba, M. L., & Pawlowski, D. (2001). *Creating culture change: An ethnographic approach to the transformation of engineering education.* Paper presented at the International Conference on Engineering Education.

Bacich, L., & Moran, J. (2018). Metodologias ativas para uma educação inovadora: uma abordagem téorico-prática. Porto Alegre: Penso.

Baeten, M., Kyndt, E., Struyven, K., & Dochy, F. (2010). Using student-centred learning environments to stimulate deep approaches to learning: Factors encouraging or discouraging their effectiveness. *Educational Research Review, 5*(3), 243–260. doi:10.1016/j.edurev.2010.06.001

Baeten, M., & Simons, M. (2014). Student teachers' team teaching: Models, effects, and conditions for implementation. *Teaching and Teacher Education, 41*, 92–110. doi:10.1016/j.tate.2014.03.010

Bailly, F., & Léné, A. (2013). The personification of the service labour process and the rise of soft skills: A French case study. *Employee Relations.*

Baily, P., Farmer, D., Crocker, B., Jessop, D., & Jones, D. (2008). *Procurement principles and management* (10th ed.). Pearson Education.

Baldo, R. F. G., Cerri, D. G. P. M., Silva, A. L., Zuim, J. B., Mergulhão, L. E. M., Adolpho, M. N., Boschi, T. F., Garcia, R. F. H., & Favalli, R. (2019). Aplicação da metodologia scrum em um estudo de caso de engenharia [Application of scrum methodology in a engineering case study]. *Revista Produção Online, Florianópolis, SC, 19*(3), 856–875. doi:10.14488/1676-1901.v19i3.3248

Ballé, M., Chaize, J., & Jones, D. (2015). Inclusive versus exclusive learning: The secret ingredient to creating a truly "lean" and "learning" culture. *Development and Learning in Organizations, 29*(1), 20–23. doi:10.1108/DLO-10-2014-0080

Ballé, M., Chaize, J., & Jones, D. (2019). Lean as a learning system: What do organizations need to do to get the transformational benefits from Toyota's method? *Development and Learning in Organizations, 33*(3), 1–4. doi:10.1108/DLO-11-2018-0147

Bals, L., Schulze, H., Kelly, S., & Stek, K. (2019). Purchasing and supply management (PSM) competencies: Current and future requirements. *Journal of Purchasing and Supply Management, 25*(5), 100572. doi:10.1016/j.pursup.2019.100572

Bardin, L. (2011). *Análise de conteúdo.* Edições 70.

Barr, M. (2017). Video games can develop graduate skills in higher education students: A randomised trial. *Computers & Education, 113*, 86–97. doi:10.1016/j.compedu.2017.05.016

Bauer, H., Brandl, F., Lock, C., & Reinhart, G. (2018). Integration of Industrie 4.0 in Lean Manufacturing Learning Factories. *Procedia Manufacturing, 23*, 147–152. doi:10.1016/j.promfg.2018.04.008

Bauters, M., Holvikivi, J., & Vesikivi, P. (2020). *An overview of the situation of Project-Based Learning in Engineering Education.* SEFI 48th Annual Conference, Enschede, The Netherlands.

Bédard, D., Lison, C., Dalle, D., Côté, D., & Boutin, N. (2012). Problem-based and project-based learning in engineering and medicine: Determinants of students' engagement and persistance. *The Interdisciplinary Journal of Problem-Based Learning, 6*(2). Advance online publication. doi:10.7771/1541-5015.1355

Belbin, R. M. (2012). *Team roles at work.* Routledge. doi:10.4324/9780080963242

Bellotti, F., Berta, R., De Gloria, A., Lavagnino, E., Antonaci, A., Dagnino, F., Ott, M., Romero, M., Usart, M., & Mayer, I. S. (2014). Serious games and the development of an entrepreneurial mindset in higher education engineering students. *Entertainment Computing, 5*(4), 357–366. doi:10.1016/j.entcom.2014.07.003

Bell, S. (2010). Project-based learning for the 21st century: Skills for the future. *The Clearing House: A Journal of Educational Strategies, Issues and Ideas, 83*(2), 39–43. doi:10.1080/00098650903505415

Benchicou, S., Aichouni, M., & Nehari, D. (2010). E-learning in engineering education: A theoretical and empirical study of the Algerian higher education institution. *European Journal of Engineering Education, 35*(3), 325–343. doi:10.1080/03043797.2010.483610

Benhima, M., & Ben, S. M. (2021). Moroccan English Department Student Attitudes Towards the Use of Distance Education During COVID-19: Moulay Ismail University as a Case Study. *International Journal of Information and Communication Technology Education, 17*(3), 105–122. Advance online publication. doi:10.4018/IJICTE.20210701.oa7

Benis, A., Nelke, S. A., & Winokur, M. (2020). *Upgrading Industrial Engineering and Management curriculum to Industry 4.0*. IEEE.

Bergmann, J., & Sams, A. (2012). Flip your classroom: reach every student in every class every day. International Society for Technology in Education.

Berns, A., Isla-Montes, J.-L., Palomo-Duarte, M., & Dodero, J.-M. (2016). Motivation, students' needs and learning outcomes: A hybrid game-based app for enhanced language learning. *SpringerPlus, 5*(1), 1305. Advance online publication. doi:10.118640064-016-2971-1 PMID:27547679

Berrocoso, J. V., Arroyo, M. D. C. G., Videla, C. B., & Cevallos, M. B. M. (2020). Trends in Educational Research about e-Learning: A Systematic Literature Review (2009–2018). *Sustainability, 12*(5153), 5153. Advance online publication. doi:10.3390u12125153

Bessant, J., & Francis, D. (1999). Developing strategic continuous improvement capability. *International Journal of Operations & Production Management, 19*(11), 1106–1119. doi:10.1108/01443579910291032

Bestiantono, D. S., Putri, Z. R. A., & Cheng, T.-H. (2020). Online Learning amid the COVID-19 Pandemic: Students' Perspectives Muhammad. *Journal of Pedagogical Sociology and Psychology, 2*(1), 45–51. doi:10.46627ilet.v1i3.46

Bhasin, S. (2013). Impact of corporate culture on the adoption of the Lean principles. *International Journal of Lean Six Sigma, 4*(2), 118–140. doi:10.1108/20401461311319329

Bhasin, S., & Found, P. (2019). Sustaining the lean ideology. *Management Decision*. Advance online publication. doi:10.1108/MD-09-2019-1254

Bhuasiri, W., Xaymoungkhoun, O., Zo, H., Rho, J. J., & Ciganek, A. P. (2012). Critical success factors for e-learning in developing countries: A comparative analysis between ICT experts and faculty. *Computers & Education, 58*(2), 843–855. doi:10.1016/j.compedu.2011.10.010

Bienhaus, F., & Haddud, A. (2018). Procurement 4.0: Factors influencing the digitisation of procurement and supply chains. *Business Process Management Journal, 24*(4), 965–984. doi:10.1108/BPMJ-06-2017-0139

Biggs, J. B. (1996). Enhancing teaching through constructive alignment. *Higher Education, 32*(3), 347–364. doi:10.1007/BF00138871

Birou, L., Lutz, H., & Zsidisin, G. A. (2016). Current state of the art and science: A survey of purchasing and supply management courses and teaching approaches. *International Journal of Procurement Management, 9*(1), 71–85. doi:10.1504/IJPM.2016.073388

Birrell, J. R., & Bullough, R. V. Jr. (2005). Teaching with a Peer: A Follow-Up Study of the 1st Year of Teaching. *Action in Teacher Education, 27*(1), 72–81. doi:10.1080/01626620.2005.10463375

Bissett-Johnson, K., & Radcliffe, D. F. (2021). Engaging engineering students in socially responsible design using global projects. *European Journal of Engineering Education, 46*(1), 4–26. doi:10.1080/03043797.2019.1674785

Bissi, W. (2007, January/June). Scrum – Metodologia de desenvolvimento ágil [Scrum – Agile development methodology]. *Campo Dig. Campo Mourão, 2*(1), 3–6.

Black, J. T., & Phillips, D. T. (2013). *Lean Engineering The Future Has Arrived.* Virtualbookworm.com Publishing.

Bolman, L. G., & Deal, T. E. (2013). *Reframing organizations: Artistry, choice and leadership* (5th ed.). Jossey-Bass.

Bonarccorsi, A., & Piccaluga, A. (1994). A Theoretical Framework for the Evaluation of University—Industry Relationships. *R & D Management, 24*(3), 229–247. doi:10.1111/j.1467-9310.1994.tb00876.x

Bonnaud, O., & Bsiesy, A. (2020). Adaptation of the Higher Education in Engineering to the Advanced Manufacturing Technologies. *Advances in Technology Innovation, 5*(2), 65–75. doi:10.46604/aiti.2020.4144

Bonwell, C. C., & Eison, J. A. (1991). *Active Learning: Creating Excitement in the Classroom.* 1991 ASHE-ERIC Higher Education Reports: ERIC.

Bonwell, C. C., & Eison, J. A. (1991). *Active Learning: Creating Excitement in the Classroom.* ASHE-ERIC Higher Education Report No 1. The George Washington University, School of Education and Human Development.

Bonwell, C., & Eison, J. (1991). *Active Learning: Creating Excitement in the Classroom.* Wiley.

Borges, K. S., Schmitt, M. A. R., & Nakle, S. M. (2014). *EduScrum – projetos de aprendizagem colaborativa baseados em scrum* [EduScrum – collaborative learning Projects based on scrum]. Instituto Federal de Educação, Ciência e Tecnologia do Rio Grande do Sul, Campus Porto Alegre.

Bortolotti, T., Boscari, S., & Danese, P. (2015). Successful lean implementation: Organizational culture and soft lean practices. *International Journal of Production Economics, 160*, 182–201. doi:10.1016/j.ijpe.2014.10.013

Bourion-Bédès, S., Tarquinio, C., Batt, M., Tarquinio, P., Lebreuilly, R., Sorsana, C., & Baumann, C. (2021b). Stress and associated factors among French university students under the COVID-19 lockdown: The results of the PIMS-CoV 19 study. *Journal of Affective Disorders, 283*, 108–114. doi:10.1016/j.jad.2021.01.041 PMID:33540333

Bourion-Bédès, S., Tarquinio, C., Batt, M., Tarquinio, P., Lebreuilly, R., Sorsana, C., Legrand, K., Rousseau, H., & Baumann, C. (2021a). Psychological impact of the COVID-19 outbreak on students in a French region severely affected by the disease: Results of the PIMS-CoV 19 study. *Psychiatry Research, 295*, 113559. doi:10.1016/j.psychres.2020.113559 PMID:33189368

Braga, E. S. O., Silva, C. S., & Rôças, G. (2019) É Pau, É Pedra... É o Fim do Caminho ou Começo de um Novo Caminhar? Reflexão sobre os Institutos Federais na Rede Federal de Educação Profissional, Científica e Tecnológica [It's Stick, It's Stone... Is it the End of the Path or the Beginning of a New Walk? Reflection on the Federal Institutes in the Federal Network for Professional, Scientific and Technological Education]. In *Educação, cultura e sociedade.* IFPB. Retrieved from http://editora.ifpb.edu.br/index.php/ifpb/catalog/book/236

Braun, J. M. (2013). Teaching Lean Six Sigma with Service-Learning. *2013 ASQ Advancing the STEM Agenda Conference.*

Brien, M. O. (2021). *Defined Learning-What is Project Based learning.* https://blog.definedlearning.com/blog/what-is-project-based-learning

Broström, T., Kaivosoja, L., & Kekkonen, M. (2019). Robotiikalla joustavaa automaatiota Oulun talousalueelle. *ePooki, Oulun ammattikorkeakoulun tutkimus- ja kehitystyön julkaisut.*

Bucciarelli, L. L. (1996). *Designing engineers.* MIT Press.

Buenaño-Fernandez, D., Luján-Mora, S., & Gil, D. (2019). A hybrid machine learning approach for the prediction of grades in computer engineering students. In *Research and Innovation Forum, Rii Forum 2019* (pp. 125–134). Springer. 10.1007/978-3-030-30809-4_13

Bulić, M., Jelaska, I., & Jelaska, P. M. (2017). The effect of E-learning on the acquisition of learning outcomes in teaching science and biology. *Croatian Journal of Education, 19*(2), 447–477. doi:10.15516/cje.v19i2.2230

Burt, D. N., Dobler, D. W., & Starling, S. L. (2003). *World class supply management: The key to supply chain management*. McGraw-Hill Irwin.

Butt, A. L., Kardong-Edgren, S., & Ellertson, A. (2018). Using Game-Based Virtual Reality with Haptics for Skill Acquisition. *Clinical Simulation in Nursing, 16*, 25–32. doi:10.1016/j.ecns.2017.09.010

Cachay, J., Wennemer, J., Abele, E., & Tenberg, R. (2012). Study on Action-Oriented Learning with a Learning Factory Approach. *Procedia: Social and Behavioral Sciences, 55*, 1144–1153. doi:10.1016/j.sbspro.2012.09.608

Callo, E. C., & Yazon, A. D. (2020). Exploring the factors influencing the readiness of faculty and students on online teaching and learning as an alternative delivery mode for the new normal. *Universal Journal of Educational Research, 8*(8), 3509–3518. doi:10.13189/ujer.2020.080826

Camosun College. (2018). *Applied Learning* https://legacy.camosun.ca/about/applied-learning/index.html

Campbell, D. T., & Stanley, J. C. (1966). *Experimental and quasi-experimental designs for research.* Houghton Mifflin Company Boston.

Campion, M. A., Fink, A. A., Ruggeberg, B. J., Carr, L., Phillips, G. M., & Odman, R. B. (2011). Doing competencies well: Best practices in competency modeling. *Personnel Psychology, 64*(1), 225–262. doi:10.1111/j.1744-6570.2010.01207.x

Cannon, R., & Newble, D. (2000). *A handbook for teachers in universities and colleges. A guide to improving teaching methods* (4th ed.). Kogan Page.

Cantoni, V., Cellario, M., & Porta, M. (2004). Perspectives and challenges in e-learning: Towards natural interaction paradigms. *Journal of Visual Languages and Computing, 15*(5), 333–345. doi:10.1016/j.jvlc.2003.10.002

Cargnin-Stieler, M., Malheiro, M. T., Alves, A. C., Lima, R. M., & Teixeira, M. C. M. (2019). Learning calculus through PBL in an industrial engineering and management program-a seven-year study. *Advances in Engineering Education, 7*(3), 1–28.

Carlson, L. E., & Sullivan, J. F. (1999). Hands-on Engineering: Learning by Doing in the Integrated Teaching and Learning Program. *International Journal of Engineering Education, 15*(1), 20–31.

Carr, A. S., & Smeltzer, L. R. (2000). An empirical study of the relationships among purchasing skills and strategic purchasing, financial performance, and supplier responsiveness. *The Journal of Supply Chain Management, 36*(3), 40–54. doi:10.1111/j.1745-493X.2000.tb00250.x

Carter, C. R., Rogers, D. S., & Choi, T. Y. (2015). Toward the theory of the supply chain. *The Journal of Supply Chain Management, 51*(2), 89–97. doi:10.1111/jscm.12073

Carter, J. R., & Narasimhan, R. (1996). Is purchasing really strategic? *International Journal of Purchasing and Materials Management, 32*(4), 20–28. doi:10.1111/j.1745-493X.1996.tb00216.x

Carvalho, A. M., Sampaio, P., Rebentisch, E., Carvalho, J. Á., & Saraiva, P. (2019). Operational excellence, organisational culture and agility: The missing link? *Total Quality Management & Business Excellence, 30*(13–14), 1495–1514. doi:10.1080/14783363.2017.1374833

Carvalho, V., Gonçalves, I. M., Lage, T., Rodrigues, R. O., Minas, G., Teixeira, S. F. C. F., Moita, A. S., Hori, T., Kaji, H., & Lima, R. A. (2021). 3D Printing Techniques and Their Applications to Organ-on-a-Chip Platforms : A Systematic Review. *Sensors (Basel), 21*(9), 3304. doi:10.339021093304 PMID:34068811

Carvalho, V., Maia, I., Souza, A., Ribeiro, J., Costa, P., Puga, H., Teixeira, S. F. C. F., & Lima, R. A. (2020). In Vitro Stenotic Arteries to Perform Blood Analogues Flow Visualizations and Measurements: A Review. *The Open Biomedical Engineering Journal, 14*(1), 87–102. doi:10.2174/1874120702014010087

Carvalho, V., Rodrigues, R. O., Lima, R. A., & Teixeira, S. (2021). Computational Simulations in Advanced Microfluidic Devices: A Review. *Micromachines, 12*(10), 1149. doi:10.3390/mi12101149 PMID:34683199

Carvalho, V., Teixeira, S., Lima, R. A., Minas, G., & Rodrigues, C. (2021). Learning Science during Summer Vacations and Its Effects on Attitude and Anxiety towards Research. *International Conference on Active Learning in Engineering Education.*

Castro, T. C., & Gonçalves, L. S. (2018). The use of gamification to teach in the nursing field. *Revista Brasileira de Enfermagem, 71*(3), 1038–1045. doi:10.1590/0034-7167-2017-0023 PMID:29924165

Catarino, S. O., Rodrigues, R. O., Pinho, D., Miranda, J. M., Minas, G., & Lima, R. (2019). Blood Cells Separation and Sorting Techniques of Passive Microfluidic Devices : From Fabrication to Applications. *Micromachines, 10*(9), 593. doi:10.3390/mi10090593 PMID:31510012

Cavinato, J. L. (1987). Purchasing performance: What makes the magic? *Journal of Purchasing and Materials Management, 23*(3), 10–16. doi:10.1111/j.1745-493X.1987.tb00186.x

CC – Creative Commons. (2021). https://creativecommons.org/

Cedefop, E. (2016). Skills and jobs survey. *Researchers and engineers: skills opportunities and challenges.*

Chamorro-Premuzic, T., Arteche, A., Bremner, A. J., Greven, C., & Furnham, A. (2010). Soft skills in higher education: Importance and improvement ratings as a function of individual differences and academic performance. *Educational Psychology, 30*(2), 221–241. doi:10.1080/01443410903560278

Chaturvedi, K., Vishwakarma, D. K., & Singh, N. (2020, July). 2021. "COVID-19 and Its Impact on Education, Social Life and Mental Health of Students : A Survey. *Children and Youth Services Review, 121*, 105866. doi:10.1016/j.childyouth.2020.105866 PMID:33390636

Chaubey, A., Bhattacharya B., & Mandal, S.K.D. (2018). *Attributes of good teaching in engineering education in Indian subcontinent.* Sadhana, Indian Academy of Sciences. doi:10.1007/s12046-018-0946-3

Chedid, M., Teixeira, L., & Rosa, M. J. (2018). The Industrial Engineers' Essential Skills: Students, teachers and industry professionals' perspectives. *Proceedings of the INTED 2018 Conference*, 1815–1825. 10.21125/inted.2018.0322

Chen, C., Jiang, B. C., & Hsu, K. (2005). An empirical study of industrial engineering and management curriculum reform in fostering students' creativity. *European Journal of Engineering Education, 30*(2), 191–202. doi:10.1080/03043790500087423

Chen, J., Kolmos, A., & Du, X. (2021). Forms of implementation and challenges of PBL in engineering education: A review of literature. *European Journal of Engineering Education, 46*(1), 90–115. doi:10.1080/03043797.2020.1718615

Chiappe, A., & Adame, S. A. (2018). *Open Educational Practices: a learning way beyond free access knowledge.* https://www.scielo.br/j/ensaio/a/jTpx6MgYFpGbzLgQJ6FY6Hh/?lang=en

Chicksand, D., Watson, G., Walker, H., Radnor, Z., & Johnston, R. (2012). Theoretical perspectives in purchasing and supply chain management: An analysis of the literature. *Supply Chain Management, 17*(4), 454–472. doi:10.1108/13598541211246611

Ching-Ter, C., Hajiyev, J., & Su, C. R. (2017). Examining the students' behavioral intention to use e-learning in Azerbaijan? The General Extended Technology Acceptance Model for E-learning approach. *Computers & Education, 111*, 128–143. doi:10.1016/j.compedu.2017.04.010

Chou, S. (2018). The Fourth Industrial Revolution: Digital fusion with internet of things. *Journal of International Affairs, 72*(1), 107–120. https://www.jstor.org/stable/26588346

Christie, M., & de Graaff, E. (2017). The philosophical and pedagogical underpinnings of Active Learning in Engineering Education. *European Journal of Engineering Education, 42*(1), 5–16. doi:10.1080/03043797.2016.1254160

Clarke, D., & Winslade, M. (2019). A school-university teacher education partnership: Reconceptualising reciprocity of learning. *Journal of Teaching and Learning for Graduate Employability, 10*(1), 138–156. doi:10.21153/jtlge2019vol10no1art797

Cohen, J. (1988). *The effect size index: d* (Vol. 2). Lawrence Erlbaum Associates, Publishers.

Condie, R., & Livingston, K. (2007). Blending online learning with traditional approaches: Changing practices. *British Journal of Educational Technology, 38*(2), 337–348. doi:10.1111/j.1467-8535.2006.00630.x

Conn, C. E. (2010). Learning the hard way (but still learning!): Using team teaching as a vehicle for pedagogical change. *Business Communication Quarterly, 73*(1), 87–91. doi:10.1177/1080569909358098

Cook, N. F., McAloon, T., O'Neill, P., & Beggs, R. (2012). Impact of a web based interactive simulation game (PULSE) on nursing students' experience and performance in life support training - A pilot study. *Nurse Education Today, 32*(6), 714–720. doi:10.1016/j.nedt.2011.09.013 PMID:22082881

Cooner, T. S. (2010). Creating opportunities for students in large cohorts to reflect in and on practice: Lessons learnt from a formative evaluation of students' experiences of a technology-enhanced blended learning design. *British Journal of Educational Technology, 41*(2), 271–286. doi:10.1111/j.1467-8535.2009.00933.x

Coronado, J.M., Moyano, A., Romero, V., Ruiz, R. & Rodríguez, J. (2021). Student Long-Term Perception of Project-Based Learning in Civil Engineering Education: An 18-Year Ex-Post Assessment. *Sustainability, 13*(1949). doi:10.3390/su13041949

Corrigan, S., Zon, G. D. R., Maij, A., McDonald, N., & Mårtensson, L. (2015). An approach to collaborative learning and the serious game development. *Cognition Technology and Work, 17*(2), 269–278. doi:10.100710111-014-0289-8

Corrocher, N., Mavilia, R., & Giorgio, M. (2020). The Sino-German alliance for the fourth industrial revolution: Dynamics and policy implications. *Journal of Economic Policy Reform, 23*(4), 426–446. doi:10.1080/17487870.2018.1547639

Costa, H. G., & Alves, A. C. (2021). *A Non-compensatory Multicriteria Model for Sorting the Influence of PBL Over Professional Skills*. Springer. doi:10.1007/978-3-030-71187-0_108

Costa, L. R. J., Honkala, M., & Lehtovuori, A. (2007). Applying the problem-based learning approach to teach elementary circuit analysis. *IEEE Transactions on Education, 50*(1), 41–48. doi:10.1109/TE.2006.886455

Costa, S., Cuzzocrea, F., & Nuzzaci, A. (2014). Uses of the internet in educative informal contexts. Implication for formal education. *Comunicar, 22*(43), 163–171. doi:10.3916/C43-2014-16

Council of the European Union. (2018). *Proposal for a Council Recommendation on Key Competences for Lifelong Learning*. Author.

Cousins, P. D., Lamming, R., Lawson, B., & Squire, B. (2008). *Strategic supply management: principles, theories and practice*. Pearson Education.

Cousins, P. D., & Spekman, R. (2003). Strategic supply and the management of inter-and intra-organisational relationships. *Journal of Purchasing and Supply Management, 9*(1), 19–29. doi:10.1016/S1478-4092(02)00036-5

Crawford, R., & Jenkins, L. (2017). Blended learning and team teaching: Adapting pedagogy in response to the changing digital tertiary environment. *Australasian Journal of Educational Technology, 33*(2), 51–72. doi:10.14742/ajet.2924

Crawley, E. F., Malmqvist, J., Ostlund, S., & Brodeur, D. (2007). *Rethinking Engineering Education: The CDIO Approach*. Springer.

Cruz, C., & Murphy, E. (1996). Purchasing's new importance requires a broader education. *Purchasing, 12*(9), 46–49.

Da Silva, G. B., Costa, H. G., & De Barros, M. D. (2015). Entrepreneurship in engineering education: A literature review. International Journal of Engineering Education.

Da Xu, L., Xu, E. L., & Li, L. (2018). Industry 4.0: State of the art and future trends. *International Journal of Production Research, 56*(8), 2941–2962. doi:10.1080/00207543.2018.1444806

Dahl, J. (2020). *Leading lean: Ensuring success and developing a framework for leadership*. O'Reilly Media.

Darby, J. L., Fugate, B. S., & Murray, J. B. (2019). Interpretative research: A complementary approach to seeking knowledge in supply chain management. *International Journal of Logistics Management*. Advance online publication. doi:10.1108/IJLM-07-2018-0187

de Araújo, J. A. R., Judson da Cruz, G., Vieira da Silva, W., Deretti, S., Dalazen, L. L., & Pereira da Veiga, C. (2016). Quality Evaluation in Post-Graduate Diploma Courses from the Students' Perspective: An Exploratory Study in Brazil. *International Journal of Management Education, 14*(3), 454–465. doi:10.1016/j.ijme.2016.10.003

De Houwer, J., Barnes-Holmes, D., & Moors, A. (2013). What is learning? On the nature and merits of a functional definition of learning. *Psychonomic Bulletin & Review, 20*(4), 631–642. doi:10.375813423-013-0386-3 PMID:23359420

De Jesus, I. R. D. I. R. D., & Costa, H. G. G. H. G. (2013). Role of production engineers in the public sector. *22nd International Conference on Production Research, ICPR 2013*. https://www.scopus.com/inward/record.uri?eid=2-s2.0-84929359638&partnerID=40&md5=0754d9a5f95b4ee82c862b6c042c7b91

De La Iglesia, J. C. F., Morante, M. C. F., & López, B. C. (2018). Influence of personal and enviroment factors on classroom ICT integration in Galicia. *Revista de Medios y Educacion, 53*, 79–91. doi:10.12795/pixelbit.2018.i53.05

De Vin, L. J., Jacobsson, L., Odhe, J. E., & Wickberg, A. (2017). Lean Production Training for the Manufacturing Industry: Experiences from Karlstad Lean Factory. *Procedia Manufacturing, 11*, 1019–1026. doi:10.1016/j.promfg.2017.07.208

Deaconu, A., Dedu, E. M., Igreţ, R. S., & Radu, C. (2018). The use of information and communications technology in vocational education and training-premise of sustainability. *Sustainability (Switzerland), 10*(5), 1466. Advance online publication. doi:10.3390u10051466

Delamare-Le Deist, F., & Winterton, J. (2005). What is competence? *Human Resource Development International, 8*(1), 27–46. doi:10.1080/1367886042000338227

Delhij, A., Solingen, R. V.., & Wijnands, W. (2016). *O Guia eduScrum* [eduScrum guide]. Time eduScrum.

Deming, W. E. (1993). *The new economics: For industry, government, education*. Massachusetts Institute of Technology.

Demir, K. A., Döven, G., & Sezen, B. (2019). Industry 5.0 and Human-Robot Co-working. *Procedia Computer Science, 158*, 688–695. doi:10.1016/j.procs.2019.09.104

Denzin, N. K. (1994). Romancing the Text: The Qualitative Researcher-Writer-as-Bricoleur on JSTOR. *Bulletin of the Council for Research in Music Education, 1*(1), 15–30. https://www.jstor.org/stable/40318652?seq=1

Devlin-Scherer, R., & Sardone, N. B. (2013). Collaboration as a Form of Professional Development: Improving Learning for Faculty and Students. *College Teaching, 61*(1), 30–37. doi:10.1080/87567555.2012.714815

Dewey, J. (1996). *Democracy and Education. An Introduction to the Philosophy of Education*. Free Press.

Dinis-Carvalho, J., Fernandes, S., & Filho, J. C. R. (2017). Combining lean teaching and learning with eduScrum. *Int. J. Six Sigma and Competitive Advantage, 10*(3/4), 221–235. doi:10.1504/IJSSCA.2017.086599

Docherty, M., & Gaubinger, K. (2018). Topic Oriented Mixed-Method System (TOMMS): A Holistic System to Integrate English Language Classes into English-Medium Instruction Courses in European Higher Education. In M. E. Auer, D. Guralnick, & I. Simonics (Eds.), *Teaching and Learning in a Digital World. Proceedings of the 20th International Conference on Interactive Collaborative Learning* (Vol. 1, pp. 840–851). Springer.

Dombrowski, U., & Mielke, T. (2013). Lean Leadership fundamental principles and their application. *Procedia CIRP, 7*, 569–574. doi:10.1016/j.procir.2013.06.034

Dombrowski, U., Wullbrandt, J., & Fochler, S. (2019). Center of Excellence for Lean Enterprise 4.0. *Procedia Manufacturing, 31*, 66–71. Advance online publication. doi:10.1016/j.promfg.2019.03.011

dos Santos, P. F., Simon, A. T., Guimarães, G. E., Amorim, M., & Vieira, M., Junior. (2017). Analyzing the competences of production engineering graduates: An industry perspective. *Production, 27*. Advance online publication. doi:10.1590/0103-6513.005317

Dowd, K. O., & Liedtka, J. (1994). What corporations seek in MBA hires: A survey. *The Magazine of the Graduation Management Admission Council, 2*.

Downe-Wambolt, B. (1992). Content analysis: Method, applications and issues. *Health Care for Women International, 13*(3), 313–321. doi:10.1080/07399339209516006 PMID:1399871

Drucker, P. F. (1977). *People and performance: The best of Peter Drucker on management*. Harper & Row, Publishers, Inc.

Du Sautoy, M. (2019). *The Creativity Code: Art and Innovation in the Age of AI*. Harvard University Press.

Duarte, L. (2019). Scrum e Métodos Ágeis – Um Guia Prático [Scrum and agile methods – a practice guide]. Gravataí. RS.

Dublin Descriptors, J. Q. I. (2004). *Dublin descriptors. The Joint Quality Initiative Working document*. Shared Dublin descriptors for the Bachelor's, Master's and Doctoral awards.

Duch, G., Duch, B. J., & Allen, D. E. (2001). *The power of problem-based learning: a practical "how to" for teaching undergraduate courses in any discipline*. Sterling: Stylus Publishing.

Dunleavy, M., & Heinecke, W. F. (2007). The Impact of 1:1 Laptop use on middle school math and science standardized test scores. *Computers in the Schools, 24*(3–4), 7–22. doi:10.1300/J025v24n03_02

Durach, C. F., Kembro, J., & Wieland, A. (2017). A new paradigm for systematic literature reviews in supply chain management. *The Journal of Supply Chain Management, 53*(4), 67–85. doi:10.1111/jscm.12145

Dutra, D. S., & Silva, J. R. (2016). Product-Service Architecture (PSA): Toward a Service Engineering perspective in Industry 4.0. *IFAC-PapersOnLine*, *49*(31), 91–96. doi:10.1016/j.ifacol.2016.12.167

Dym, C. L., Agogino, A. M., Eris, O., Frey, D. D., & Leifer, L. J. (2005). Engineering design thinking, teaching, and learning. *Journal of Engineering Education*, *94*(1), 103–120. doi:10.1002/j.2168-9830.2005.tb00832.x

Egbert, S. C., & Camp, S. (2016). Practical strategies for team teaching in the distance education (IVC) context. In Interdisciplinary Approaches to Distance Teaching: Connecting Classrooms in Theory and Practice (pp. 141-153). doi:10.4324/9781315694412

Eisenhardt, K. M., & Graebner, M. E. (2007). Theory building from cases: Opportunities and challenges. *Academy of Management Journal*, *50*(1), 25–32. doi:10.5465/amj.2007.24160888

Elberse, A. (2008, July). The Long Tail Debate: A Response to Chris Anderson. *Harvard Business Review*.

Ellen MacArthur Foundation. (2012). *Towards the Circular Economy - Economic and Business Rationale for an Accelerated Transition*. Ellen MacArthur Foundation.

Elsayed, E. A. (1999). Industrial Engineering Education: A Prospective. *European Journal of Engineering Education*, *24*(4), 415–421. doi:10.1080/03043799908923576

Elsevier. (2021). *University-industry collaboration*. https://www.elsevier.com/research-intelligence/university-industry-collaboration

Eltantawy, R. A., Giunipero, L. C., & Fox, G. L. (2009). A strategic skill based model of supplier integration and its effect on supply management performance. *Industrial Marketing Management*, *38*(8), 925–936. doi:10.1016/j.indmarman.2008.12.022

Emiliani, B. (2008). *Practical lean leadership: A strategic leadership guide for executives*. The Center for Lean Business Management, LLC.

Enke, J., Glass, R., Kreß, A., Hambach, J., Tisch, M., & Metternich, J. (2018). Industrie 4.0 - Competencies for a modern production system: A curriculum for Learning Factories. *Procedia Manufacturing*, *23*, 267–272. Advance online publication. doi:10.1016/j.promfg.2018.04.028

Erpenbeck, J., & Scharnhorst, A. (2005). Modellierung von Kompetenzen im Licht der Selbstorganisation. In T. Meynhardt & E. J. Brunner (Eds.), *Selbstorganisation Managen - Beiträge zur Synergetik der Organisation*. Waxmann Verlag GmbH.

ESG Report. (2015). *Standards and Guidelines for Quality Assurance in the European Higher Education Area (ESG)*. Retrieved from https://www.enqa.eu/index.php/home/esg/

EU Science Hub. (2019). *Learning and Skills for Digital Era*. https://ec.europa.eu/jrc/en/research-topic/learning-and-skills

EU. (2020). *Skills for Industry Curriculum Guidelines 4.0: Future-proof education and training for manufacturing in Europe*. doi:10.2826/69418

Eurofound. (2019). *The future of manufacturing in Europe*. Author.

European Commission. (2008). *Communication from the Commission to the European Parliament and the Council — The raw materials initiative: meeting our critical needs for growth and jobs in Europe*. COM (2008) 699.

European Commission. (2011). *Communication from the Commission to the European Parliament, the Council, the European Economic and Social Committee and the Committee of the Regions, Tackling The Challenges in Commodity Markets and on Raw Materials*. COM (2011) 25.

European Commission. (2015). *Communication from the Commission to the European Parliament, the Council, the European Economic and Social Committee and the Committee of the Regions, Closing the Loop - An EU Action Plan for the Circular Economy*. COM(2015) 614.

European Commission. (2018). *Report on Critical Raw Materials and the Circular Economy*. Commission Staff Working Document.

European Union (2000). *Directive 2000/53/EC of 18 September 2000 on end-of life vehicles*. Official Journal of the European Communities, (21.10.2000) 269/34–269/43.

European Union (2005). *Directive 2005/64/EC of 26 October 2005 on the type-approval of motor vehicles with regard to their reusability, recyclability and recoverability and amending*. Official Journal of the European Communities, (25.11.2005) 310/10–310/27.

European Union. (2019). *Key competences for lifelong learning*. doi:10.2766/569540

Faes, W., Knight, L., & Matthyssens, P. (2001). Buyer profiles: An empirical investigation of changing organizational requirements. *European Journal of Purchasing & Supply Management*, *7*(3), 197–208. doi:10.1016/S0969-7012(00)00031-9

Faisal, R. A., Jobe, M. C., Ahmed, O., & Sharker, T. (2021). Mental Health Status, Anxiety, and Depression Levels of Bangladeshi University Students During the COVID-19 Pandemic. *International Journal of Mental Health and Addiction*, •••, 1–16. doi:10.100711469-020-00458-y

Faustino, Catarino, Lima, & Minas. (2016). Biomedical Microfluidic Devices by Using Low-Cost Fabrication Techniques: A Review. *Journal of Biomechanics, 49*(11), 2280–92. . doi:10.1016/j.jbiomech.2015.11.031

Fauvel, A. M., Miller, L. K., Lane, P., & Farris, J. (2010). Reflections on an interdisciplinary, community-based, team-taught adventure. *The Journal of Continuing Higher Education*, *58*(1), 40–46. doi:10.1080/07377360903528089

Fava-de-Moraes, F., & Fava, M. (2000). A Iniciação Científica: Muitas Vantagens e Poucos Riscos. *São Paulo em Perspectiva*, *14*(1), 73–77. doi:10.1590/S0102-88392000000100008

Fawcett, S. E., & Rutner, S. M. (2014). A longitudinal view of supply chain education: Assessing the challenge of retaining relevance in today's dynamic marketplace. *International Journal of Logistics Management*, *25*(1), 180–201. doi:10.1108/IJLM-07-2012-0061

Feisel, E., Hartmann, E., & Giunipero, L. C. (2011). The importance of the human aspect in the supply function: Strategies for developing PSM proficiency. *Journal of Purchasing and Supply Management*, *17*(1), 54–67. doi:10.1016/j.pursup.2010.07.001

Feld, W. M. (2000). Lean Manufacturing: Tools, Techniques, and How to Use Them (1st ed.). St. Lucie Press.

Felder, R. M., Stice, J. E., & Rugarcia, A. (2000). The future of Engineering Education VI. Making Reform Happen. *Chemical Engineering Education*, *34*(3), 208–215.

Ferguson, R., Barzilai, S., Ben-zvi, D., Chinn, C. A., Herodotou, C., Hod, Y., Kali, Y., Kukulska-Hulme, A., Kupermintz, H., McAndrew, P., Rienties, B., Sagy, O., Scanlon, E., Sharples, M., Weller, M., & Whitelock, D. (2017). *Innovating Pedagogy 2017: Open University Innovation Report 6*. The Open University.

Fernandes, S., Alves, A. C., & Uebe-Mansur, A. F. (2020). Student-Centered Assessment Practices: An Integrated Approach With Project-Based Learning (PBL). In Determining the Reliability of Online Assessment and Distance Learning (p. 340). IGI Global.

Fernandes, S., Cunha, R., Torres, D. P. C., Fernandes, S., Cunha, R., Torres, D., & Pimentel, C. (2020). JELA: An Alternative Approach to Industrial Engineering and Management student's Lean Management Education. In *Proceedings of 6th European Lean Educator Conference (ELEC2019)* (pp. 79–87). Springer International Publishing.

Fernandes, S., Flores, M. A., & Lima, R. M. (2007). Project-led education in engineering: Monitoring and assessing the learning process. *Proceedings of SEFI and IGIP Joint Annual Conference 2007: Joining Forces in Engineering Educations Towards Excellence*, 345–346.

Fernandes, S. R. G. (2014). Preparing Graduates for Professional Practice: Findings from a Case Study of Project-based Learning (PBL). *Procedia: Social and Behavioral Sciences, 139*, 219–226. doi:10.1016/j.sbspro.2014.08.064

Fernandes, S., Mesquita, D., Flores, M. A., & Lima, R. M. (2014). Engaging students in learning: Findings from a study of project-led education. *European Journal of Engineering Education, 39*(1), 55–67. doi:10.1080/03043797.2013.833170

Ferrari, R. (2015). Writing narrative style literature reviews. *Medical Writing, 24*(4), 230–235. doi:10.1179/2047480615Z.000000000329

Ferreira, C. M., & Serpa, S. (2018). Society 5.0 and Social Development: Contributions to a Discussion. *Management and Organizational Studies, 5*(4), 26. doi:10.5430/mos.v5n4p26

Ferreira, E. P., & Martins, A. (2016). Educrum – The empowerment of students in engineering education? In *Proceedings of the 12th International CDIO Conference*. Turku University of Applied Sciences.

Field, A. (2009). *Discovering statistics using SPSS* (Vol. 3). Sage publications.

Filella Guiu, G., Cabello, E., Pérez-Escoda, N., & Ros-Morente, A. (2016). Evaluation of the Emotional Education program "Happy 8-12" for the assertive resolution of conflicts among peers. *Electronic Journal of Research in Educational Psychology, 14*(3), 582–601. doi:10.14204/ejrep.40.15164

Filho, G. E., Sauer, L. Z., Almeida, N. N., & Villas-Boas, V. (2019). *Uma nova sala de aula é possível* [A new classroom is possible]. LTC.

Flinders, D. J. (1988). Teacher isolation and the new reform. *Journal of Curriculum and Supervision, 4*(1), 17–29.

Flint, R. W. Jr, & Dorr, N. (2010). Social neuroscience at the college of saint rose: The art of team teaching in emerging areas of psychological science. *Journal of Undergraduate Neuroscience Education, 8*(2), 122–127. https://www.scopus.com/inward/record.uri?eid=2-s2.0-77956286723&partnerID=40&md5=62cb459ca0248b76923d46df794d0d4e PMID:23493798

Flöthmann, C., Hoberg, K., & Wieland, A. (2018). Competency requirements of supply chain planners & analysts and personal preferences of hiring managers. *Supply Chain Management, 23*(6), 480–499. doi:10.1108/SCM-03-2018-0101

Flumerfelt, B. S. (2008). *Lean Thinking for schools.* Academic Press.

Flumerfelt, S. & Soma, P. (n.d.). *Transforming the way we do business.* Destin, FL: Charactership Lean Publishing.

Flumerfelt, S. (2017). *Lean Leadership at Oakland University* [online video]. Retrieved from: https://drive.google.com/drive/u/3/search?q=owner:haughey%40oakland.edu

Flumerfelt, S., Alves, A. C., & Kahlen, F.-J. (2013). What Lean Teaches Us About Ethics in Engineering. *Education and Globalization, 5.* doi:10.1115/IMECE2013-62393

Flumerfelt, S., Kahlen, F.-J. F.-J., Alves, A., Calvo-Amodio, J., & Hoyle, C. (2014). Systems Competency for Engineering Practice. *Systems, Design, and Complexity, 11.* doi:10.1115/IMECE2014-40142

Flumerfelt, S., Alves, A. C., Kahlen, F.-J., & Siriban Manalang, A. B. (2012). Why Require Ethics in Engineering? *Volume 5: Education and Globalization. General Topics*, *5*, 495–502. doi:10.1115/IMECE2012-89392

Flumerfelt, S., Alves, A. C., & Khalen, F.-J. (2014). Lean Engineering Education: The DNA of Content and Competency. *Proceedings of the 2014 IIE Engineering Lean and Six Sigma Conference, 2014 Lean Educators Conference.*

Flumerfelt, S., Alves, A. C., Leão, C. P., & Wade, D. L. (2016). What do organizational leaders need from Lean graduate programming. *European Journal of Training and Development*, *40*(5), 302–320. doi:10.1108/EJTD-01-2015-0005

Flumerfelt, S., Kahlen, F. J., Alves, A. C., & Siriban-Manalang, R. (2015). *Lean Engineering Education: Driving Content and Competency Mastery*. The American Society of Mechanical Engineers. doi:10.1115/1.860502

Flumerfelt, S., Ross, W., & Wade, D. (2017). Lean learning at Oakland University. In A. C. Alves, S. Flumerfelt, & F. J. Kahlen (Eds.), *Lean Education: An Overview of Current Issues*. Springer. doi:10.1007/978-3-319-45830-4_6

Flumerfelt, S., Schwartz, K. G., Mavris, E., & Briceno, S. (2019). *Complex systems engineering: Theory and practice*. AIAA Press. doi:10.2514/4.105654

Foden, F. E. (1963). Herbert Schofield and Loughborough College. *The Vocational Aspect of Secondary and Further Education*, *15*(32), 231–246. doi:10.1080/03057876380000271

Forrest, C. J., & Swanton, T. (2021). Longitudinal associations between soft skills, education and labour market outcomes: evidence from a survey of young Australians. *Education + Training*. doi:10.1108/ET-10-2020-0325

Forza, C., & Salvador, F. (2006). HRM policies for mass customization: understanding individual competence requirements and training needs for the mass customizing industrial company. In Mass Customization: Challenges and Solutions (pp. 251–269). Springer.

Frank, M., Lavy, I., & Elata, D. (2003). Implementing the project-based learning approach in an academic engineering course. *International Journal of Technology and Design Education*, *13*(3), 273–288. doi:10.1023/A:1026192113732

Freeman, S., Eddy, S. L., McDonough, M., Smith, M. K., Okoroafor, N., Jordt, H., & Wenderoth, M. P. (2014). Active learning increases student performance in science, engineering, and mathematics. *Proceedings of the National Academy of Sciences of the United States of America*, *111*(23), 8410–8415. doi:10.1073/pnas.1319030111 PMID:24821756

Freitas, I. (2021). More Checks for Less Waste in the Lamination Process of a Shipbuilding Company Pursuing Lean Thinking. In A. Dolgui, A. Bernard, D. Lemoine, G. von Cieminski, & D. Romero (Eds.), *Advances in Production Management Systems. Artificial Intelligence for Sustainable and Resilient Production Systems. APMS 2021. IFIP Advances in Information and Communication Technology* (Vol. 630). Springer. doi:10.1007/978-3-030-85874-2_69

Fundação Oswaldo Cruz. (2021). *Análise do impacto da Pandemia causada pelo SARS-Cov-2 sobre PPGBB* [Analysis of the impact of the pandemic caused by SARS-Cov-2 on PPGBB]. Retrieved from https://www.posgraduacao.icc.fiocruz.br/analise-do-impacto-da-pandemia-causada-pelo-sars-cov-2-sobre-o-ppgbb/

Gaebel, B. M., & Zhang, T. (2018). *Learning and teaching in the European Higher Education Area*. E. U. Association. https://eua.eu/downloads/publications/trends-2018-learning-and-teaching-in-the-european-higher-education-area.pdf

Galati, F., & Bigliardi, B. (2019). Industry 4.0: Emerging themes and future research avenues using a text mining approach. *Computers in Industry*, *109*, 100–113. doi:10.1016/j.compind.2019.04.018

García, I., & Cano, E. (2018). A computer game for teaching and learning algebra topics at undergraduate level. *Computer Applications in Engineering Education*, *26*(2), 326–340. doi:10.1002/cae.21887

Garcia, J. M., Homkes, R., Carnes, M. T., & Taylor, K. D. (2015, June 14-17). *Lessons Learned from Team-Teaching a PBL Robotics Course with Multi- Disciplinary Instructors and Students. 122nd ASEE Annual Conference & Exposition*, Seattle, WA.

Gast, I., Schildkamp, K., & van der Veen, J. T. (2017). Team-Based Professional Development Interventions in Higher Education: A Systematic Review. *Review of Educational Research*, *87*(4), 736–767. doi:10.3102/0034654317704306 PMID:28989192

Gehrke, L., Kühn, A. T., Rule, D., Moore, P., Bellmann, C., Siemes, S., Dawood, D., Singh, L., Kulik, J., & Standley, M. (2015). *A Discussion of Qualifications and Skills in the Factory of the Future: A German and American Perspective*. Academic Press.

Gento, A. M., de Benito, J. J., Sanz, P., & Pascual, J. A. (2016). Lean School: A Practical Space of Cooperative Learning from the Factory to the University. In C. Hernández (Ed.), *Advances in Management Engineering. Lecture Notes in Management and Industrial Engineering* (pp. 209–220). Springer International Publishing.

Gento, A. M., Pimentel, C., & Pascual, J. A. (2021). Lean school: An example of industry-university collaboration. *Production Planning and Control*, *32*(6), 473–488. doi:10.1080/09537287.2020.1742373

George, M. L., Maxey, J., Rowlands, D., & Price, M. (2004). The Lean Six Sigma Pocket Toolbook: A Quick Reference Guide to 100 Tools for Improving Quality and Speed (1st ed.). McGraw-Hill.

Gero, J. S. (1996). Creativity, emergence and evolution in design. *Knowledge-Based Systems*, *9*(7), 435–448. doi:10.1016/S0950-7051(96)01054-4

Giunipero, L. C., Denslow, D., & Eltantawy, R. (2000). *A skills-based analysis of the world class purchaser*. Academic Press.

Giunipero, L. C., Denslow, D., & Eltantawy, R. (2005). Purchasing/supply chain management flexibility: Moving to an entrepreneurial skill set. *Industrial Marketing Management*, *34*(6), 602–613. doi:10.1016/j.indmarman.2004.11.004

Giunipero, L. C., & Handfield, R. B. (2004). Purchasing. *Education + Training*, 2.

Giunipero, L. C., Handfield, R. B., & Eltantawy, R. (2006). Supply management's evolution: Key skill sets for the supply manager of the future. *International Journal of Operations & Production Management*, *26*(7), 822–844. doi:10.1108/01443570610672257

Giunipero, L. C., & Pearcy, D. H. (2000). World-class purchasing skills: An empirical investigation. *The Journal of Supply Chain Management*, *36*(3), 4–13. doi:10.1111/j.1745-493X.2000.tb00081.x

Gladman, A. (2015). Team teaching is not just for teachers! Student perspectives on the collaborative classroom. *TESOL Journal*, *6*(1), 130–148. doi:10.1002/tesj.144

Glick, D. F., & Kulbok, P. A. (2001). Program revision: A dynamic outcome of evaluation. *Quality Management in Health Care*, *10*(1), 37–44. doi:10.1097/00019514-200110010-00005 PMID:11702469

Goerke, M., Schmidt, M., Busch, J., & Nyhuis, P. (2015). Holistic Approach of Lean Thinking in Learning Factories. *Procedia CIRP*, *32*, 138–143. doi:10.1016/j.procir.2015.02.221

Goodridge, D., Westhorp, G., Rotter, T., Dobson, R., & Bath, B. (2015). Lean and leadership practices: Development of an initial realist program theory. *BMC Health Services Research*, *15*(1), 362. doi:10.118612913-015-1030-x PMID:26345184

Gowripeddi, V. V., Bijjahalli, M. C., Janardhan, N., & Bhimavaram, K. R. (2021). Role of Education 4.0 Technologies in Driving Industry 4.0. *17th International Conference on Remote Engineering and Virtual Instrumentation, REV 2020: Advances in Intelligent Systems and Computing*, 576–587. 10.1007/978-3-030-52575-0_48

GPIMEM UNESP. (2020, May 31). *Mathnews*. https://youtu.be/sz_8EKAuh5Q

Graafland, M., Bemelman, W. A., & Schijven, M. P. (2015). Appraisal of face and content validity of a serious game improving situational awareness in surgical training. *Journal of Laparoendoscopic & Advanced Surgical Techniques. Part A.*, *25*(1), 43–49. doi:10.1089/lap.2014.0043 PMID:25607899

Graafland, M., Bemelman, W. A., & Schijven, M. P. (2017). Game-based training improves the surgeon's situational awareness in the operation room: A randomized controlled trial. *Surgical Endoscopy*, *31*(10), 4093–4101. doi:10.100700464-017-5456-6 PMID:28281117

Graafland, M., Vollebergh, M. F., Lagarde, S. M., Van Haperen, M., Bemelman, W. A., & Schijven, M. P. (2014). A serious game can be a valid method to train clinical decision-making in surgery. *World Journal of Surgery*, *38*(12), 3056–3062. doi:10.100700268-014-2743-4 PMID:25159118

Graham, R. (2010). *UK approaches to engineering project-based learning*. White Paper sponsored by the Bernard M. Gordon/MIT Engineering Leadership Program.

Graham, R. (2018). *The global state of the art in engineering education. Massachusetts Institute of Technology*. MIT.

Grasso, D., & Martinelli, D. (2007). Holistic engineering. *The Chronicle of Higher Education. The Chronicle Review*, *53*(28), B8–B9.

Gregory, A. T., & Denniss, A. R. (2018). An Introduction to Writing Narrative and Systematic Reviews— Tasks, Tips and Traps for Aspiring Authors. *Heart Lung and Circulation*, *27*(7), 893–898. doi:10.1016/j.hlc.2018.03.027 PMID:29857977

Gross, H. (1966). Make or buy decisions in growing firms. *The Accounting Review*, *41*(4), 745–753.

Gudmundsdottir, S. (2001). Narrative research on school practice. In V. Richardson (Ed.), *Fourth handbook for research on teaching* (pp. 26–240). MacMillan.

Guenaga, M., Arranz, S., Florido, I. R., Aguilar, E., De Guinea, A. O., Rayón, A., Bezanilla, M. J., & Menchaca, I. (2013). Serious games for the development of employment oriented competences. *Revista Iberoamericana de Tecnologias Del Aprendizaje*, *8*(4), 176–183. doi:10.1109/RITA.2013.2285024

Guerra, A., & Rodriguez-Mesa, F. (2021). Educating engineers 2030 – PBL, social progress and sustainability. *European Journal of Engineering Education*, *46*(1), 1–3. doi:10.1080/03043797.2020.1828678

Guerra, A., Ulseth, R., & Kolmos, A. (2017). *PBL in Engineering Education* (A. Guerra, R. Ulseth, & A. Kolmos, Eds.). SensePublishers., doi:10.1007/978-94-6300-905-8

Gunel, M. (2006). *Investigating the Impact of Teachers' Implementation Practices on Academic Achievement in Science during a Long-Term Professional Development Program on the Science Writing Heuristic*. Iowa State University. doi:10.31274/rtd-180813-12062

Gupta, K., Mashinini, M., Mukhawana, D., & Louw, A. (2019). Development of industry 4.0 virtual lab for manufacturing engineering education. *Proceedings of the International Conference on Industrial Engineering and Operations Management*, 1857–1867.

Gupta, K., Mukhawana, D., & Mashinini, M. (2019). A project based learning tool for industry 4.0 manufacturing engineering education. *Proceedings of the International Conference on Industrial Engineering and Operations Management*, 978–984.

Gürel, E. (2017). SWOT analysis: A theoretical review. *Journal of International Social Research, 87*(1-2), 149–200. doi:10.17719/jisr.2017.1832

Hackathorn, J., Solomon, K. L., Blankmeyer, E. D., Tennial, A. M., & Garczynski, R. E. (2011). Learning by Doing: An Empirical Study of Active Teaching Techniques. *The Journal of Effective Teaching, 11*(2), 40–54.

Hacking, I. (1984). Experimentation and scientific realism. In *Science and the Quest for Reality* (pp. 162–181). Springer. doi:10.1007/978-1-349-25249-7_7

Hadgraft, R. G., & Kolmos, A. (2020). Emerging learning environments in engineering education. *Australasian Journal of Engineering Education, 25*(1), 3–16. doi:10.1080/22054952.2020.1713522

Hamari, J., Shernoff, D. J., Rowe, E., Coller, B., Asbell-Clarke, J., & Edwards, T. (2016). Challenging games help students learn: An empirical study on engagement, flow and immersion in game-based learning. *Computers in Human Behavior, 54*, 170–179. doi:10.1016/j.chb.2015.07.045

Hannafin, M. J., Hannafin, K. M., Land, S. M., & Oliver, K. (1997). Grounded practice and the Design of Constructivist Learning Environments. *Educational Technology Research and Development, 45*(3), 101–117. doi:10.1007/BF02299733

Hannafin, M. J., & Land, S. M. (2000). Technology and student-centered learning in higher education: Issues and practices. *Journal of Computing in Higher Education, 12*(1), 3–30. doi:10.1007/BF03032712

Hartmann, E. A., & Bovenschulte, M. (2013). Skills Needs Analysis for Industry 4.0 Based on Roadmaps for Smart Systems. *Using Technology Foresights for Identifying Future Skills Needs. Global Workshop Proceedings*, 24–36.

Hecklau, F., Galeitzke, M., Flachs, S., & Kohl, H. (2016). Holistic Approach for Human Resource Management in Industry 4.0. *Procedia CIRP, 54*, 1–6. doi:10.1016/j.procir.2016.05.102

Heckman, J. J., & Kautz, T. (2012). Hard evidence on soft skills. *Labour Economics, 19*(4), 451–464. doi:10.1016/j.labeco.2012.05.014 PMID:23559694

Heinonen, H., Törmänen, J., Soukkio, J., & Riihinen, M. (2020). *Projektin HRS: loppuraportti*. Academic Press.

Heinrich, C. J., Darling-Aduana, J., & Martin, C. (2020). The potential and prerequisites of effective tablet integration in rural Kenya. *British Journal of Educational Technology, 51*(2), 498–514. doi:10.1111/bjet.12870

Helariutta, A., Fred, M., Kangastie, H., Merimaa, M., & Päällysaho, S. (2021). *Avoin TKI-integroitu oppiminen - toteuttamistapoja ja oppimisympäristöjä. Laurea julkaisut.* Academic Press.

Helle, L., Tynjälä, O., & Olkinuora, P. (2003). Project-based learning in post-secondary education – theory, practice and rubber sling shots. *Higher Education, 51*, 87–314.

Hellier, S., & Davidson, L. (2018). Team teaching in nursing education. *Journal of Continuing Education in Nursing, 49*(4), 186–192. doi:10.3928/00220124-20180320-09 PMID:29596706

Henshaw, R. (1991). Desirable attributes for professional engineers. In J. B. Agnew & C. Creswell (Eds.), *Broadening Horizons of Engineering Education, 3rd Annual conference of Australasian Association for Engineering Education* (pp. 199–204). Academic Press.

Hernandez, B., Montaner, T., Sese, F. J., & Urquizu, P. (2011). The role of social motivations in e-learning: How do they affect usage and success of ICT interactive tools? *Computers in Human Behavior*, *27*(6), 2224–2232. doi:10.1016/j.chb.2011.07.001

Hernandez-de-Menendez, M., & Morales-Menendez, R. (2019). Technological innovations and practices in engineering education: A review. *International Journal on Interactive Design and Manufacturing*, *13*(2), 713–728. doi:10.100712008-019-00550-1

Hernandez-Linares, R., Sánchez, H., Agudo, J. E., & Rico, M. (2017). Chronos: A tool to develop the time management competence among engineering students. *Computer Applications in Engineering Education*, *25*(1), 79–89. doi:10.1002/cae.21780

Hernez-Broome, G., & Hughes, R. J. (2004). Leadership development: Past, present, and future. *Human Resource Planning, 27*(1).

Heyse, V., Erpenbeck, J., & Max, H. (2004). *Kompetenzen erkennen, bilanzieren und entwickeln.* Waxmann Verlag.

Heywood, J. (2005). *Engineering education: Research and development in curriculum and instruction.* John Wiley and Sons. doi:10.1002/0471744697

Higgs, M., & Rowland, D. (2005). All changes great and small: Exploring approaches to change and its leadership. *Journal of Change Management*, *5*(2), 121–151. doi:10.1080/14697010500082902

Hines, P., Holweg, M., & Rich, N. (2004). Learning to evolve a review of contemporary lean thinking. *International Journal of Operations & Production Management*, *24*(10), 994–1011. doi:10.1108/01443570410558049

Hodges, C., Moore, S., Lockee, B., Trust, T., & Bond, A. (2020). The difference between emergency remote teaching and online learning. *EDUCAUSE Review*, *12*, •••. https://er.educause.edu/articles/2020/3/the-difference-between-emergency-remote-teaching-and-online-learning

Hoffman, J. (2014). *Avaliação mediadora: uma prática em construção da pré-escola à Universidade* [Mediator evaluation: a practice in construction from preschool to university]. Porto Alegre.

Hoidn, S. (2017). *Student-centered learning environments in higher education classrooms.* Palgrave Macmillan. doi:10.1057/978-1-349-94941-0

Hong, J.-H. (2007). The Comparison of Problem-based Learning (PmBL) Model and project-based learning (PtBL) model. *International Conference on Engineering Education–ICEE.*

Hrast, S., & Savec, V. F. (2018). ICT-supported inquiry-based learning. *World Transactions on Engineering and Technology Education, 16*(4), 398–403. https://www.scopus.com/inward/record.uri?eid=2-s2.0-85062687538&partnerID=40&md5=4649f1374ae29596070d5c0da0be43e7

Huang, R. H., Liu, D. J., Guo, J., Yang, J. F., Zhao, J. H., Wei, X. F., Knyazeva, S., Li, M., Zhuang, R. X., Looi, C. K., & Chang, T. W. (2020). *Guidance on flexible learning during campus closures: Ensuring course quality of higher education in COVID-19 outbreak.* Smart Learning Institute of Beijing Normal University. https://iite.unesco.org/wp-content/uploads/2020/05/Guidance-on-Flexible-Learning-during-Campus-Closures-in-COVID-19-outbreak-SLIBNU-V1.2_0508.pdf

Huang, R. H., Liu, D. J., Tlili, A., Yang, J. F., & Wang, H. H. (2020). *Handbook on facilitating flexible learning during educational disruption: The Chinese experience in maintaining undisrupted learning in COVID-19 outbreak.* Smart Learning Institute of Beijing Normal University. https://iite.unesco.org/wp-content/uploads/2020/03/Handbook-on-Facilitating-Flexible-Learning-in-COVID-19-Outbreak-SLIBNU_V2.0_20200324.pdf

Hua, W., Xiaofeng, L., Zhenqiang, B., Jun, R., Ban, W., & Liming, L. (2020). Consideration on the strategies during epidemic stage changing from emergency response to continuous prevention and control. *Zhonghua Liu Xing Bing Xue Za Zhi*, *41*(2), 297–300. doi:10.3760/cma.j.issn.0254-6450.2020.02.003 PMID:32088947

Huerta, D., & McMillan, V. (2012). Reflections on collaborative teaching of science information literacy and science writing: Plans, processes and pratfalls. In *Libraries within their Institutions* (pp. 19–28). Creative Collaborations. doi:10.1300/J121v17n01_03

Hughes, D. L., Whitburn, L. Y., Colasante, M., McDonald, A. C., McGowan, H., Zacharias, A., & Green, R. A. (2020). Developing anatomy demonstrators of the future: The role of team-teaching. *Journal of University Teaching and Learning Practice, 17*(5), 1-16. https://www.scopus.com/inward/record.uri?eid=2-s2.0-85098052551&partnerID=40&md5=824aed685d13d46534990e0c8a74849a

Hummel, H. G. K., Joosten-ten Brinke, D., Nadolski, R. J., & Baartman, L. K. J. (2017). Content validity of game-based assessment: Case study of a serious game for ICT managers in training. *Technology, Pedagogy and Education*, *26*(2), 225–240. doi:10.1080/1475939X.2016.1192060

Hunt, I., O'Sullivan, D., Rolstadas, A., Horan, M., & Precup, L. (2004). Survey of manufacturing curricula from around the world. *Production Planning and Control*, *15*(1), 71–79. doi:10.1080/09537280410001662583

Hurlburt, G., Voas, J., Miller, K., Laplante, P., & Michael, B. (2010). A nonlinear perspective on higher education. *Computer*, *43*(12), 90–92. doi:10.1109/MC.2010.338

Icek, A. (2005). *Attitudes, Personality and Behaviour* (2nd ed.). Milton-Keynes: Open University Press.

Imai, M. (1986). *Kaizen: The Key to Japan's Competitive Success*. McGraw-Hill.

Institute of Industrial and Systems Engineers. (2020a). *Origins of IISE*. https://www.iise.org/details.aspx?id=295

Institute of Industrial and Systems Engineers. (2020b). *The industrial engineering body of knowledge*. IISE Body of Knowledge. https://www.iise.org/

Instituto Federal do Rio de Janeiro. (2021, May 15). *Programa de Pós-graduação (Mestrado e Doutorado Profissional) em Ensino de Ciências* [Postgraduate Program (Master's and Professional Doctorate) in Science Teaching]. https://portal.ifrj.edu.br/cursos-pos-graduacao/stricto-sensu/propec

Irwin, M., & Supplee, L. H. (2012). Directions in implementation research methods for behavicral and social science. *The Journal of Behavioral Health Services & Research*, *39*(4), 339–342. doi:10.100711414-012-9293-z PMID:22907687

Isoherranen, V., & Ratnayake, R. C. (2018). Performance assessment of microenterprises operating in the Nordic Arctic region. *Journal of Small Business and Entrepreneurship*, *30*(5), 431–449. doi:10.1080/08276331.2018.1438769

Jasondavies. (2021). Retrieved May 19, 2021 from https://www.jasondavies.com/wordcloud

Jauch, L. R., & Wilson, H. K. (1979). A strategic perspective for make or buy decisions. *Long Range Planning*, *12*(6), 56–61. doi:10.1016/0024-6301(79)90056-6

Johnsen, H. M., Fossum, M., Vivekananda-Schmidt, P., Fruhling, A., & Slettebø, Å. (2016). Teaching clinical reasoning and decision-making skills to nursing students: Design, development, and usability evaluation of a serious game. *International Journal of Medical Informatics*, *94*, 39–48. doi:10.1016/j.ijmedinf.2016.06.014 PMID:27573310

Johnson, B., Ulseth, R., Smith, C., & Fox, D. (2015, October 21-24). The impacts of project based learning on self-directed learning and professional skill attainment: A comparison of project based learning to traditional engineering education. 2015 IEEE Frontiers in Education Conference (FIE).

Johnson, D. W., Johnson, R. T., & Smith, K. A. (1998a). Active Learning: Cooperation in the College Classroom (2nd ed.). Interaction Book Co.

Johnson, D. W., Johnson, R. T., & Smith, K. A. (1998b). Cooperative Learning Returns to College: What Evidence is There That it Works? *Change: The Magazine of Higher Learning*, 26–35.

Johnson, L., Adams Becker, S., Estrada, V., & Martín, S. (2013). *The Technology Outlook for STEM+ Education 2013-2018: An NMC Horizon Project Sector Analysis* (Vol. 148). The New Media Consortium. https://files.eric.ed.gov/fulltext/ED559376.pdf

Jones, F., & Harris, S. (2012). Benefits and Drawbacks of Using Multiple Instructors to Teach Single Courses. *College Teaching*, *60*(4), 132–139. doi:10.1080/87567555.2012.654832

Jones, J. W., & Mezo, M. (2014). Capstone = team teaching + team learning + industry. *ASEE Annual Conference and Exposition, Conference Proceedings.*

Joshi, M., & Vaidya, R. (2013). A review of paradigm shift from conventional to personalized e-learning. *2013 2nd International Conference on Advances in Computing, Communications and Informatics, ICACCI 2013*, 1263–1269. 10.1109/ICACCI.2013.6637359

Junco, R., & Cotten, S. R. (2011). Perceived academic effects of instant messaging use. *Computers & Education*, *56*(2), 370–378. doi:10.1016/j.compedu.2010.08.020

Kaczmarczyk, J., Davidson, R., Bryden, D., Haselden, S., & Vivekananda-Schmidt, P. (2016). Learning decision making through serious games. *The Clinical Teacher*, *13*(4), 277–282. doi:10.1111/tct.12426 PMID:26213136

Kagermann, H., Wahister, W., & Helbig, J. (2013). *Recommendations for implementing the strategic initiative INDUSTRIE 4.0: Securing the Future of German Manufacturing Industry*. Academic Press.

Kagermann, H., Wahlster, W., & Helbig, J. (2013). *Recommendations for Implementing the Strategic Initiative INDUSTRIE 4.0*. Academic Press.

Kahlen, F.-J., Flumerfelt, S., Sinban-Manalang, A. B., & Alves, A. (2011). Benefits of lean teaching. *ASME 2011 International Mechanical Engineering Congress and Exposition, IMECE 2011, 5*, 351–358.

Kaliappen, N. (2019). Educational benefits of using Business Strategy Game (BSG) in teaching and learning strategic management. *International Journal of Emerging Technologies in Learning*, *14*(7), 209–215. doi:10.3991/ijet.v14i07.9792

Kanungo, S., Sharma, D., & Aggarwal, A. (2021). Impact and New Normal. *International Journal of Security and Privacy in Pervasive Computing*, *13*(1), 57–68. Advance online publication. doi:10.4018/IJSPPC.2021010103

Kanwal, F., & Rehman, M. (2014). E-learning adoption model: A case study of Pakistan. *Life Science Journal, 11*(4), 78–86. https://www.scopus.com/inward/record.uri?eid=2-s2.0-84899152280&partnerID=40&md5=08de3e65655d108cf64f5cfdd905d4d3

Karanicolas, S., Snelling, C., & Winning, T. (2018). *Flipped Learning in Adelaide*. University of Adelaide. https://www.adelaide.edu.au/flipped-classroom/about/

Karim, N. S. A., & Heckman, R. (2005). Group communication media choice and the use of information and communication technology to support learning: A case study. *Campus-Wide Information Systems*, *22*(1), 28–42. doi:10.1108/10650740510574401

Katz, D., Zerillo, J., Kim, S., Hill, B., Wang, R., Goldberg, A., & DeMaria, S. (2017). Serious gaming for orthotopic liver transplant anesthesiology: A randomized control trial. *Liver Transplantation*, *23*(4), 430–439. doi:10.1002/lt.24732 PMID:28133947

Kaya, R. (2019). The Conceptual Framework of the Industry 4.0 Revolution. In F. Ayhan (Ed.), *Several Dimensions of Innovation, Technology and Industry 4.0* (pp. 13–271). Peter Lang. doi:10.3726/b16202

Keidanren. (2018). *Society 5.0: co-creating the future*. www.keidanren.or.jp/en/policy/2018/095_outline.pdf

Kekkonen, M., & Juntunen, T. (2018). Tuottava tapa oppia projektityöskentelyä. *Toolilainen, ammattikorkeakoulujen tekniikan ja liikenteen alan järjestölehti.*

Kekkonen, M., Jokinen, T., & Isoherranen, V. (2021) Potkua Pohjois-Suomen PKT-yrityksiin Lean-toimintamallia soveltamalla. Avoin TKI-integroitu oppiminen - toimintamallit ja hyvät käytänteet. (pp. 43–48). Laurea julkaisut.

Keough, M. (1993). Buying your way to the top. *The McKinsey Quarterly*, *1993*(3), 41–63.

Kern, D., Moser, R., Sundaresan, N., & Hartmann, E. (2011). Purchasing Competence: A Stakeholder-Based Framework for Chief Purchasing Officers. *Journal of Business Logistics*, *32*(2), 122–138. doi:10.1111/j.2158-1592.2011.01011.x

Kersten, S. (2018). Approaches of Engineering Pedagogy to Improve the Quality of Teaching in Engineering Education. In J. Drummer, G. Hakimov, M. Joldoshov, T. Köhler, & S. Udartseva (Eds.), *Vocational Teacher Education in Central Asia. Technical and Vocational Education and Training: Issues, Concerns and Prospects* (Vol. 28, pp. 129–139). Springer.

Khan, A. H., Sultana, M. S., Hossain, S., Hasan, M. T., Ahmed, H. U., & Sikder, M. T. (2020). The Impact of COVID-19 Pandemic on Mental Health & Wellbeing among Home-Quarantined Bangladeshi Students: A Cross-Sectional Pilot Study. *Journal of Affective Disorders*, *277*(May), 121–128. doi:10.1016/j.jad.2020.07.135 PMID:32818775

Khanna, D., & Prasad, A. (2020). *Problems Faced by Students & Teachers During Online Education Due to COVID-19 and How to Resolve Them* [Paper presentation]. *6th International Conference on Education and Technology (ICET)*, Indonesia.

Khan, T. (2021). How COVID-19 has impacted my PhD research. *Journal of Health and Social Care Improvement*, *3*(2), 4–9. Retrieved April 10th, 2021, from https://www.researchgate.net/publication/341727378_How_COVID-19_has_impacted_my_PhD_research

Khanzadi, M., Shahbazi, M. M., Arashpour, M., & Ghosh, S. (2019). Lean design management using a gamified system. *Scientia Iranica*, *26*(1), 15–25. doi:10.24200ci.2018.20325

Khasawneh, A. M. (2010). Using internet technology to support flexible learning in jordan: A case of Hashemite university. *International Journal of Innovation and Learning*, *8*(1), 1–10. doi:10.1504/IJIL.2010.034010

Khattar, A., Jain, P. R., & Quadri, S. M. K. (2020). *Effects of the Disatroue Pandemic COVID 19 on Learning Styles, Activities and mental Health of Young Indian Students- A Machine Learning Approach* [Paper presentation]. International Conference on Intelligent Computing and Control Systems (ICICCS 2020), Secunderabad.

Khoshnodifar, Z., Abbasi, E., Farhadian, H., Sadighi, H., & Pouratashi, M. (2020). Comparative comparison of lecture and team member teaching design methods in agricultural higher education system of Iran. *Journal of Agricultural Science and Technology*, *22*(4), 891–904. https://www.scopus.com/inward/record.uri?eid=2-s2.0-85090734994&partnerID=40&md5=af92d3714e28f235ccf6309cdf93b16d

Kidd, P. T. (1994). *Agile Manufacturing forging new frontiers*. Addison Wesley Publishers.

Killen, K. H., & Kamauff, J. W. (1995). *Managing purchasing: Making the supply team work*. Irwin Publishing.

Killingsworth, B. L., & Xue, Y. (2015). Investigating factors influencing students' learning in a team teaching setting. *International Journal of Cognitive Research in Science, Engineering and Education, 3*(2), 9-16. https://www.scopus.com/inward/record.uri?eid=2-s2.0-85010411070&partnerID=40&md5=f81164f563a66e1ae216bfa8fc88cbd6

Kilpatrick, W. H. (1918). The project method. *Teachers College Record, 19*(4), 319–335.

Kilpatrick, W. H. (1921). Dangers and difficulties of the project method and how to overcome them. *Teachers College Record, 22*(4), 283–287.

King, S. (2006). Promoting Paired Placements in Initial Teacher Education. *International Research in Geographical and Environmental Education, 15*(4), 370–386. doi:10.2167/irg201.0

Kiratli, N., Rozemeijer, F., Hilken, T., de Ruyter, K., & de Jong, A. (2016). Climate setting in sourcing teams: Developing a measurement scale for team creativity climate. *Journal of Purchasing and Supply Management, 22*(3), 196–204. doi:10.1016/j.pursup.2016.04.006

Kirchherr, J., Reike, D., & Hekkert, M. (2017). Conceptualizing the circular economy: An analysis of 114 definitions. *Resources, Conservation and Recycling, 127*, 221–232. doi:10.1016/j.resconrec.2017.09.005

Kleyn, T., & Valle, J. (2014). Modeling collaborative teaching in teacher education: Preparing pre-service teachers to teach all students. In Advances in Research on Teaching (Vol. 21, pp. 137-164). Academic Press.

Knight, L., Meehan, J., Tapinos, E., Menzies, L., & Pfeiffer, A. (2020). Researching the future of purchasing and supply management: The purpose and potential of scenarios. *Journal of Purchasing and Supply Management, 100624*. Advance online publication. doi:10.1016/j.pursup.2020.100624

Knight, L., Tu, Y.-H., & Preston, J. (2014). Integrating skills profiling and purchasing portfolio management: An opportunity for building purchasing capability. *International Journal of Production Economics, 147*, 271–283. doi:10.1016/j.ijpe.2013.06.013

Knoll, M. (1997). The project method: Its vocational education origin and international development. *Journal of Industrial Teacher Education, 34*(3), 59–80.

Kodkanon, K., Pinit, P., & Murphy, E. (2018). High-school teachers' experiences of interdisciplinary team teaching. *Issues in Educational Research, 28*(4), 967–989. https://www.scopus.com/inward/record.uri?eid=2-s2.0-85056302680&partnerID=40&md5=c9af039af6d318345af03df05938c2cc

Kokostaki, D., Menzies, V., & Wiggins, A. (2016). Project-Based Leaning. *Improving Schools, 19*(3), 267–27. doi:10.1177/1365480216659733

Kolb, D. A. (1984). *Experiential Learning: Experience as the Source of Learning and Development.* Prentice Hall.

Kolchin, M. G., & Giunipero, L. C. (1993). *Purchasing education and training: Requirements and resources.* Center for Advanced Purchasing Studies.

Kolmos, A., & de Graaff, E. (2014). Problem-based and project-based learning in engineering education. Cambridge handbook of engineering education research, 1, 141-161.

Kolmos, A. (1996). Reflections on project work and problem-based learning. *European Journal of Engineering Education, 21*(2), 141–148. doi:10.1080/03043799608923397

Komiyama, H., & Yamada, K. (2018). *New Vision 2050.* Springer Japan. doi:10.1007/978-4-431-56623-6

Kostko, A. (2019). The impact of team teaching on student attitudes and classroom performance in introductory philosophy courses. *Teaching Philosophy, 42*(4), 329–354. doi:10.5840/teachphil20191017111

Kowalewski, K.-F., Hendrie, J. D., Schmidt, M. W., Proctor, T., Paul, S., Garrow, C. R., Kenngott, H. G., Müller-Stich, B. P., & Nickel, F. (2017). Validation of the mobile serious game application Touch Surgery™ for cognitive training and assessment of laparoscopic cholecystectomy. *Surgical Endoscopy*, *31*(10), 4058–4066. doi:10.100700464-017-5452-x PMID:28281111

Krafcik, J. F. (1988). Triumph Of The Lean Production System. *Sloan Management Review*, *30*(1), 41–52.

Krajcik, J. S., & Blumenfeld, P. C. (2006). Project-based learning. The Cambridge handbook of the learning sciences, 317-333.

Kriz, W. C., & Auchter, E. (2016). 10 Years of Evaluation Research Into Gaming Simulation for German Entrepreneurship and a New Study on Its Long-Term Effects. *Simulation & Gaming*, *47*(2), 179–205. doi:10.1177/1046878116633972

Kröber, C., & Münster, S. (2016). Educational App Creation for the Cathedral in Freiberg. In J. Spector, D. Ifenthaler, D. Sampson, & P. Isaias (Eds.), *Competencies in Teaching, Learning and Educational Leadership in the Digital Age* (pp. 303–318). Springer. doi:10.1007/978-3-319-30295-9_19

Kropsu-Vehkapera, H., Isoherranen, V., & Kess, P. (2013). Company assessment in production management education. Education conference, Tampere Finland.

Kruger, J., & Dunning, D. (1999). Unskilled and unaware of it: How difficulties in recognizing one's own incompetence lead to inflated self-assessments. *Journal of Personality and Social Psychology*, *77*(6), 1121–1134. doi:10.1037/0022-3514.77.6.1121 PMID:10626367

Kulski, M. M., & Kerr, R. (2012). Managing teaching teams effectively. *International Journal of Learning*, *18*(9), 149–160. doi:10.18848/1447-9494/CGP/v18i09/47730

Kumar, R. (2020). *Industrial Engineering* (2nd ed.). Jyothis.

Kumar, H., & Raghavendran, S. (2015). Gamification, the finer art: Fostering creativity and employee engagement. *The Journal of Business Strategy*, *36*(6), 3–12. doi:10.1108/JBS-10-2014-0119

Kurdve, M., & Bellgran, M. (2021). Green lean operationalisation of the circular economy concept on production shop floor level. *Journal of Cleaner Production*, *278*, 1–11. doi:10.1016/j.jclepro.2020.123223

Lagro, J., van de Pol, M. H. J., Laan, A., Huijbregts-Verheyden, F. J., Fluit, L. C. R., & Olde Rikkert, M. G. M. (2014). A Randomized Controlled Trial on Teaching Geriatric Medical Decision Making and Cost Consciousness With the Serious Game GeriatriX. *Journal of the American Medical Directors Association*, *15*(12), 957.e1–957.e6. doi:10.1016/j.jamda.2014.04.011 PMID:24913210

Lai, K.-W., & Pratt, K. (2007). Positive to a degree: The effects of ICT use in New Zealand secondary schools. *Computers in the Schools*, *24*(3–4), 95–109. doi:10.1300/J025v24n03_07

Laker, D. R., & Powell, J. L. (2011). The differences between hard and soft skills and their relative impact on training transfer. *Human Resource Development Quarterly*, *22*(1), 111–122. doi:10.1002/hrdq.20063

Lamancusa, J. S., Zayas, J. L., Soyster, A. L., Morell, L., & Jorgensen, J. (2008). The Learning Factory: Industry-Partnered Active Learning. *Journal of Engineering Education*, *97*(1), 5–11. doi:10.1002/j.2168-9830.2008.tb00949.x

Lamb, L. C., DiFiori, M. M., Jayaraman, V., Shames, B. D., & Feeney, J. M. (2017). Gamified Twitter Microblogging to Support Resident Preparation for the American Board of Surgery In-Service Training Examination. *Journal of Surgical Education*, *74*(6), 986–991. doi:10.1016/j.jsurg.2017.05.010 PMID:28545826

Land, S., & Hannafin, M. (2000). Student-Centered Learning Environments. In D. Jonassen & S. Land (Eds.), *Theoretical foundations of learning environments* (pp. 1–24). Lawrence Erlbaum Associates, Inc. Publishers.

Lang, J. D., Cruise, S., McVey, F. D., & McMasters, J. (1999). Industry expectations of new engineers: A survey to assist curriculum designers. *Journal of Engineering Education, 88*(1), 43–51. doi:10.1002/j.2168-9830.1999.tb00410.x

Langroudi, Z. K. (2015). Effective factors on information and communication technology usage by university teachers; a case study of kharazmi university. *International Journal of Information Science and Management, 13*(2), 51–59.

Lasagabaster, D. (2018). Fostering team teaching: Mapping out a research agenda for English-medium instruction at university level. *Language Teaching, 51*(3), 400–416. doi:10.1017/S0261444818000113

Lattuca, L. R., Knight, D. B., Ro, H. K., & Novoselich, B. J. (2017). Supporting the Development of Engineers' Interdisciplinary Competence. *Journal of Engineering Education, 106*(1), 71–97. doi:10.1002/jee.20155

Lei 11892/2008. (2008). Retrieved April 28, 2021, from http://www.planalto.gov.br/ccivil_03/_ato2007-2010/2008/lei/l11892.htm

Lemberger, M., Webb, L., & Moore, M. (2011). Student Success Skills : An Evidence-Based Cognitive and Social Change Theory for Student Achievement Linda Webb, Florida State University, and Molly m. Moore, University of Northern Colorado. *Journal Od Education, 192*(2), 88–99. doi:10.1177/0022057412192002-311

Leoni, R. (2014). Graduate employability and the development of competencies. The incomplete reform of the "Bologna Process". *International Journal of Manpower, 35*(4), 448–469. doi:10.1108/IJM-05-2013-0097

Leuven/Louvain-la-Neuve Communiqué. (2009). *The Bologna Process 2020: the European Higher Education Area in the new decade*. Communiqué of the Conference of European Ministers Responsible for Higher Education, Leuven and Louvain-la-Neuve.

Lewis, L. H., & Williams, C. J. (1994). Experiential Learning: Past and Present. *New Directions for Adult and Continuing Education, 62*(62), 5–16. doi:10.1002/ace.36719946203

Li, R., & Antiohos, A. (2021). *Problem Based Learning (PBL) in four-week term block mode teaching.* Paper presented at the 2021 IEEE Global Engineering Education Conference (EDUCON). 10.1109/EDUCON46332.2021.9453991

Liker, J. K., & Convis, G. L. (2012). *The Toyota way to lean leadership: Achieving and sustaining excellence through leadership development.* McGraw-Hill.

Lima, R. M., Carvalho, J. D., Sousa, R.M., Alves, A.C., Moreira, F., Fernandes, S., & Mesquita, D. (2017). Ten Years of Project-Based Learning (PBL) in Industrial Engineering and Management at the University of Minho in PBL in Engineering Education - International Perspectives on Curriculum Change. Sense Publishers.

Lima, R. M., da Silva, J. M., van Hattum-Janssen, N., Monteiro, S. B. S., & de Souza, J. C. (2012). *Project-based learning course design : A service design approach.* Academic Press.

Lima, R. M., Dinis-Carvalho, J., Sousa, R. M., Alves, A. C., Moreira, F., Fernandes, S., & Mesquita, D. (2017). Ten Years of Project-Based Learning (PBL) in Industrial Engineering and Management at the University of Minho. In PBL in Engineering Education (pp. 33–51). SensePublishers. doi:10.1007/978-94-6300-905-8_3

Lima, R. M., Dinis-Carvalho, Sousa, R., Arezes, P., & Mesquita, D. (2018) Project-based learning as a bridge to the industrial practice. In Closing the gap between practice and research in industrial engineering, lecture notes in management and industrial engineering. doi:10.1007/978-3-319-58409-6_41

Lima, R. M., Mesquita, D., Sousa, R. M., Monteiro, M. T. T., & Cunha, J. (2019). Curriculum Analysis Process: analysing fourteen Industrial Engineering programs. *PAEE/ALE2019*, 92–101.

Lima, Teixeira, Minas, Rodrigues, & Carvalho. (2021). *I9MASKS A Micro Step towards the Development of PDMS Masks for the Protection of COVID-19*. Universidade do Minho – MEtRICs.

Lima, R. M., Mesquita, D., Amorim, M., Jonker, G., & Flores, M. A. (2012). An analysis of knowledge areas in industrial engineering and management curriculum. *International Journal of Industrial Engineering and Management*, *3*(2), 75–82.

Lima, R. M., Andersson, P. H., & Saalman, E. (2017). Active Learning in Engineering Education: A (re)introduction. *European Journal of Engineering Education*, *42*(1), 1–4. doi:10.1080/03043797.2016.1254161

Lima, R. M., Carvalho, D., Flores, M. A., & Van, N. (2007). A case study on project led education in engineering : Students ' and teachers ' perceptions. *European Journal of Engineering Education*, *32*(3), 337–347. doi:10.1080/03043790701278599

Lima, R. M., & Mesquita, D. C. R., & M., R. (2017). Defining the Industrial and Engineering Management Professional Profile: A longitudinal study based on job advertisements. *Production*, •••, 27.

Lin, D. T., Park, J., Liebert, C. A., & Lau, J. N. (2015). Validity evidence for Surgical Improvement of Clinical Knowledge Ops: A novel gaming platform to assess surgical decision making. *American Journal of Surgery*, *209*(1), 79–85. doi:10.1016/j.amjsurg.2014.08.033 PMID:25454955

Liuta, A. V., Perig, A. V., Afanasieva, M. A., & Skyrtach, V. M. (2019). Didactic games as student-friendly tools for learning hydraulics in a technical university's undergraduate curriculum. Industry and Higher Education. doi:10.1177/0950422218824507

Li, Y. (2020). Language–content partnership in higher education: Development and opportunities. *Higher Education Research & Development*, *39*(3), 500–514. doi:10.1080/07294360.2019.1685947

Logan, J. (2011). *The new Chinese city: globalization and market reform* (Vol. 52). John Wiley & Sons.

Longenecker, C. O., Papp, G., & Stansfield, T. (2009). Quarterbacking real and rapid organizational improvement. *Leader to Leader*, *2009*(51), 17–23. doi:10.1002/ltl.313

Longenecker, C. O., & Yonker, R. D. (2013). Leadership deficiencies in rapidly changing organizations: Multisource feedback as a needs assessment tool–Part I. *Industrial and Commercial Training*, *45*(3), 159–165. doi:10.1108/00197851311320577

Longo, F., Nicoletti, L., & Padovano, A. (2017). Smart operators in industry 4.0: A human-centered approach to enhance operators' capabilities and competencies within the new smart factory context. *Computers & Industrial Engineering*, *113*, 144–159. doi:10.1016/j.cie.2017.09.016

López-Fernández, D., Ezquerro, J. M., Rodríguez, J., Porter, J., & Lapuerta, V. (2019). Motivational Impact of Active Learning Methods in Aerospace Engineering Students. *Acta Astronautica*, *165*, 344–354. doi:10.1016/j.actaastro.2019.09.026

López-Pérez, M. V., Pérez-López, M. C., & Rodríguez-Ariza, L. (2011). Blended learning in higher education: Students' perceptions and their relation to outcomes. *Computers & Education*, *56*(3), 818–826. doi:10.1016/j.compedu.2010.10.023

LucasJ. (2014). https://www.livescience.com/47499-what-is-engineering.html

Lutsenko, G. V., & Lucenko, G. V. (2018). Project-based learning in automation engineering curriculum. Proceedings of the 46th SEFI Annual Conference 2018: Creativity, Innovation and Entrepreneurship for Engineering Education Excellence.

Luzzini, D., & Ronchi, S. (2016). Cinderella purchasing transformation: Linking purchasing status to purchasing practices and business performance. *Production Planning and Control*, *27*(10), 787–796. doi:10.1080/09537287.2015.1137986

Lytras, M. D., Visvizi, A., Daniela, L., Sarirete, A., & De Pablos, P. O. (2018). Social networks research for sustainable smart education. *Sustainability (Switzerland), 10*(9), 2974. Advance online publication. doi:10.3390u10092974

Mahdi, A. E. (2006). Introducing peer-supported learning approach to tutoring in engineering and technology courses. *International Journal of Electrical Engineering Education, 43*(4), 277–287. doi:10.7227/IJEEE.43.4.1

Ma, K., Teng, H., Du, L., & Zhang, K. (2014). Project-Driven Learning-by-Doing Method for Teaching Software Engineering Using Virtualization Technology. *International Journal of Emerging Technologies in Learning, 9*(9), 26–31. doi:10.3991/ijet.v9i9.4006

Malathi, M., Paramasivam, C., Sangeetha, S., Gospeline Christiana, S., & Nagalakshmi, K. (2019). Effective implementation of padlet ICT tool for engineering education. *International Journal of Innovative Technology and Exploring Engineering, 8*(8), 1586–1590.

Maletina, L., Karmanova, Y., & Kashpur, V. (2015). Implication of interdisciplinary team teaching of an ESP course. *Mediterranean Journal of Social Sciences, 6*(5), 556–561. doi:10.5901/mjss.2015.v6n5p556

Mann, H. B., & Whitney, D. R. (1947). On a Test of Whether one of Two Random Variables is Stochastically Larger than the Other. *Annals of Mathematical Statistics, 18*(1), 50–60. doi:10.1214/aoms/1177730491

Mann, L., Chang, R., Chandrasekaran, S., Coddington, A., Daniel, S., Cook, E., Crossin, E., Cosson, B., Turner, J., Mazzurco, A., Dohaney, J., O'Hanlon, T., Pickering, J., Walker, S., Maclean, F., & Smith, T. D. (2021, January 2). From problem-based learning to practice-based education: A framework for shaping future engineers. *European Journal of Engineering Education, 46*(1), 27–47. Advance online publication. doi:10.1080/03043797.2019.1708867

Mano, H., & Oliver, R. L. (1993). Assessing the Dimensionality and Structure of the Consumption Experience: Evaluation, Feeling, and Satisfaction. *The Journal of Consumer Research, 20*(3), 451–466. doi:10.1086/209361

Manrique-Losada, B., Gasca-Hurtado, G. P., & Álvarez, M. C. G. (2015). Assessment proposal of teaching and learning strategies in software process improvement. *Revista de la Facultad de Ingeniería, 2015*(77), 105–114. doi:10.17533/udea.redin.n77a13

Maphalala, M. C., & Adigun, O. T. (2021). Academics' Experience of Implementing E-Learning in a South African Higher Education Institution. *International Journal of Higher Education, 10*(1), 1–13. doi:10.5430/ijhe.v10n1p1

Markauskaite, L., & Goodyear, P. (2018). *Universities should take stronger leadership on knowledge and how it matters.* University of Sidney. https://theconversation.com/universities-should-take-stronger-leadership-on-knowledge-and-how-it-matters-89849

Martínez-Cerdá, J.-F., Torrent-Sellens, J., & González-González, I. (2018). Promoting collaborative skills in online university: Comparing effects of games, mixed reality, social media, and other tools for ICT-supported pedagogical practices. *Behaviour & Information Technology, 37*(10–11), 1055–1071. doi:10.1080/0144929X.2018.1476919

Martínez-Cerdá, J.-F., Torrent-Sellens, J., & González-González, I. (2020). Socio-technical e-learning innovation and ways of learning in the ICT-space-time continuum to improve the employability skills of adults. *Computers in Human Behavior, 107*, 105753. Advance online publication. doi:10.1016/j.chb.2018.10.019

Martin, P., & Tate, K. (2001). *Getting Started in Project Management.* John Wiley & Sons, Inc.

Masetto, M. T. (2011). *O professor na hora da verdade: a prática docente no Ensino Superior* [T he teacher at the moment of truth: teaching practice in Higher Education] (1st ed.). Avercamp.

Mason, J., & Loader, K. (2019). Using a Serious Game to Train Violence Risk Assessment and Management Skills. *Simulation & Gaming, 50*(2), 124–135. Advance online publication. doi:10.1177/1046878119837314

Masters, K. (2013). Edgar Dale's Pyramid of Learning in medical education: A literature review. *Medical Teacher*, *35*(11), e1584–e1593. doi:10.3109/0142159X.2013.800636 PMID:23758180

McComas, W. F. (2014). *The Language of Science Education: An Expanded Glossary of Key Terms and Concepts in Science Teaching and Learning*. Springer., doi:10.1007/978-94-6209-497-0

McDonald, A. C., Green, R. A., Zacharias, A., Whitburn, L. Y., Hughes, D. L., Colasante, M., & McGowan, H. (2021). Anatomy Students That are "Team-Taught" May Achieve Better Results Than Those That are "Sole-Taught". *Anatomical Sciences Education*, *14*(1), 43–51. doi:10.1002/ase.1954 PMID:32145155

McKeefry, H. (1998, Aug.). Opportunity Knocks-Education and Technical Skills Will Open the Door to Lucrative Job Offers. *CMP Net*, 3.

McKenzie, S., Hains-Wesson, R., Bangay, S., & Bowtell, G. (2020). A team-teaching approach for blended learning: An experiment. *Studies in Higher Education*, 1–15. Advance online publication. doi:10.1080/03075079.2020.1817887

McKinsey. (2019). Industry 4.0: Capturing value at scale in discrete manufacturing. *Advanced Industries*. doi:10.1007/s12599-014-0334-4

MEC. (2019). *Diretrizes Curriculares Nacionais dos Cursos de Engenharia*. Author.

MEC/CNE. (2019). *Diretrizes Curriculares Nacionais do Curso de Graduação em Engenharia*. http://portal.mec.gov.br/docman/marco-2019-pdf/109871-pces001-19-1/file

Mehta, R., Hoegg, J., & Chakravarti, A. (2011). Knowing Too Much: Expertise-Induced False Recall Effects in Product Comparison. *The Journal of Consumer Research*, *38*(3), 535–554. doi:10.1086/659380

Meletiou-Mavrotheris, M., & Prodromou, T. (2016). Pre-Service Teacher Training on Game-Enhanced Mathematics Teaching and Learning. *Technology. Knowledge and Learning*, *21*(3), 379–399. doi:10.100710758-016-9275-y

Mellors-Bourne, R., May, T., Haynes, K., & Talbot, M. (2017). *The state of engineering*. Retrieved from https://www.engineeringuk.com/media/1355/enguk-report-2017.pdf

Mengist, W., Soromessa, T., & Legese, G. (2020). Method for conducting systematic literature review and meta-analysis for environmental science research. *MethodsX*, *7*, 100777. doi:10.1016/j.mex.2019.100777 PMID:31993339

Merriam, S. B., & Tisdell, E. J. (2015). *Qualitative research: A guide to design and implementation*. John Wiley & Sons.

Mesquita, D. I. A. (2015). O currículo da formação em engenharia no âmbito do processo de Bolonha: Desenvolvimento de competências e perfil profissional na perspectiva dos docentes, dos estudantes e dos profissionais [The engineering training curriculum under the Bologna process: Development of skills and professional profile from the perspective of teachers, students and professionals] (Doutorado em Ciências da Educação Especialidade em Desenvolvimento Curricular). Instituto de Educação, Universidade do Minho, Portugal.

Mesquita, D., Lima, R. M., & Flores, M. A. (2012). *Developing professional competencies through projects in interaction with companies : A study in Industrial Engineering and Management Master Degree Project Based Learning (PBL) and Professional Practice*. Academic Press.

Mesquita, D., Salimova, T., Soldatova, E., Atoev, S., & Lima, R. M. (2019). What can be recommended to engineering teachers from the analysis of 16 European teaching and learning best practices? *Proceedings of the SEFI 47th Annual Conference*, 770–779.

Messing, J., Altas, I., & Howarth, J. (2008). Implementing Mmicrosoft PiL in Australian schools. *International Journal of Emerging Technologies in Learning*, *3*(2), 63–68. https://www.scopus.com/inward/record.uri?eid=2-s2.0-84862905363&partnerID=40&md5=da7c88fcf192b826630e018d3b55e048

Méxas, M. P., Costa, H. G., & Quelhas, O. L. G. (2013). Avaliação da importância relativa dos critérios para a seleção de Sistemas Integrados de Gestão (ERP) para uso em empresas da construção civil. *Gestão & Produção*, *20*(2), 337–356. doi:10.1590/S0104-530X2013000200007

Méxas, M. P., Quelhas, O. L. G., Costa, H. G., & De Jesus Lameira, V. (2013). A set of criteria for selection of enterprise resource planning (ERP). *International Journal of Enterprise Information Systems*, *9*(2), 44–69. doi:10.4018/jeis.2013040103

Miles, M. B., & Huberman, A. M. (1994). *An Expanded Sourcebook Qualitative Data Analysis*. SAGE Publications.

Mills, J. E., & Treagust, D. F. (2003). Engineering Education – Is Problem-based or Project-based Learning the Answer? *Australasian Journal of Engineering Education*. http://www.aaee.com.au/journal/2003/mills_treagust03.pdf

Mills, J. E., & Treagust, D. F. (2003). Engineering Education: Is Problem-Based or Project-Based Learning the answer? *Australasian Journal of Engineering Education*, *3*(2), 2–16. https://www.researchgate.net/publication/246069451_

Minett-Smith, C., & Davis, C. L. (2020). Widening the discourse on team-teaching in higher education. *Teaching in Higher Education*, *25*(5), 579–594. doi:10.1080/13562517.2019.1577814

Minichiello, A., Blake, T. D., Goodridge, W. H., & Sam, D. D. (2011). Team teaching that goes the distance: Team instruction for a broadcast introductory engineering course. *ASEE Annual Conference and Exposition, Conference Proceedings*.

Miranda, J., Pérez-Rodríguez, R., Borja, V., Wright, P. K., & Molina, A. (2019). Sensing, smart and sustainable product development (S 3 product) reference framework. *International Journal of Production Research*, *57*(14), 4391–4412. doi:10.1080/00207543.2017.1401237

Mitchell, P., & Forer, P. (2010). *Blended Learning: The Perceptions of First-year Geography Students*. doi:10.1080/03098260902982484

Mitchell, J. E., & Rogers, L. (2020). Staff perceptions of implementing project-based learning in engineering education. *European Journal of Engineering Education*, *45*(3), 349–362. doi:10.1080/03043797.2019.1641471

Moeed, A. (2013). Science Investigation That Best Supports Student Learning: Teachers Understanding of Science Investigation. *International Journal of Environmental and Science Education*, *8*(4), 537–559. doi:10.12973/ijese.2013.218a

Moen, T. (2006). Reflections on the Narrative Research Approach. *International Journal of Qualitative Methods*, *5*(4), 56–69. doi:10.1177/160940690600500405

Moen, T., Gudmundsdottir, S., & Flem, A. (2003). Inclusive practice: A biographical approach. *Teaching and Teacher Education*, *19*(3), 359–370. doi:10.1016/S0742-051X(03)00020-9

Moher, D., Liberati, A., Tetzlaff, J., & Altman, D. G. (2009). Preferred Reporting Items for Systematic Reviews and Meta-Analyses: The PRISMA Statement. *PLoS Medicine*, *6*(7), e1000097. doi:10.1371/journal.pmed.1000097 PMID:19621072

Monden, Y. (1998). Toyota Production System: An Integrated Approach to Just-In-Time (3rd ed.). Engineering and Management Press.

Money, A., & Coughlan, J. (2016). Team-taught versus individually taught undergraduate education: A qualitative study of student experiences and preferences. *Higher Education*, *72*(6), 797–811. doi:10.1007/s10734-015-9976-5

Monitoria de Matemática IFRJ-BCDUC [IFRJ-BCDUC Mathematics Monitoring]. (2021, March 26). *Behind the smoke screen*. https://www.youtube.com/watch?v=w9eL9WyynR0

Monteblanco, A. D. (2020). Power Dynamics, Common Pitfalls, and Successful Strategies Associated with Co-Teaching. *College Teaching*, *69*(2), 63–68. doi:10.1080/87567555.2020.1810610

Monteiro, E., & Reis, M. C. G. (2019). Afro-brazilian heritage in the context of quilombola school education. *Educação e Realidade*, *44*(2). Advance online publication. doi:10.1590/2175-623688369

Moran, S. (2019). *An applied Guide to process and plant design*. Elsevier.

Moreira, F., Mesquita, D., & van Hattum-Janssen, N. (2011). The importance of the project theme in Project-Based Learning: a study of student and teacher perceptions. *Proceedings of the 2011 Project Approaches in Engineering Education, 53*(9). doi:10.1017/CBO9781107415324.004

Moreira, M. C. A., Rôças, G., Anjos, M. B., & Pereira, M. V. (2016) Analysis of educational products in a professional science education master's degree course. *Proceedings of XVII International Organization for Science and Technology Education Symposium.*

Moreira, F., Alves, A. C., & Sousa, R. M. (2010). Towards Eco-efficient Lean Production Systems. In *IFIP Advances in Information and Communication Technology* (Vol. 322, pp. 100–108). Balanced Automation Systems for Future Manufacturing Networks. doi:10.1007/978-3-642-14341-0_12

Moretto, V. P. (2017). Planejamento: planejando a educação para o desenvolvimento de competências [Planning: planning education for skills development]. Petrópolis, RJ: Vozes.

Mourtzis, D., & Doukas, M. (2014). Design and Planning of Manufacturing Networks for Mass Customisation and Personalisation: Challenges and Outlook. *Procedia CIRP*, *19*, 1–13. doi:10.1016/j.procir.2014.05.004

Muhammad, A. H., Siddique, A., Youssef, A. E., Saleem, K., Shahzad, B., Akram, A., & Al-Thnian, A.-B. S. (2020). A hierarchical model to evaluate the quality of web-based E-learning systems. *Sustainability (Switzerland)*, *12*(10), 4071. Advance online publication. doi:10.3390u12104071

Mulder, M., Wesselink, R., & Bruijstens, H. C. (2005). Job profile research for the purchasing profession. *International Journal of Training and Development*, *9*(3), 185–204. doi:10.1111/j.1468-2419.2005.00230.x

Muller, E. W. (2001). *Report on the job analysis to update the certified purchasing manager and accredited purchasing practitioner examinations*. Center for Advanced Purchasing Studies.

Murphy, E. (1995). Half the battle is knowing what skills to acquire. *Purchasing*, *119*(9), 49–54.

Mursion. (2021). *From Skill to Instinct: How Higher Education Can Bridge the Gap Between Classroom and Career.* Academic Press.

Mustata, I. C., Alexe, C. G., & Alexe, C. M. (2017). Developing competencies with the general management II business simulation game. *International Journal of Simulation Modelling*, *16*(3), 412–421. doi:10.2507/IJSIMM16(3)4.383

Nakayama, R. S., de Mesquita Spínola, M., & Silva, J. R. (2020). Towards I4.0: A comprehensive analysis of evolution from I3.0. *Computers & Industrial Engineering*, *144*, 106453. doi:10.1016/j.cie.2020.106453

Narlikar, A. (2003). *International Trade and Developing Countries: Bargaining coalitions in the GATT & WTO* (Vol. 13). Taylor & Francis.

Nepomuceno, L. D. de O., & Costa, H. G. (2015). Analyzing perceptions about the influence of a master course over the professional skills of its alumni: A multicriteria approach. *Pesquisa Operacional*. Advance online publication. doi:10.1590/0101-7438.2015.035.01.0187

Neves, Cunha, Arezes, Leao, Teixeira, Lobarinhas, & Teixeira. (2006). *IN2TEC: A Multidisciplinary Research Project Involving Researchers, Students and Industry*. Academic Press.

Nino, M., & Evans, M. A. (2015). Fostering 21st-Century Skills in Constructivist Engineering Classrooms with Digital Game-Based Learning. *Revista Iberoamericana de Tecnologias Del Aprendizaje*, *10*(3), 143–149. doi:10.1109/RITA.2015.2452673

Nonaka, I., & Takeuchi, H. (1995). *The knowledge-creating Company: How Japanese companies create the dynamics of innovation*. Oxford University Press.

OER. (2019). *UNESCO Recommendation on OER*. https://en.unesco.org/themes/building-knowledge-societies/oer/recommendation PDI - Plano de Desenvolvimento Institucional da 2019/2023

Ohno, T. (1988). *Toyota Production System: Beyond large-scale production*. Productivity Press.

Oliveira, A. F. (2021). *Museus, Centros de Ciências e Formação Inicial de professores: contribuições da prática de mediação realizada no Museu Ciência* [Museums, Science Centers and Initial Teacher Training: contributions from the mediation practice carried out at the Science Museum] (Dissertação de Mestrado). Programa de Pós-graduação em Ensino de Ciência, Instituto Federal de Educação, Ciência e Tecnologia do Rio de Janeiro, City-C, Rio de Janeiro, Brasil.

Oliveira, A. S. S., de Barros, M. D. D., de Carvalho Pereira, F., Gomes, C. F. S. F. S., & da Costa, H. G. G. (2018). Prospective scenarios: A literature review on the Scopus database. *Futures*, *100*, 20–33. doi:10.1016/j.futures.2018.03.005

Osborne, W., & Case, K. (2013). Instructional Factories and Training Courses: The Inception and Implementation of Training Courses for Un-Skilled and Semi-Skilled Munitions Workers during the Great War. *Proceedings of the 11th International Conference on Manufacturing Research (ICMR2013)*, 331–336.

Osguthorpe, R. T., Osguthorpe, R. T., & Graham, C. R. (2003). Blended Learning Environments: Definitions and Directions. *Quarterly Review of Distance Education*, *4*(3), 227–233.

Ott, C., & Meek, N. (2019). Pair Teaching in Action. *ACM International Conference Proceeding Series*.

Oyetade, K. E., Zuva, T., & Harmse, A. (2020). Technology adoption in education: A systematic literature review. *Advances in Science. Technology and Engineering Systems*, *5*(6), 108–112. doi:10.25046/aj050611

Pagell, M., Das, A., Curkovic, S., & Easton, L. (1996). Motivating the purchasing professional. *The Journal of Supply Chain Management*, *32*(3), 27.

Pahl, G., & Beitz, W. (2013). *Engineering design: a systematic approach*. Springer Science & Business Media.

Pais-Montes, C., Freire-Seoane, M. J., & López-Bermúdez, B. (2019). Employability traits for engineers: A competencies-based approach. *Industry and Higher Education*, *33*(5), 308–326. doi:10.1177/0950422219854616

Papaioannou, P., & Charalambous, K. (2011). Principals' attitudes towards ICT and their perceptions about the factors that facilitate or inhibit ICT integration in primary schools of Cyprus. *Journal of Information Technology Education*, *10*(1), 349–369. doi:10.28945/1530

Papanastasiou, E. C. (2005). Factors Structure of The Attitudes Toward Research Scale. *Statistics Education Research Journal*, *4*(1), 16–26. doi:10.1037/t64085-000

Park, S., Hironaka, S., Carver, P., & Nordstrum, L. (2013). *Continuous Improvement In Education*. papers3://publication/uuid/2260843A-153F-4D8F-9206-FC474ED2972C

Pascual, J. A., Hoyuelos, I., Mateo, M., & Gento, A. M. (2019). Lean school: A learning factory for training lean manufacturing in a physical simulation environment. *Management and Production Engineering Review*, *10*(1), 4–13.

Pascual, J. A., Pimentel, C., Mateo, M., Hoyuelos, I., Matias, J., & Gento, A. M. (2020). A Learning Factory for Remanufacturing: A New Configuration at Valladolid Lean School. In *Proceedings of 6th European Lean Educator Conference (ELEC2019)*. Springer International Publishing. 10.1007/978-3-030-41429-0_1

Pecore, J. L. (2015). From Kilpatrick's Project Method to Project-Based Learning. In M. Y. Eryaman & B. C. Bruce (Eds.), International Handbook of Progressive Education (pp. 155–171). Academic Press.

Pekkanen, P., Niemi, P., Puolakka, T., Pirttilä, T., & Huiskonen, J. (2020). Building integration skills in supply chain and operations management study programs. *International Journal of Production Economics*, *225*, 107593. doi:10.1016/j.ijpe.2019.107593

Pereira, M. A. C., Barreto, M. A. M., & Pazeti, M. (2017). Application of Project-Based Learning in the first year of an Industrial Engineering Program: Lessons learned and challenges. *Production*, *27*.

Pereira, A. G., Lima, T. M., & Charrua-Santos, F. (2020). Society 5.0 as a Result of the Technological Evolution: Historical Approach. In T. Ahram, R. Taiar, S. Colson, & A. Choplin (Eds.), *Advances in Intelligent Systems and Computing* (pp. 700–705). Springer., doi:10.1007/978-3-030-25629-6_109

Pereira, F. C., Costa, H. G., & Pereira, V. (2017). Patent filings versus articles published: A review of the literature in the context of Multicriteria Decision Aid. *World Patent Information*, *50*, 17–26. doi:10.1016/j.wpi.2017.07.003

Pereira, F. C., Costa, H. G., & Pereira, V. (2019). Stimulating and inhibiting factors of patent filing with Brazilian universities. *International Journal of Entrepreneurship and Innovation Management*, *23*(3), 261–280. doi:10.1504/IJEIM.2019.099844

Pereira, M. A. C., Barreto, M. A. M., & Pazeti, M. (2017). Application of Project-Based Learning in the first year of an Industrial Engineering Program: Lessons learned and challenges. *Production*, *27*, 1–13. doi:10.1590/0103-6513.223816

Perkins, D. N. (2013). *Knowledge as design*. Routledge. doi:10.4324/9781315803067

Perrenoud, P. (1999). Construir as competências desde a escola [Building skills from school]. Porto Alegre. *The Art of Medication*.

Pettigrew, A. M. (2001). Management research after modernism. *British Journal of Management*, *12*(s1), S61–S70. doi:10.1111/1467-8551.12.s1.8

Pine, B. J. (1993). *Mass Customization: The New Frontier in Business Competition*. Harvard Business School Press.

Pinzone, M., Fantini, P., Perini, S., Garavaglia, S., Taisch, M., & Miragliotta, G. (2017). Jobs and Skills in Industry 4.0: An Exploratory Research. *IFIP International Conference on Advances in Production Management Systems*, 282-288. 10.1007/978-3-319-66923-6_33

Pittich, D., Tenberg, R., & Lensing, K. (2020). Learning Factories for Complex Competence Acquisition. *European Journal of Engineering Education*, *45*(2), 196–213. doi:10.1080/03043797.2019.1567691

Piwowar-Sulej, K. (2021). Human resources development as an element of sustainable HRM – with the focus on production engineers. *Journal of Cleaner Production*, *278*, 124008. doi:10.1016/j.jclepro.2020.124008 PMID:32901179

Plataforma Nilo Peçanha. (2021, September 20). Retrieved from http://plataformanilopecanha.mec.gov.br/

Poh, M.-Z., Swenson, N. C., & Picard, R. W. (2010). A wearable sensor for unobtrusive, long-term assessment of electrodermal activity. *IEEE Transactions on Biomedical Engineering*, *57*(5), 1243–1252. doi:10.1109/TBME.2009.2038487 PMID:20172811

Pope-Ruark, R., Motley, P., & Moner, W. (2019). Creative innovation takes a (team-teaching) family. *Teaching & Learning Inquiry*, *7*(1), 120–135. doi:10.20343/teachlearninqu.7.1.8

Powell, P. (2000). From classical to project-led education. In A. S. Pouzada (Ed.), *Project Based Learning. Project-led Education and Group Learning* (pp. 11–42). Universidade do Minho.

Powell, P., & Weenk, W. (2003). *Project-led Engineering Education*. Lemma.

Prahalad, C. K., & Hamel, G. (1990). The core competence of the corporation. *Harvard Business Review*, *68*(3), 79–91.

Pratama, U. N., & Surahman, E. (2020) *Investigating Student Responses of online learning during the Covid Pandemic in Art Education* [Paper presentation]. *6th International Conference on Education and Technology (ICET)*, Indonesia.

Prince, M. (2004). Does active learning work? A review of the research. *Journal of Engineering Education*, *93*(3), 223–231. doi:10.1002/j.2168-9830.2004.tb00809.x

Prince, M. J., & Felder, R. M. (2006). Inductive teaching and learning methods: Definitions, comparisons, and research bases. *Journal of Engineering Education*, *95*(2), 123–138. doi:10.1002/j.2168-9830.2006.tb00884.x

Prince, M., & Felder, R. (2006). Inductive Teaching and Learning Methods: Definitions, Comparisons and Research Bases. *J. Engr. Education*, *95*(2), 123–138.

Project Management Institute (PMI). (2017). *A Guide to the Project Management Body of Knowledge (PMBOK Guide)*. Project Management Institute.

Ptacek, R., Coats, R., & Ptacek, T. (2012). *Today's lean leader*. MCS Media, Inc.

Qin, J., Chui, Y.-P., Pang, W.-M., Choi, K.-S., & Heng, P.-A. (2010). Learning blood management in orthopedic surgery through gameplay. *IEEE Computer Graphics and Applications*, *30*(2), 45–57. doi:10.1109/MCG.2009.83 PMID:20650710

Radovan, M., & Kristl, N. (2020). Learning and teaching in a virtual learning environment – the importance of community building and collaboration. *Sodobna Pedagogika/Journal of Contemporary Educational Studies, 71*(2), 10–23. https://www.scopus.com/inward/record.uri?eid=2-s2.0-85089198731&partnerID=40&md5=ee91a39cbaf3314f6669d90304fcae70

Rajamma, R. K., & Sciandra, M. R. (2020). Planning and Implementing a Graduate Online Team-Taught Marketing Course. *Journal of Marketing Education*, *42*(2), 108–122. doi:10.1177/0273475318786336

Rajarapollu, P. R., & Bhagwatkar, S. A. (2018). ICT - An Effective Way for Active and Collaborative Learning Teaching Process in Engineering Education - A Review. *2017 International Conference on Transforming Engineering Education, ICTEE 2017*. 10.1109/ICTEED.2017.8585697

Randazzo, M., Priefer, R., & Dakwar, R. K. (2021). Project-Based Learning and Traditional Online Teaching of Research Methods During COVID-19: An Investigation of Research Self-Efficacy and Student Satisfaction. *Frontiers in Education.*, *6*(662850). Advance online publication. doi:10.3389/feduc.2021.66285

Ravi, V. (2015). *Industrial Engineering and Management*. PHI Learning Private Limited.

Redecker, C., & Punie, Y. (2017). *European Framework for the Digital Competence of Educators: DigCompEdu*. https://ec.europa.eu/jrc/en/publication/eur-scientific-and-technical-research-reports/european-framework-digital-competence-educators-digcompedu

Regueras, L. M., Verdú, E., Muñoz, M. F., Perez, M. A., de Castro, J. P., & Verdú, M. J. (2009). Effects of competitive E-learning tools on higher education students: A case study. *IEEE Transactions on Education*, *52*(2), 279–285. doi:10.1109/TE.2008.928198

Reichheld, F. F. (2003). The One Number You Need to Grow. *Harvard Business Review*. https://hbr.org/2003/12/the-one-number-you-need-to-grow

Reichheld, F. (2006). *The Ultimate Question: Driving Good Profits and True Growth*. Harvard Business School Publishing Corporation.

Reinalda, B., & Kulesza, E. (2006). *The Bologna Process – Harmonizing Europe's Higher Education*. Verlag Barbara Budrich. doi:10.2307/j.ctvbkk39v

Renuka, G. (2021). Innovative Teaching and Learning Strategies for Materials Engineering Education. *Materials Today: Proceedings*. Advance online publication. doi:10.1016/j.matpr.2021.04.350

Rezaie, M., Nosratabadi, H. E., & Fazlollahtabar, H. (2012). Applying KANO model for users' satisfaction assessment in e-learning systems: A case study in iran virtual higher educational systems. *International Journal of Information and Communication Technology Education*, *8*(3), 1–12. doi:10.4018/jicte.2012070101

Ribeiro, E. A. (2008). *A perspectiva da entrevista na investigação qualitativa*. Academic Press.

Ricaurte, M., & Viloria, A. (2020). Project-Based Learning as a Strategy for Multi-Level Training Applied to Undergraduate Engineering Students. *Education for Chemical Engineers*, *33*, 102–111. doi:10.1016/j.ece.2020.09.001

Rickard, A., & Walsh, T. (2019). Policy, practice and process in team teaching: A pilot project with co-operating teachers and student teachers on school placement. *Irish Educational Studies*, *38*(3), 309–326. doi:10.1080/03323315.2019.1625798

Riordan, C. M. (2013, June 4). Work-life balance isn't the point. *Harvard Business Review*. Retrieved from: https://hbr.org/2013/06/work-life-balance-isnt-the-poi

Rodrigues, R. O., Sousa, P. C., Gaspar, J., Bañobre-López, M., Rui, L., & Minas, G. (2020). Organ-on-a-Chip: A Preclinical Microfluidic Platform for the Progress of Nanomedicine. *Small*, *16*(51), 2003517. doi:10.1002mll.202003517 PMID:33236819

Rodriguez, D. S. S., Costa, H. G., & Do Carmo, L. F. R. R. S. (2013). Métodos de auxílio multicritério à decisão aplicados a problemas de PCP: Mapeamento da produção em periódicos publicados no Brasil. In Gestao e Producao. doi:10.1590/S0104-530X2013000100010

Rolfes, K., Perry, L., & Shunk, D. (2019). *Industry 4.0 and people-centric leadership*. AME. https://www.ame.org/target/articles/2019/industry-40-and-people-centric-leadership

Romi, I. M. (2017). A Model for e-learning Systems Success: Systems, Determinant, and Performance. *International Journal of Emerging Technologies in Learning*, *12*(10), 4–20. doi:10.3991/ijet.v12i10.6680

Rosero-Zambrano, C. A., Avila, A., Osorio, L. A., & Aguirre, S. (2018). Impact of Adding Internet Technology on Student Performance and Perception of Autonomy in Fundamentals of Electronics Course. *Journal of Science Education and Technology*, *27*(2), 177–187. doi:10.100710956-017-9716-y

Rugarcia, A., Felder, R. M., Woods, D. R., & Stice, J. E. (2000). The future of Engineering Education I. A vision for a new century. *Chemical Engineering Education*, *34*(1), 16–25.

Rutz, E., Eckart, R., Wade, J. E., Maltbie, C., Rafter, C., & Elkins, V. (2003). Student performance and acceptance of instructional technology: Comparing technology-enhanced and traditional instruction for a course in statics. *Journal of Engineering Education*, *92*(2), 133–140. doi:10.1002/j.2168-9830.2003.tb00751.x

Rybnicek, R., & Königsgruber, R. (2019). What Makes Industry–University Collaboration Succeed? A Systematic Review of the Literature. *Journal of Business Economics*, *89*(2), 221–250. doi:10.100711573-018-0916-6

Rychen, D. S., & Salganik, L. H. (2000). *Definition and selection of Key competencies*. https://www.oecd.org/edu/skills-beyond-school/41529556.pdf

Sackey, S. M., & Bester, A. (2016). Industrial engineering curriculum in Industry 4.0 in a South African context. *South African Journal of Industrial Engineering*, *27*(4), 101–114. doi:10.7166/27-4-1579

Sadek, S. H., Rubio, M., Rui, L., & Vega, E. J. (2021). Blood Particulate Analogue Fluids: A Review. *Materials (Basel)*, *14*(9), 2451. Advance online publication. doi:10.3390/ma14092451 PMID:34065125

Saeed, A. A., Mutashar, A. M., & Aldakheel, A. (2018). The impact of applying collaborative team teaching method on students' outcomes. *Indian Journal of Forensic Medicine and Toxicology*, *12*(4), 139–143. doi:10.5958/0973-9130.2018.00212.8

Salehi, H., Shojaee, M., & Sattar, S. (2014). Using e-learning and ICT courses in educational environment: A review. *English Language Teaching*, *8*(1), 63–70. doi:10.5539/elt.v8n1p63

Saliba, T. A., Borges de Mattos Custodio, L., Adas Saliba, N., & Adas Saliba Moimaz, S. (2019). The Impact of Scientific Initiation on the Scientific Production of Graduate Students. *Revista Da ABENO*, *19*(1), 127–133. doi:10.30979/rev.abeno.v19i1.873

Salifu, I. (2021). Exploring Coteaching as a Trend in Higher Education. *College Teaching*, *69*(3), 150–160. doi:10.1080/87567555.2020.1838426

Salonen, A. O., & Savander-Ranne, C. (2015). Teachers' Shared Expertise at a Multidisciplinary University of Applied Sciences. *SAGE Open*, *5*(3). Advance online publication. doi:10.1177/2158244015596206

Salvendy, G. (2001). Handbook of Industrial Engineering: Technology and Operations Management (3rd ed.). John Wiley & Sons.

Sanchez-Franco, M. J. (2010). WebCT - The quasimoderating effect of perceived affective quality on an extending Technology Acceptance Model. *Computers & Education*, *54*(1), 37–46. doi:10.1016/j.compedu.2009.07.005

Sánchez-Polo, M. T., Cegarra-Navarro, J.-G., Cillo, V., & Wensley, A. (2019). Overcoming knowledge barriers to health care through continuous learning. *Journal of Knowledge Management*, *23*(3), 508–526. doi:10.1108/JKM-10-2018-0636

Sang, G., Valcke, M., van Braak, J., Tondeur, J., & Zhu, C. (2011). Predicting ICT integration into classroom teaching in Chinese primary schools: Exploring the complex interplay of teacher-related variables. *Journal of Computer Assisted Learning*, *27*(2), 160–172. doi:10.1111/j.1365-2729.2010.00383.x

Santandreu-Mascarell, C., Canós-Darós, L., & Pons-Morera, C. (2011). Competencies and skills for future Industrial Engineers defined in Spanish degrees. *Journal of Industrial Engineering and Management*, *4*(1). Advance online publication. doi:10.3926/jiem.2011.v4n1.p13-30

Sá, P., Costa, A. P., & Moreira, A. (2021). *Reflexões em torno de Metodologias de Investigação: recolha de dados* (Vol. 2). UA Editora.

Saravanan, S., Mahesh, V., Kumar, D. G., & Kshatri, S. S. (2021, March). Improving Student's Learning with Efficient Learning Techniques: A Case Study of First Year Basic Electrical Engineering Course. *Materials Today: Proceedings*. Advance online publication. doi:10.1016/j.matpr.2021.02.784

Sawhney, R., & Chason, S. (2005). Human behavior based exploratory model for successful implementation of lean enterprise in industry. *Performance Improvement Quarterly*, *18*(2), 76–96. doi:10.1111/j.1937-8327.2005.tb00334.x

Sawilowsky, S. S. (2009). New Effect Size Rules of Thumb. *Journal of Modern Applied Statistical Methods; JMASM*, *8*(2), 26. doi:10.22237/jmasm/1257035100

Scerri, M., Presbury, R., & Goh, E. (2020). An Application of the Mentoring Framework to Investigate the Effectiveness of Mentoring Programs between Industry Mentors and Student Mentees in Hospitality. *Journal of Hospitality and Tourism Management*, *45*(August), 143–151. doi:10.1016/j.jhtm.2020.08.010

Schmulian, A., & Coetzee, S. A. (2019). To Team or Not to Team: An Exploration of Undergraduate Students' Perspectives of Two Teachers Simultaneously in Class. *Innovative Higher Education*, *44*(4), 317–328. doi:10.100710755-019-9466-2

Schoenherr, T. (2010). Outsourcing decisions in global supply chains: An exploratory multi-country survey. *International Journal of Production Research*, *48*(2), 343–378. doi:10.1080/00207540903174908

Scholten, K., & Dubois, A. (2017). Advancing the skill set of SCM graduates–An active learning approach. *International Journal of Operations & Production Management*, *37*(11), 1683–1699. doi:10.1108/IJOPM-11-2015-0724

Schulze, H., Bals, L., & Johnsen, T. E. (2019). Individual competences for sustainable purchasing and supply management (SPSM) A literature and practice perspective. *International Journal of Physical Distribution & Logistics Management*, *49*(3), 287–304. doi:10.1108/IJPDLM-01-2018-0036

Scopus. (n.d.). Retrieved May 3, 2019, from https://www-scopus.ez24.periodicos.capes.gov.br/term/analyzer.uri?sid=a318f65953790c5f3e34092c19bad69c&origin=resultslist&src=s&s=EID%282-s2.0-78049514934+OR+2-s2.0-79251568005+OR+2-s2.0-80052818916+OR+2-s2.0-84890889040+OR+2-s2.0-84897615891+OR+2-s2.0-84

Scucuglia, R. (2012). *On the nature of students' digital mathematical performance: When elementary school students produce mathematical multimodal artistic narratives*. Alemanha: Verlag/LAP Lambert Academic Publishing.

Seager, W., Ruskov, M., Sasse, M. A., & Oliveira, M. (2011). Eliciting and modelling expertise for serious games in project management. *Entertainment Computing*, *2*(2), 75–80. doi:10.1016/j.entcom.2011.01.002

Seif, E. (2014). *Seven Types of Projects that Foster Powerful Learning*. https://www.ascd.org/blogs/seven-types-of-projects-that-foster-powerful-learning

Selkrig, M., & Keamy, K. (2015). Promoting a willingness to wonder: Moving from congenial to collegial conversations that encourage deep and critical reflection for teacher educators. *Teachers and Teaching*, *21*(4), 421–436. doi:10.1080/13540602.2014.969104

Serafim, A. P., Durães, R. S. S., Rocca, C. C. A., Gonçalves, P. D., Saffi, F., Cappellozza, A., … Lotufo-Neto, F. (2021). Exploratory study on the psychological impact of COVID-19 on the general Brazilian population. *PLoS ONE, 16*(2). doi:10.1371/journal.pone.0245868

Sewilam, H., Nacken, H., Breuer, R., & Pyka, C. (2017). Competence-based and game-based capacity development for sustainable water management in Germany. *Environmental Earth Sciences*, *76*(3), 131. Advance online publication. doi:10.100712665-017-6416-0

Seymour, M. W., & Seymour, D. (2013). Are two professors better than one? Student and faculty perceptions of co-teaching. *International Journal of Learning*, *20*(1), 39–52. https://www.scopus.com/inward/record.uri?eid=2-s2.0-85016634914&partnerID=40&md5=b695b3a8588e5f115bf26a82947cc565

Sharp, H. (2019). Bricolage research in history education as a scholarly mixed-methods design - CORE Reader. *History Education Research Journal, 16*(1), 50–62.

Sheppard, S., Colby, A., Macatangay, K., & Sullivan, W. (2006). What is engineering practice? *International Journal of Engineering Education, 22*(3), 429–438.

Shingo Institute. (2020). *The Shingo Model for Operational Excellence*. https://shingo.org/model

Shingo. (1989). *A study of Toyota Production System from an Industrial Engineering*. Academic Press.

Shorfuzzaman, M., Alelaiwi, A., Masud, M., Hassan, M. M., & Hossain, M. S. (2015). Usability of a cloud-based collaborative learning framework to improve learners' experience. *Computers in Human Behavior*, *51*, 967–976. doi:10.1016/j.chb.2014.10.002

SHRM (Society for Human Resource Management). (2008, March 1). *Leadership competencies*. Retrieved from: https://www.shrm.org/ResourcesAndTools/hr-topics/behavioral-competencies/leadership-and-navigation/Pages/leadershipcompetencies.aspx

Shute, V. J., Wang, L., Greiff, S., Zhao, W., & Moore, G. (2016). Measuring problem solving skills via stealth assessment in an engaging video game. *Computers in Human Behavior*, *63*, 106–117. doi:10.1016/j.chb.2016.05.047

Silva, G. B., & Costa, H. G. (2015). Mapeamento de um núcleo de partida de referências em Data Mining a partir de periódicos publicados no Brasil. *Gestão & Produção*, *22*(1), 107–118. doi:10.1590/0104-530X792-13

Siqueira, B. J. S., & Aparecida, A. E. (2020). Socially responsible investments (SRIs) – mapping the research field. In Social Responsibility Journal. doi:10.1108/SRJ-06-2019-0204

Smith, K. E. (2001). The concept and importance of productivity. In *Maynard's Industrial Engineering Handbook* (p. 2567). McGraw Hill Standard Handbooks.

Smith, K. K., & Winn, V. G. (2017). Co-teaching in the college classroom. *Teaching Education*, *28*(4), 435–448. doi:10.1080/10476210.2017.1325863

Smith, T. (2017). Gamified Modules for an Introductory Statistics Course and Their Impact on Attitudes and Learning. *Simulation & Gaming*, *48*(6), 832–854. doi:10.1177/1046878117731888

Snyder, H. (2019). Literature review as a research methodology: An overview and guidelines. *Journal of Business Research*, *104*, 333–339. doi:10.1016/j.jbusres.2019.07.039

Soares, C. (2021). *Metodologias ativas: uma nova experiência de aprendizagem*. Editora Cortez.

Soares, F. O., Sepúlveda, M. J., Monteiro, S., Lima, R. M., & Dinis-Carvalho, J. (2013). An integrated project of entrepreneurship and innovation in engineering education. *Mechatronics*, *23*(8), 987–996. doi:10.1016/j.mechatronics.2012.08.005

Sousa, R. M., Moreira, F., & Alves, A. C. (2013). Active Learning using Physical Prototypes and Serious Games. *Proceedings of the Fifth International Symposium on Project Approaches (PAEE2013)*, 1-9.

Souza, D. G., Vergottini, V. S., Bernini, D. S. D. (2018). Educação dos tempos modernos através da aprendizagem colaborativa: uma abordagem sobre EDUSCRUM [Modern time education through collaborative learning: an eduScrum approach]. *Anais do XXIX Simpósio Brasileiro de Informática na Educação (SBIE)*.

Spear, S. J. (2004). Learning to lead at Toyota. *Harvard Business Review*, *82*(5), 78–86, 151. https://www.ncbi.nlm.nih.gov/pubmed/15146738 PMID:15146738

Sperl, T. (2017). *Lean Leadership at Oakland University (online video)*. Oakland University. Retrieved from https://drive.google.com/drive/u/3/search?q=owner:haughey%40oakland.edu

Spitz-Oener, A. (2006). Technical change, job tasks, and rising educational demands: Looking outside the wage structure. *Journal of Labor Economics*, *24*(2), 235–270. doi:10.1086/499972

Squire, B., Brown, S., Readman, J., & Bessant, J. (2006). The impact of mass customisation on manufacturing trade-offs. *Production and Operations Management*, *15*(1), 10–21. doi:10.1111/j.1937-5956.2006.tb00032.x

Starčič, A. I., & Turk, Z. (2010). Powerful learning environments in engineering education. *7th WSEAS International Conference on Engineering Education, EDUCATION'10, International Conference on Education and Educational Technologies*, 409–416. https://www.scopus.com/inward/record.uri?eid=2-s2.0-79958763750&partnerID=40&md5=e8188bff062cfb8cfe6aa6ba889ffafb

Stefanovic, D., Drapsin, M., Nikolic, J., Scepanovic, D., Radjo, I., & Drid, P. (2011). Empirical study of student satisfaction in e-learning system environment. *Technics Technologies Education Management*, *6*(4), 1152–1164. https://www.scopus.com/inward/record.uri?eid=2-s2.0-84855521462&partnerID=40&md5=a344298b5a59d03fa3af02b5ea26a6e5

Stek, K., & Schiele, H. (2021). (in press). How to train supply managers – necessary and sufficient purchasing skills leading to success. *Journal of Purchasing and Supply Management*, *100700*. Advance online publication. doi:10.1016/j.pursup.2021.100700

Stinenbosch, M., & Stek, K. (2017, July). The gap between academic purchasing education and the market needs. *In-Procurement*, 24-25.

Stinenbosch, M. (2017). *Analysis of represented purchasing skills in academic literature and in current Dutch education provision*. University of Twente.

Stone, R., Caird-Daley, A., & Bessell, K. (2009). SubSafe: A games-based training system for submarine safety and spatial awareness (Part 1). *Virtual Reality (Waltham Cross)*, *13*(1), 3–12. doi:10.100710055-008-0110-1

Suárez, L. M. C. (2016). Investigating website appearance and usability effects on student satisfaction with the website: A descriptive analysis in three countries. *International Review on Public and Nonprofit Marketing*, *13*(3), 223–238. doi:10.100712208-015-0147-2

Sudjimat, D. A., Nyoto, A., & Romlie, M. (2021). Implementation of Project-Based Learning Model and Workforce Character Development for the 21st Century in Vocational High School. *International Journal of Instruction*, *14*(1), 181–198. doi:10.29333/iji.2021.14111a

Sugimori, Y., Kusunoki, K., Cho, F., & Uchikawa, S. (1977). Toyota production system and Kanban system Materialization of just-in- time and respect-for-human system. *International Journal of Production Research*, *15*(6), 553–564. doi:10.1080/00207547708943149

Suki, N. M., & Suki, N. M. (2019). Structural Relationships in the Embedding of Role-Play Games in a Class for Japanese Language Proficiency: Towards a Unified View. *Technology. Knowledge and Learning*, *24*(1), 65–87. doi:10.100710758-016-9294-8

Sulisworo, D., Ummah, R., Nursolikh, M., & Rahardjo, W. (2020). The analysis of the critical thinking skills between blended learning implementation: Google Classroom and Schoology. *Universal Journal of Educational Research*, *8*(3B), 33–40. doi:10.13189/ujer.2020.081504

SuperData Research, Inc. (2019). *Worldwide digital games market: September 2019*. SuperData Research.

Sutherland, J. (2014). *Scrum: The art of doing twice the work in half the time*. LeYa.

Suzaki, K. (1993). *The new shop floor management: Empowering people for continuous improvement*. Academic Press.

Swedberg, R. (2020). Exploratory Research. In C. Elman, J. Gerring, & J. Mahoney (Eds.), The Production of Knowledge: Enhancing Progress in Social Science (pp. 17-41). Cambridge University Press. doi:10.1017/9781108762519.002

Takeuchi, H., Osono, E., & Shimizu, N. (2008). The Contradictions That Drive Toyota's Success. *Harvard Business Review*.

Tan, H. R., Chng, W. H., Chonardo, C., Ng, M. T. T., & Fung, F. M. (2020). How Chemists Achieve Active Learning Online during the COVID-19 Pandemic: Using the Community of Inquiry (CoI) Framework to Support Remote Teaching. *Journal of Chemical Education*, *97*(9), 2512–2518. doi:10.1021/acs.jchemed.0c00541

Tan, H.-S., Ivander, Oktarina, R., Reynaldo, V., & Sharina, C. (2020). Conceptual development of learning factory for industrial engineering education in Indonesia context as an enabler of students' competencies in industry 4.0 era. *IOP Conference Series. Earth and Environmental Science*, *426*, 012123. doi:10.1088/1755-1315/426/1/012123

Tassabehji, R., & Moorhouse, A. (2008). The changing role of procurement: Developing professional effectiveness. *Journal of Purchasing and Supply Management*, *14*(1), 55–68. doi:10.1016/j.pursup.2008.01.005

Tatham, P., Wu, Y., Kovács, G., & Butcher, T. (2017). Supply chain management skills to sense and seize opportunities. *International Journal of Logistics Management*, *28*(2), 266–289. doi:10.1108/IJLM-04-2014-0066

Tawafak, R. M., College, A. B., Sohail, O., Malik, I., Oman, G., Alfarsi, A. B., & College, O. (2021, July). Impact of Technologies During the COVID-19 Pandemic for Improving Behavioral Intention to Use E-Learning. *International Journal of Information and Communication Technology Education*, *17*(3), 137–150. Advance online publication. doi:10.4018/IJICTE.20210701.oa9

Teixeira, A. L. O. (2021). *"Estamira" e Educação Ambiental Crítica: inspirações na formação de professors* ["Estamira" and Critical Environmental Education: inspirations in teacher education] (Dissertação de Mestrado). Programa de Pós-graduação em Ensino de Ciência, Instituto Federal de Educação, Ciência e Tecnologia do Rio de Janeiro, City-C, Rio de Janeiro, Brasil.

Teixeira, I., Correia, T., Carvalho, V., Rodrigues, N., Lima, R. A., & Teixeira, S. (2021). Numerical Simulation: Study on the Path of Expelled and Exhaled Water Vapor Particles. *5th International Conference on Numerical and Symbolic Computation Developments and Applications*.

Tejada, R. V. (2019). *Self-Learning Techniques to Contribute to the Learning of the English Language of the Students*. https://files.eric.ed.gov/fulltext/ED595374.pdf

Teotônio, I., Hecht, M., Castro, L. C., Gandolfi, L., Pratesi, R., Nakano, E. Y., Puppin Zandonadi, R., & Pratesi, C. B. (2020). Repercussion of COVID-19 pandemic on brazilians' quality of life: A nationwide cross-sectional study. *International Journal of Environmental Research and Public Health*, *17*(22), 1–11. doi:10.3390/ijerph17228554 PMID:33218087

Thakker, S.V., Parab, J., & Kaisare, S., (2020). *Systematic research of e-learning platforms for solving challenges faced by Indian engineering students*. doi:10.1108/AAOUJ-09-2020-0078

Thakur, P., Dutt, S., & Chauhan, A. (2021 April) Problem Based Learning in Education– its Need for Sustainable Development. *Journal of Engineering Education Transformations, 34*(4).

Thomas, J. (2000). *A review of research on project-based learning*. Autodesk Foundation. http://www.ydae.purdue.edu/LCT/HBCU/documents/AReviewofResearchofProject-BasedLearning.pdf

Thomas, J. W. (2000). *A review of research on Project-Based learning*. The Autodeks Foundation.

Thomassen, A. O., & Stentoft, D. (2020). Educating students for a complex future–why integrating a problem analysis in problem-based learning has something to offer. *The Interdisciplinary Journal of Problem-Based Learning, 14*(2). Advance online publication. doi:10.14434/ijpbl.v14i2.28804

Thompson, A., Strickland III, A. J., & Gamble, J. (2007). *Crafting and Executing Strategy: Text and Reading* (15th ed.). McGraw Hill-Irwin.

Tien, J. M. (2020). Toward the Fourth Industrial Revolution on Real-Time Customization. *Journal of Systems Science and Systems Engineering, 29*(2), 127–142. doi:10.100711518-019-5433-9

Tisch, M., Hertle, C., Abele, E., Metternich, J., & Tenberg, R. (2016). Learning Factory Design: A Competency-Oriented Approach Integrating Three Design Levels. *International Journal of Computer Integrated Manufacturing, 29*(12), 1355–1375. doi:10.1080/0951192X.2015.1033017

Tisch, M., Hertle, C., Cachay, J., Abele, E., Metternich, J., & Tenberg, R. (2013). A Systematic Approach on Developing Action-Oriented, Competency-Based Learning Factories. *Procedia CIRP, 7*, 580–585. doi:10.1016/j.procir.2013.06.036

Todd, R. H., & Magleby, S. P. (2004). Evaluation and rewards for faculty involved in engineering design education. *International Journal of Engineering Education, 20*(3), 333–340.

Trentin, A., Somià, T., Sandrin, E., & Forza, C. (2019). Operations managers' individual competencies for mass customization. *International Journal of Operations & Production Management, 39*(9/10), 1025–1052. doi:10.1108/IJOPM-10-2018-0592

Trent, R. J., & Monczka, R. M. (2003). Understanding integrated global sourcing. *International Journal of Physical Distribution & Logistics Management, 33*(7), 607–629. doi:10.1108/09600030310499286

Trevelyan, J. (2019). Transitioning to engineering practice. *European Journal of Engineering Education, 44*(6), 821–837. doi:10.1080/03043797.2019.1681631

Tripp, D. (2005, September/December). Pesquisa-ação: Uma introdução metodológica [Action Research: a metodology introduction]. *Educação e Pesquisa, São Paulo, 31*(3), 443–466. doi:10.1590/S1517-97022005000300009

Tsybulsky, D. (2019). The team teaching experiences of pre-service science teachers implementing PBL in elementary school. *Journal of Education for Teaching, 45*(3), 244–261. doi:10.1080/09589236.2019.1599505

Tsybulsky, D., & Muchnik-Rozanov, Y. (2019). The development of student-teachers' professional identity while team-teaching science classes using a project-based learning approach: A multi-level analysis. *Teaching and Teacher Education, 79*, 48–59. doi:10.1016/j.tate.2018.12.006

Tuomivaara, A., & Laakko, M. (2020). *5S Project JMC Engine Oy* [Final report]. Academic Press.

Tuononen, T., Parpala, A., & Lindblom-Ylänne, S. (2019). Graduates' evaluations of usefulness of university education, and early career success–a longitudinal study of the transition to working life. *Assessment & Evaluation in Higher Education, 44*(4), 581–595. doi:10.1080/02602938.2018.1524000

Tüzün, H., Yilmaz-Soylu, M., Karakuş, T., Inal, Y., & Kizilkaya, G. (2009). The effects of computer games on primary school students' achievement and motivation in geography learning. *Computers & Education, 52*(1), 68–77. doi:10.1016/j.compedu.2008.06.008

Tyni, P. (2021). *New standards for development of laser cutting machines.* Oulu University of Applied Sciences. https://www.theseus.fi/handle/10024/356966

Uébe-Mansur, A. F., Leite, F. C., & Bastos, H. P. P. (2016). Lean education for applied science universities: A proposal by federal institutes of applied sciences in Brazil. In Lean Education: An Overview of Current Issues. doi:10.1007/978-3-319-45830-4_2

Uébe-Mansur, A. F., Alves, A. C., & Torres, R. B. (2019). Trello as virtual learning environment and active learning organiser for PBL classes: An analysis under Bloom's Taxonomy. *International Symposium on Project Approaches in Engineering Education*, 9.

Umek, L., Keržič, D., Aristovnik, A., & Tomaževič, N. (2017). An assessment of the effectiveness of moodle e-learning system for undergraduate Public Administration education. *International Journal of Innovation and Learning, 21*(2), 165–177. doi:10.1504/IJIL.2017.081939

UNDP – United Nations Development Programme. (2008). *Competency Framework.* https://procurement-notices.undp.org/view_file.cfm?doc_id=10206

UNESCO. (2010). *Engineering: Issues, Challenges and Opportunities for Development.* UNESCO.

UNESCO. (2020). *International Commission on the Future of Education, Education in a Post-COVID World: Nine ideas for public action.* https://unesdoc.unesco.org/ark:/48223/pf0000373717/PDF/373717eng.pdf

UNESCO. (2021). *Engineering for Sustainable Development.* UNESCO.

UNESCO. (2021). *Engineering for Sustainable Development.* United Nations.

United Nations. (2015). *2030 Agenda.* https://nacoesunidas.org/pos2015/agenda2030/

Universities of the Future. (2019). *Industry 4.0 Implications For Higher Education Institutions: State of Maturity and Competency Needs.* Author.

Unnikrishnan, B. (2014). Medical Students' Research - Facilitators and Barriers. *Journal of Clinical and Diagnostic Research, 8*(12), XC01–XC04. doi:10.7860/JCDR/2014/10223.5291 PMID:25654016

Usmani, R. S., Saeed, A., & Tayyab, M. (2021). Role of ICT for Community in Education During COVID-19. In N. Zaman, K. Rafique, & V. Ponnusamy (Eds.), *ICT Solutions for Improving Smart Communities in Asia* (pp. 125–150). IGI Global. doi:10.4018/978-1-7998-7114-9.ch006

Valério M., Moreira A., Braz B. (2019). A sala de aula invertida na universidade pública Brasileira: evidências da prática em uma licenciatura em ciências exatas [The flipped classroom in the Brazilian public university: evidence of practice in a degree in exact sciences]. *Revista Thema, 16*(1).

Van der Valk, W., Sumo, R., Dul, J., & Schroeder, R. G. (2016). When are contracts and trust necessary for innovation in buyer-supplier relationships? A necessary condition analysis. *Journal of Purchasing and Supply Management, 22*(4), 266–277. doi:10.1016/j.pursup.2016.06.005

van Dun, D. H., & Wilderom, C. P. M. (2016). Lean-team effectiveness through leader values and members' informing. *International Journal of Operations & Production Management, 36*(11), 1530–1550. doi:10.1108/IJOPM-06-2015-0338

van Hattum-Janssen, N., & Vasconcelos, R. M. (2008). The Role of the Tutor in Project-Led Education: the development of an evaluation instrument. *International Conference on Engineering and Technology Education*, 27–31.

Vangrieken, K., & Kindt, E. (2019). The teacher as an island? A mixed method study on the relationship between autonomy. and collaboration. *European Journal of Psychology of Education, 35*(1), 177–204. doi:10.100710212-019-00420-0

Vasiliene-Vasiliauskiene, V., Vasiliauskas, A. V., Meidute-Kavaliauskiene, I., & Sabaityte, J. (2020). Peculiarities of educational challenges implementing project-based learning. *World Journal on Educational Technology: Current Issues, 12*(2), 136–149.

Veile, J. W., Kiel, D., Müller, J. M., & Voigt, K.-I. (2019). Lessons learned from Industry 4.0 implementation in the German manufacturing industry. *Journal of Manufacturing Technology Management*. doi:10.1108/JMTM-08-2018-0270

Verkuyl, M., Atack, L., Mastrilli, P., & Romaniuk, D. (2016). Virtual gaming to develop students' pediatric nursing skills: A usability test. *Nurse Education Today, 46*, 81–85. doi:10.1016/j.nedt.2016.08.024 PMID:27614548

Veselov, G., Pljonkin, A., & Fedotova, A. Y. (2019). Project-based learning as an effective method in education. *ICMET 2019: Proceedings of the 2019 International Conference on Modern Educational Technology*, 54-57, 10.1145/3341042.3341046

Vesikivi, P., Holvikivi, J., Lakkala, M., & Bauters, M. (2016). Teacher collaboration in IT project courses: Resistance and success. *44th Annual Conference of the European Society for Engineering Education - Engineering Education on Top of the World: Industry-University Cooperation, SEFI 2016.*

Vesikivi, P., Lakkala, M., Holvikivi, J., & Muukkonen, H. (2019). Team teaching implementation in engineering education: Teacher perceptions and experiences. *European Journal of Engineering Education, 44*(4), 519–534. doi:10.1080/03043797.2018.1446910

Von der Gracht, H., Giunipero, L. C., & Schueller, M. (2016). *Future-proof procurement; Now or never: the big procurement transformation.* Retrieved from https://advisory.kpmg.us/articles/2017/future-proof-procurement.html

Von der Gracht, H., Huber-Straßer, A., Schueller, M., Müller, N., Lichtenau, P. E., & Zühlke, H. (2018). *Rethinking the value chain. A study on AI, humanoids and robots - Artificial Intelligence: Possible business application and development scenarios to 2040.* Retrieved from https://assets.kpmg/content/dam/kpmg/xx/pdf/2018/09/rethinking-the-value-chain.pdf

von Tunzelmann, N. (2003). Historical Coevolution of Governance and Technology in the Industrial Revolutions. *Structural Change and Economic Dynamics, 14*(4), 365–384. doi:10.1016/S0954-349X(03)00029-8

Vrchota, J., Mařiková, M., Řehoř, P., Rolínek, L., & Toušek, R. (2019). Human Resources Readiness for Industry 4.0. *Journal of Open Innovation, 6*(1), 3. doi:10.3390/joitmc6010003

Vybornov, N. A., Levchenko, A. A., Likhter, A. M., & Smirnov, V. V. (2014). Research practice as a constituent of the worldwide CDIO initiative. In *Proceedings of the 5th International methodological conference on "Changes in Education in the 21st century: Best International Practices and Russian Experience. How to Form Business Mindset and Innovative Thinking"*. Astrakhan: Astrakhan State University Publishing House.

Vygotsky, L. S. (1986). *Thought and language.* MIT Press.

Wahid, K. A., Ismail, W. S. A. W., & Numprasertchai, H. (2019). The role of social media in collective learning. *International Journal of Innovation and Learning, 25*(4), 363–376. doi:10.1504/IJIL.2019.099981

Walsh, E. O., & Davis, E. C. (2017). The geology and sociology of consumption: Team-teaching sustainability in an interdisciplinary first-year seminar. *Journal of Geoscience Education, 65*(2), 126–135. doi:10.5408/16-172.1

Wang, T., & Yang, K. (2016). Technology-enhanced science teaching and learning: Issues and trends. In Science Education Research and Practice in Asia: Challenges and Opportunities (pp. 461–481). Springer Singapore. doi:10.1007/978-981-10-0847-4_25

Wang, X. (2010). eLearning in Europe: Action plans and implications. *Proceedings - 2010 International Conference on Artificial Intelligence and Education, ICAIE 2010*, 240–245. 10.1109/ICAIE.2010.5641445

Wang, Y. S. (2003). Assessment of learner satisfaction with asynchronous electronic learning systems. *Information & Management, 41*(1), 75–86. doi:10.1016/S0378-7206(03)00028-4

Wang, Y., Jing, X., Han, W., Jing, Y., & Xu, L. (2020). Positive and negative affect of university and college students during COVID-19 outbreak: A network-based survey. *International Journal of Public Health, 65*(8), 1437–1443. doi:10.100700038-020-01483-3 PMID:33063142

Wasserman, E., & Migdal, R. (2019). Professional development: Teachers' attitudes in online and traditional training courses. *Online Learning Journal, 23*(1), 132–143. doi:10.24059/olj.v23i1.1299

Weidman, J. C., Twale, D. J., & Stein, E. L. (2001). *Socialization of Graduate and Professional Students in Higher Education A Perilous Passage?* ASHE-ERIC Higher Education Report.

Wertsch, J. V. (1991). *Voices of the mind: A sociocultural approach to mediated action.* Harvard University Press.

WHO. (2020, March 29). *WHO/Europe | Coronavirus disease (COVID-19) outbreak - WHO announces COVID-19 outbreak a pandemic.* Retrieved from https://www.euro.who.int/en/health-topics/health-emergencies/coronavirus-covid-19/news/news/2020/3/who-announces-covid-19-outbreak-a-pandemic

Williams, B., Figueiredo, J., & Trevelyan, J. (2018). Engineering Practice in a Global Context. CRC Press.

Williams, B., & Carvalho, L. C. (2020). Trajectory of a fourth-generation female entrepreneur as seen through a monomyth lens – Casa Ermelinda Freitas. *Journal of Small Business and Entrepreneurship*, 1–21. doi:10.1080/08276331.2020.1789827

Willingham, D. T. (2008). Critical thinking: Why is it so hard to teach? *Arts Education Policy Review, 109*(4), 21–32. doi:10.3200/AEPR.109.4.21-32

Witte, A. E., & Daly, P. (2014). Proverbial wisdom - A "serious" international business game. *Journal of International Education in Business, 7*(1), 2–13. doi:10.1108/JIEB-12-2012-0029

Wöllner, C., & Ginsborg, J. (2011). Team teaching in the conservatoire: The views of music performance staff and students. *British Journal of Music Education, 28*(3), 301–323. doi:10.1017/S0265051711000222

Womack, J. P., Jones, D. T., & Ross, D. (1990). The Machine That Changed the World. New York: Free Press.

Womack, J. P., & Jones, D. T. (1996). *Lean Thinking: Banish Waste and Create Wealth in Your Corporation.* Free Press.

Wong, C. Y., Grant, D. B., Allan, B., & Jasiuvian, I. (2014). Logistics and supply chain education and jobs: A study of UK markets. *International Journal of Logistics Management, 25*(3), 537–552. doi:10.1108/IJLM-01-2013-0003

Wong, G., Greenhalgh, T., Westhorp, G., Buckingham, J., & Pawson, R. (2013). RAMESES publication standards: Meta-narrative reviews. *BMC Medicine, 11*(1), 20. doi:10.1186/1741-7015-11-20 PMID:23360661

Wood, D., Underwood, J., & Avis, P. (1999). Integrated learning systems in the classroom. *Computers & Education, 33*(2–3), 91–108. doi:10.1016/S0360-1315(99)00027-5

World Economic Forum. (2014). *Towards the Circular Economy: Accelerating the scale-up across global supply chains.* https://www3.weforum.org/docs/WEF_ENV_TowardsCircularEconomy_Report_2014.pdf

World Economic Forum. (2015). *"The skills needed in the 21st century," New Vision for Education - Unlocking the Potential of Technology.* https://widgets.weforum.org/nve-2015/chapter1.html

World Economic Forum. (2015). *The skills needed in the 21st century. New Vision for Education - Unlocking the Potential of Technology.* https://widgets.weforum.org/nve-2015/chapter1.html

Xie, Y., Qui, Y., Huang, Y., Li, Y., Liu, A., & Wu, J. (2020), *The Construction and Effect of the MOOC-based One Plus Three Teaching model During the covid 19 pandemic* [Paper presentation]. *2020 Ninth International Conference of Educational Innovation through Technology (EITT)*. 10.1109/EITT50754.2020.00042

Yee, C. P., Ismail, Z. I. A., & Seman, E. A. A. (2019). A check on planning access for blended learning. *International Journal of Recent Technology and Engineering, 8*(2), 1338–1345. doi:10.35940/ijrte.B1064.0882S819

Yin, R. K. (2009). Case study research: Design and methods. *The Canadian Journal of Action Research, 14*(1), 69–71.

Zanasi, A., Ruini, F., & Bonzio, A. (2017). Intelligence analysts' training through serious games: The leila project. *International Journal of Safety and Security Engineering, 7*(3), 380–389. doi:10.2495/SAFE-V7-N3-380-389

Zandin. (2001). *Maynard's Industrial Engineering Handbook* (5th ed.). McGraw Hill Standard Handbooks.

Zawawi, N. A. W. A., Umar, A. A., Aziz, A. R. A., Crouzier, O., Khamidi, M. F., & Idrus, A. (2014). PFI procurement skills: A cross-sector survey of practitioners. *Procedia Engineering, 77*, 170–178. doi:10.1016/j.proeng.2014.07.016

Zidoun, Y., Dehbi, R., Talea, M., & El Arroum, F. (2019). Designing a theoretical integration framework for mobile learning. *International Journal of Interactive Mobile Technologies, 12*(12), 152–170. doi:10.3991/ijim.v13i12.10841

Zimmerling, E., Höflinger, P. J., Sandner, P. G., & Welpe, I. M. (2016). A system framework for gamified Cost Engineering. *Information Systems Frontiers, 18*(6), 1063–1084. doi:10.100710796-016-9624-3

Zonnenshain, A., & Kenett, R. S. (2020). Quality 4.0—The challenging future of quality engineering. *Quality Engineering, 32*(4), 1–13. doi:10.1080/08982112.2019.1706744

Zunk, B. (2018). Positioning "Techno-Economics" as an Interdisciplinary Reference Frame for Research and Teaching at the Interface of Applied Natural Sciences and Applied Social Sciences: An Approach Based on Austrian IEM Study Programmes. *International Journal of Industrial Engineering and Management, 9*(1), 17–23.

Zunk, B. M., & Sadei, C. (2015). Sharpening the Industrial Engineering and Management qualification profile: Research findings from Austria. *International Journal of Industrial Engineering and Management, 6*(3), 109–120. https://www.scopus.com/inward/record.uri?eid=2-s2.0-84947928261&partnerID=40&md5=b0f42724b97636d4a62cfbdbc4defdd2

About the Contributors

Anabela Carvalho Alves's main research interests are in the areas of Production Systems Design and Operation; Lean Production (Lean Healthcare, Lean Services, Lean Product Development, Lean & TRIZ, Lean-Green, and Lean & Ergonomics); Production Planning and Control, Project Management and Engineering Education, with a particular interest in active learning methodologies, Project-Led Education (PLE), Project/Problem-Based Learning (PBL) and Lean Education. She is Author/co-author of more than 130 publications in conferences publications or communications, 2 books, 2 co-edited books, 4 editions of conference proceedings, 21 book chapters, and 30 international journal articles. She supervised two Ph.D. and supervising 5 PhD, supervised one post-doc, 61 Master dissertations in the areas of Industrial Engineering and Management. Member of the Scientific and Organizing Committee of the International Symposium on Project Approaches in Engineering Education (PAEE), (http://paee.dps.uminho.pt). She organizes of the topic of Curriculum Innovations, Pedagogy, and Learning Methodologies in the ASME conference IMECE. She has been organizing workshops about Lean Education and PBL.

Natascha van Hattum-Janssen works as a quality assurance officer for the Research & graduate School of Saxion University of Applied Sciences in Enschede, the Netherlands. She holds a PhD in Education Sciences (2004), specialisation in Curriculum Development and an Msc in Educational Science and Technology from the University of Twente. She has been a researcher in Engineering Education for more than 20 years and currently works on professional identity of engineering students and engineering professionals as well as professional socialisation in engineering curricula. She is founding member and former president of the Project Approaches in Engineering Education Association. Her research interests are in curriculum development, engineering education, professional identity, narrative research, project-based learning, and faculty development.

* * *

André Luiz Aquere holds a degree in Civil Engineering from the University of Brasília (1984), a Master's in Civil Engineering from the Pontifical Catholic University of Rio de Janeiro (1989), a Ph.D. in Industrial and Systems Engineering from the University of Minho (2010) and held a post-doctorate degree at the ALGORITMI Center of the University of Minho (2019-2020). He is currently an Associate Professor at the University of Brasília, which he joined in 1986, working in the Undergraduate Courses in Civil Engineering and in the Postgraduate Program in Structures and Civil Construction (PECC-FT), as a Master and Doctoral advisor. He was one of the founders of the Production Engineering Course at UnB, introducing Project-Based Learning, in which he also served as a professor in the disciplines of

Production Systems Design. From 1991 to 1995, he worked at Kyoto University (Japan) in the field of Computational Modeling of Inverse Problems in Electrodynamics. From 2001 to 2003 he worked on updating the earthquake-resistant design of the Angra 3 Nuclear Power Plant, with an emphasis on soil-structure interaction models. From 2005 to 2010 he created and coordinated the Project Laboratory at the University of Brasília. From June 2013 to November 2016 he served as Director of Infrastructure Management, at the Rectory of UnB, with the attribution of implementing and managing a system for managing the production and maintenance of civil infrastructure at the university. Has experience in Civil Engineering, with an emphasis on Building Design, Seismic Analysis and Project Management, currently working on the following topics: Project Management in Engineering, Project Based Learning (PBL) and Lean Production (Lean Production).

Elaine Aparecida Araújo received his PhD in Management from Universidade Federal de Lavras (UFLA, Brasil) in 2012. She is currently an associate professor and member of the Decision Analysis Group at Universidade Federal Fluminense (UFF), Brazil, acting on the following subjects: multicriteria decision making, clustering, and performance evaluation.

Ariana Araújo holds a Master degree in Industrial Engineering and Management from the University of Minho, Portugal (2011). She has 10 years of experience in industrial engineering and Lean Production in a Multinational Company. Currently, she is a PhD researcher in Lean Production and Pull Systems.

Daniel Beneroso is an Associate Professor of Engineering Teaching & Learning at the University of Nottingham. He completed a PGCHE (Postgraduate Certificate in Higher Education), awarded by the University of Nottingham in 2019. Since then, he has developed a passion for understanding what makes learning happen in the context of Higher Education. Currently, his research interests are in engineering education, evidence-based teaching and learning, gender and ethnicity equity, and culture of empowerment and collaboration. He has experience as an educational facilitator and leader working with academics, engineers and students in different roles (course director and accreditation director) to create enhanced learning environments able to help future engineers to thrive in the rapidly changing world of work. These have been recognised by a Senior Fellowship of the Higher Education Academy and a number of awards ('HE Innovate Awards' for the most innovative approach to supporting students in the UK (2021), 'Lord Dearing Award' for his outstanding contribution to the development of teaching and student learning (2021), 'Excellence in Teaching' recognition by the Nottingham Faculty of Engineering for his excellent contribution in teaching (2018), and a Highly Commendation award in the Best Research Supervisor category at the University of Nottingham Staff Oscars (2019).

Eduardo Braga is a doctorate student in the Vocacional Programme of Science Education at Instituto Federal do Rio de Janeiro, Brazil and a professor of Mathematics.

Violeta Meneses Carvalho completed the Integrated Master's in Biomedical Engineering at the University of Minho and her area of expertise is Biomaterials, Rehabilitation, and Biomechanics. Currently, she is a Ph.D. student of Mechanical Engineering at the University of Minho in Mechanical Engineering and Resource Sustainability Center (MEtRICs), and an invited teacher in the Production and Systems Department. She is the author and co-author of 19 publications in conferences and international journals, and 2 book chapters and received 7 awards and/or honors. Her main research lies in computational simulations, blood flow, biomedical devices, organs-on-a-chip, and nanomedicine.

Maria João Castro is twenty-two years old and is currently on the Industrial Engineering and Management Master's Degree, at the University of Minho.

Helder Costa has Doctorate and Master's Degrees in Mechanical Engineering from the Pontifical Catholic University of Rio de Janeiro and Graduation in Mechanical Engineering from the Fluminense Federal University. Scientific and Events Director of the National Association of Postgraduate and Research Programs in Production Engineering (ANPEPRO). Full Professor of the Post-Graduate Program (Masters and Doctorate) in Production Engineering at UFF. Consultant for project analysis (CNPQ, CAPES and FAPERJ) and article reviewer (EJOR, IJPE and Ecological Indicators, among other indexed journals). Member of the CAPES team of consultants to review the evaluation form indicators (ENG III). Received a commendation from the Brazilian Navy and PMERJ, in recognition of his support for the scientific development of these institutions and was nominated as the author of an outstanding paper (Journal of Modeling Management-EMERALD, 2016) and as an outstanding reviewer of indexed journals.

Shannon Flumerfelt, Ph.D., is a leading international consultant in the use lean for all aspects of education and government. She is the Founder of Charactership Lean Consulting and serves as an Endowed Professor of Lean, Pawley Lean Institute, and Associate Professor, Organizational Leadership at Oakland University in Rochester, MI. She has worked as an administrator and has conducted educational transformation projects for institutions. She is an accomplished presenter and author, with engagements from the Shingo Institute, the American Society of Mechanical Engineers, and the American Institutes of Research. Her books include: Transforming the Way We Do Business, Lean Engineering Education, Lean Education: An Overview of Current Issues, Trans-disciplinary Perspectives on Complex Systems.

Inês Freitas is a 22 years old student from Guimarães, Portugal, who is currently on the Industrial Engineering and Management Master's Degree at the University of Minho.

Angel-Manuel Gento-Municio is a PhD Industrial Engineering and Assistant Professor at the Department of Management Science at the Faculty of Industrial Engineering, University of Valladolid. He is a researcher at the Process Design and Improvement group (DIMEPRO) of University of Valladolid. His fields of research are lean manufacturing, supply chain management, and risk management by simulation and fuzzy sets.

Margarida Pires Gonçalves is twenty-three years old and is currently attending the Industrial Engineering and Management Master's Degree, at the University of Minho.

Calandra Green is an established leader in healthcare and education. Her experiences include executive consulting and coaching and clinical healthcare leadership. She completed her doctorate on the topic of Lean Leadership.

H. S. Guruprasad received his B.E. degree in Computer Science and Engineering from Malnad College of Engineering. He received his post graduation M.S. CSE from BITS, Pilani. He holds Ph.D.

in Computer Science from M.G.R. University, Chennai. He is working as Professor in the Department of Information Science and Engineering at B.M.S. College of Engineering, Bengaluru, India. He possesses teaching experience of over two decades. Internet of Things, Cloud Computing and Communication Networks are his research interests.

Cristine Martins Gomes de Gusmão has a PhD in Computer Science from the Informatics Center of the Federal University of Pernambuco (CIn-UFPE) in 2007. Professor at the Federal University of Pernambuco (UFPE) since 2010, (i) Deputy Head of the Department of Biomedical Engineering, with undergraduate and graduate activities postgraduate courses; (ii) General Coordinator of the Open University of the Unified Health System at UFPE (UNA-SUS UFPE); (iii) General Coordinator of the Educational and Social Research Group at SABER Technologies; (iv) Member of the ICDE OER Advocacy Committee. Member of the International Council for Open and Distance Education (ICDE), since 2017. Permanent professor of the Postgraduate Programs in Biomedical Engineering and in Mathematics and Technological Education at UFPE. Associate Researcher at the Laboratory of Technological Innovation in Health (LAIS) at the Federal University of Rio Grande do Norte. Has experience in coordinating development, research and innovation projects funded by CNPq, FACEPE, FINEP, Ministry of Health - Federal Government of Brazil. Main areas of expertise: Technology-mediated education, Open Educational Practices, Educational Content Production and Project Management. Interests: Micro Credentials and Microlearning, Learning Analytics, Open Educational Resources, Skills and Learning Paths, Sustainable Development Goals 3 and 4 - Agenda 2030, Connected Health and Health Education.

Ville Isoherranen (Dr.Sc. Tech) is the Director of School of Engineering and Natural Resources at the Oulu University of Applied Sciences, Oulu, Finland. Dr. Isoherranen is also Adjunct Professor at the Faculty of Technology, Industrial Engineering and Management at the University of Oulu, Oulu, Finland. Dr. Isoherranen's research interests are strategic management, operational excellence, knowledge based management, and customer-focused enterprises.

Neila Jesus is a master student in the Master Vocational Programme in Teaching and Its Technologies Instituto Federal do Rio de Janeiro and Supervisor/Teacher in the Rio das Ostras Municipality, Brazil.

Isabelle Leão is twenty-three years old and is currently on the Industrial Engineering and Management Master's Degree, at the University of Minho.

Rui A. Lima is an Associate Professor at the Department of Mechanical Engineering, University of Minho (UMinho), and a researcher at the Mechanical Engineering and Resource Sustainability Center (MEtRICs, UMinho) and Transport Phenomena Research Center (CEFT), FEUP, University of Porto. He received a Dipl. Eng. in Mechanical Engineering from UMinho (Portugal) in 1998, a MSc-Eng from Sheffield Univ. (U.K.) in 2000 and a PhD in Engineering from Tohoku Univ. (Japan) in 2007. His main research interests include microfluidics, nanofluidics and blood flow in biomedical microdevices.

Erik Lopes holds a degree in Civil Engineering from the University of Brasília. He worked in the extension activities of the Consulting Club of the University of Brasília and in the junior company Concreta, where he was project manager and, later, organizational president. He is interested in project management, with a focus on agile management, active learning methodologies, entrepreneurship, and business management.

Heidi Manninen holds a Master degree in Industrial Engineering and Management from the University of Minho, Portugal (2011). After 10 years as an Industrial Engineer, she has a wide experience in logistics processes in a Multinational Company.

Andre Uébe-Mansur is head of Master Vocation Programme in Teaching and Its Technologies at Instituto Federal Fluminense, Brazil. He works as professor and researcher in Education and Business Management focused on active learning methodologies and computing in education. Andre has developed his doctoral research on Complex Thinking and Digital Social Network for student collaborative learning and continues studying about new ways for students skills development for the digital economy. His research was awarded by the Brazilian administration council. He holds a PhD in Computing in Education from the Rio Grande do Sul Federal University, Brazil and also graduated in Business Administration since 1998 . He believes that computing technologies and active learning methodologies can make difference for the innovative student skills to the new digital economy.

Eduardo Marinho is twenty-three years old and is currently on the Industrial Engineering and Management Master's Degree, at the University of Minho.

Graca Minas is an Associate Professor at the Department of Electronics Engineering and a researcher at the MicroElectroMechanical Systems Center (CMEMS-UMinho), both at University of Minho (UMinho), Portugal. She received a Dipl. Eng. in Industrial Electronics Engineering in 1994, a MSc-Eng in 1998 and a PhD in Electronics Engineering in 2004 from UMinho, in the field of Lab-on-a-chip devices, working in cooperation with the Laboratory for Electronic Instrumentation, Delft University of Technology, The Netherlands. Her main research interests include lab-on-a-chip, organ-on-a-chip, on-chip integration of electronic circuits, optics and biosensors.

Rafael José Sousa Moreira is twenty-two years old and is currently on the Industrial Engineering and Management Master's Degree, at the University of Minho.

Rui Manuel Ribeiro da Mota received his Graduation in Management and Finance from ISAG-Instituto Superior de Administração e Gestão in 1998. In 2004 completed the Executive Master and Business Administration in Porto Business School. In 2006 completed a twelve-year career as Finance Director in businesses like textile, metalworking and electronics industries. In 2006 is hired by Sonae as a controller for the hospitality business and in 2007 is invited to lead process improvement team of the finance department of the food retail business. In 2012 participates in the creation of the Sonae's Lean Management System called IOW – Improving Our Work and starts its implementation in non-food retail business unit. In 2016 is invited to lead the coordination of the Lean program for all Sonae business units. He is also the Director of the IOW School.

Leonor Pacheco is twenty-two years old and is currently on the Industrial Engineering and Management Master's Degree, at the University of Minho.

José A. Pascual is a Lecturer in the Department of Management Science at the Faculty of Industrial Engineering, University of Valladolid. His fields of research are the simulation of financial markets through ABM, educational games, role play applied to teaching, learning factories, educational practices, lean manufacturing and logistics. It belongs to the European Social Simulation Association (ESSA) and the Association for the Development of Organisational Engineering (ADINGOR).

Carina Pimentel is an Industrial Engineering and Management Assistant Professor at the Department of Economics, Management, Industrial Engineering and Tourism of the University of Aveiro. She is a researcher at the Decision Support Systems group of the Research Unit on Governance, Competitiveness and Public Policies (GOVCOPP) and of the Research and Development Unit for Mechanical and Industrial Engineering (UNIDEMI). Her recent research interests include smart supply chain management, production planning and control, lean manufacturing, and urban logistics.

Michelle Ramos is a doctoral student in Production Engineering (UFF) in the Operational Research line with focus on Multi-criteria Decision Making methods (MCDM), Master in Production Engineering and Computer Systems (UFF), MBA in Business Management with emphasis on Logistics (FGV), international specialization in Strategic Management and International Marketing (La Verne), graduated in Administration from Centro Universitário Serra dos Órgãos. Works as a Researcher at the Laboratory of Innovative Enterprises - LEI (UFF) for 10 years, currently working as part of the Project of Development Methodology for Condition Based Maintenance in Blowout Preventers (BOPs). Worked as a researcher in other projects in organizational management area in the 2010-2021 period, including Strategic Planning projects, Mapping Management Practices, Information and Knowledge Management Model, Energy use of Waste with rotary drum pyrolysis technology and Organizational Innovation. She taught classes in undergraduate courses in Administration and Production Engineering. Has experience in the areas of Production Engineering and Administration with an emphasis on Logistics, Project Management, Strategies, and Innovation.

John Robinson is an Associate Professor in Chemical & Environmental Engineering at the University of Nottingham. He is an expert in microwave processing technologies, with 15 years of experience which spans the scale-up of continuous process that operate in challenging industrial environments through to establishing a core scientific understanding of microwave-material interactions. He oversees the process engineering design curriculum, leading a team of staff to deliver design teaching across 4 years of the Chemical Engineering degree programmes.

Giselle Roças is PhD in Ecology and is Professor/Advisor at the postgraduate programme of Science Education at Instituto Federal do Rio de Janeiro, Brazil.

Cristina S. Rodrigues is an Assistant Professor at the Department of Production and Systems Engineering, University of Minho (UMinho), and a researcher at the Centro Algoritmi (Algoritmi, UMinho). She has a PhD in Industrial and Systems Engineering., University of Minho, in the area of Numerical and Statistical Methods, and is responsible for courses in applied statistics at undergraduate and postgraduate studies in engineering. Her current research interests include structural equation modeling, research by questionnaire and multivariate statistics in engineering.

Frederico Henrichs Sheremetieff received his MSc in Production Engineering from Universidade Federal Fluminense (UFF) (2020), Brazil. He is currently a member of the Decision Analysis Group at Universidade Federal Fluminense (UFF), Brazil, acting on the following subjects: game-based learning methods and multicriteria decision making.

Pedro Silva is twenty-two years old and is currently on the Industrial Engineering and Management Master's Degree, at the University of Minho.

Klaas Stek is a lecturer and researcher at the University of Twente in the Netherlands and the Graz University of Technology in Austria and works in educational Erasmus+ projects to design improved educational methods in the field of purchasing and supply management. The dissertation of Dr Stek regarded Purchasing Skills Leading to Success and studied the competencies that facilitate innovation sourcing.

Senhorinha Teixeira is an associate professor at the Production and System Engineering Department at University of Minho, Portugal and she was the Head of Department from 2012 to 2015. She obtained her first degree in System Engineering and Informatics at the University of Minho, Portugal, and she was awarded a PhD in Chemical Engineering at the University of Birmingham, U.K. Her PhD research work took place at the AERE, Harwell, U.K. Currently, she teaches numerical methods and CFD courses. She supervised and co-supervised 10 PhD and 50 Master dissertations. She is author and/or coauthor of more than 200 publications in conferences, 9 book chapters and 50 international journal articles. Her main research areas include modelling and computer simulation of industrial processes. She is a researcher at the Centre ALGORITMI at University of Minho, Portugal.

Sheetal V. A. received the Bachelor of Engineering degree from Visvesvaraya Technological University, Belagavi, India, in 2005 and Master of Technology from Visvesvaraya Technological University, Belagavi, India, in 2012. She is currently working as Assistant Professor in department of Computer Science and Engineering, B.M.S. College of Engineering Bangalore. Her area of interests include Internet of Things, Teaching and Learning process.

Nandhini Vineeth received her B.E. Degree in Computer Science and Engineering from Tamil Nadu College of Engineering. She received her Post Graduation M.Tech in Computer Network Engineering from B.M.S. College of Engineering under Visveshwaraya Technological University. She holds Ph.D. degree in Computer and Information sciences from Visveshwaraya Technological University. Vehicular Adhoc Networks, Network Coding and Video Streaming are her research interests.

Index

F

G

H

I

K

L

M

O

P

Q

R

S

T

U

V

W

Y

Premier Reference Source
Stabilizing Currency and Preserving Economic Sovereignty Using the Grondona System

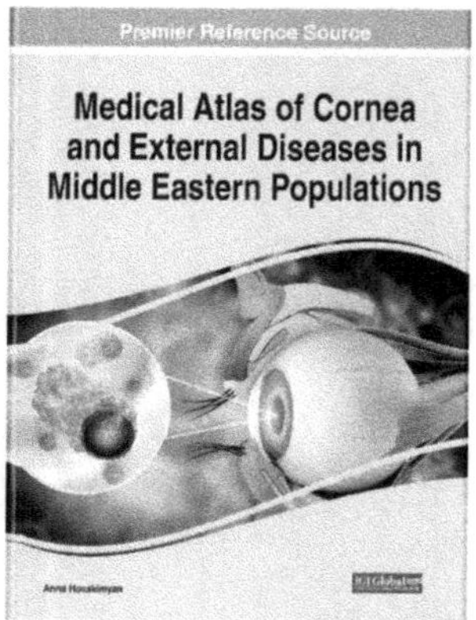

Premier Reference Source
Medical Atlas of Cornea and External Diseases in Middle Eastern Populations

Premier Reference Source
Examining Biophilia and Societal Indifference to Environmental Protection

Premier Reference Source
Using Narratives and Storytelling to Promote Cultural Diversity on College Campuses

www.ingramcontent.com/pod-product-compliance
Lightning Source LLC
LaVergne TN
LVHW062108300926
849161LV00010B/2399
* 9 7 8 1 7 9 9 8 8 8 1 6 1 *